Metathesis in Natural Product Synthesis

Edited by
Janine Cossy, Stellios Arseniyadis,
and Christophe Meyer

Related Titles

Christmann, M., Bräse, S. (eds.)

Asymmetric Synthesis – The Essentials

395 pages with 85 figures and 3 tables
2008
Softcover
ISBN: 978-3-527-32093-6

Hudlicky, T., Reed, J. W.

The Way of Synthesis
Evolution of Design and Methods for Natural Products

1018 pages with 285 figures and 15 tables
2007
Softcover
ISBN: 978-3-527-31444-7

Sheldon, R. A., Arends, I., Hanefeld, U.

Green Chemistry and Catalysis

448 pages with 476 figures and 21 tables
2007
Hardcover
ISBN: 978-3-527-30715-9

Enders, D., Jaeger, K.-E. (eds.)

Asymmetric Synthesis with Chemical and Biological Methods

470 pages with 84 figures and 57 tables
2007
Hardcover
ISBN: 978-3-527-31473-7

Dalko, P. I. (ed.)

Enantioselective Organocatalysis
Reactions and Experimental Procedures

559 pages with 364 figures and 115 tables
2007
Hardcover
ISBN: 978-3-527-31522-2

Cornils, B., Herrmann, W. A., Muhler, M., Wong, C.-H. (eds.)

Catalysis from A to Z
A Concise Encyclopedia

approx. 1560 pages in 3 volumes with 2778 figures and 104 tables
2007
Hardcover
ISBN: 978-3-527-31438-6

de Vries, J. G., Elsevier, C. J. (eds.)

The Handbook of Homogeneous Hydrogenation

1641 pages in 3 volumes with 422 figures and 254 tables
2007
Hardcover
ISBN: 978-3-527-31161-3

Metathesis in Natural Product Synthesis

Strategies, Substrates and Catalysts

Edited by
Janine Cossy, Stellios Arseniyadis, and Christophe Meyer

With a Foreword by Robert H. Grubbs

WILEY-VCH
WILEY-VCH Verlag GmbH & Co. KGaA

The Editors

Prof. Janine Cossy
Laboratorie de Chimie
Organique, ESPCI
10 Rue Vauquelin
75231 Paris Cedex 05
France

Dr. Stellios Arseniyadis
Laboratoire de Chimie
Organique, ESPCI
10 Rue Vauquelin
75231 Paris Cedex 05
France

Dr. Christophe Meyer
Laboratoire de Chimie
Organique, ESPCI
10 Rue Vauquelin
75231 Paris Cedex 05
France

Cover
The gryffon painting being part of the front cover picture has been kindly provided by Dominique Escortell

Library of Congress Card No.: applied for

British Library Cataloguing-in-Publication Data
A catalogue record for this book is available from the British Library.

Bibliographic information published by the Deutsche Nationalbibliothek
The Deutsche Nationalbibliothek lists this publication in the Deutsche Nationalbibliografie; detailed bibliographic data are available on the Internet at http://dnb.d-nb.de.

Cover Adam Design, Weinheim
Typesetting Laserwords Private Limited, Chennai, India
Printing and Binding Strauss GmbH, Mörlenbach

Printed in the Federal Republic of Germany
Printed on acid-free paper

ISBN: 978-3-527-32440-8

Foreword

In the last few decades, metathesis has been among the key reactions that have revolutionized the synthesis of complex molecules. Many organic chemists in academic and industrial laboratories, in the field of natural products, have used this reaction as a very practical, versatile, and selective synthetic tool. Olefin metathesis has helped to elevate the art and science of chemical synthesis to its present high level.

The examples in this book will demonstrate that organic chemists, with the metathesis reaction in hand, have a new way to consider the connections that are required for efficient access to natural products. This book assembles the most important and interesting examples in the synthesis of natural products using metathesis. Owing to the possibilities opened by olefin and acetylenic metathesis, a great variety of carbocyclic – nitrogen-, oxygen-, sulfur-containing heterocycles – natural products with small-, medium-, and macrocyclic size can be obtained rapidly. The synthetic transformations that couple metathesis steps in cascade reactions are particularly elegant. Emphasis has been put on the metathesis step showing the importance of the catalysts that are tolerant of a large variety of functional groups, very regio-, stereoselective, and even enantioselective. The power of the catalysts and of the metathesis reaction can be appreciated when alternative pathways are considered.

Every reaction and catalyst can always be improved. In the area of metathesis, the development of more active and robust catalysts, catalysts that can control the *E* and *Z* stereoselectivity of the formed olefins, particularly the stereoselectivity of trisubstituted olefins, or catalysts that can control the enantioselectivity remains a challenge. As has been demonstrated in the past, improvements of the catalyts give rise to increasingly exciting applications in the field of complex molecules and particularly in the field of natural products synthesis. This book will be a good source of inspiration for those planning future developments of metathesis reactions in the field of natural and non-natural products.

Robert H. Grubbs

Metathesis in Natural Product Synthesis: Strategies, Substrates and Catalysts. Edited by Janine Cossy, Stellios Arseniyadis, and Christophe Meyer

ISBN: 978-3-527-32440-8

Contents

Metathesis in Natural Product Synthesis: Strategies, Substrates and Catalysts. Edited by Janine Cossy, Stellios Arseniyadis, and Christophe Meyer

ISBN: 978-3-527-32440-8

Preface

In the 1960s, the ring-opening polymerization of cycloalkenes and the disproportionation of linear alkenes, both used by the polymer and the petroleum industry, were the first reported examples of "olefin metathesis reactions." Whereas those transformations were generally carried out with ill-defined catalysts, the mechanism of olefin metathesis proposed by Chauvin and Hérisson in 1971 identified metal carbenes as catalytically active species with reactions proceeding through metallacyclobutane intermediates. The mid-1970s saw the emergence of the first well-defined alkylidene–metal complexes for olefin metathesis initially based on tantalum and tungsten. However, in the late 1980s, the quest for higher functional group tolerance resulted in the development of the molybdenum complex, also known as *Schrock's catalyst*, which was later used by Grubbs and Fu in ring-closing metathesis (RCM) to access oxygen and nitrogen heterocycles. Up to now, several applications of RCM to natural products synthesis have been reported using Schrock's catalyst as the initiator; however, its air- and moisture sensitivity, which implies the use of a glove box or Schlenk techniques, has certainly hampered its more widespread use by organic chemists. In 1992, Grubbs and coworkers reported the first stable vinylidene ruthenium catalyst to be active in both ring-opening metathesis (ROM) and RCM. In 1995, further refinements led to the development of an air- and moisture-stable as well as highly functional group-tolerant benzylidene ruthenium complex also known as *Grubbs first-generation catalyst*. The latter became the first user-friendly metathesis catalyst and has allowed numerous synthetic applications. The replacement of one phosphine by a strongly σ-donating *N*-heterocyclic carbene ligand to further improve the stability of the active species and accelerate the initiation phase stimulated the discovery of the second-generation catalysts. To date, many catalysts have been devised with the goal of improving the rate of initiation and the stability of the catalytic propagating species to enable the metathesis of sterically hindered substrates. This was attained by fine-tuning the steric and/or electronic properties of the benzylidene part or the *N*-heterocyclic carbene of the ruthenium complexes, and/or other subtle ligand exchange. For the tremendous impact of metathesis in the science of organic synthesis, Chauvin, Grubbs, and Schrock received the Nobel Prize in Chemistry in 2005.

The aim of the book is to emphasize the impact of metathesis on the synthesis of natural products and/or biologically active compounds, and highlight how they

Metathesis in Natural Product Synthesis: Strategies, Substrates and Catalysts. Edited by Janine Cossy, Stellios Arseniyadis, and Christophe Meyer

ISBN: 978-3-527-32440-8

have provided new and elegant solutions to many synthetic puzzles. As RCM has been the first class of metathesis reactions routinely used in natural products chemistry, the first three chapters of the book will highlight its applications to the synthesis of small- to medium-size carbocycles (Chapter 1, N. Blanchard and J. Eustache), nitrogen heterocycles (Chapter 2, L. van Delft and Floris P. J. T. Rutjes), and oxygen heterocyles (Chapter 3, J. D. Rainier). Phosphorus and sulfur heterocycles synthesized *via* RCM also deserved a section since they have found useful applications in the stereoselective synthesis of acyclic subunits found in various natural products (Chapter 4, C.D. Thomas and P.R. Hanson). The use of RCM for the synthesis of macrocyclic compounds has also been covered (Chapter 5, A. Gradillas and J. Pérez-Castells) since it constitutes an attractive alternative to traditional routes such as macrolactonization or macrolactamization. Alkynes can also be used as reacting partners in metathesis reactions as illustrated in the two following chapters of the book. Indeed, while ene–yne metathesis catalyzed by alkylidene ruthenium complexes allows a convenient access to conjugated dienes (Chapter 6, M. Mori), ring-closing alkyne metathesis using a well-defined tungsten–alkylidyne complex or molybdenum precatalysts activated *in situ* offers a convenient route toward cycloalkynes (Chapter 7, P. Davies). As for many reactions, there are situations where a planned metathesis event was found to be either unsuccessful or did not operate with high efficiency, stereoselectivity, and/or chemoselectivity. Silicon-tethered metathesis (Chapter 8, P. A. Evans) and the use of an unsaturated relay allowing initiation of metathesis at an appropriate reactive site (Chapter 9, T. R. Hoye and J. Jeon) are two strategies that have been used to circumvent some of these problems. More recently, cross-metathesis (CM) has emerged as a useful catalytic and chemoselective alternative to traditional olefination methods. Applications in the context of natural product synthesis have therefore been covered (Chapter 10, J. Prunet and L. Grimaud). After disclosing the synthetic potential of each of the different metathesis reactions, it appeared important to illustrate how their combination in cleverly designed cascades has led to some impressive and elegant synthesis of structurally complex natural products (Chapter 11, M. Porta and S. Blechert). The development of chiral molybdenum or ruthenium catalyst has also enabled the achievement of enantioselective metathesis reactions whose applications yet reported to the synthesis of natural products have been listed in one chapter (Chapter 12, A. H. Hoveyda, S. J. Malcolmson, S. J. Meek, and A. R. Zhugralin). Finally, the last section of the book is devoted to solid-phase metathesis, which constitutes a useful tool in diversity-oriented synthesis for chemical biology while also simplifying the purification stages (Chapter 13, S. Barluenga, P.-Y. Dakas, R. Jogireddy, G. Valot, and N. Winssinger).

We would like to warmly thank all the authors for contributing to this book and acknowledge their expertise on the different topics that have been covered. We also thank the team at Wiley-VCH and especially Stefanie Volk for her helpful assistance during the preparation of this book.

We sincerely hope that this book will be a valuable source of information for researchers working in both academic and industrial laboratories and that it will stimulate new applications and developments of metathesis in the field of natural product synthesis.

Janine Cossy, Stellios Arseniyadis, and Christophe Meyer.

List of Catalysts

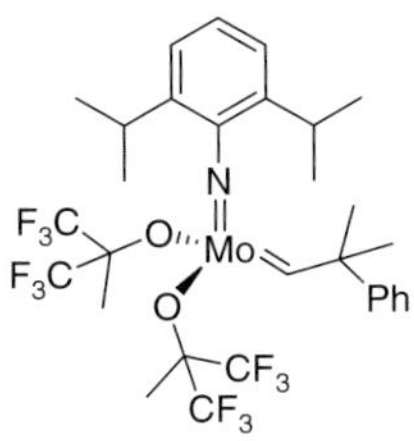

[Mo]-**I**

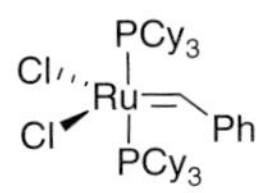

[Ru]-**I**

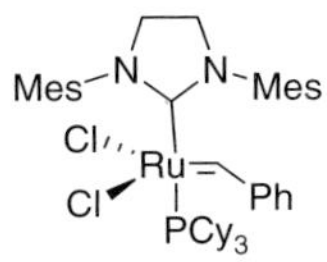

[Ru]-**II**

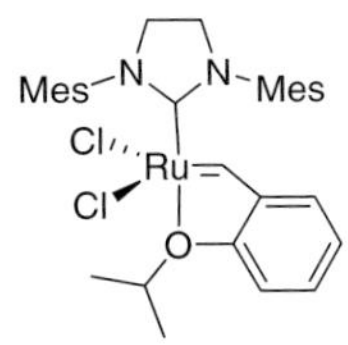

[Ru]-**III**

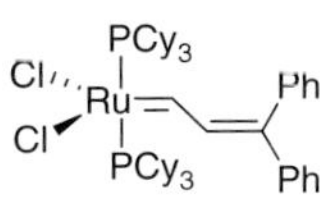

[Ru]-**IV**

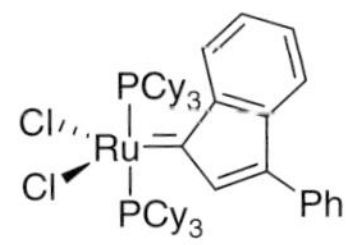

[Ru]-**V**

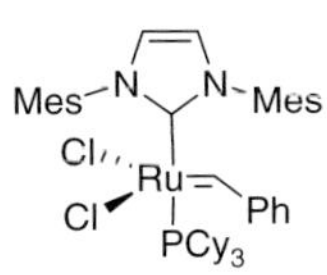

[Ru]-**VI**

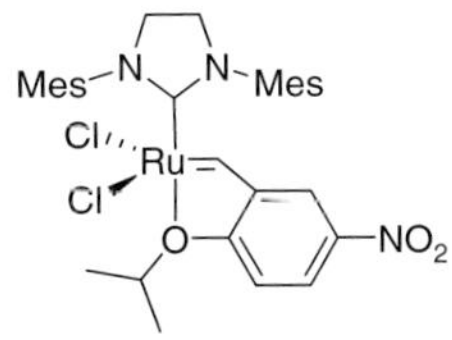

[Ru]-**VII**

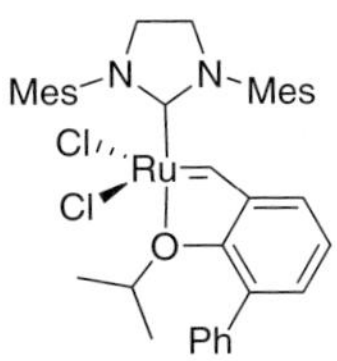

[Ru]-**VIII**

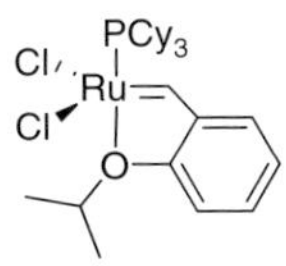

[Ru]-**IX**

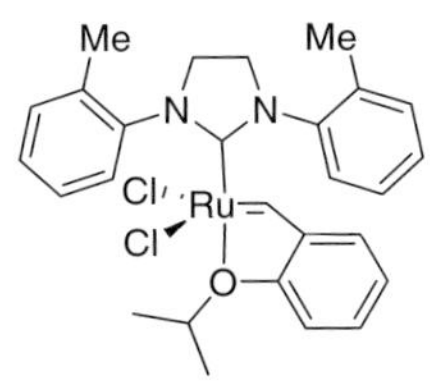

[Ru]-**X**

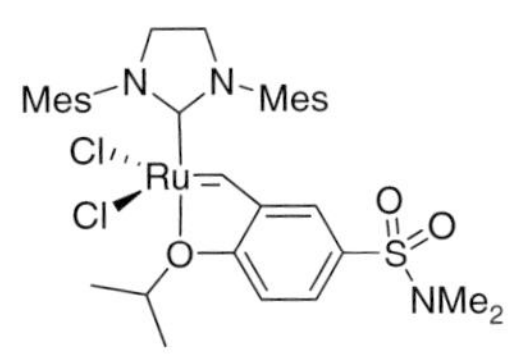

[Ru]-**XI**

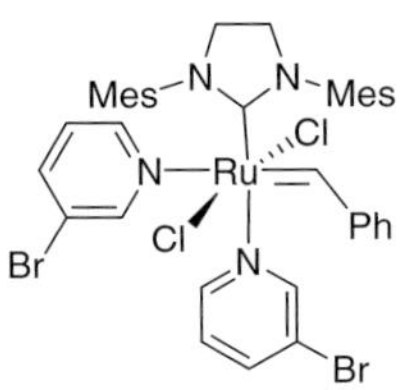

[Ru]-**XII**

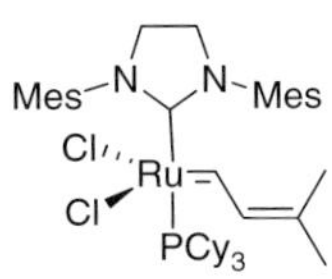

[Ru]-**XIII**

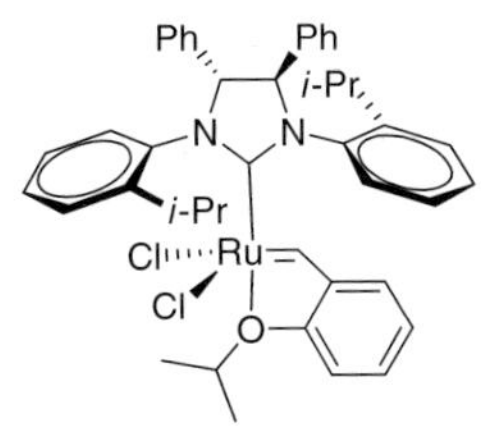

(*R*,*R*)-[Ru]-**XIV**

Metathesis in Natural Product Synthesis: Strategies, Substrates and Catalysts. Edited by Janine Cossy, Stellios Arseniyadis, and Christophe Meyer

ISBN: 978-3-527-32440-8

List of Contributors

Sofia Barluenga
Université Louis Pasteur de Strasbourg
Organic and Bioorganic Chemistry Laboratory
Institut de Science et Ingénierie Supramoléculaire
8 Allée Gaspard Monge
67000 Strasbourg
France

Nicolas Blanchard
Université de Haute-Alsace
Ecole Nationale Supérieure de Chimie de Mulhouse
Laboratoire de Chimie Organique et Bioorganique associé au CNRS
3 Rue Alfred Werner
68093 Mulhouse Cedex
France

Siegfried Blechert
Technische Universität Berlin
Institute of Chemistry
Straße des 17. Juni 135
10623 Berlin
Germany

Pierre-Yves Dakas
Université Louis Pasteur de Strasbourg
Organic and Bioorganic Chemistry Laboratory
Institut de Science et Ingénierie Supramoléculaire
8 Allée Gaspard Monge
67000 Strasbourg
France

Paul W. Davies
University of Birmingham
School of Chemistry
Birmingham
B15 2TT
United Kingdom

Jacques Eustache
Université de Haute-Alsace
Ecole Nationale Supérieure de Chimie de Mulhouse
Laboratoire de Chimie Organique et Bioorganique associé au CNRS
3 Rue Alfred Werner
68093 Mulhouse Cedex
France

Metathesis in Natural Product Synthesis: Strategies, Substrates and Catalysts. Edited by Janine Cossy, Stellios Arseniyadis, and Christophe Meyer

ISBN: 978-3-527-32440-8

P. Andrew Evans
The University of Liverpool
Department of Chemistry
Crown Street
Liverpool L69 7ZD
UK

Ana Gradillas
Universidad CEU-San Pablo
Departamento de Química
Facultad de Farmacia
Urb. Montepríncipe
28668 Boadilla del Monte
Madrid
Spain

Laurence Grimaud
Ecole Nationale Supérieure des
Techniques Avancées
Unité Chimie et Procédés
32 boulevard Victor
75739 Paris Cedex 15
France

Paul R. Hanson
University of Kansas
Department of Chemistry
1251 Wescoe Hall Drive
Malott Hall
Lawrence, KS 66045
USA

Amir H. Hoveyda
Boston College
Department of Chemistry
Eugene F. Merkert
Chemistry Center
Chestnut Hill
MA 02467
USA

Thomas R. Hoye
University of Minnesota
Department of Chemistry
207 Pleasant Street
SE
Minneapolis
Minnesota 55455
USA

Junha Jeon
University of Minnesota
Department of Chemistry
207 Pleasant Street
SE
Minneapolis
Minnesota 55455
USA

Rajamalleswaramma Jogireddy
Université Louis Pasteur de
Strasbourg
Organic and Bioorganic
Chemistry Laboratory
Institut de Science et Ingénierie
Supramoléculaire
8 Allée Gaspard Monge
67000 Strasbourg
France

Steven J. Malcolmson
Boston College
Department of Chemistry
Eugene F. Merkert
Chemistry Center
Chestnut Hill
MA 02467
USA

Simon J. Meek
Boston College
Department of Chemistry
Eugene F. Merkert
Chemistry Center
Chestnut Hill
MA 02467
USA

Silvie A. Meeuwissen
Radboud University Nijmegen
Institute for Molecules and
Materials
Heyendaalseweg 135
6525 ED Nijmegen
The Netherlands

Miwako Mori
Health Sciences University of
Hokkaido
Ishikari-Tobetsu
Hokkaido
061-0293
Japan

Marta Porta
Technische Universität Berlin
Institute of Chemistry
Straße des 17. Juni 135
10623 Berlin
Germany

Joëlle Prunet
Ecole Polytechnique
Laboratoire de Synthèse
Organique
UMR CNRS 7652
DCSO
91128 Palaiseau
France

Javier Pérez-Castells
Universidad CEU-San Pablo
Departamento de Química
Facultad de Farmacia
Urb. Montepríncipe
28668 Boadilla del Monte
Madrid
Spain

Jon D. Rainier
University of Utah
Department of Chemistry
315 East 1400 South
Salt Lake City
UT 84112
USA

Floris P. J. T. Rutjes
Radboud University Nijmegen
Institute for Molecules and
Materials
Heyendaalseweg 135
6525 ED Nijmegen
The Netherlands

Christopher D. Thomas
University of Kansas
Department of Chemistry
1251 Wescoe Hall Drive
Malott Hall
Lawrence, KS 66045
USA

Sebastiaan (Bas) A. M. W. van den Broek
Radboud University Nijmegen
Institute for Molecules and
Materials
Heyendaalseweg 135
6525 ED Nijmegen
The Netherlands

Floris L. van Delft
Radboud University Nijmegen
Institute for Molecules and Materials
Heyendaalseweg 135
6525 ED Nijmegen
The Netherlands

Gaële Valot
Université Louis Pasteur de Strasbourg
Organic and Bioorganic Chemistry Laboratory
Institut de Science et Ingénierie Supramoléculaire
8 Allée Gaspard Monge
67000 Strasbourg
France

Nicolas Winssinger
Université Louis Pasteur de Strasbourg
Organic and Bioorganic Chemistry Laboratory
Institut de Science et Ingénierie Supramoléculaire
8 Allée Gaspard Monge
67000 Strasbourg
France

Adil R. Zhugralin
Boston College
Department of Chemistry
Eugene F. Merkert Chemistry Center
Chestnut Hill
MA 02467
USA

Abbreviations

Ac	acetyl
acac	acetylacetonato
AIBN	azobisiisobutyronitrile
Ar	aryl
AROM	asymmetric ring-closing metathesis
BBN	9-borabicyclononane
9-BBN	9-borabicyclo[3.3.0]nonan-9-yl
Bn	benzyl
Boc	*tert*-butyloxycarbonyl
BOM	benzyloxymethyl
BQ	1,4-benzoquinone
Bz	benzoyl
CAN	ceric ammonium nitrate
Cat.	catalytic
Cbz	benzyloxycarbonyl
CM	cross-metathesis
COD	cycloocta-1,5-diene
Cp	cyclopentadienyl
Cp*	1,2,3,4,5-pentamethylcyclopentadienyl
CSA	10-camphorsulfonic acid
Cy	cyclohexyl
dba	dibenzylideneacetone
DBU	1,8-diazabicyclo[5.4.0]undec-7-ene
DCC	N,N′-dicyclohexylcarbodiimide
DCE	1,2-dichloroethane
DDQ	2,3-dichloro-5,6-dicyanobenzoquinone
DIAD	diisopropyl azodicarboxylate
DIPEA	diisopropylethylamine
DMAP	*N,N*-dimethylaminopyridine
DMB	3′,5′-dimethoxybenzoin
dmdba	bis(3,5-dimethoxybenzylidene)acetone
DMDO	dimethyldioxirane
DMF	*N,N*-dimethylformamide

Metathesis in Natural Product Synthesis: Strategies, Substrates and Catalysts. Edited by Janine Cossy, Stellios Arseniyadis, and Christophe Meyer

ISBN: 978-3-527-32440-8

DMP	Dess-Martin periodinane
DMP	dimethoxypropane
DMPU	1,3-dimethyl-3,4,5,6-tetrahydro-2-(1H)-pyrimidinone
DMSO	dimethylsulfoxyde
DPPA	diphenylophosphoryl azide
dppp	1,3-bis(diphenylphosphino)propane
EDC	1-ethyl-3-(3-dimethylaminopropyl)carbodiimide
Et	ethyl
Fmoc	9-fluorenylmethoxycarbonyl
Grubbs II	benzylidene[1,3-bis(2,4,6-trimethylphenyl)-2-imidazolidinylidene] dichloro(tricyclohexylphosphine) ruthenium
HMDS	hexamethyldisilazane
HMPA	hexamethylphosphoroamide
HOBt	hydroxybenzotriazole
i-pr	isopropyl
IBX	2-iodoxybenzoic acid
dIcr	(+)-?2-caranyl
Im	imidazole
Ipc	diisopinocampheyl
KHMDS	potassium hexamethyldisilazide
LDA	lithium N,N-diisopropylamide
LiHMDS	lithium bistrimethylsilylamide
m-CPBA	*m*-chloroperbenzoic acid
Me	methyl
MEM	2-methoxyethoxymethoxy
Mes	mesityl
MOM	methoxymethyl
Ms	methanesulfonyl
MS	molecular sieves
NaHMDS	sodium hexamethyldisilazide
NBS	N-bromosuccinimide
NCS	N-chlorosuccinimide
NIS	N-iodosuccinimide
NMO	N-methylmorpholine N-oxide
PCC	pyridinium chlorochromate
Ph	phenyl
Phth	phthalyl
Piv	pivaloyl
PMB	*p*-methoxybenzyl
PMP	4-methoxyphenyl
PPTS	pyridinium p-toluenesulfonate
*p*Ts	*para*-toluensulfonyl
Py	pyridine
Quant.	quantitative
RCAM	ring-closing alkyne metathesis

RCEYM	ring-closing ene-yne metathesis
RCM	ring-closing metathesis
ROM	ring-opening metathesis
ROMP	ring-opening metathesis polymerization
RRCM	relay ring-closing metathesis
RRM	ring-rearrangement metathesis
rt	room temperature
t-Bu	tertiary butyl
TBAF	tetrabutylammonium fluoride
TBDPS	*tert*-butyldiphenylsilyl
TBS	*tert*-butyldimethylsilyl
TCE	trichloroethoxymethoxy
Teoc	2-(trimethylsilyl)ethoxycarbonyl
TES	triethylsilyl
Tf	trifluoromethanesulfonyl
TFA	trifluoroacetic acid
2-Th	2-thienyl
THF	tetrahydrofuran
THP	tetrahydropyran
TIPS	triisopropylsilyl
TMEDA	N,N,N′,N′-tetramethylethylenediamine
TMS	trimethylsilyl
Tol-BINAP	2,2′-bis(ditolylphosphino)-1,1′-binaphthalene
TPAP	tetra(n-propyl)ammonium perruthenate
Tr	triphenylmethyl (trityl)
Trt	trityl
Ts	*p*-toluenesulfonyl
TS	transition state

1
Synthesis of Natural Products Containing Medium-size Carbocycles by Ring-closing Alkene Metathesis

Nicolas Blanchard and Jacques Eustache

1.1
Introduction

This chapter deals with the synthesis of naturally occurring molecules (or related models) and focuses on the construction of medium-size carbocycles by ring-closing metathesis (RCM). We have arbitrarily chosen to organize this chapter by increasing ring size. Strategic aspects and potential problems are discussed.

1.2
Formation of Five-membered Carbocycles by RCM

Strategic positioning of the double bonds of a 1,6-diene prior to RCM can be efficiently accomplished via [3,3]-sigmatropic rearrangements. Only selected examples are discussed below, based on originality and efficiency criteria. Catalytic asymmetric Claisen rearrangement was reported by Hiersemann *et al.* in the enantioselective synthesis of the C10–C18 segment of ecklonialactone B (**1**), a C18-oxylipin isolated from the brown algae *Ecklonia stolonifera* and *Egregia menziessi* [1]. The [3,3]-sigmatropic rearrangement of **2** using catalyst **3** led to the acyclic α-keto ester **4** that can be reduced and cyclized to the corresponding disubstituted cyclopentene **5** using [Ru]-**II**. Further functional group transformations led to the desired C10–C18 fragment **6** of the natural product (Scheme 1.1).

A [3,3]-sigmatropic rearrangement/RCM sequence [2] was also used as a key step in the total synthesis of a bioactive spirobenzofuran **7** isolated from the mycelium cultures of *Acremonium* sp. HKI 0230 [3]. The two consecutive quaternary centers embedded in the 1,6-diene **9** are worthy of note in the context of cyclopentene formation by RCM, and the five-membered ring compound **10** was obtained in high yield (97%) using catalyst [Ru]-**I**. The same type of strategy was applied for the efficient construction of the two vicinal quaternary carbon atoms present in the herbertanes sesquiterpenes (Scheme 1.2) [4].

Besides the Ireland–Claisen [3,3]-sigmatropic rearrangement/RCM sequence, some examples of Johnson–Claisen/RCM were reported by Ghosh and Maity [5].

Metathesis in Natural Product Synthesis: Strategies, Substrates and Catalysts. Edited by Janine Cossy, Stellios Arseniyadis, and Christophe Meyer

ISBN: 978-3-527-32440-8

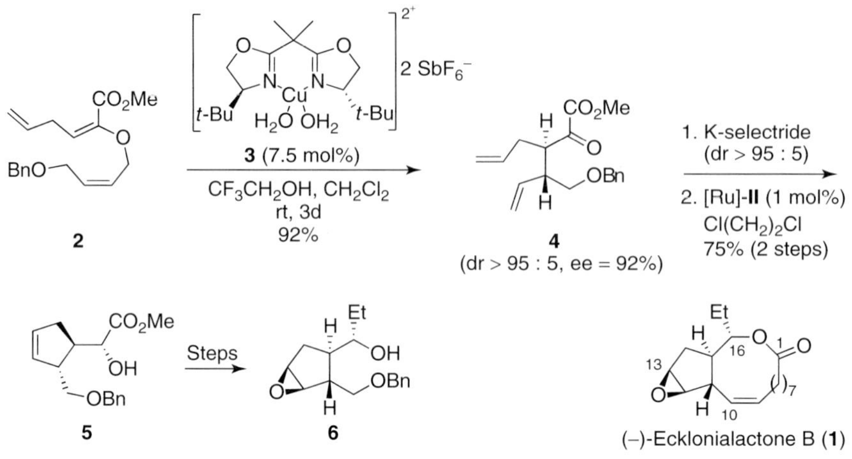

Scheme 1.1

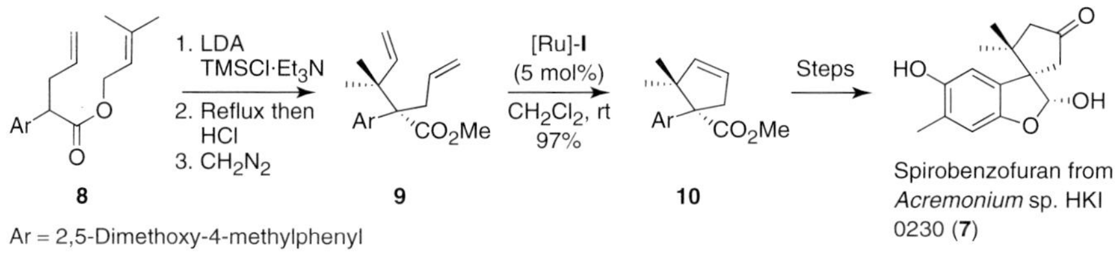

Scheme 1.2

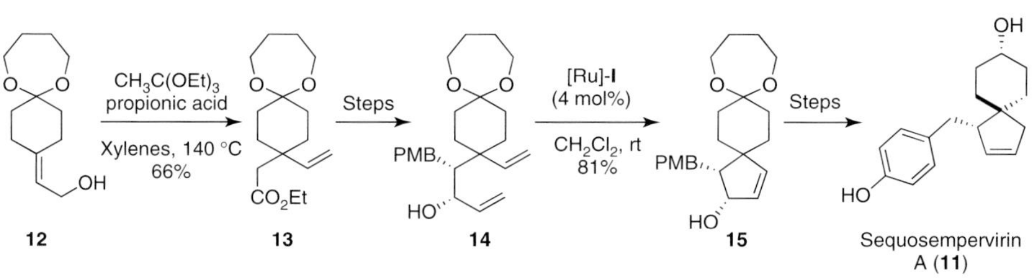

Scheme 1.3

In the first total synthesis of sequosempervirin A (**11**), a norlignan isolated from *Sequoia sempervirens*, the γ,δ-unsaturated carbonyl derivative **13** was prepared from **12** by Johnson–Claisen rearrangement. Further steps led to **14**, the precursor of the key RCM reaction catalyzed by [Ru]-**I**, which established the desired spiro structure of compound **15** (Scheme 1.3).

An analogous Johnson–Claisen rearrangement/RCM sequence was used by Ghosh *et al.* in the synthesis of the carbocyclic core of the nucleoside (−)-BCA (**16**), a potential inhibitor of HIV reverse transcriptase (Scheme 1.4) [6]. The chiral cyclopentene **20** was obtained in excellent yield through a [Ru]-**I**-mediated RCM of 1,6-diene **19**. Further epoxidation and functional group transformations led to (−)-BCA core structure **23**.

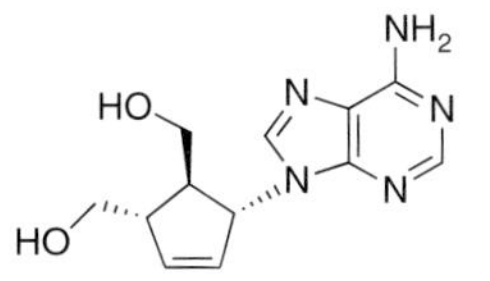

Bis(hydroxymethyl)cyclopentenyl adenine, (–)-BCA (**16**)

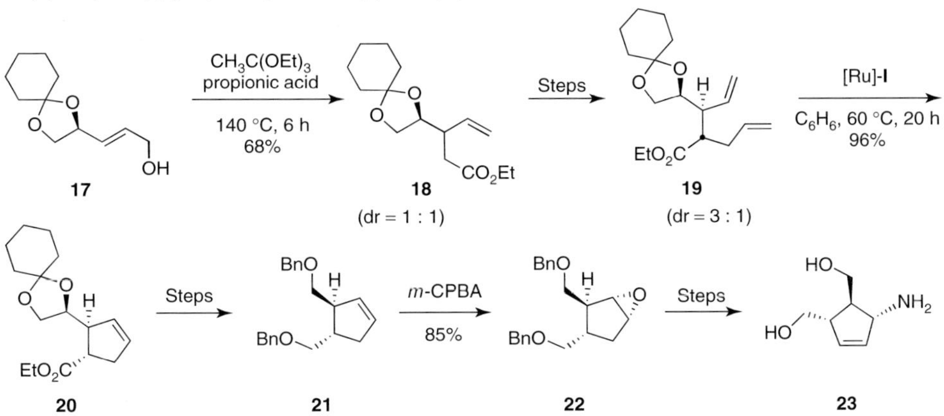

Scheme 1.4

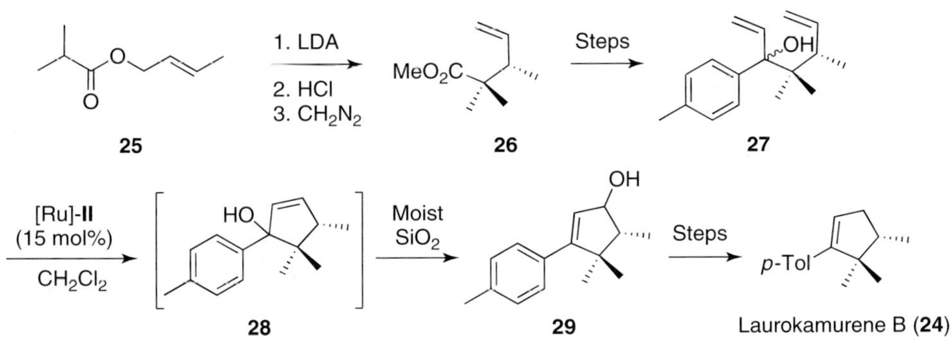

Scheme 1.5

Another [3,3]-sigmatropic rearrangement/RCM sequence was developed by Srikrishna and coworkers in the first total synthesis of the aromatic sesquiterpene (±)-laurokamurene B (**24**) (Scheme 1.5) [7]. RCM of **27** led to cyclopentenol **28** bearing a labile hydroxyl moiety, both benzylic and allylic. The latter compound was found to be unstable and when treated with silica gel delivered the rearranged alcohol **29** in a one-pot operation.

Derivatives of nitiol (**30**), a novel complex sesterterpenoid that acts as a modulator of IL-2 gene expression, were prepared using RCM of vinyl ketone **33** as a key step (Scheme 1.6) [8]. Worthy of note is the use of catalyst [Ru]-**XIII** [9] as [Ru]-**II** was found to increase intermolecular cross-metathesis at the expense of the desired

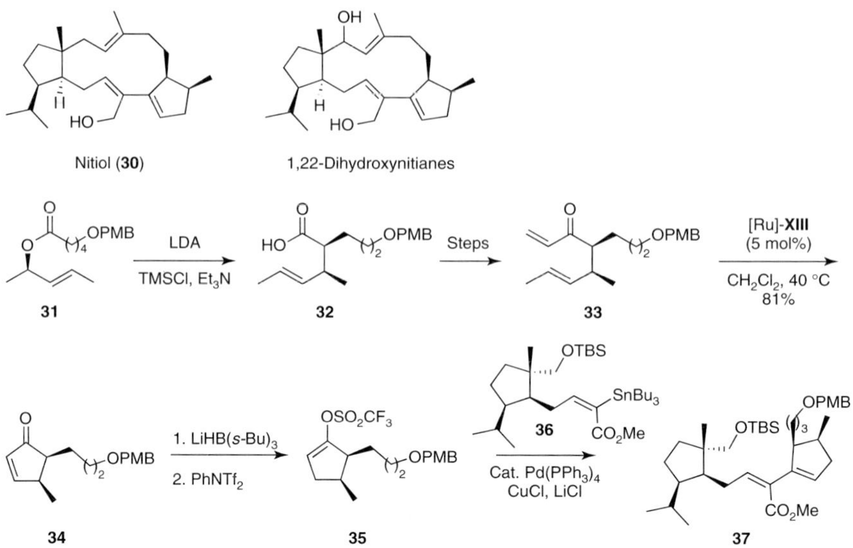

Scheme 1.6

RCM. Conjugate reduction of enone **34** using L-selectride followed by trapping of the resulting enolate with $PhNTf_2$ delivered enol triflate **35**. The latter was then submitted to a Stille cross-coupling reaction with vinyl stannane **36** leading to the advanced intermediate **37** in the synthesis of 1,22-dihydroxynitianes.

In the total synthesis of (−)-allosamizoline (**38**), the carbocyclic fragment of the chitinase inhibitor allosamidin, Donohoe and Rosa reported the efficient elaboration of the cyclopentenyl moiety via the RCM of a 1,6-diene (Scheme 1.7) [10]. Wittig olefination of aldehyde **39**, readily obtained from D-glucosamine, proved troublesome when basic phosphorus ylide **40** was used and the 1,6-diene **42** was isolated in 18% yield. Since the terminal substituents of the alkenes are not transferred into the cyclized products during RCM reactions, the reactivity of acrylate **43** was evaluated as the latter was obtained in high yield (91%) using the less basic ylide **41**. RCM proceeds equally well, delivering cyclopentene **44** in 88% yield.

Formation of cyclopentenes through the RCM of 1,6-dienes can prove itself difficult in very crowded environment. During synthetic efforts toward the tetracyclic skeleton of the sesquiterpenoid tashironin (**47**), Mehta and Maity reported the efficient RCM of **50** using [Ru]-**I** that led to **51** in 86% yield [11]. However, the RCM of **52** having a methyl substituent in allylic position led to **53** in considerably lower yield (25%) even in the presence of the more active [Ru]-**II** catalyst (Scheme 1.8).

A rapid elaboration of the ABC core structure of the tetranortriterpene dumsin (**54**) was reported by Srikrishna *et al.* starting from the readily available (*R*)-carvone [12]. Two independent efficient RCM reactions, catalyzed by [Ru]-**I**, were used as

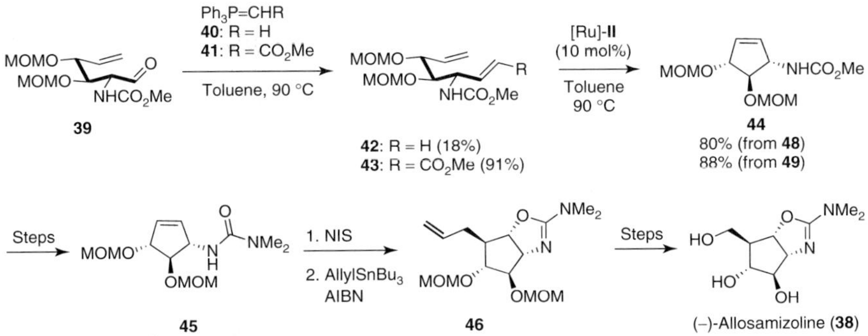

Scheme 1.7

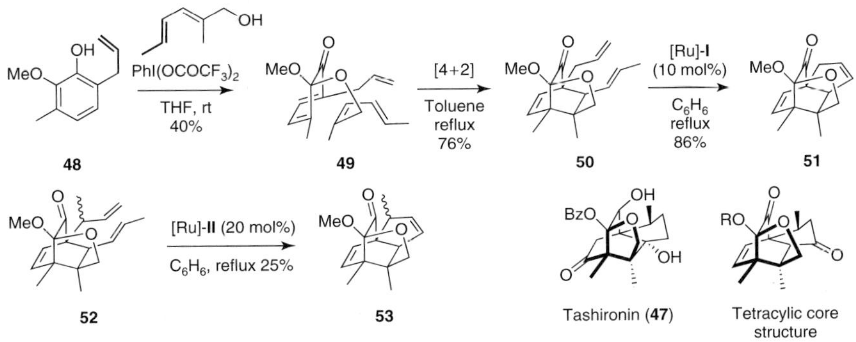

Scheme 1.8

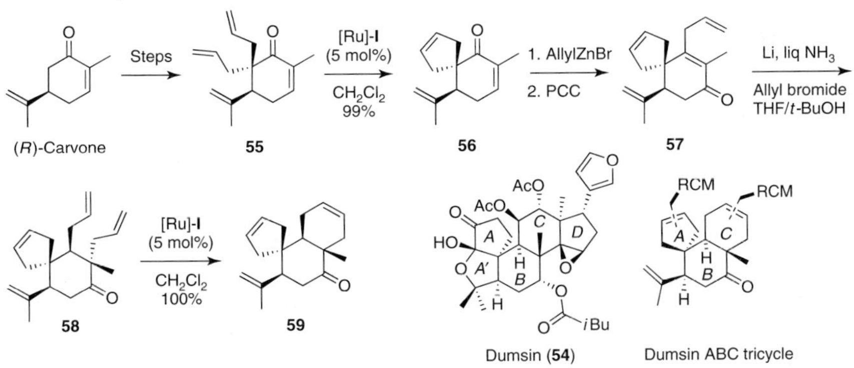

Scheme 1.9

key steps to create the spiro cyclopentene in compound **56** and the trans-fused BC rings in compound **59** (Scheme 1.9).

In challenging cases, simple transformations of the allylic substituents may promote a productive RCM of 1,6-dienes to the corresponding cyclopentenes as shown by Eustache *et al.* in the stereoselective synthesis of spiroepoxide **61**, a

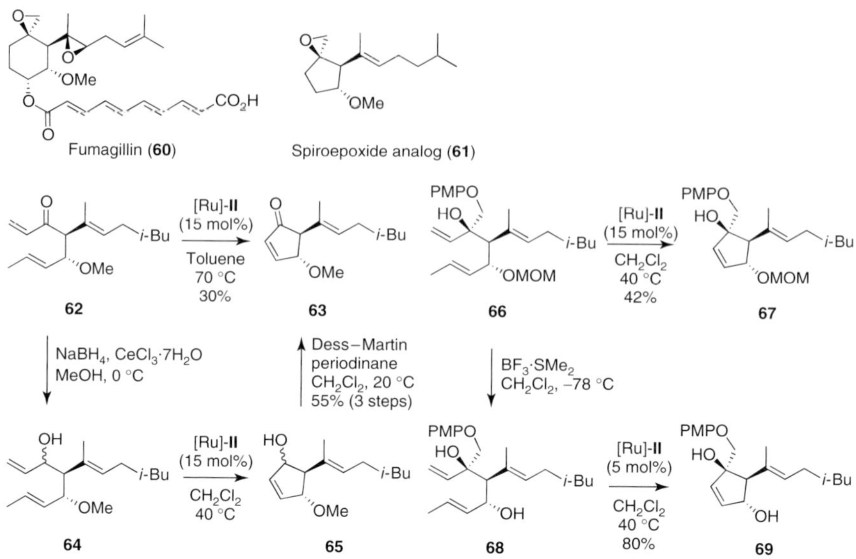

Scheme 1.10

potential methionine aminopeptidase-2 inhibitor related to the natural product fumagillin (**60**) (Scheme 1.10) [13]. Actually, the RCM of vinyl ketone **62** was found to be sluggish using catalyst [Ru]-**II** (toluene, 70 °C) yielding only 30% of the cyclopentenone **63**. The enhanced reactivity of the corresponding allylic alcohol **64** allowed a significant yield increase (55%, over three steps from **62**). However, the presence of a potential coordination site for the catalyst could also be detrimental to the efficiency of the RCM. Thus, RCM of the Methoxymethyl (MOM) ether **66** gave cyclopentene **67** in moderate yield (42%), whereas the corresponding free alcohol **68** cyclized to **69** in 80% yield with a threefold decrease in catalyst loading (Scheme 1.10).

Cyclopentenes bearing a trisubstituted double bond could be efficiently prepared via the RCM of the corresponding 1,6-diene. Catalyst [Ru]-**II** is well suited for this transformation since [Ru]-**I** usually leads to inferior yields. The prochiral trisubstituted olefin could then be further reduced or oxidized, thus creating a new tertiary or quaternary stereogenic center.

Trauner *et al.* used RCM as a key step to elaborate the cyclopentyl core of (−)-guanacastepene E (**70**) and (−)-heptemerone B (**71**) (Scheme 1.11) [14]. The cyclization precursor was prepared via an asymmetric ene reaction of glyoxylate **72**. Further transformations led to 1,6-diene **74** that could be cyclized to cyclopentenone **75** in 86% yield using [Ru]-**II** (5 mol%). Conjugate addition of the organocuprate **76** then established the C11 quaternary stereocenter. The central seven-membered ring was then closed with an electrochemical oxidation.

Fomannosin (**79**) is a sesquiterpene metabolite isolated from the wood-destroying Basidiomycetes fungus *Fomes annonsus* (Fr.) Karst. Its inherent instability and very congested structure have elicited a great deal of interest from the synthetic

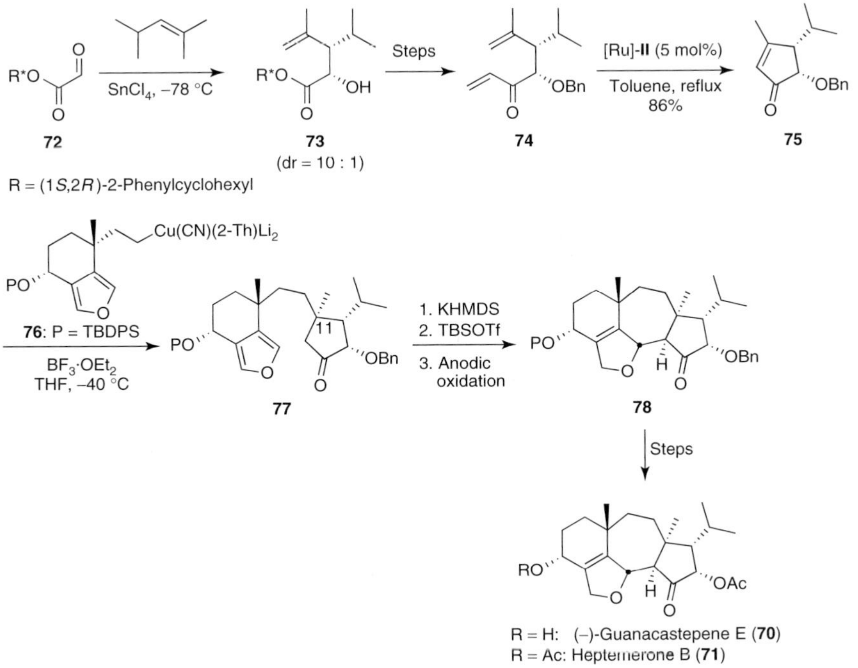

Scheme 1.11

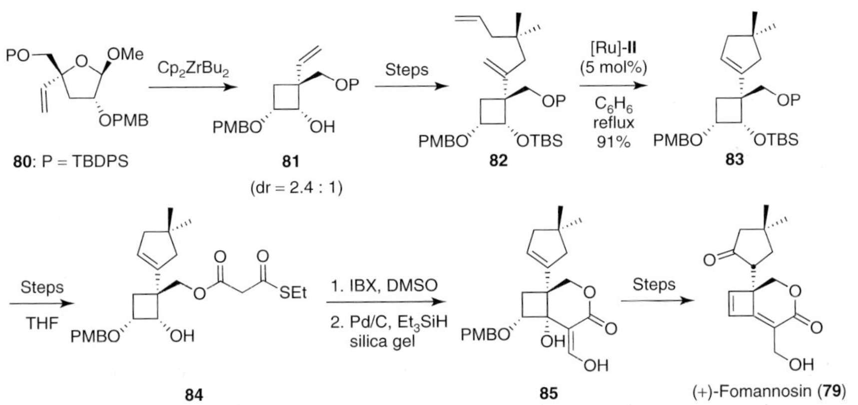

Scheme 1.12

community. Paquette *et al.* reported a straightforward strategy for the elaboration of both antipodes of fomannosin. Compound **80** (prepared from α-D-glucose) underwent a zirconium-promoted ring contraction providing the four-membered ring **81**. The challenging RCM of **82** catalyzed by [Ru]-**II** was used as a key step, and led to cyclopentene derivative **83** featuring a trisubstituted double bond adjacent to

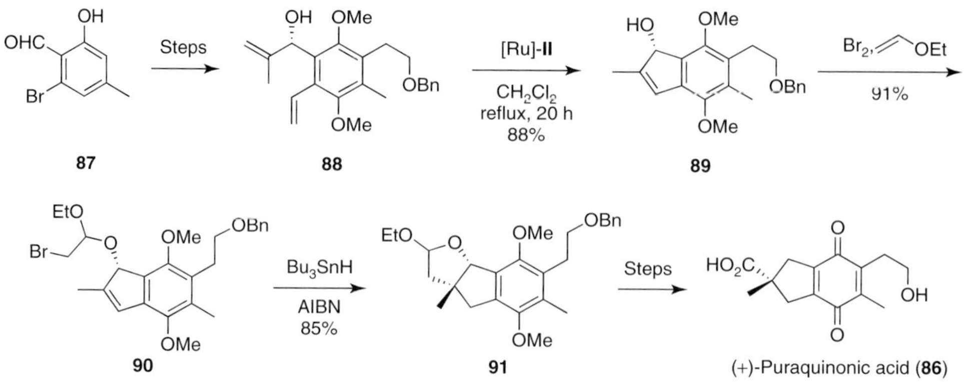

Scheme 1.13

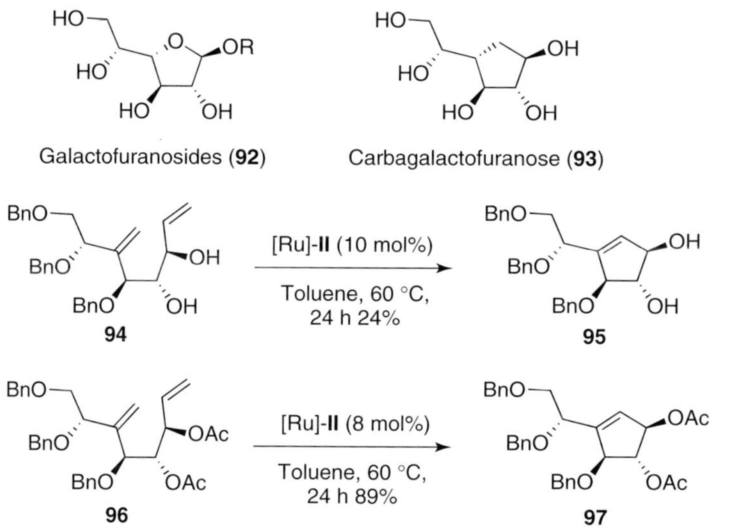

Scheme 1.14

a quaternary stereocenter (Scheme 1.12) [15]. The lactone ring was then elaborated by an intramolecular Knoevenagel reaction.

RCM was used as a key step in the total synthesis of (+)-puraquinonic acid (**86**), a fungal metabolite that induces differentiation in HL-60 cells [16]. Starting from aromatic aldehyde **87**, the RCM precursor **88** was prepared in a few steps. Following RCM leading to **89** (88%), the newly formed trisubstituted alkene was involved in a radical cyclization. Bromo acetal **90** was treated with Bu_3SnH and AIBN thus triggering a stereoselective 5-*exo* trig cyclization reaction. Further steps led to the natural product **86** (Scheme 1.13).

Carbahexofuranoses and carbapentofuranoses syntheses have been intensively investigated and RCM reaction emerged as one of the most practical and efficient methods to this effect. Among the numerous syntheses of such derivatives, the synthesis of 4α-carba-β-D-galactofuranose **93** is worthy of note [17]. A dramatic protecting group effect was observed during the RCM of **94** and **96**. Whereas free allylic alcohols are known to enhance reactivity, RCM of **94** leads to the five-membered ring **95** in poor yield (24%). Under the same conditions, the diacetate **96** undergoes an efficient RCM providing **97** in 89% yield (Scheme 1.14).

Two simultaneous RCM have been used in efforts toward the core structure of the elisabethin diterpenoids [18]. Trisallylcarveol (**99**), accessible from (*R*)-carvone, can be transformed in a single step under mild conditions albeit with a moderate yield into the tricyclic compound **100** possessing the core structure of the desired target molecule (Scheme 1.15).

Tetrasubstituted double bonds embedded in a five-membered ring can also be efficiently prepared via RCM. In the total synthesis of (±)-spirotenuipesines A (**101**)

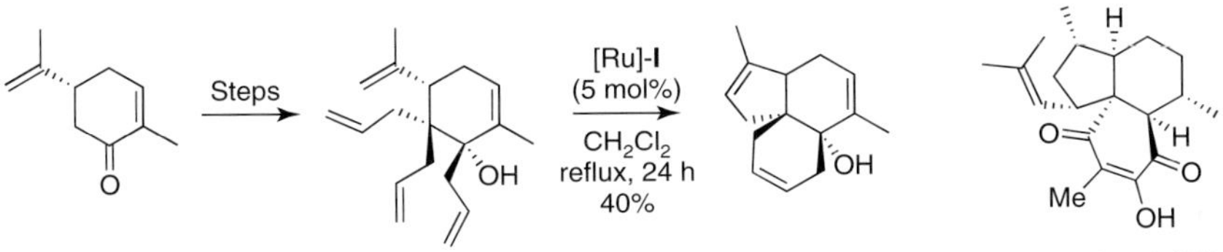

Scheme 1.15

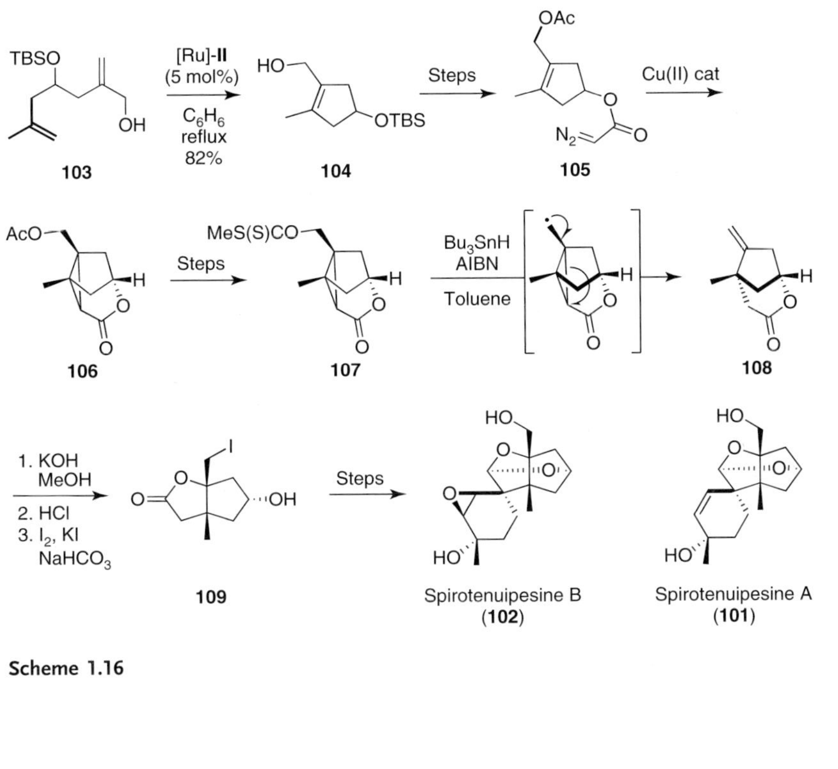

Scheme 1.16

and B (**102**), Danishefsky *et al.* simplified the preparation of a key intermediate, compound **104**, from nine steps [19] to three steps when an RCM strategy was adopted (Scheme 1.16) [20]. The resulting tetrasubstituted double bond in **104** was then submitted to an intramolecular diazoester cyclopropanation reaction. Radical opening of the resulting strained cyclopropyl group in **107** delivered bicyclic lactone **108**, which was converted to **109** by iodolactonization. The latter compound could be further transformed to (±)-spirotenuipesines A (**101**) and B (**102**).

Another example of elaboration of tetrasubstituted double bonds by RCM was disclosed by Kotora *et al.* in a formal total synthesis of estrone (Scheme 1.17). Starting from **110**, two zirconium-mediated cyclizations enabled the preparation of compound **112** whose RCM catalyzed by [Ru]-**II** (20 mol%) (toluene, 90 °C) provided estrone precursor **113** in 82% yield [21].

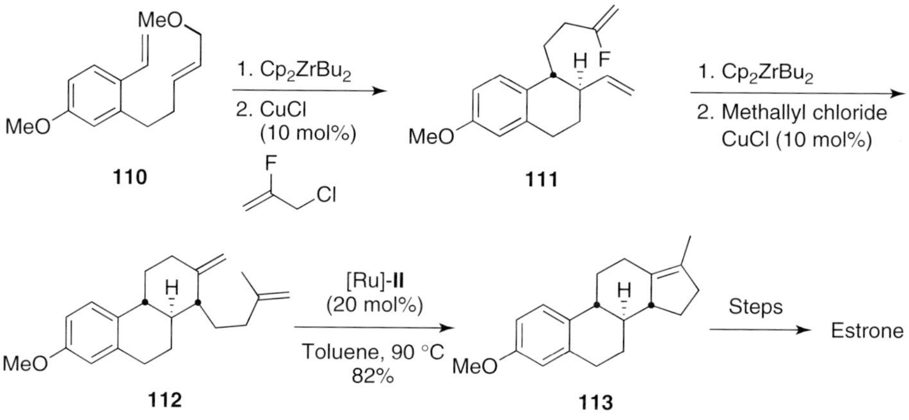

Scheme 1.17

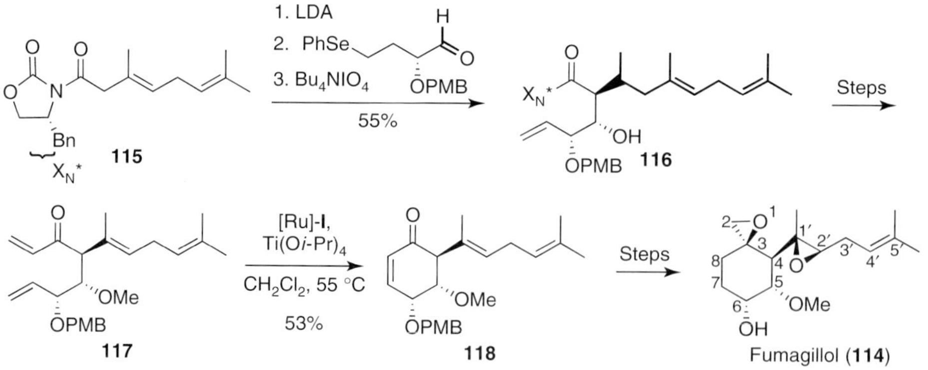

Scheme 1.18

1.3 Formation of Six-membered Carbocycles by RCM

Among the recently reported synthetic approaches to angiogenesis inhibitors such as ovalicin, fumagillin, and synthetic analogs, several use RCM to construct the six-membered ring characteristic of this family of molecules [22–24]. In the first metathesis-based synthesis of fumagillol (**114**) (Scheme 1.18) [22a], the nonfunctionalized C7–C8 bond was established through an RCM/hydrogenation sequence. In agreement with Fürstner's observations, adding $Ti(Oi\text{-}Pr)_4$ was required for the RCM of **117** to proceed with catalyst [Ru]-**I** [25]. The resulting advanced intermediate **118** was easily converted to fumagillol. The approach was used for the preparation of fumagillin analogs **121–123**, (Scheme 1.19) [22b] Using the more active [Ru]-**II** eliminated the need for $Ti(Oi\text{-}Pr)_4$ in the RCM of **119** to **120**.

The second approach to fumagillol (**114**) relies on a Claisen/RCM sequence (Scheme 1.20) [23c]. Starting from 1,2;5,6-di-*O*-isopropylidene D-mannitol (**124**), the ester **125** was prepared and subjected to a glycolate-Claisen rearrangement leading to **126** (77%). RCM of **126** afforded cyclohexene **127** in high yield (94%), which was converted to an advanced intermediate previously used by Sorensen *et al.* in their synthesis of fumagillol [26].

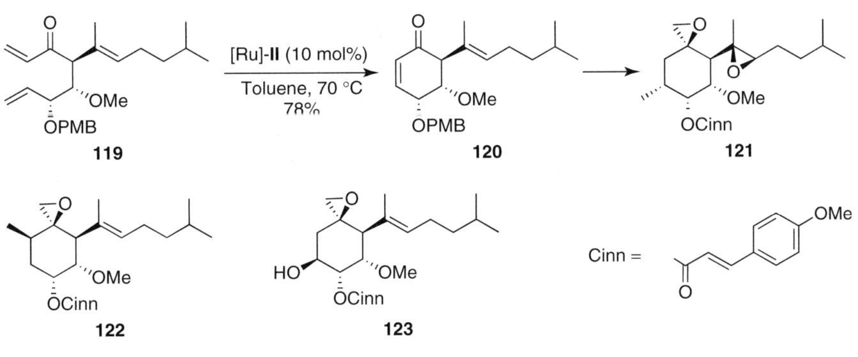

Scheme 1.19

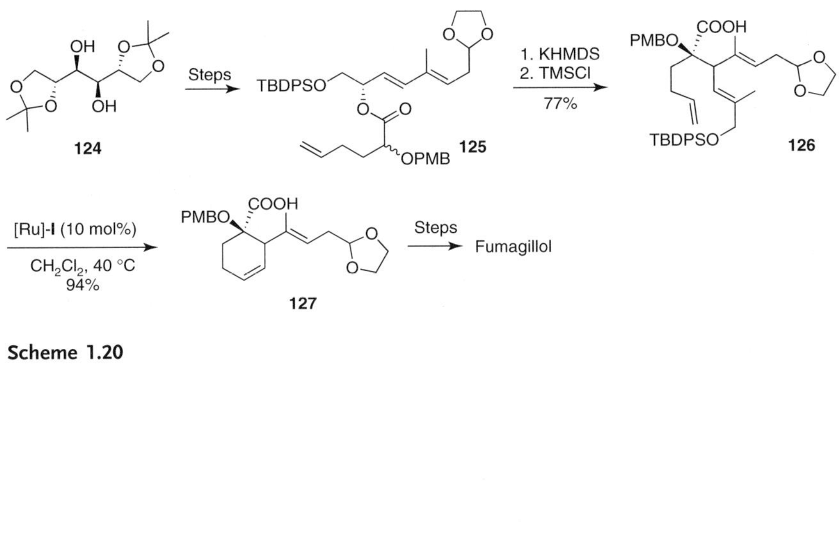

Scheme 1.20

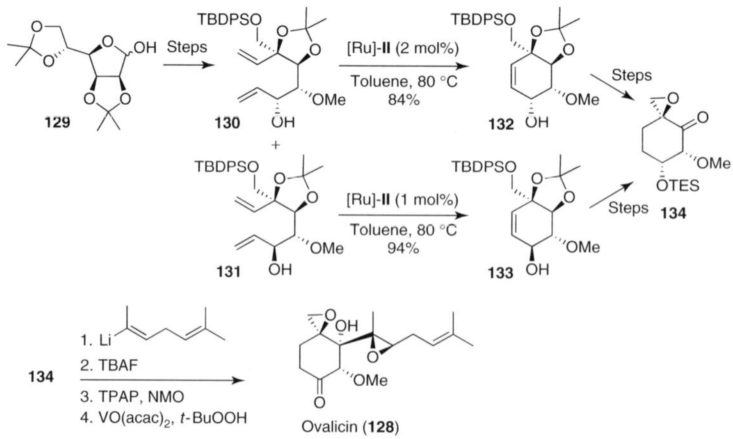

Scheme 1.21

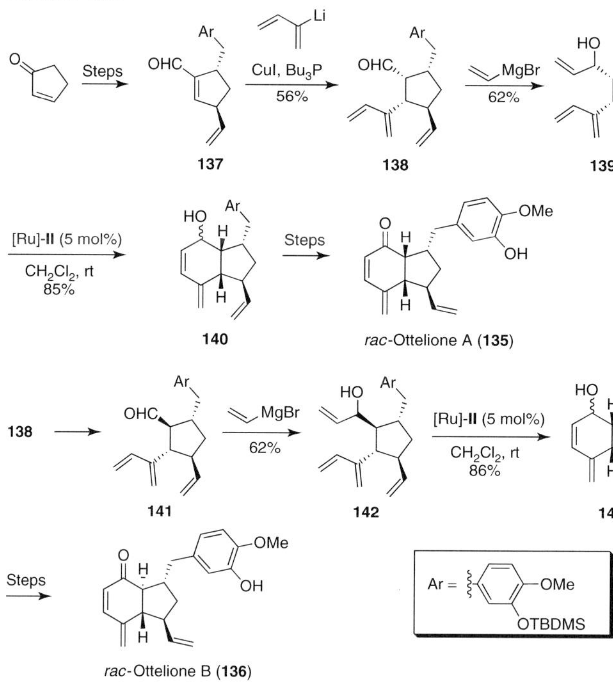

Scheme 1.22

In contrast to the two previous syntheses, the six-membered ring was constructed first in Takahashi's *et al.* synthesis of ovalicin (**128**) (Scheme 1.21) [24]. Thus, 2,3;5,6-di-*O*-isopropylidene-D-mannose (**129**) was converted to the diastereomeric RCM substrates **130** and **131**, which smoothly cyclized to compounds **132** and **133** in the presence of [Ru]-**II** catalyst.

Clive *et al.* designed a synthesis of both otteliones A (**135**) and B (**136**), featuring an efficient cuprate conjugate addition/Grignard reaction/RCM sequence

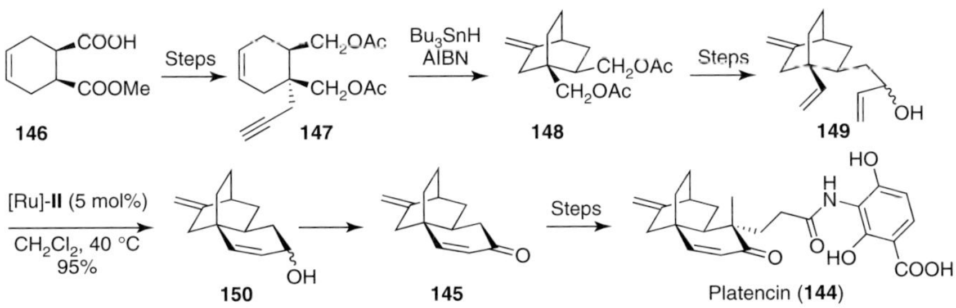

Scheme 1.23

(Scheme 1.22) [27]. Starting from cyclopentenone, the cyclopentene carboxaldehyde **137** was prepared. Conjugate addition of the organocopper derived from 2-butadienyllithium generated **138** and subsequent treatment with vinylmagnesium bromide afforded the RCM substrate **139**. This compound was cyclized using catalyst [Ru]-**II** to furnish intermediate **140**, which was converted to *rac*-ottelione A in two steps. Using the same sequence, the epimeric *trans*-aldehyde **141** (obtained by equilibration of **138**) afforded *rac*-ottelione B via **142** and **143**. An asymmetric version of these syntheses was also developed.

Very recently, two RCM-based but conceptually different formal syntheses of platencin **144** were disclosed [28, 29]. In both cases, the target was intermediate **145** that has been previously converted to platencin by Nicolaou *et al.* [30a] and Rawal *et al.* [30c].

In Lee's *et al.* synthesis [28], the bicyclo[2.2.2]octane moiety was assembled through a radical cyclization/skeletal rearrangement [31]. Standard chemical manipulation then led to triene **149** whose RCM ([Ru]-**II**) afforded cyclohexenol **150** in 95% yield (Scheme 1.23).

Mulzer's *et al.* approach implies a nonobvious construction of the strained bicyclo[2.2.2]octane skeleton by RCM (Scheme 1.24) [29]. Starting from (−)-perillaldehyde (**151**), the decaline **152** was prepared. Wittig reaction afforded triene **153**, which was submitted to RCM conditions ([Ru]-**II**, CH_2Cl_2, reflux). RCM completion required more than 24 hours and 8 mol% of catalyst but the yield of **154** was excellent (90%) and no side-products were reported. Compound **154b** was then converted to intermediate **145** in two steps (Scheme 1.24).

Polyprenylated acylphloroglucinols, such as garsubellin A (**155**) and hyperforin A (**156**), are bioactive molecules characterized by a polycyclic bridged system [32]. In 2005, Shibasaki *et al.* described a RCM-based total synthesis of (±)-garsubellin A (Scheme 1.25) [33].

The α,β-unsaturated ketone **157** was converted to cyclohexanone **158** and a Claisen rearrangement/RCM sequence was used to construct the characteristic bicyclo[3.3.1]nonane skeleton.[1] Thus, enol ether **159** cleanly rearranged at high temperature to diene **160**.[2] Treatment of **160** by [Ru]-**III** [(20 mol%), toluene, reflux,

1) The Claisen/RCM sequence has been used repeatedly for constructing various ring systems.

2) Ethylene carbonate was added to the reaction mixture to compensate for carbonate hydrolysis in 183.

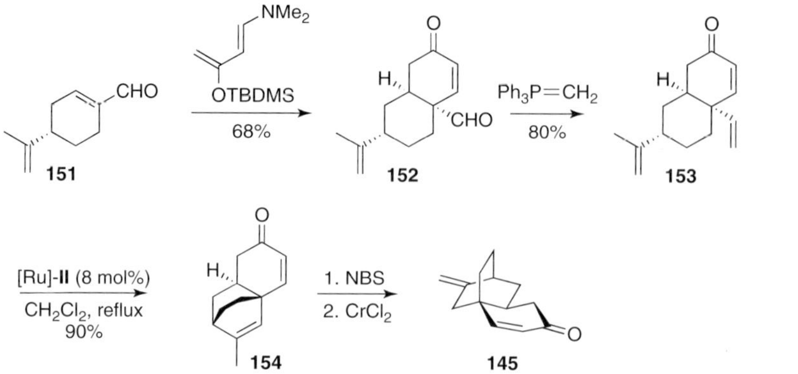

Scheme 1.24

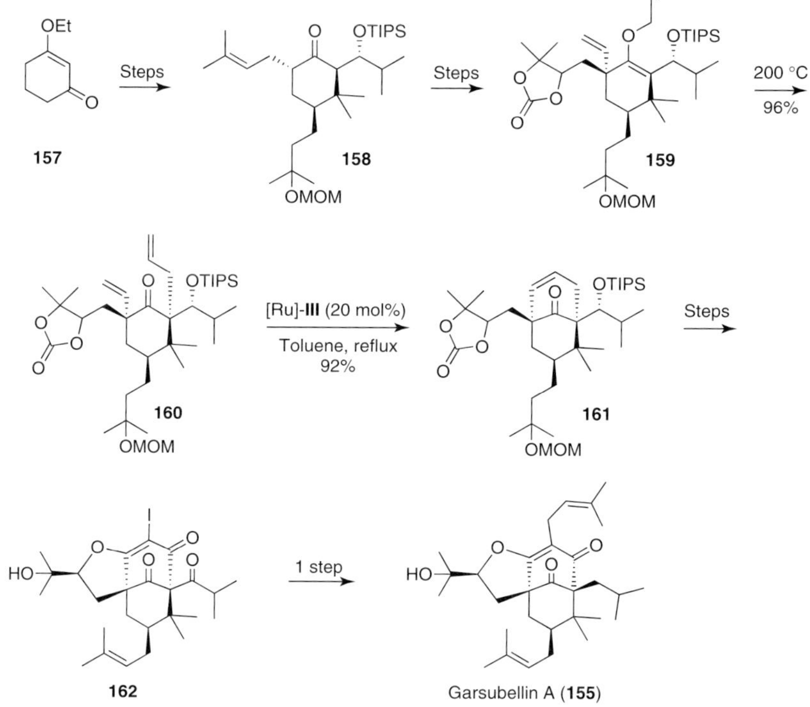

Scheme 1.25

48 hours] led to intermediate **161** in excellent yield [34]. Finally, **161** was converted to **162** and then to (±)-garsubellin A (**155**). As shown in Scheme 1.26, the method could not be extended to hyperforin (**156**) as RCM of **163** led to **164** possessing a seven-membered ring (Scheme 1.26).

Eight to ten-membered-rings, which are common in natural products [35], can be prepared by heterolytic fragmentation of bicyclic systems [36]. The latter can be formed by several methods among which RCM is one of the most efficient.

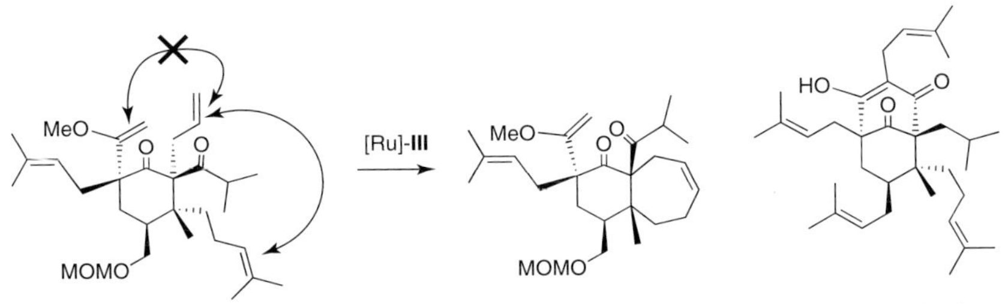

Scheme 1.26

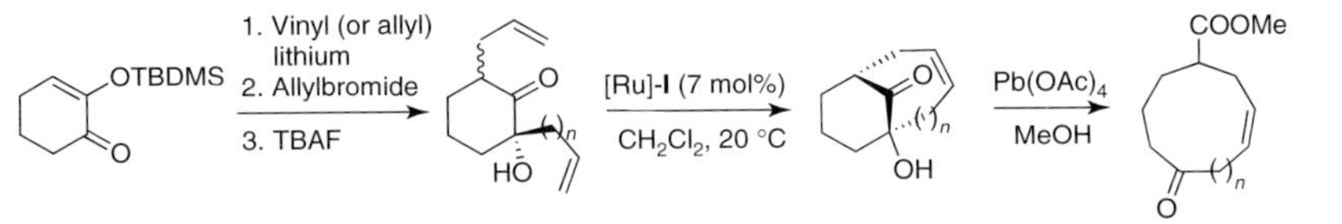

Scheme 1.27

Mascareñas *et al.* reported an RCM/ring fragmentation strategy to eight- and nine-membered carbocycles (Scheme 1.27) [37]. Starting from 1,2-cyclohexanedione, the dienes **166a** and **166b** were obtained. When submitted to RCM conditions ([Ru]-**I**, 20 °C), only the diastereomers having the two unsaturated side chains in a *syn* relationship reacted. The resulting bridged compounds **167a** and **167b** were cleaved by lead tetraacetate to afford ketoesters **168a** and **168b**. By contrast, direct construction of the eight-membered ring **170** by RCM of **169** was unsuccessful (Scheme 1.28) [37a].

A related RCM/fragmentation protocol was used for preparing germacranolides that may exist as *E,E*-, *E,Z*-, or *Z,Z*-10-membered rings (Figure 1.1) [38].

Starting from (*R*)-carvone, the trienes **171** and **172** were prepared and RCM provided decalines **173** and **174**, respectively (Scheme 1.29). Selective conversion of the secondary alcohols in **173** and **174** to the corresponding mesylates and Wharton fragmentation [39] led to (*Z,Z*)- and (*E,Z*)-germacratrienes **177** and **178**. The more substituted germacrenes **179–182** were also prepared using this strategy [38].

A similar strategy was reported for a synthesis of (±)-periplanone C (**183**) (Scheme 1.30) [40]. The triene **185** was prepared in a few steps from 3-isopropylanisole and smoothly cyclized using [Ru]-**I** (3 mol%) to the *trans*-decaline

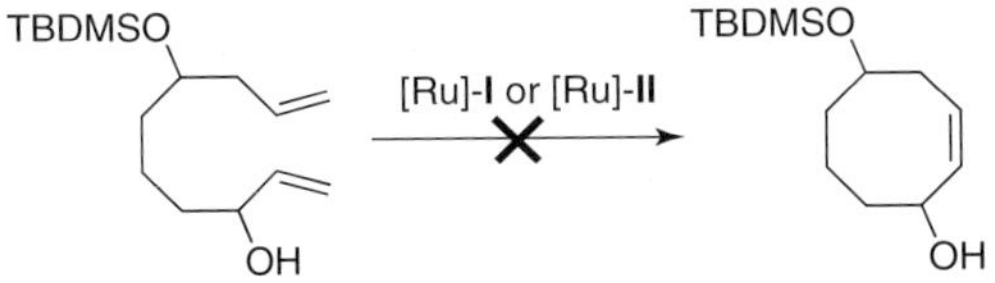

Scheme 1.28

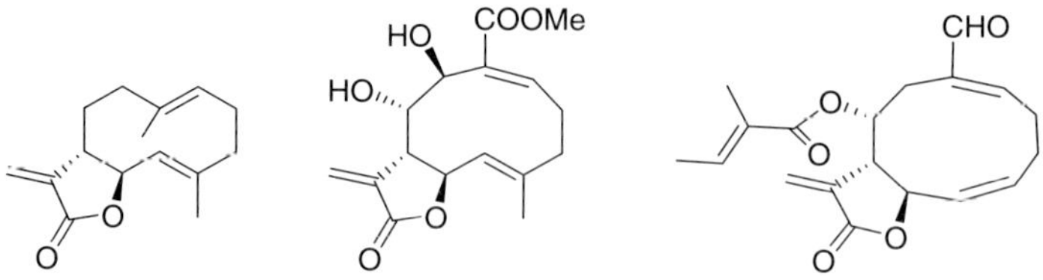

Figure 1.1 Various types of germacranolides.

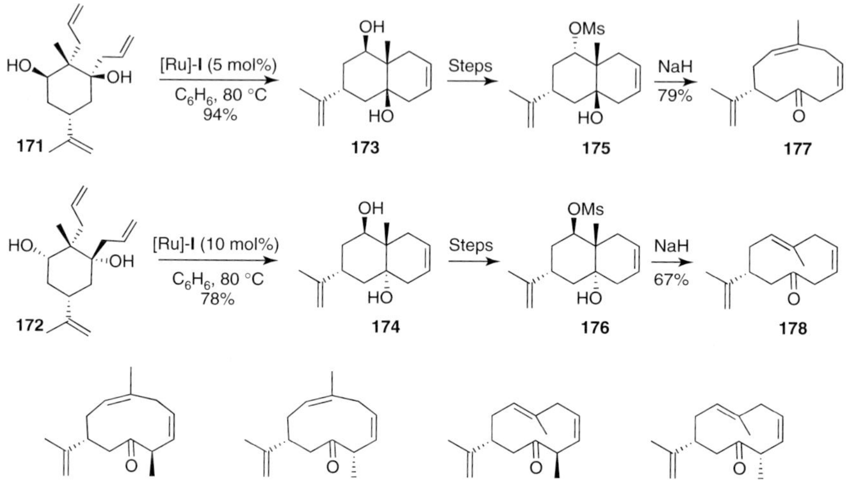

Scheme 1.29

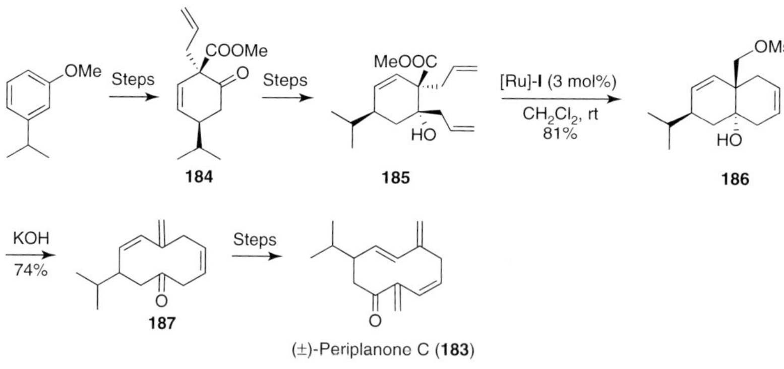

Scheme 1.30

186 (81%). Mono-*O*-mesylation and treatment with KOH afforded cyclodecanone **187**, which was converted to (±)-periplanone C (**183**) in a few steps. Interestingly, using the same conditions, *epi*-**185** did not cyclize. This was attributed to the favorable preorganization of **185**, favoring RCM, whereas the more stable conformation of *epi*-**185** places the two olefins away from each other (Scheme 1.31) [41].

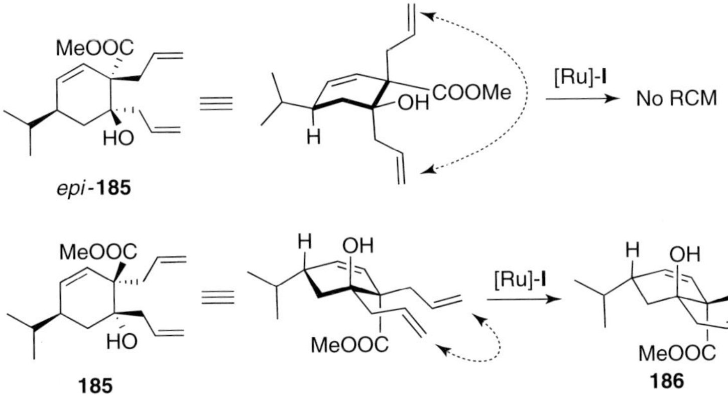

Scheme 1.31

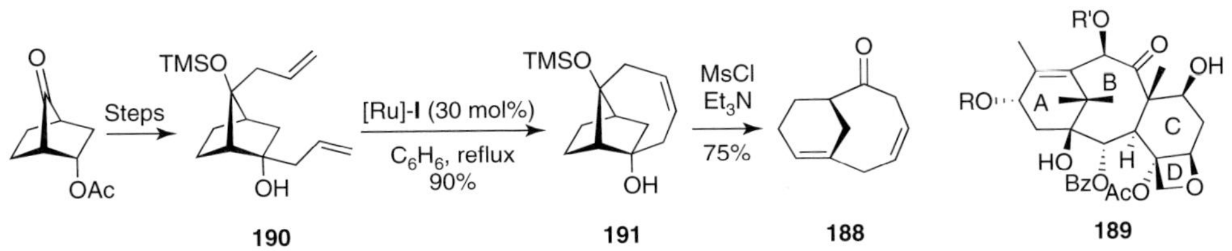

Scheme 1.32

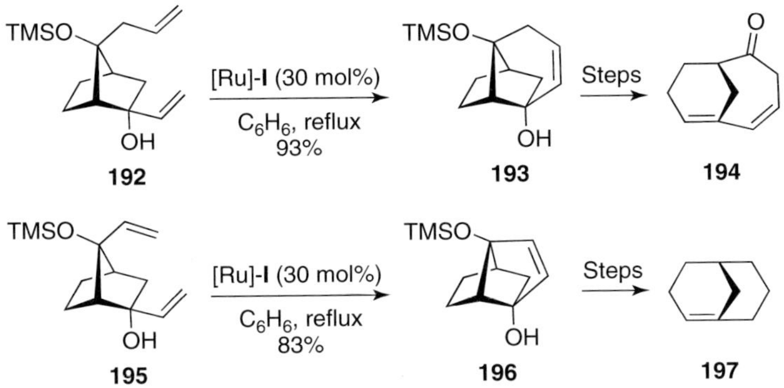

Scheme 1.33

Access to anti-Bredt alkenes is possible using an RCM/fragmentation sequence [42]. The strategy has been used for preparing the bicyclic ketone **188** having a structure similar to the AB rings of taxanes **189**. RCM of **190** catalyzed by [Ru]-**I** afforded **191** (90%) and subsequent fragmentation provided ketone **188** (Scheme 1.32). Application to the construction of other bridged ring systems has been examined. RCM of **192** delivered **193** in high yield (93%) and the cyclization of **195** to afford the strained cycloalkene **196** (83%) is noteworthy (Scheme 1.33).

In 2006, Robichaud and Tremblay reported the first RCM-based formal synthesis of (+)-compactin (**198**) [43]. The functionalized cyclohexene **199** (ring A) was prepared and submitted to RCM conditions to construct the six-membered B

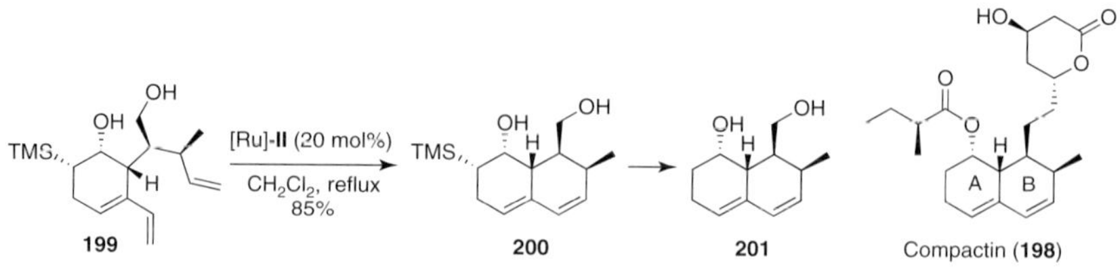

Scheme 1.34

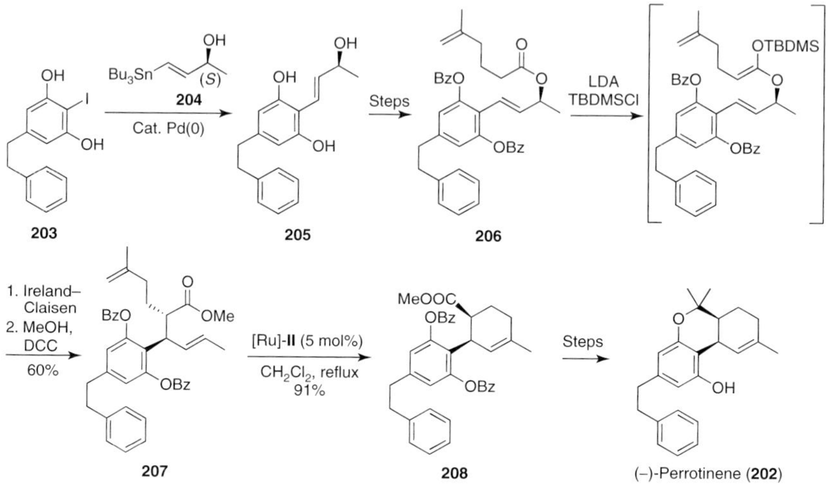

Scheme 1.35

ring, thus completing the elaboration of the fully functionalized decaline system **200**. Removal of the trimethylsilyl (TMS) group was difficult, and provided the target compound **201** in 31% yield, whose transformation to compactin had been previously described (Scheme 1.34).

The first synthesis of (−)-perrotinene **202** was reported in 2008 [44]. The strategy is again characterized by an Ireland–Claisen rearrangement/RCM key sequence (Scheme 1.35). Alcohol **205** was prepared by Stille coupling between the aryl iodide **203** and the enantiomerically pure vinyl stannane **204**. The ensuing, highly stereoselective, Ireland–Claisen rearrangement of the silyl ketene acetal derived from **206** allows a complete chirality transfer and the control of the configuration of the two stereogenic centers in **207**. RCM of the latter substrate proceeded smoothly and gave cyclohexene **208** in high yield (91%).

Recently, Li *et al.* described a synthesis of (−)-10α- and (−)-10β-hydroxy-4-muurolen-3-one (**209**) [45]. Starting from (*R*)-carvone, compound **210** was prepared in a few steps and its RCM proceeded smoothly using catalyst [Ru]-**II** to afford intermediate **211** in quantitative yield (Scheme 1.36).

Vannusal A (**212**) is a complex triterpene characterized by a pentacyclic bridged core. Nicolaou *et al.* have devised a synthetic approach to compound **216** possessing the core structure of vannusal A (Scheme 1.37) [46]. A Mn(III)-induced radical

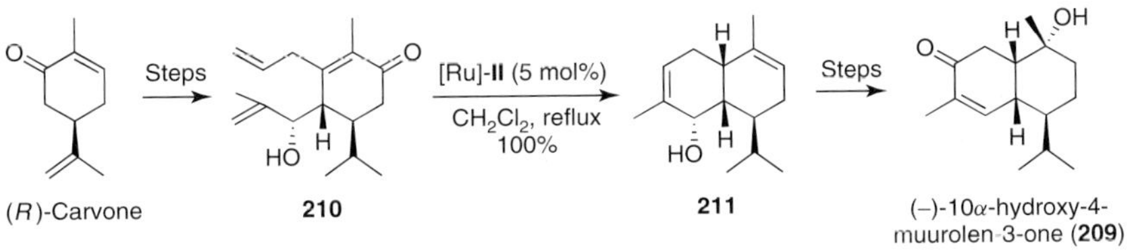

Scheme 1.36

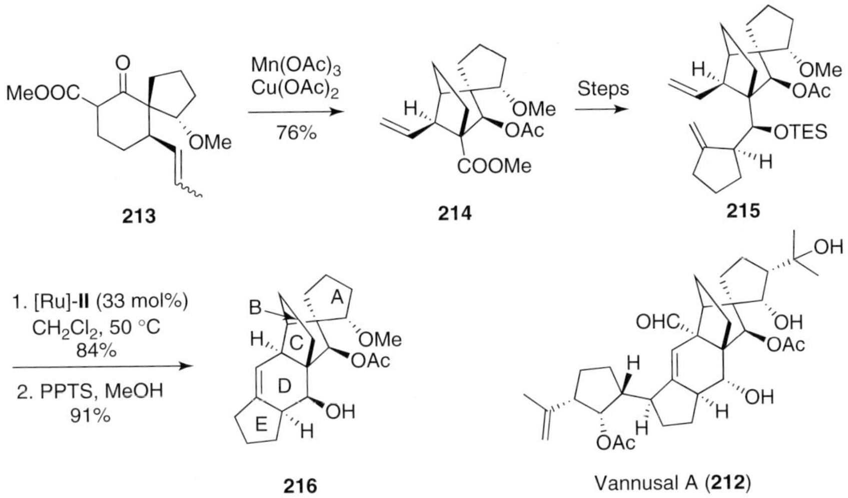

Scheme 1.37

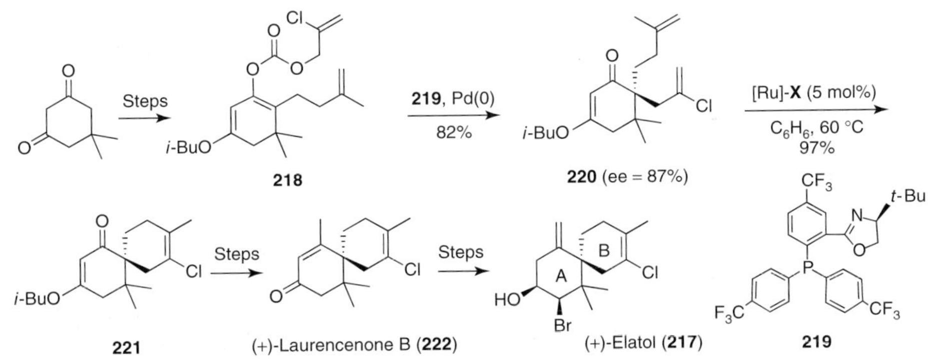

Scheme 1.38

cyclization of the β-ketoester **213** was used to prepare the bridged tricyclic system ABC **214**. This compound was converted in a few steps to the RCM substrate **215** whose cyclization proceeded cleanly and in good yield (84%) using catalyst [Ru]-**II** (33 mol%).

In 2008, Stoltz *et al.* reported the first synthesis of elatol **217** (Scheme 1.38) [47]. Starting from dimedone, the mixed carbonate **218** was prepared. Decarboxylative oxygen-to-carbon migration was effected using a Pd(0) complex based on the chiral

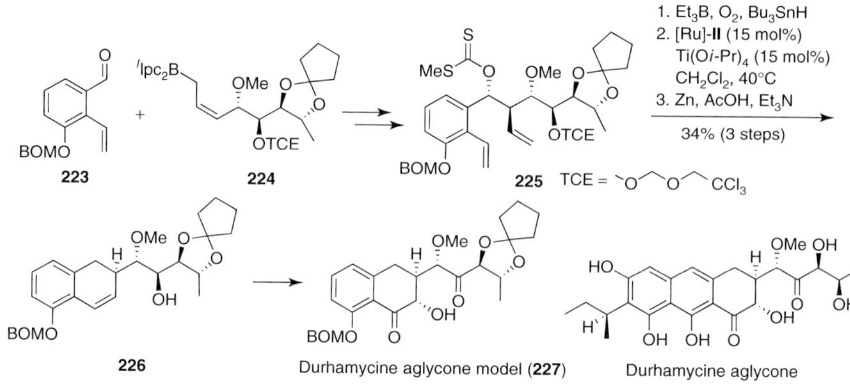

Scheme 1.39

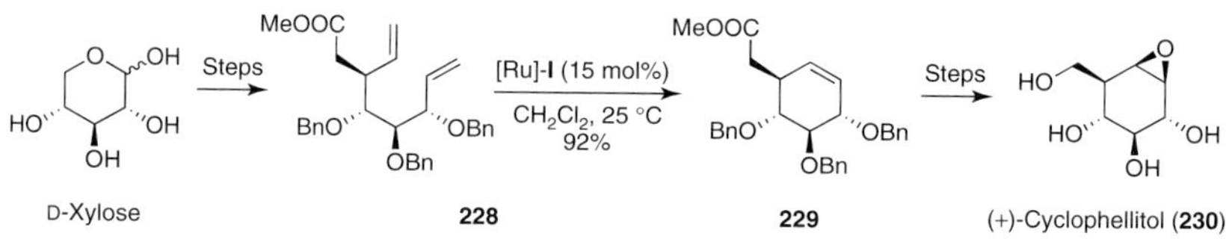

Scheme 1.40

phosphine **219** [48] to afford the optically enriched ketone **220** (ee = 87%) in good yield (82%). The challenging RCM of **220**, leading to a tetrasubstituted double bond, proceeded exceedingly well using catalyst [Ru]-**X** [49]. The RCM product **221** was then converted to (+)-laurencenone (**222**) then (+)-elatol (**217**) in a few steps.

Roush *et al.* have reported the synthesis of a durhamycin aglycon model (Scheme 1.39) [50]. Allylation of aldehyde **223** with the allylic borane **224** provided **225** after the formation of a xanthate. Deoxygenation gave the requisite RCM precursor whose cyclization proceeded reasonably well provided that $Ti(Oi\text{-}Pr)_4$ was used as an additive. As suggested by model studies, this additive may avoid the coordination of the ruthenium carbene by either the aryl ether (OBOM) or the neighboring ethers (OMe, OTCE).

The mild conditions of the olefin metathesis reaction make it an ideal method for the preparation of cyclitols and carbasugars [51, 52]. In general, the syntheses start from readily available carbohydrate derivatives that are converted to dienes and then submitted to RCM conditions. The resulting functionalized cyclohexenes are then converted to the target molecules in a few straightforward steps [53b]. According to this principle, the syntheses of (+)-cyclophellitol (**230**) (Scheme 1.40) [54], valienamine (**233**) (Scheme 1.41) [53, 55], valiolamine (**236**) (Scheme 1.42) [56], conduritol derivatives **238**–**240** (Scheme 1.43) [57], phosphatidyl inositol [58], and analogs (**243**) (Scheme 1.44) [59] have been reported.

Noncarbohydrate starting materials may also be used for preparing carbasugars. For example, Roush's *et al.* asymmetric allylation of chiral aldehyde **244** gave diene **245**. Subsequent RCM catalyzed by [Ru]-**II** and dihydroxylation allowed access to

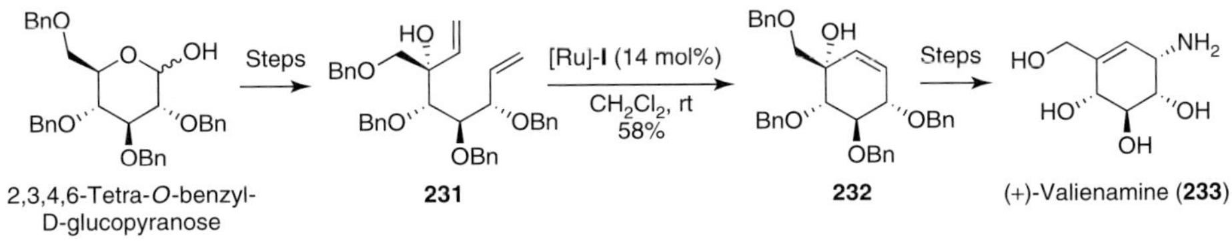

Scheme 1.41

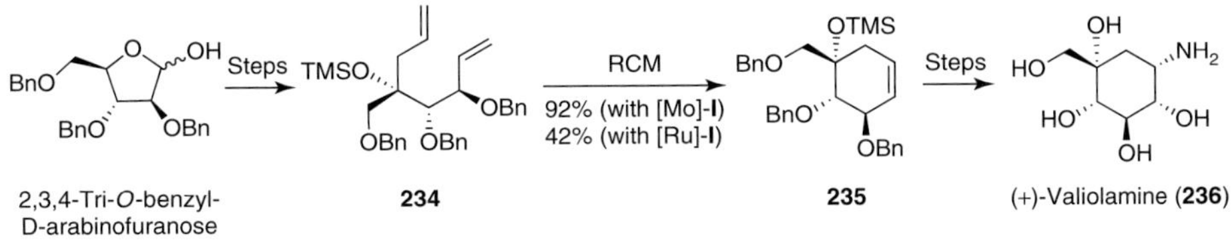

Scheme 1.42

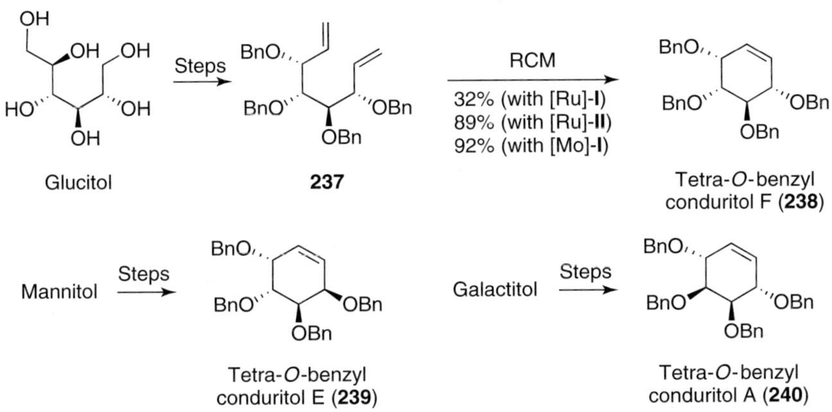

Scheme 1.43

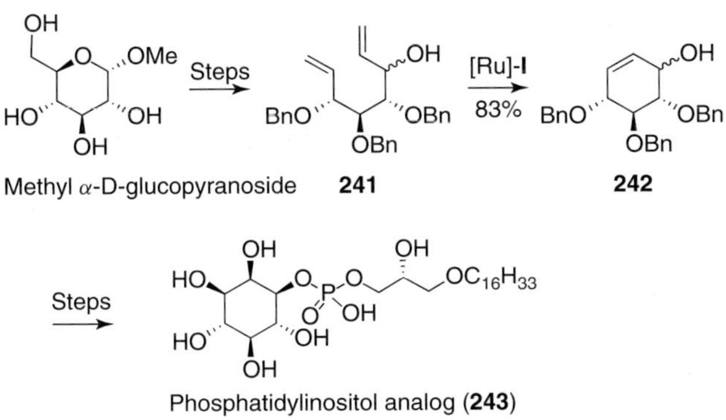

Scheme 1.44

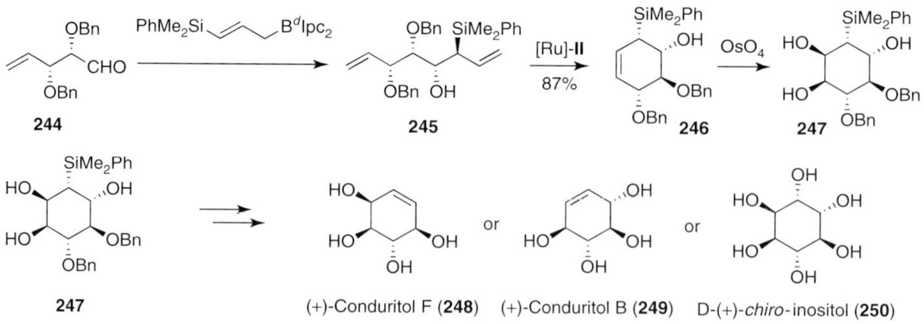

Scheme 1.45

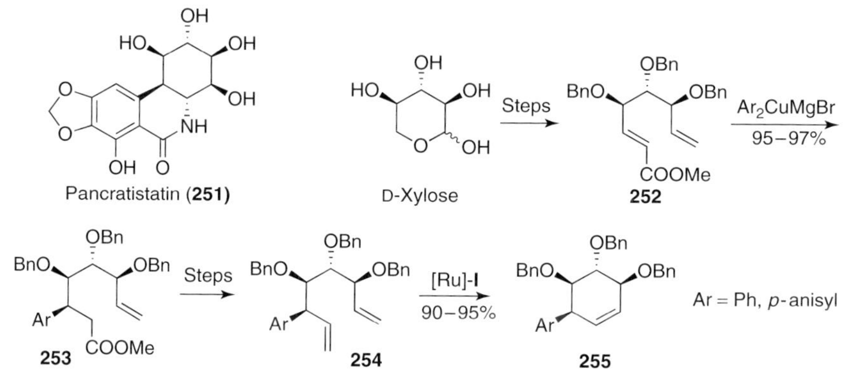

Scheme 1.46

the pivotal intermediate **247**. This compound was converted to conduritols F (**248**) and B (**249**) by stereospecific *syn* (basic conditions) or *anti* (acidic conditions) Peterson elimination, respectively, and debenzylation. Alternatively, Tamao–Fleming oxidation and debenzylation provided D-(+)-*chiro*-inositol (**250**) (Scheme 1.45) [60].

Pancratistatin (**251**), an important anticancer product, is structurally related to carbasugars. In 2004, Kornienko and Nadein reported a synthesis of 1-aryl-1-deoxyconduritols that were used as synthetic intermediates to prepare a series of simplified pancratistatin analogs. Compound **252**, prepared from D-xylose, underwent conjugate addition of magnesium diarylcuprates to afford **253**. Further steps led to **254** whose RCM catalyzed by [Ru]-**I** provided aryl deoxyconduritols **255** (Scheme 1.46) [61, 62].

1.4
Formation of Seven-membered Carbocycles by RCM

RCM of 1,8-dienes constitutes an efficient access to cycloheptenes since di-, tri-, and tetrasubstituted olefins could be elaborated in good yields with catalyst loading as low as 1 mol%. In this context, the highly demanding field of natural product synthesis constitutes an efficient test for the development of more active and

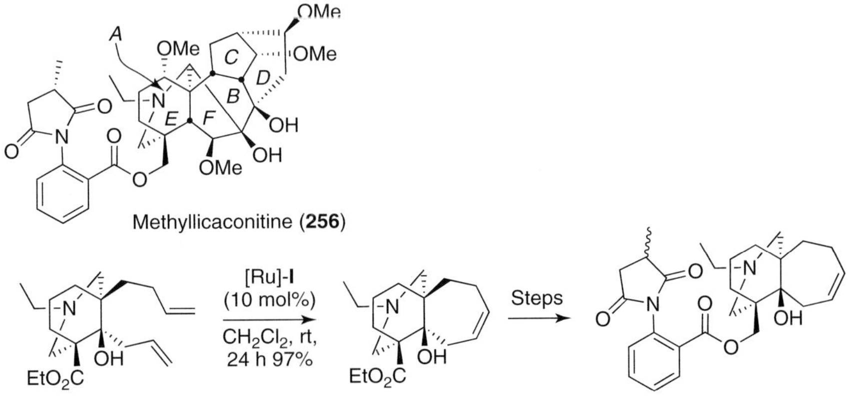

Methyllicaconitine (**256**)

257 **258** **259**

Scheme 1.47

functional group tolerant catalysts as shown by Stoltz in the (−)-cyanthiwigin F synthesis. The following section, by no mean exhaustive, covers the more recent applications in RCM of 1,8-dienes applied to the synthesis of biologically relevant targets. When possible, comparison between various catalysts and experimental conditions is also discussed.

Challenging RCM was reported during the synthesis of the ABE tricyclic analogs **259** of the alkaloid methyllicaconitine (**256**), a competitive antagonist of nicotine acetylcholine receptors [63]. The B ring system was synthesized via a [Ru]-**I** catalyzed RCM of the 1,8-diene **257** possessing two contiguous quaternary stereogenic centers. Compound **258** having the characteristic *trans* AB ring fusion of methyllicaconitine was obtained in excellent yield (97%) (Scheme 1.47).

A key step in the total synthesis of (±)-frondosin B (**260**) reported by Mehta and Likhite is RCM of 1,8-diene **262** using [Ru]-**I** (C_6H_6, reflux) [64]. The corresponding bicyclic 6,7-fused ring system **263** was obtained in good yield (80%). Cyclization of **264** mediated by boron trifluoride led to the corresponding benzofuran **265**, and subsequent acid-catalyzed isomerization of the trisubstituted olefin then provided (±)-frondosin B (**260**) (Scheme 1.48).

In synthetic studies toward rameswaralide (**266**), a promising anti-inflammatory agent, the cycloheptenone tricyclic core **269** was efficiently prepared through RCM of the acyclic ω-alkenyl acrylate **268** under very mild conditions using [Ru]-**II** (10 mol%) (CH_2Cl_2, rt) in high yield (90%) (Scheme 1.49) [65].

A related ω-alkenyl acrylate RCM was used in the first total synthesis of (+)-sundiversifolide (**270**) reported by Shishido *et al.* [66]. An excellent yield of chiral cycloheptenone **273** was obtained from **272** using catalyst [Ru]-**II**. Enone **273** was then chemoselectively reduced to the corresponding allylic alcohol **274**, thus setting the stage for an Eschenmoser–Claisen rearrangement. Further iodolactonization/radical reduction led to bicyclic lactone **275**, a late intermediate in the synthesis of the natural product (Scheme 1.50).

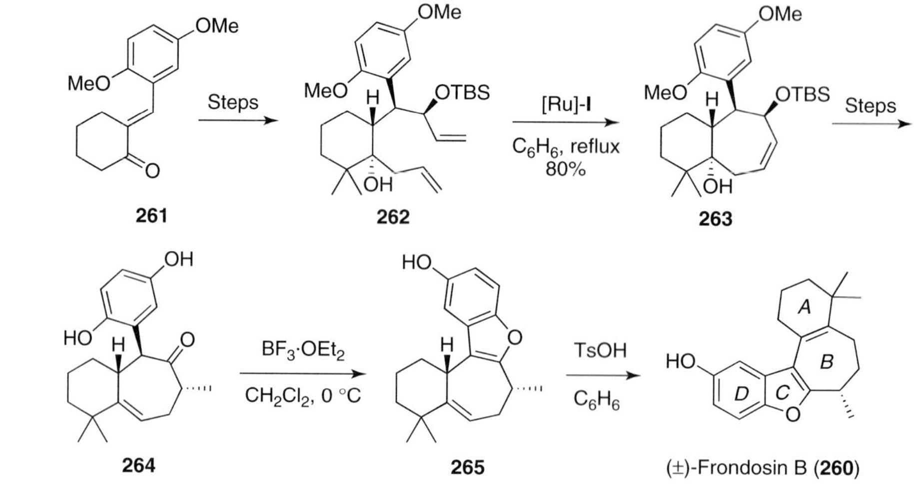

Scheme 1.48

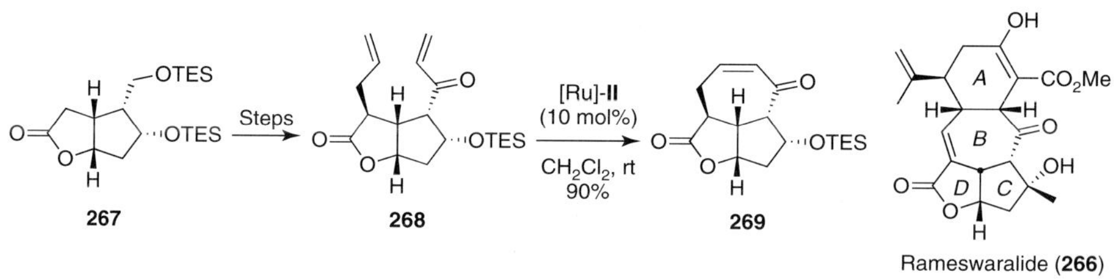

Scheme 1.49

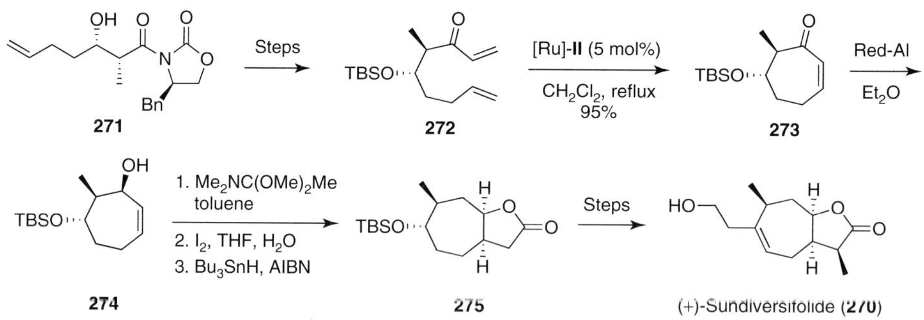

Scheme 1.50

Trisubstituted olefins embedded in seven-membered carbocycles are also synthesized in high yields from the corresponding 1,8-dienes. In their elegant total synthesis of (−)-cyanthiwigin F (**276**), Stoltz and Enquist used for the first time catalyst [Ru]-**X** [49] possessing an increased activity for the generation of trisubstituted olefins. A double Pd-catalyzed decarboxylative allylic alkylation of **277** in

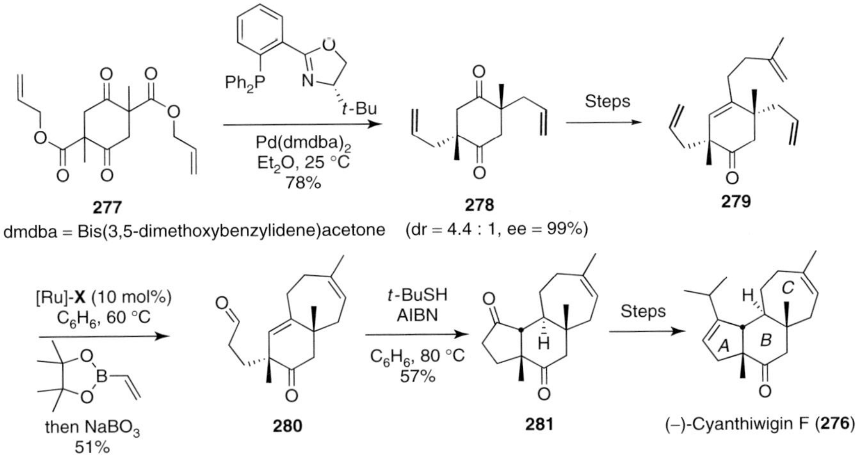

Scheme 1.51

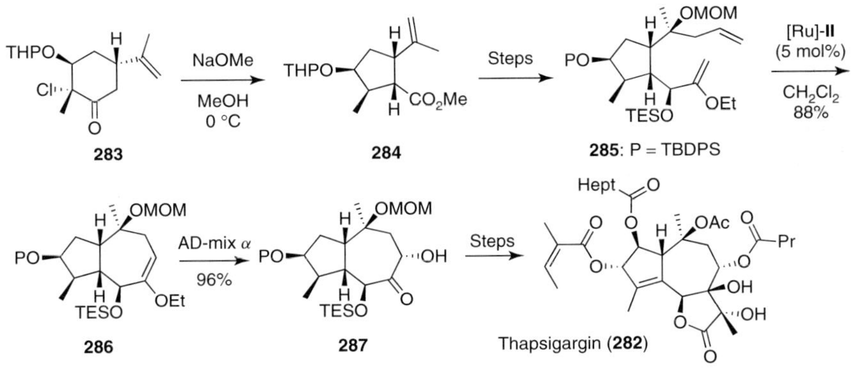

Scheme 1.52

the presence of a chiral ligand gave **278**, which was converted to tetraene **279** [67]. RCM of **279** followed by a one-pot cross-metathesis with vinylpinacolboronate led after oxidation to the bicyclic aldehyde **280** possessing the BC ring system of the natural product. Radical cyclization of the latter compound (*t*-BuSH, AIBN) established the desired ABC tricyclic motif **281** that could be further transformed to (−)-cyanthiwigin F (**276**) (Scheme 1.51).

The guaianolide class of natural products is characterized by a tricyclic 5,7,5-ring system [68]. Among these promising sesquiterpene lactones were identified the thapsigargins (such as **282**, Scheme 1.52), potent inhibitors of sarco/endoplasmic reticulum ATPases (SERCAs). Recently, Ley reported a successful strategy to access these complex natural products based on the construction of the seven-membered ring by RCM. Compound **283**, prepared from (*S*)-carvone, underwent Favorskii ring contraction and **284** was converted to enol ether **285**. RCM of **285**, using catalyst [Ru]-**II** (5 mol%) allows the formation of the bicyclic enol ether **286** in excellent yield (88%) [69, 70]. The latter was then efficiently converted into the corresponding α-hydroxyketone **287** via asymmetric dihydroxylation. The same key step was used in the total synthesis of closely related guaianolides, trilobolide, nortrilobolide and thapsivillosin F [71].

Ingenol (**288**) is a diterpenoid possessing a fascinating bicyclo[4.4.1]undecane skeleton, with a highly strained intrabridgehead topology. RCM-based strategies for the construction of the B ring of ingenol were reported by Wood [72] and Kigoshi [73]. In Wood's total synthesis, compound **289** (synthesized from 3-carene) underwent Diels–Alder cycloaddition with cyclopentadiene to afford **290** (Scheme 1.53). Ring-opening metathesis (ROM) with ethylene gave **291**, which was further elaborated to the RCM precursors **292** and **293**. Only 45% of

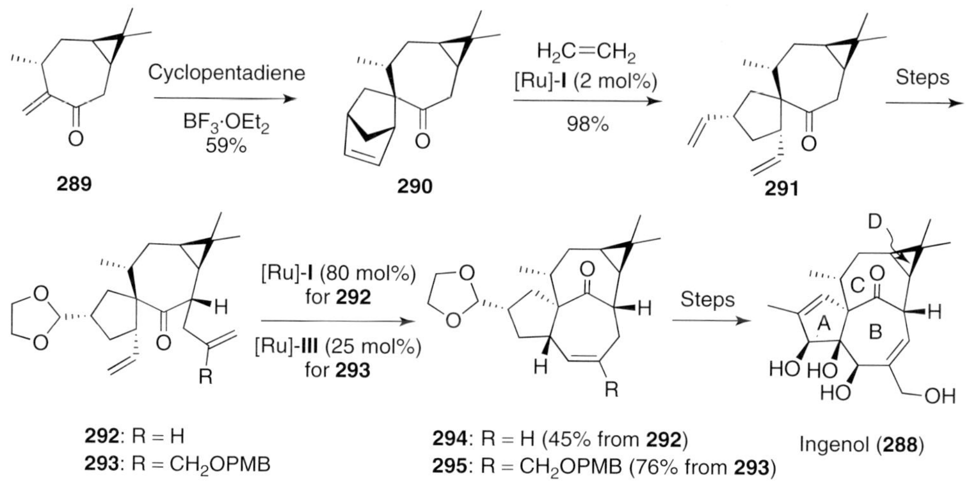

Scheme 1.53

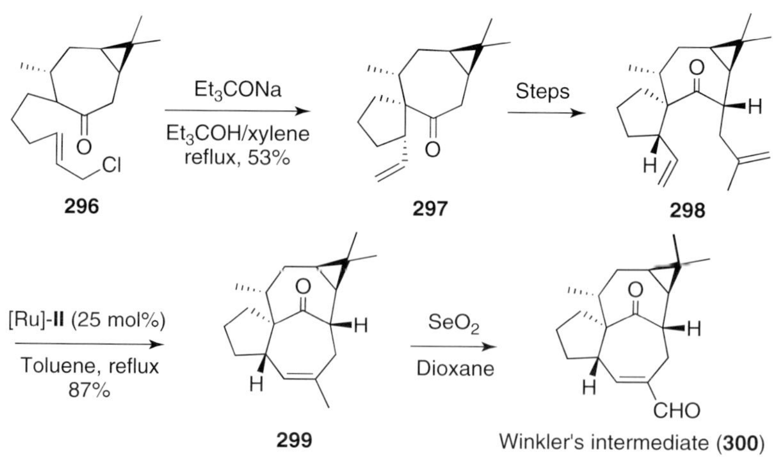

Scheme 1.54

the desired cycloheptenone **294** was obtained from **287** using [Ru]-**I** (80 mol%), whereas an excellent 87% yield of **295** was obtained from **293** with [Ru]-**III** (25 mol%).

Kigoshi's formal synthesis (Scheme 1.54) relies on the efficient spirocyclization reaction of ketone **296** also prepared from 3-carene. The AC ring system was then further elaborated into 1,8-diene **298** that could be cyclized to **299** in 87% yield using [Ru]-**II** (25 mol%). A dramatic temperature and solvent effect was observed as no reaction took place in dichloromethane. Allylic oxidation then delivered aldehyde **300**, an intermediate in Winkler's total synthesis of ingenol [74].

A rare example of formation of cycloheptadiene from 1,3,8-triene was reported by Hiemstra *et al.* in the enantioselective synthesis of the tetracyclic left-hand substructure **301** of solanoeclepin A (**302**) (Scheme 1.55) [75]. Starting from compound **303**, a [4+2]-cycloaddition was used to elaborate the oxabicyclic motif. Nolan's catalyst [Ru]-**VI** (15 mol%) was used to cyclize the 1,3,8-triene **305** to the desired cycloheptadiene **306** in quantitative yield. It is worth mentioning that [Ru]-**I** was quite inefficient in this case even if a stoichiometric quantity was used.

A concise asymmetric total synthesis of (+)-cyanthiwigin U (**307**) was reported by Phillips and Pfeiffer relying on the clever use of tandem metathesis of bicyclo[2.2.2]octenes (Scheme 1.56) [76]. The two-directional tandem ROM–RCM of compound **308**, using catalyst [Ru]-**II** under an atmosphere of ethylene, simultaneously generated the cyclopentenone and the cycloheptenone rings of

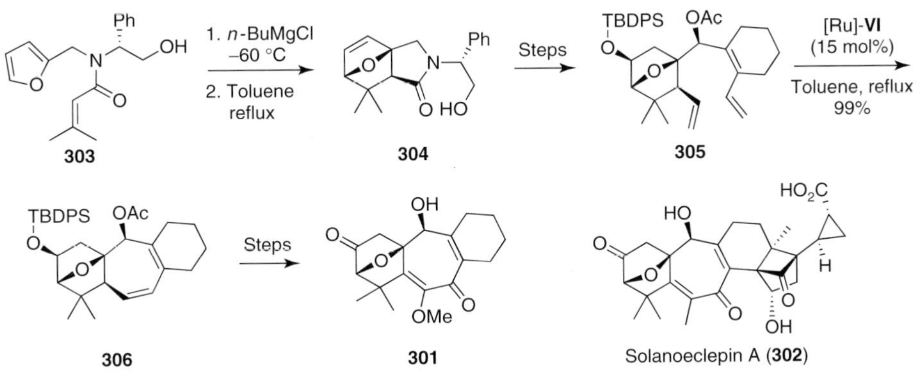

Scheme 1.55

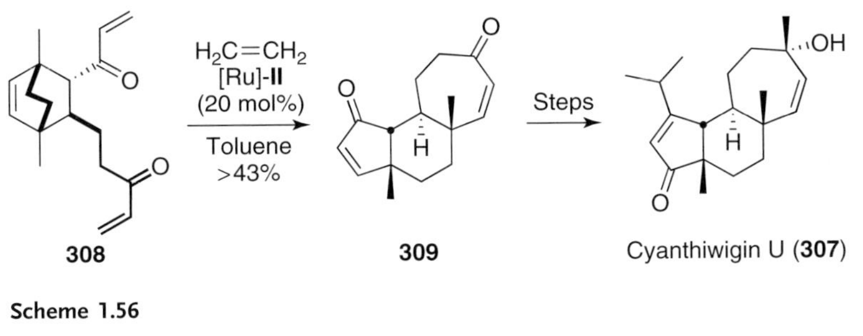

Scheme 1.56

309 with great efficiency. This synthetic strategy was later extended to the synthesis of cyanthiwigin W and Z [77].

An RCM was reported by Wicha *et al.* in a synthetic approach to the diterpenes guanacastepene A (**310**) and heptemerone G (**311**) (Scheme 1.57) [78]. The 1,8-diene **312** possessing a 1,1-disubstituted olefin was cyclized efficiently to **313** in high yield (96%) using catalyst [Ru]-**II**. Further functional group transformation including epoxidation and Dauben oxidation [79] led to the AB core structure **314** of the targeted diterpenes.

This very efficient RCM should not mask the fact that trisubstituted double bonds of terpenoids, imposed by the biosynthesis pathway from isoprene units, could be challenging to prepare through the RCM of 1,8-dienes as subtle differences in the metathesis precursor can have a dramatic impact on the reaction rate and yields [80].

For example, [Ru]-**I** could efficiently catalyze the RCM of the substituted 1,8-diene **315** to the corresponding trisubstituted cycloheptene **316** as demonstrated by Tori *et al.* in the total synthesis of two sphenolobane-type diterpenes (*Z*,*E*)- and (*E*,*E*)-**320** isolated from the liverwort *Anastrophyllum aurium* (Scheme 1.58) [81].

However, during a synthesis of (−)-tormesol (**321**), a diterpene isolated from *Halimium viscosum*, the cyclization of the related nonenolizable β-ketoester **322** proved sluggish as only 50% conversion was observed using catalyst [Ru]-**I** in combination with Ti(O*i*-Pr)$_4$ (1 equiv.). The more active [Ru]-**II** overcame this limitation and allowed the synthesis of cycloheptene **323** in excellent yield with a low

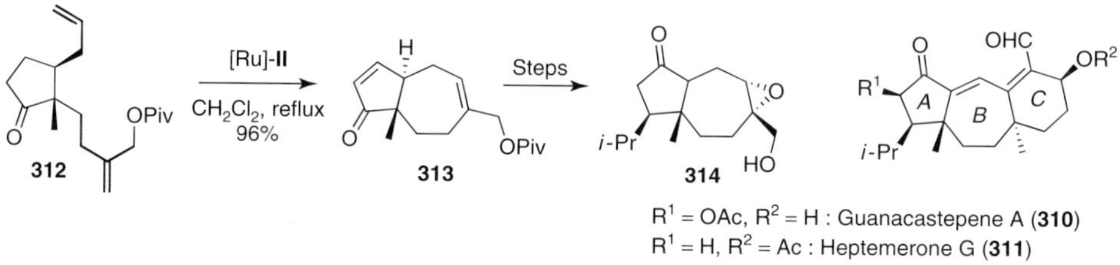

Scheme 1.57

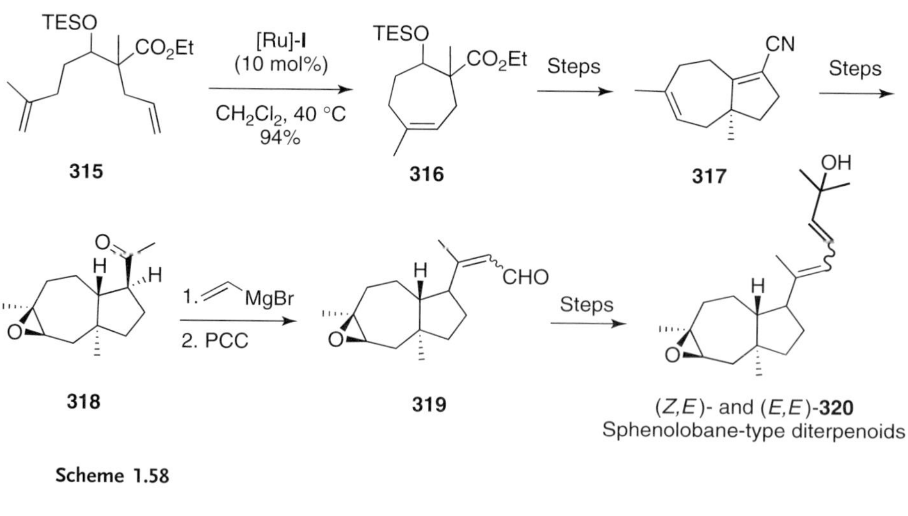

Scheme 1.58

catalyst loading (1 mol%) [82]. Compound **323** was then converted to (−)-tormesol through intermediates **324** and **325** (Scheme 1.59) [81, 82].

A very challenging RCM was reported by Reiser *et al.* during the enantioselective synthesis of arglabin (**326**), a promising antitumor agent isolated from *Artemisia glabella* [83]. Condensation of cyclopropanecarbaldehyde **327** with allylic silane **328** led, after treatment under basic conditions, to adduct **329**, which was further elaborated to **330**. The tetrasubstituted double bond embedded into the seven-membered ring of **331** was efficiently prepared via the RCM of compound **330**. Worthy of note is that an inert gas sparging of the reaction combined with catalyst [Ru]-**II** (15 mol%) had to be used to ensure complete conversion. Epoxidation from the α-face of the seven-membered ring proved troublesome since the more stable β-epoxide was obtained using dimethyldioxirane or peracids. A hydroxyl-directed vanadium-catalyzed epoxidation of the homoallylic alcohol **331** overcame the intrinsic bias of the system, leading to the desired epoxide **332** in good yield and selectivity (Scheme 1.60).

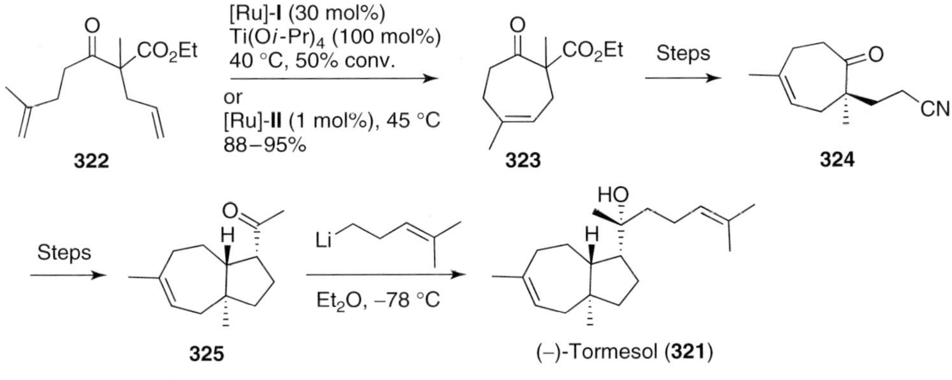

Scheme 1.59

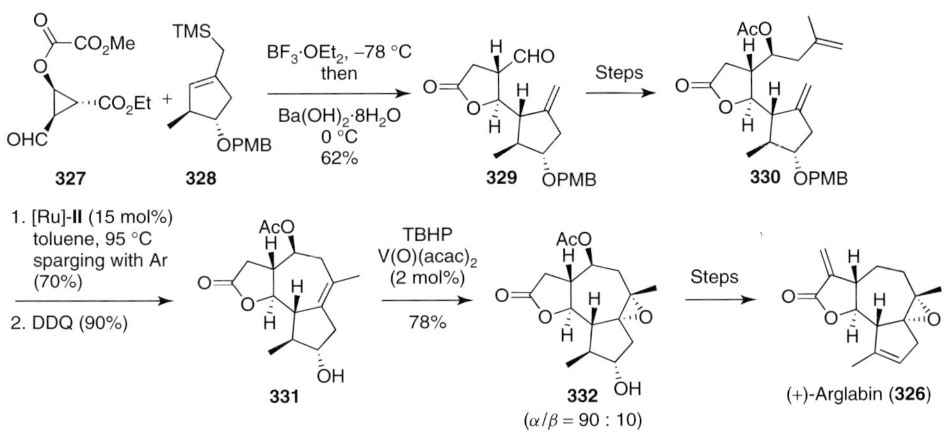

Scheme 1.60

1.5
Formation of Eight-membered Carbocycles by RCM

Cyclooctanoids are among the most difficult carbocycles to elaborate from acyclic precursors due to severe developing ring strain and restricted bond rotation in the precyclization conformer [84]. In the mid 1990s, RCM of 1,9-dienes emerged as a powerful strategy for the synthesis of carbocyclic eight-membered ring [85]. Steric and conformational effects were found to have a dramatic impact on the rate and efficiency of cyclization reactions. High catalyst loading (up to 50 mol%), long reaction times, and elevated temperatures were usually reported. Up to 2006, an excellent review compiles the recent advances in the synthesis of challenging di- [86] and trisubstituted [87] double bonds embedded in cyclooctenyl rings [88].

The dicyclo[*a*,*b*]cyclooctane ring systems found in ophiobolin M (**333**) and fusicoccin A (**334**) terpenes have stimulated a great deal of interest. The groups of Williams [89] and Wicha [90] reported two different RCM approaches aiming at the rapid elaboration of the eight-membered B ring system. The first approach relied on a [Ru]-**I** catalyzed RCM of triene **335** leading to cyclooctene **336** in moderate yield (Scheme 1.61) [89]. Worthy of note is the fact that [Ru]-**II** proved inferior as a catalyst leading to lower conversion and partial isomerization of the terminal olefin(s).

The second approach [90] reports the synthesis of the AB ring system of serpendione (**337**) via RCM reaction of **338** (Equation 1, Scheme 1.62) that contains

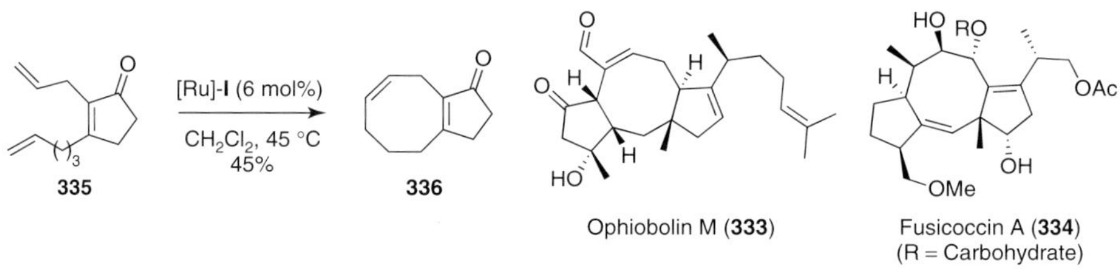

Scheme 1.61

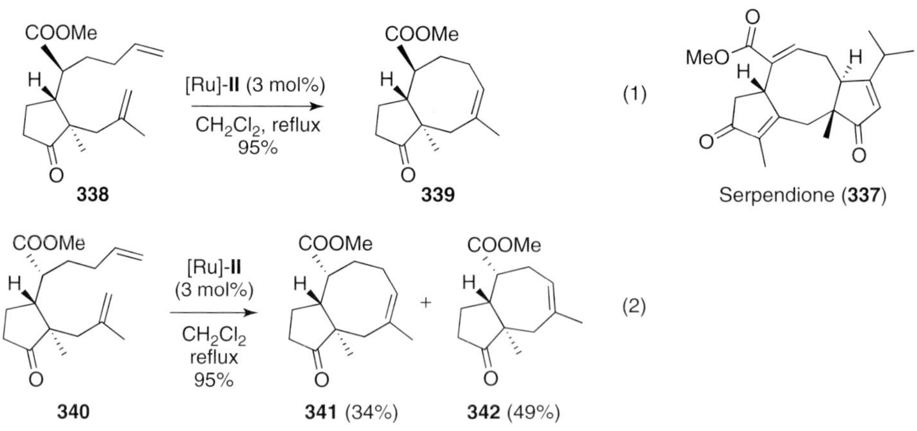

Scheme 1.62

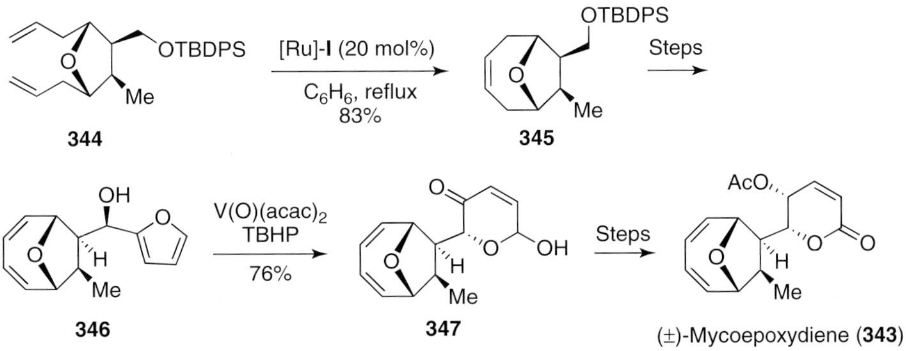

Scheme 1.63

both a geminal disubstituted and a monosubstituted olefin. The [Ru]-**II** catalyzed transformation was extremely efficient and delivered compound **339** possessing the AB ring system in 95% yield. When exposed to [Ru]-**II**, the epimeric compound **340** could be transformed to the desired cyclooctene **341**, albeit in low yield (34%). Isomerization of the allylic double bond prior to cyclization accounts for the major cycloheptene product **342** (49%) (Equation 2, Scheme 1.62), thus once again stressing the importance of subtle steric and conformational effects in the formation of medium rings by RCM. By contrast, the use of [Ru]-**I** led only to dimerization.

Imposing a conformational constraint that favors cyclization is a useful strategy for the RCM of 1,9-dienes. In the total synthesis of (±)-mycoepoxydiene (**343**), Tadano *et al.* reported the RCM of the diallylated tetrahydrofuran **344** using catalyst [Ru]-**I** under diluted conditions to prevent oligomerization [91]. A good yield of the desired oxygen-bridged cyclooctene **345** was obtained (83%). A dramatic solvent dependence was noted as CH_2Cl_2 induced only decomposition of the starting diene **344**. Compound **345** was converted to (±)-mycoepoxydiene **343** in several steps including oxidation of furylcarbinol **346** to dihydropyranone **347** (Scheme 1.63).

Vinigrol (**348**) is a metabolite of the fungal strain *Virgaria nigra* that possesses a fascinating tricyclo[4.4.4.0^{4a,8a}]tetradecane skeleton [92]. Its biological activity, including antihypertensive, platelet aggregation-inhibiting, and tumor necrosis factor antagonist properties, has fueled a considerable synthetic interest. The groups of Barriault [93] and Paquette [94] investigated two different approaches involving disconnections of the carbocyclic eight-membered ring through RCM at the C9–C10 or C10–C11 positions, respectively. Unfortunately, these two strategies met with failure. As observed in previous studies, [Ru]-**I** was not active enough to promote the RCM of 1,9-dienes **349**, **350**, and **351**, and [Ru]-**II** led to double bond migration (Scheme 1.64).

In order to increase conformational flexibility in the cyclization precursor, Paquette *et al.* considered the relocation of the double bond from the $\Delta^{3,4}$ to the $\Delta^{2,3}$ site to project axially the two C1 and C5 substituents. As shown by MM3 calculations on model systems, the closer proximity of these two substituents in the $\Delta^{2,3}$ isomer could allow a more RCM. Diene **352** was then prepared and

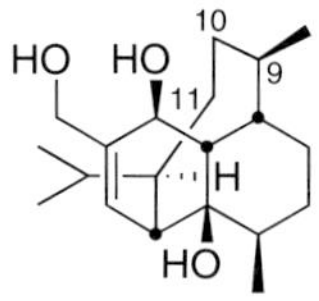

Vinigrol (**348**)

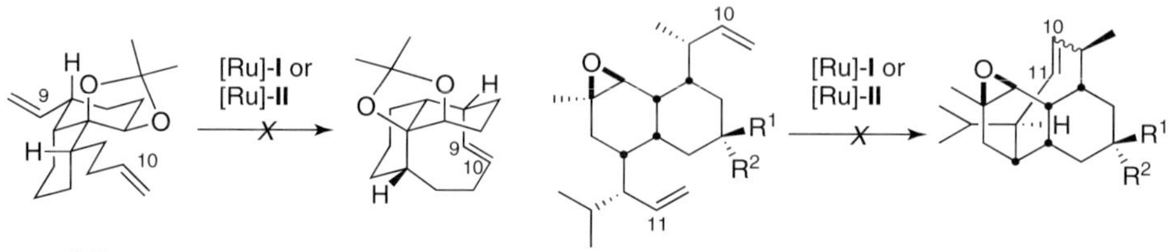

349

350: R^1, R^2 = OCH_2CH_2O
351: R^1 = OTBDPS, R^2 = H

Scheme 1.64

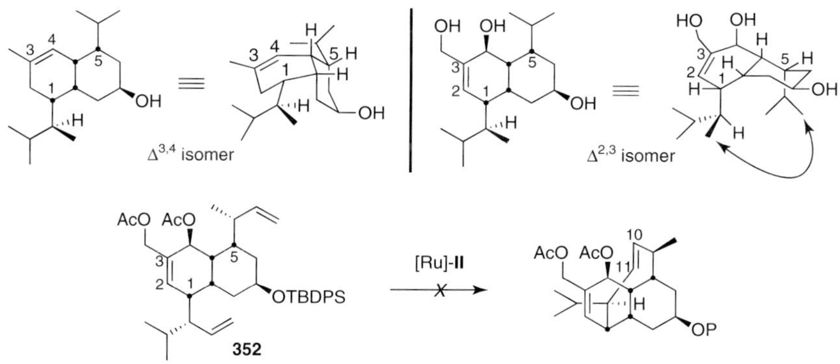

Scheme 1.65

subjected to catalyst [Ru]-**II** but unfortunately, only starting material was recovered (Scheme 1.65).

Prunet *et al.* have extensively investigated the possibility of synthesizing the B ring system of Taxol® (**353**) via the RCM of 1,9-diene **355** (Scheme 1.66) [95]. When a diastereomeric mixture of the latter compound and *epi*-**355** was subjected to catalyst [Ru]-**I**, only compound **355** cyclized to cyclooctene **356** (possessing the C9–C10 unsaturation) albeit slowly, emphasizing once again that RCM is highly dependent upon the conformational and steric effects in the cyclization precursor (Equation 1, Scheme 1.66). Recently, the formation of the C10–C11 unsaturation by RCM was reported and proved more efficient in terms of protecting group compatibility and catalyst efficiency (Equation 2, Scheme 1.66). Unprotected diol **357** led to the desired cyclooctene ring **358** in an impressive quantitative yield, thus suggesting that the BC ring system of taxol may be synthesized according to this strategy. These recent results are in stark contrast to the previous C9–C10

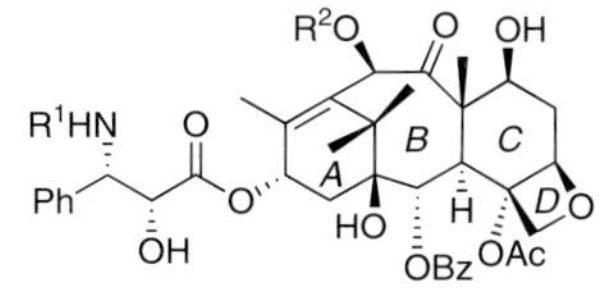

R^1 = Bz, R^2 = Ac: Taxol (paclitaxel) (**353**)
R^1 = Boc, R^2 = H: Taxotere (docetaxel) (**354**)

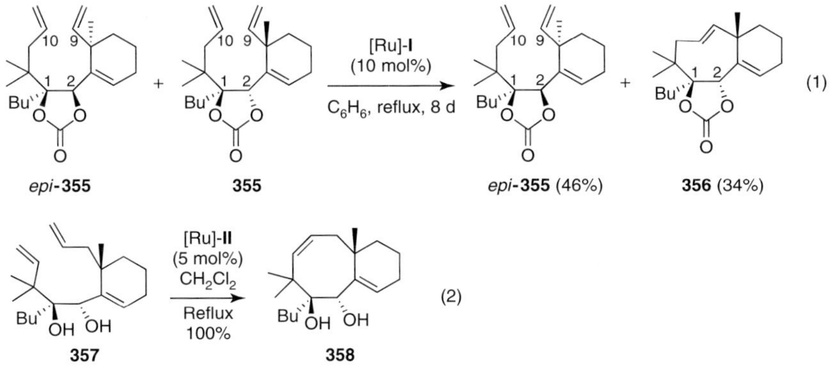

Scheme 1.66

RCM strategy that was thwarted by conformational constraints imposing a cyclic carbonate protecting group of the C1–C2 diol motif.

1.6
Formation of Nine-membered Carbocycles by RCM

The failed metathesis-based synthesis of pestalotiopsin A **359**, compared to the successful synthesis of cornexistins, reported by Paquette *et al.* [96] and Clark *et al.* [97] illustrates well the uncertainty linked to RCM toward 8 to 10-membered rings. RCM of several substrates **360–362** was evaluated (Figure 1.2). For **360** and **361**, regardless of the catalyst used ([Ru]-**I**, **II**, **III**, **VII**, and **XI**), the starting diene was either recovered (when the reaction was performed below 80 °C) or decomposed (in refluxing toluene). The MOM ether **362** was isomerized to **363**.

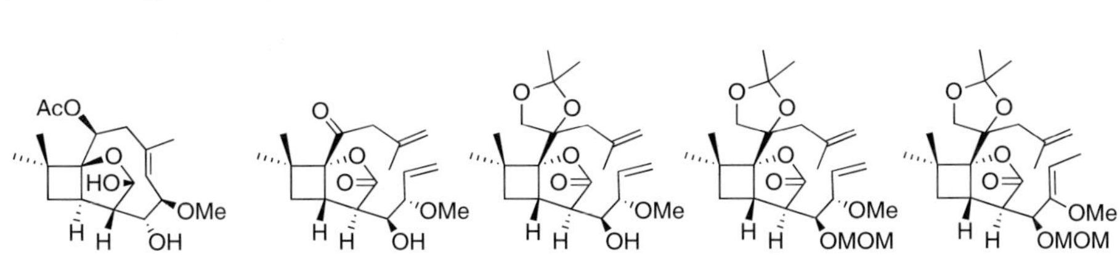

Pestalotiopsin A (**359**) **360** **361** **362** **363**

Figure 1.2 Pestalotiopsin a synthesis:unteactive RCM substrates.

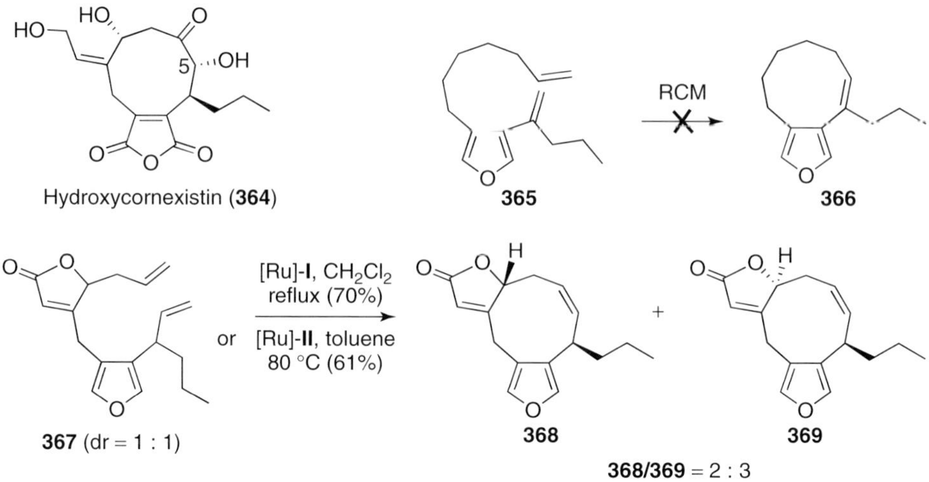

Scheme 1.67

In 2003, Clark *et al.* reported a novel strategy for the preparation of cornexistins. In model studies, attempts to form intermediate **364**, featuring a trisubstituted olefin, by the RCM of **365** were unsuccessful regardless of the catalyst used ([Ru]-**I** or [Ru]-**II**). In contrast, the cyclization of **367** (a racemic mixture of two diastereomers) led to the isomeric mixtures **368** and **369** in good yield. In subsequent studies, the RCM reaction was examined more closely and shown to be significantly stereoselective (**368**/**369** = 2 : 3) [98].[3)] Compound **368** was converted to (±)-5-*epi*-hydroxycornexistin in a few additional steps (Scheme 1.67).

1.7
Formation of 10-membered Carbocycles by RCM

In 2001 and 2002, Koskinen *et al.* reported what appeared to be the first examples of 10-membered ring carbocycle syntheses via RCM [99]. As can be seen in Table 1.1, the reaction proved difficult to perform and yields remained moderate even under optimized conditions. Using alcohol **370a** (*syn*/*anti* = 2 : 1), a single cyclized product (**371a**) was obtained in low yield and some starting material was recovered. With the silyl ether **370b** as substrate, the reaction was cleaner, affording a mixture of three diastereomers in fair yield. Addition of Ti(O*i*-Pr)$_4$ was slightly beneficial. The roughly 2 : 1 *trans*/*cis* ratio of the major RCM products **371b** and **372b** compares well with the corresponding 2 : 1 *syn*/*anti* ratio of the substrate **370b** [99a]. When

3) [Ru]-I catalyst was used for this transformation (18 hours, CH_2Cl_2, reflux, 77%). [Ru]-II did not bring any significant improvement (personal communication from Dr Clark).

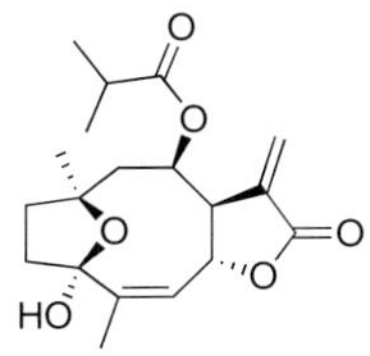

Diversifolin (**374**)

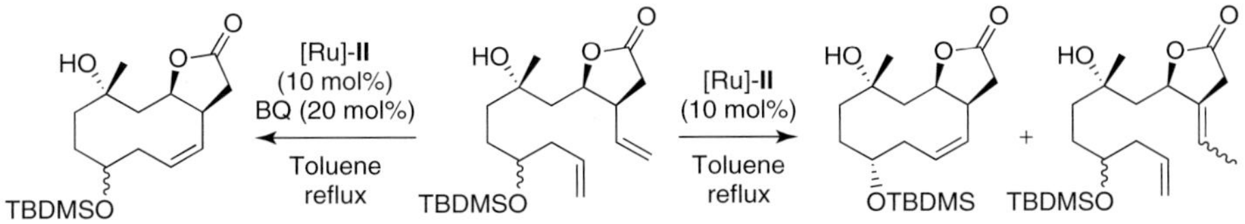

(*R*)-**376** (31%)
(*S*)-**376** (40%)

375
dr = 1 : 1

(*R*)-**376** (27%)
(*S*)-**376** (trace)

377 (35%)

Scheme 1.68

Table 1.1 Cyclization of alcohol 370a and TBDMS ether 370b.

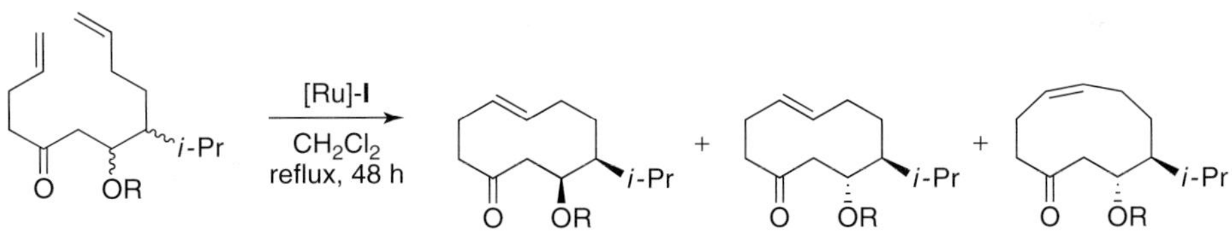

370a : R = H
370b : R = TBDMS
(*syn/anti* : 2 : 1)

371a **371b**

372a **372b**

373a **373b**

Substrate	Substrate recovery (%)	Products		
370a	25	**371a** (11%)	–	–
370b	6	**371b** (16%)	**372b** (35%)	**373b** (11%)
370b + Ti(O*i*-Pr)$_4$ (20 mol%)	0	**371b** (21%)	**372b** (47%)	**373b** (8%)

the more active [Ru]-**II** catalyst was used, only the *syn*-**370b** precursor cyclized to afford a mixture of **371b** and **373b** in a 1 : 6 ratio. The *anti*-**370b** isomer was not recovered and was probably converted to polymeric material [99b].

A synthetic study toward diversifolin (**374**) was published in 2007 by Kobayashi *et al.* (Scheme 1.68) [100]. First, a long sequence (17 steps) led to the key RCM substrate **375**, obtained as a mixture of epimers. RCM was difficult and required significant optimization. While using [Ru]-**I** no reaction was observed, whereas [Ru]-**II** in 1,2-dichloroethane (DCE) led to the formation of **377** in low yield (15%) in which one of the terminal olefin had shifted. When toluene was used as the solvent,

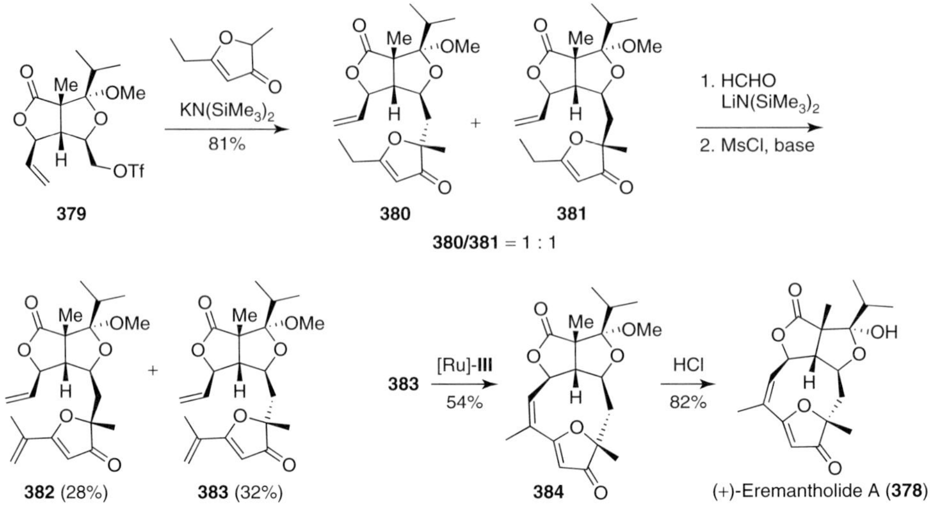

Scheme 1.69

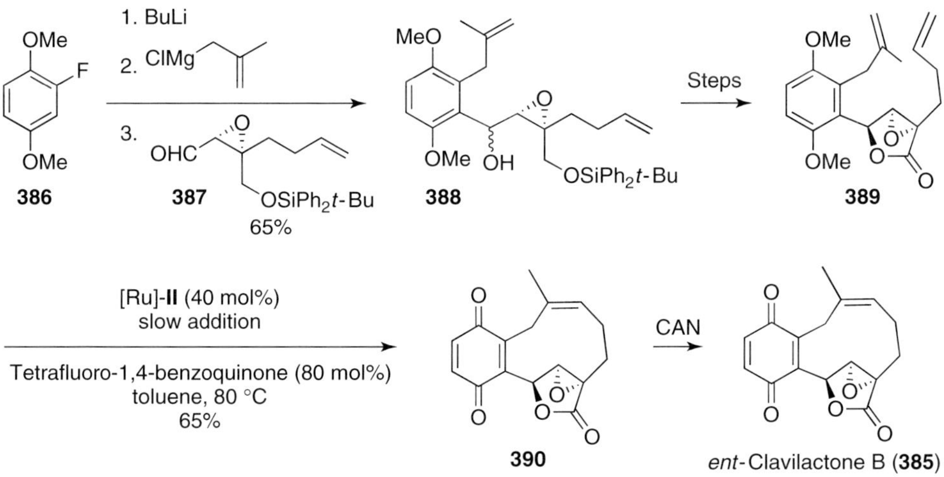

Scheme 1.70

377 was obtained in 35% yield along with (3*R*)-**376** (27%), and some starting material was recovered. Only (3*R*)-**375** cyclizes rapidly and no (3*S*)-**376** was formed. RCM of (3*S*)-**375** is more difficult allowing the competing olefin isomerization to prevail. This may be avoided by adding 1,4-benzoquinone (BQ), whose role is probably to prevent the accumulation of ruthenium hydrides (decomposition products of [Ru]-**II**) that are thought to be responsible for olefin migration [101].

In Hale's and Li's synthesis of (+)-eremantholide A (**378**) (Scheme 1.69), the RCM step takes place almost at the end of the synthesis and involves a fairly hindered substrate. Epimers **382** and **383** were prepared by alkylation of triflate **379** followed by hydroxymethylation and subsequent dehydration. Using [Ru]-**III** as catalyst, the reaction works well. Remarkably, only epimer **383** reacted, affording the desired bicyclic system **384** in 54% yield and (+)-eremantholide A, while the other diastereomer **382** cross-metathesized [102].

In Barrett's synthesis of *ent*-clavilactone B **385** (Scheme 1.70), the benzyne intermediate generated from **386** was allowed to react with methallylmagnesium chloride and epoxy-aldehyde **387**, leading to compound **388**, which was converted to the RCM precursor **389**. As expected, RCM proved to be difficult and had to be performed under special conditions: the catalyst [Ru]-**II** (40 mol%) was slowly added, along with tetrafluoro-1,4-BQ (to prevent olefin isomerization), and the formed ethylene was removed during the reaction. Under these conditions, the RCM product **390** was obtained in a satisfactory yield (65%) and converted to *ent*-clavilactone B [103].

During the period 2001–2006, Gennari *et al.* published a series of accounts describing the synthesis, via RCM, of simplified eleuthesides as potential anticancer agents [104–106]. In particular, these authors recently reported a novel formal synthesis of eleutherobin **391** (Scheme 1.71) [105, 106].

Starting from (*R*)-(−)-carvone, the half-protected dialdehyde **392** was prepared. Diastereoselective reagent-controlled oxyallyltitanations (with TADDOL as

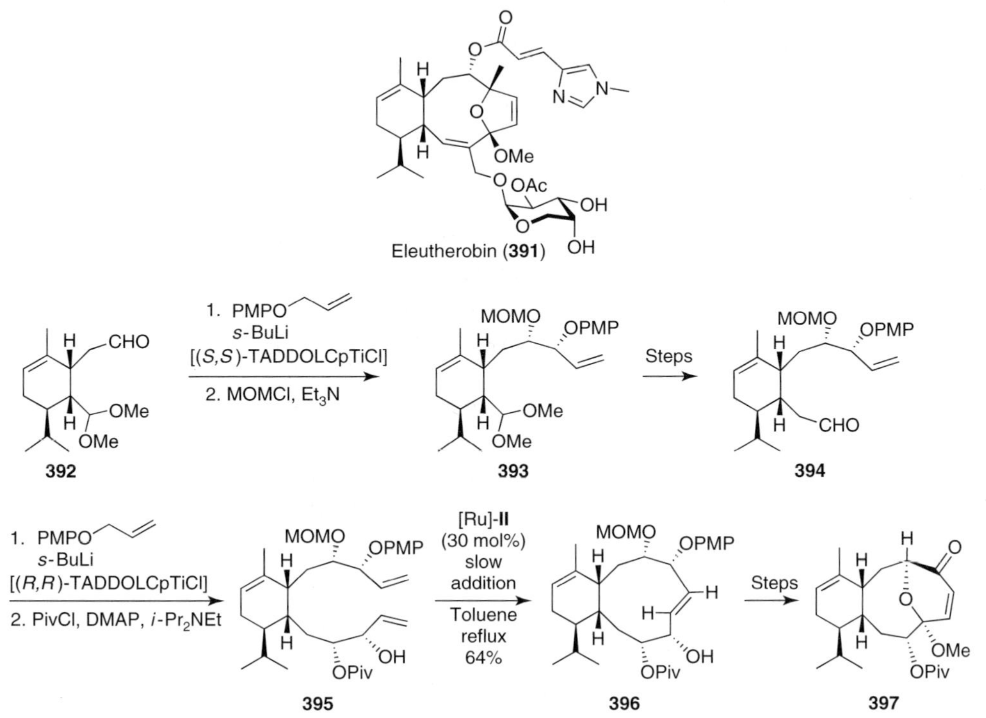

Scheme 1.71

the chiral ligand) of both aldehydic functions led to the RCM precursor **395**, which was cyclized using [Ru]-**II** catalyst to afford the 10-membered carbocycle **396**. The latter was converted to the advanced intermediate **397** that had previously been converted to eleutherobine. Forcing RCM conditions were necessary ([Ru]-**II** [30 mol%], slow addition, toluene, reflux), and surprisingly, only the (*E*)-cycloalkene **396** was obtained [25, 106, 107]. Using simplified models of **395**, Gennari *et al.* have shown by a variety of methods (molecular mechanics, DFT) that the *trans*-four-membered ruthenacycle intermediates **398** leading to (*E*)-cyclodecenes are significantly more stable than the corresponding *cis*-ruthenacycles **399**, leading to (*Z*)-cyclodecenes (Scheme 1.72). RCM of diastereomer **400** was more difficult than that of **395**, affording the ring-cyclized product **401** in low yield alongside with regioisomer **402** resulting from a metathetic ring rearrangement (RCM–ROM–RCM cascade).

Excentricine **403** belongs to a rare class of isoquinoline alkaloids featuring a bridged, bicyclic ether moiety. A possible approach to this class of compounds was illustrated by the synthesis of the simplified model compound **408** (Scheme 1.73) [108]. The strategy relies on a Bischler–Napieralski isoquinoline synthesis and an RCM as key reactions. The site of the RCM reaction was chosen so as to avoid the formation of inactive ruthenium complexes (likely to be observed for γ,δ- or δ,ε-unsaturated amides or esters). RCM of diene **406** proceeded smoothly using catalyst [Ru]-**II** to afford the desired 10-membered ring **407** in good yield (68%). However, the last deprotection step was unsatisfactory and proceeded only in low yield. The solution to this problem was provided by simply inverting the reaction sequence. Thus, a facile RCM in the presence of [Ru]-**I** afforded the 14-membered macrocyclic lactam **409**. After hydrogenation of the olefin, a Bischler–Napieralski reaction led to the 2,3-dihydroisoquinoline **410** that was reduced to **408**.

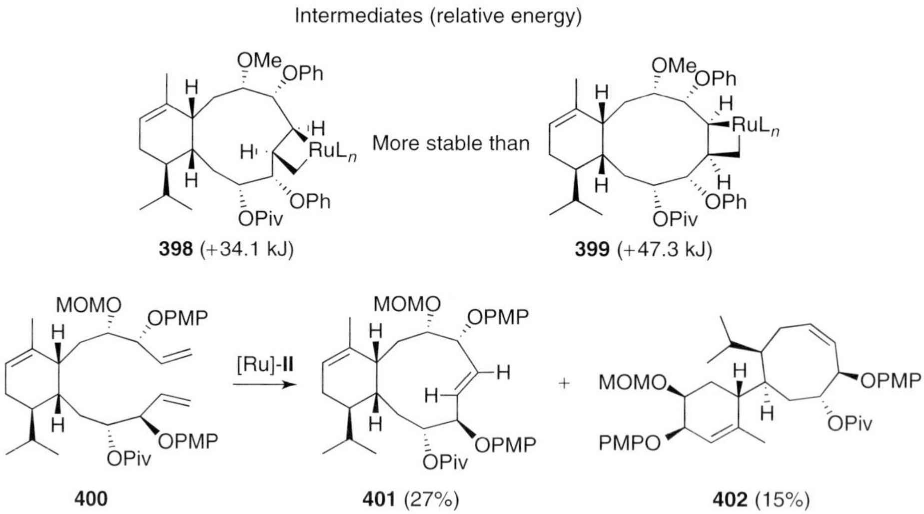

Scheme 1.72

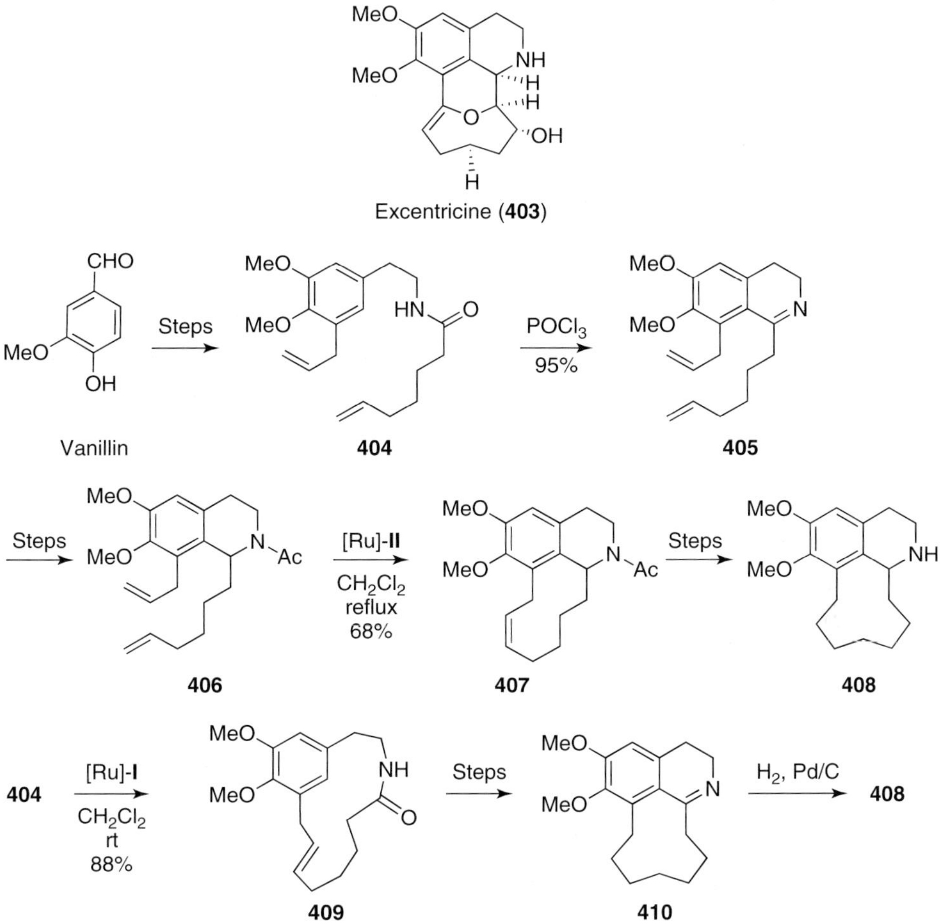

Scheme 1.73

1.8 Conclusion

In the last decade, the alkene/alkyne metathesis reaction has become a major synthetic tool for the preparation of complex organic molecules and, not surprisingly, RCM has proved to be particularly useful for the synthesis of medium-size (5–10-membered) rings. For ring sizes of five to seven, the reaction works well but for larger (8–10-membered) rings, the metathesis is less predictable and its outcome depends on still imperfectly understood factors. Remarkable sequences, in particular [3,3]-sigmatropic rearrangement/RCM or RCM/fragmentation combinations, have been classically used as synthetic strategies.

References

1 Wang, Q., Millet, A., and Hiersemann, M. (2007) *Synlett*, 1683–1686.

2 Probst, N.P., Haudrechy, A., and Plé, K. (2008) *J. Org. Chem.*, **73**, 4338–4341 and references cited therein.

3 Srikrishna, A. and Laksmi, B.V. (2005) *Tetrahedron Lett.*, **46**, 7029–7031.

4 (a) Srikrishna, A. and Lakshmi, B.V. (2005) *Tetrahedron Lett.*, **46**, 4879–4881; (b) Srikrishna, A. and Lakshmi, B.V. (2005) *Synlett*, 1173–1175.

5 Maity, S. and Ghosh, S. (2007) *Tetrahedron Lett.*, **48**, 3355–3358.

6 Banerjee, S., Ghosh, S., Sinha, S., and Ghosh, S. (2005) *J. Org. Chem.*, **70**, 4199–4202.

7 Srikrishna, A., Khan, I.A., Babu, R.R., and Sajjanshetty, A. (2007) *Tetrahedron*, **63**, 12616–12620.

8 Wilson, M.S., Woo, J.C.S., and Dake, G.R. (2006) *J. Org. Chem.*, **71**, 4237–4245.

9 Chatterjee, A.K., Morgan, J.P., Scholl, M., and Grubbs, R.H. (2000) *J. Am. Chem. Soc.*, **122**, 3783–3784.

10 Donohoe, T.J. and Rosa, C.P. (2007) *Org. Lett.*, **9**, 5509–5511.

11 Mehta, G. and Maity, P. (2007) *Tetrahedron Lett.*, **48**, 8865–8868.

12 Srikrishna, A., Pardeshi, V.H., and Thriveni, P. (2008) *Tetrahedron: Asymmetry*, **19**, 1392–1396.

13 Rodeschini, V., van de Weghe, P., Tarnus, C., and Eustache, J. (2005) *Tetrahedron Lett.*, **46**, 6691–6695.

14 Miller, A.K., Hughes, C.C., Kennedy-Smith, J.J., Gradl, S.N., and Trauner, D. (2006) *J. Am. Chem. Soc.*, **128**, 17057–17062.

15 Paquette, L.A., Peng, X., Yang, J., and Kang, H.-J. (2008) *J. Org. Chem.*, **73**, 4548–4558.

16 Clive, D.L.J. and Yu, M. (2002) *Chem. Commun.*, 2380–2381.

17 Frigell, J. and Cumpstey, I. (2007) *Tetrahedron Lett.*, **48**, 9073–9076.

18 Srikrishna, A., Pardeshi, V.H., and Satyanarayana, G. (2007) *Tetrahedron Lett.*, **48**, 4087–4090.

19 Ziegler, F.E., Metcalf, C.A.III, Nangia, A., and Schulte, G. (1993) *J. Am. Chem. Soc.*, **115**, 2581–2589.

20 (a) Dai, M. and Danishefsky, S.J. (2007) *J. Am. Chem. Soc.*, **129**, 3498–3499; (b) Dai, M., Krauss, I.J., and Danishefsky, S.J. (2008) *J. Org. Chem.*, **73**, 9576–9583.

21 Herrmann, P., Buděšínský, M., and Kotora, M. (2008) *J. Org. Chem.*, **73**, 6202–6206.

22 (a) Boiteau, J.G., van de Weghe, P., and Eustache, J. (2001) *Org. Lett.*, **3**, 2737–2740; (b) Rodeschini, V., Boiteau, J.-G., van de Weghe, P., Tarnus, C., and Eustache, J. (2004) *J. Org. Chem.*, **69**, 357–373.

23 (a) Picoul, W., Urchegui, R., Haudrechy, A., and Langlois, Y. (1999) *Tetrahedron Lett.*, **40**, 4797–4800; (b) Picoul, W., Bedel, O., and Langlois, Y. (2003) *Pure Appl. Chem.*, **75**, 235–249; (c) Bedel, O., Haudrechy, A., and Langlois, Y. (2004) *Eur. J Org. Chem.*, 3813–3819.

24 Takahashi, S., Hishinuma, N., Koshino, H., and Nakata, T. (2005) *J. Org. Chem.*, **70**, 10162–10165.

25 Fürstner, A. and Langemann, K. (1997) *Synthesis*, 792–803.

26 (a) Vosburg, D.A., Weiler, S., and Sorensen, E.J. (2003) *Chirality*, **15**, 156–166; (b) Vosburg, D.A., Weiler, S., and Sorensen, E.J. (1999) *Angew. Chem. Int. Ed.*, **38**, 971–974.

27 Clive, D.L.J. and Liu, D. (2008) *J. Org. Chem.*, **73**, 3078–3087.

28 Yun, S.Y., Zheng, J.-C., and Lee, D. (2008) *Angew. Chem. Int. Ed.*, **47**, 6201–6203.

29 Tiefenbacher, K. and Mulzer, J. (2008) *Angew. Chem. Int. Ed.*, **47**, 6199–6200.

30 (a) Nicolaou, K.C., Tria, G.S., and Edmonds, D.J. (2008) *Angew. Chem. Int. Ed.*, **47**, 1780–1783; (b) Nicolaou, K.C., Tria, G.S., and Edmonds, D.J. (2008) *Angew. Chem. Int. Ed.*, **47**, 1780–1783; (c) Hayashida, J. and Rawal, V.H. (2008) *Angew. Chem. Int. Ed.*, **47**, 4373–4376.

31 Toyota, M., Yokota, M., and Ihara, M. (2001) *J. Am. Chem. Soc.*, **123**, 1856–1861.

32 Ciochina, R. and Grossman, R.B. (2006) *Chem. Rev.*, **106**, 3963–3986.

33 Kuramochi, A., Usuda, H., Yamatsugu, K., Kanai, M., and Shibasaki, M. (2005) *J. Am. Chem. Soc*, **127**, 14200–14201.

34 Kingsbury, J.S., Harrity, J.P.A., Bonitatebus, P.J., Hoveyda, A.H., Yokota, M., and Ihara, M. (1999) *J. Am. Chem. Soc.*, **121**, 791–799.

35 Mehta, G. and Singh, V. (1999) *Chem. Rev.*, **99**, 881–930.

36 Grob, C.A. and Schiess, P.W. (1967) *Angew. Chem.*, **79**, 1–14.

37 (a) Rodrìguez, J.R., Castedo, L., and Mascareñas, J.L. (2002) *Chem. Eur J.*, **8**, 2923–2930; (b) Rodrìguez, J.R., Castedo, L., and Mascareñas, J.L. (2000) *Org. Lett.*, **2**, 3209–3212.

38 Mehta, G. and Kumaran, R.S. (2005) *Tetrahedron Lett.*, **46**, 8831–8835.

39 Wharton, P.S. (1961) *J. Org. Chem.*, **26**, 4781–4782.

40 Ivkovic, A., Matovic, R., and Saicic, R.N. (2004) *Org. Lett.*, **6**, 1221–1224.

41 Matovic, R., Ivkovic, A., Manojlovic, M., Tokic-Vujosevic, Z., and Saicic, R.N. (2006) *J. Org. Chem.*, **71**, 9411–9419.

42 Mehta, G. and Kumaran, R.S. (2002) *Chem.Commun.*, 1456–1457.

43 Robichaud, J. and Tremblay, F. (2006) *Org. Lett.*, **8**, 597–600.

44 Song, Y., Hwang, S., Gong, P., Kim, D., and Kim, S. (2008) *Org. Lett.*, **10**, 269–271.

45 Bi, F.Q., Fang, L.J., Zhang, C.X., and Li, Y.L. (2008) *Chin. Chem. Lett.*, **19**, 1265–1267.

46 Nicolaou, K.C., Jennings, M.P., and Dagneau, P. (2002) *Chem.Commun.*, 2480–2481.

47 White, D.E., Stewart, I.C., Grubbs, R.H., and Stoltz, B.M. (2008) *J. Am. Chem. Soc*, **130**, 810–811.

48 Tani, K., Behenna, D.C., McFadden, R.M., and Stoltz, B.M. (2007) *Org. Lett.*, **9**, 2529–2531.

49 Stewart, I.C., Ung, T., Pletnev, A.A., Berlin, J.M., Grubbs, R.H., and Schrodi, Y. (2007) *Org. Lett.*, **9**, 1589–1592.

50 Pragani, R. and Roush, W.R. (2008) *Org. Lett.*, **10**, 4613–4616.

51 For a recent review about carbasugars see: Arjona, O., Gómez, A.M., López, J.C., and Plumet, J. (2007) *Chem. Rev.*, **107**, 1919–2036.

52 For reviews about olefin metathesis in carbohydrate chemistry see: (a) Roy, R. and Das, S.K. (2000) *Chem. Commun.*, 519–529; (b) Madsen, R. (2007) *Eur. J Org. Chem.*, 399–415.

53 Kapferer, P., Sarabia, F., and Vasella, A. (1999) *Helv. Chim. Acta*, **82**, 645–656.

54 Ziegler, F.E. and Wang, Y. (1998) *J. Org. Chem.*, **63**, 426–427.

55 Chang, Y.-K., Lee, B.-Y., Kim, D.J., Lee, G.S., Jeon, H.B., and Kim, K.S. (2005) *J. Org. Chem.*, **70**, 3299–3302.

56 Sellier, O., van de Weghe, P., Le Nouen, D., Strehler, C., and Eustache, J. (1999) *Tetrahedron Lett.*, **40**, 853–856.

57 Ackermann, L., El Tom, D., and Fürstner, A. (2000) *Tetrahedron*, **56**, 2195–2202.

58 Nishikawa, A., Saito, S., Hashimoto, Y., Koga, K., and Shirai, R. (2001) *Tetrahedron Lett.*, **42**, 9195–9198.

59 Andresen, T.L., Skytte, D.M., and Madsen, R. (2004) *Org. Biomol. Chem.*, **2**, 2951–2957.

60 Heo, J.-N., Holson, E.B., and Roush, W.R. (2003) *Org. Lett.*, **5**, 1697–1700.

61 Nadein, O.N. and Kornienko, A. (2004) *Org. Lett.*, **6**, 831–834.

62 Kireev, A.S., Nadein, O.N., Agustin, V.J., Bush, N.E., Evidente, A., Manpadi, M., Ogasawara, M.A., Rastogi, S., Togelj, S., Shors, S.T., and Kornienko, A. (2006) *J. Org. Chem.*, **71**, 5694–5707.

63 Barker, D., Brimble, M.A., McLeod, M.D., and Savage, G.P. (2004) *Org. Biomol. Chem.*, **2**, 1659–1669.

64 Mehta, G. and Likhite, N.S. (2008) *Tetrahedron Lett.*, **49**, 7113–7116.

65 Mehta, G. and Lakshminath, S. (2006) *Tetrahedron Lett.*, **47**, 327–330.

66 Yokoe, H., Sasaki, H., Yoshimura, T., Shindo, M., Yoshida, M., and Shishido, K. (2007) *Org. Lett.*, **9**, 969–971.

67 Enquist, J.A. Jr. and Stoltz, B.M. (2008) *Nature*, **453**, 1228–1231.

68 Schall, A. and Reiser, O. (2008) *Eur. J. Org. Chem.*, 2353–2364.

69 Andrews, S.P., Ball, M., Wierschem, F., Cleator, E., Oliver, S., Högenauer, K., Simic, O., Antonello, A., Hünger, U., Smith, M.D., and Ley, S.V. (2007) *Chem. Eur. J.*, **13**, 5688–5712.

70 van de Weghe, P., Bisseret, P., Blanchard, N., and Eustache, J. (2006) *J. Organomet. Chem.*, **691**, 5078–5108.

71 Oliver, S.F., Högenauer, K., Simic, O., Antonello, A., Smith, M.D., and Ley, S.V. (2003) *Angew. Chem. Int. Ed.*, **42**, 5996–6000.

72 Nickel, A., Maruyama, T., Tang, H., Murphy, P.D., Greene, B., Yusuff, N., and Wood, J.L. (2004) *J. Am. Chem. Soc.*, **126**, 16300–16301.

73 Watanabe, K., Suzuki, Y., Aoki, K., Sakakura, A., Suenaga, K., and Kigoshi, H. (2004) *J. Org. Chem.*, **69**, 7802–7808.

74 Winkler, J.D., Rouse, M.B., Greaney, M.F., Harrison, S.J., and Jeon, Y.T. (2002) *J. Am. Chem. Soc.*, **124**, 9726–9728.

75 Benningshof, J.C.J., Blaauw, R.H., van Ginkel, A.E., Rutjes, F.P.J.T., Fraanje, J., Goubitz, K., Schenk, H., and Hiemstra, H. (2000) *Chem. Commun.*, 1465–1466.

76 Pfeiffer, M.W.B. and Phillips, A.J. (2005) *J. Am. Chem. Soc.*, **127**, 5334–5335.

77 Pfeiffer, M.W.B. and Phillips, A.J. (2008) *Tetrahedron Lett.*, **49**, 6860–6861.

78 Michalak, K., Michalak, M., and Wicha, J. (2008) *Tetrahedron Lett.*, **49**, 6807–6809.

79 Dauben, W.G. and Michno, D.M. (1977) *J. Org. Chem.*, **42**, 682–685.

80 Kirkland, T.A. and Grubbs, R.H. (1997) *J. Org. Chem.*, **62**, 7319–7318.

81 Nakashima, K., Inoue, K., Sono, M., and Tori, M. (2002) *J. Org. Chem.*, **67**, 6034–6040.

82 Nakashima, K., Fujisaki, N., Inoue, K., Minami, A., Nagaya, C., Sono, M., and Tori, M. (2006) *Bull. Chem. Soc. Jpn.*, **79**, 1955–1962.

83 Kalidindi, S., Jeong, W.B., Schall, A., Bandichhor, R., Nosse, B., and Reiser, O. (2007) *Angew. Chem. Int. Ed.*, **46**, 6361–6363.

84 (a) Miller, S.J., Kim, S.-H., Chen, Z.-R., and Grubbs, R.H. (1995) *J. Am. Chem. Soc.*, **117**, 2108–2109; (b) Galli, C. and Mandolini, L. (2000) *Eur. J. Org. Chem.*, 3117–3125; (c) Illuminati, G. and Mandolini, L. (1981) *Acc. Chem. Res.*, **14**, 95–102; (d) Mitchell, L., Parkinson, J.A., Percy, J.M., and Singh, K. (2008) *J. Org. Chem.*, **73**, 2389–2395.

85 Maier, M.E. (2000) *Angew. Chem. Int. Ed.*, **39**, 2073–2077.

86 (a) Wenz, M., Groβbach, D., Beitzel, M., and Blechert, S. (1999) *Synthesis*, 607–614; (b) Krafft, M.E., Cheung, Y.Y., and Abboud, K.A. (2001) *J. Org. Chem.*, **66**, 7443–7448; (c) Paquette, L.A. and Efremov, I. (2001) *J. Am. Chem. Soc.*, **123**, 4492–4501.

87 (a) Paquette, L.A., Tae, J., Arrington, M.P., and Sadoun, A.H. (2000) *J. Am. Chem. Soc.*, **122**, 2742–2748; (b) Fürstner, A. and Langemann, K. (1996) *J. Org. Chem.*, **61**, 8746–8749.

88 Michaut, A. and Rodriguez, J. (2006) *Angew. Chem. Int. Ed.*, **45**, 5740–5750.

89 Ruprah, P.K., Cros, J.-P., Pease, J.E., Whittingham, W.G., and Williams, J.M.J. (2002) *Eur. J. Org. Chem.*, 3145–3152.

90 Michalak, K., Michalak, M., and Wicha, J. (2005) *Tetrahedron Lett.*, **46**, 1149–1153.

91 Takao, K.-I., Watanabe, G., Yasui, H., and Tadano, K.-I. (2002) *Org. Lett.*, **4**, 2941–2943.

92 Uchida, I., Ando, T., Fukami, N., Yoshida, K., Hashimoto, M., Tada, T., Koda, S., and Morimoto, Y. (1987) *J. Org. Chem.*, **52**, 5292–5293.

93 Morency, L. and Barriault, L. (2005) *J. Org. Chem.*, **70**, 8841–8853.

94 Paquette, L.A., Efremov, I., and Liu, Z. (2005) *J. Org. Chem.*, **70**, 505–509.

95 (a) Ma, C., Schiltz, S., Le Goff, X.F., and Prunet, J. (2008) *Chem. Eur. J.*, **14**, 7314–7323; (b) Schiltz, S., Ma, C., Ricard, L., and Prunet, J. (2006) *J. Organomet. Chem.*, **691**, 5438–5443.

96 Paquette, L.A., Dong, S., and Parker, G.D. (2007) *J. Org. Chem.*, **72**, 7135–7147.

97 Clark, J.S., Marlin, F., Nay, B., and Wilson, C. (2003) *Org. Lett.*, **5**, 89–92.

98 Clark, J.S., Northall, J.M., Marlin, F., Nay, B., Wilson, C., Blake, A.J., and Waring, M.J. (2008) *Org. Biomol. Chem.*, **6**, 4012–4025.

99 (a) Nevalainen, M. and Koskinen, A.M.P. (2001) *Angew. Chem. Int. Ed.*, **40**, 4060–4062; (b) Nevalainen, M. and

Koskinen, A.M.P. (2002) *J. Org. Chem.*, **67**, 1554–1560.

100 Nakamura, T., Oshida, M., Nomuira, T., Nakazaki, A., and Kobayashi, S. (2007) *Org. Lett.*, **9**, 5533–5536.

101 Hong, S.H., Sanders, D.P., Lee, C.W., and Grubbs, R.H. (2005) *J. Am. Chem. Soc.*, **127**, 17160–17161.

102 Li, Y. and Hale, K.J. (2007) *Org. Lett.*, **9**, 1267–1270.

103 Larrosa, I., Da Silva, M.I., Gómez, P.M., Hannen, P., Ko, E., Lenger, S.R., Linke, S.R., White, A.J.P., Wilton, D., and Barrett, A.G.M. (2006) *J. Am. Chem. Soc.*, **128**, 14042–14043.

104 Castoldi, D., Caggiano, L., Bayón, P., Costa, A.M., Cappella, P., Sharon, O., and Gennari, C. (2005) *Tetrahedron*, **61**, 2123–2139.

105 Castoldi, D., Caggiano, L., Panigada, L., Sharon, O., Costa, A.M., and Gennari, C. (2005) *Angew. Chem.*, **117**, 594–597.

106 Castoldi, D., Caggiano, L., Panigada, L., Sharon, O., Costa, A.M., and Gennari, C. (2006) *Chem. Eur. J.*, **124**, 51–62.

107 Bourgeois, D., Mahuteau, J., Pancrazi, A., Nolan, S.P., and Prunet, J. (2000) *Synthesis*, 869–882.

108 Larghi, E.L. and Kaufman, T.S. (2008) *Tetrahedron*, **64**, 9921–9927.

2
Natural Products Containing Medium-sized Nitrogen Heterocycles Synthesized by Ring-closing Alkene Metathesis

Sebastiaan (Bas) A. M. W. van den Broek, Silvie A. Meeuwissen, Floris L. van Delft, and Floris P. J. T. Rutjes

2.1
Introduction

The formation of carbon–carbon bonds is a crucial issue in organic synthesis. Out of all methodologies available, the metathesis reaction has, over the past decade, evolved as one of the most useful tools in this field to rapidly construct the skeleton of carba- and heterocyclic molecules [1]. Key step of the metathesis reaction entails the metal-catalyzed redistribution of two carbon–carbon double bonds by a scission–recombination process [2]. Among various applications of the metathesis reaction, ring-closing metathesis (RCM) has emerged as one of the most powerful tools for the construction of natural and unnatural cyclic compounds (Scheme 2.1) [3]. Other types of metathesis reactions developed so far include cross-metathesis (CM), ring-opening metathesis (ROM), ring rearrangement metathesis (RRM, combining ROM with RCM in a tandem sequence), and ene–yne metathesis [3].

The metathesis reaction has already been used for several decades in polymer chemistry for ring-opening metathesis polymerization (ROMP) [4]. Only since the development of well-defined molybdenum and ruthenium carbene complexes by Schrock [5] and Grubbs [6] in 1990 and 1992, respectively, an explosion of their application in organic synthesis has been witnessed. Initially, it was demonstrated that the Schrock catalyst [Mo]-**I** could serve as a homogeneous catalyst for olefin RCM [7]. Unfortunately, metal carbene complex [Mo]-**I** suffers from extreme sensitivity to air and moisture, which precluded widespread use among organic chemists. Not long thereafter, the more stable yet less reactive benzylidene ruthenium complex [Ru]-**IV** was developed [8]. In 1995, additional research in this field led to the ruthenium carbene catalyst [Ru]-**I** [9], which is currently commercially available as the Grubbs-I or first-generation Grubbs catalyst. This catalyst still appeared to be somewhat less reactive than catalyst [Mo]-**I** in olefin metathesis, but yet retained the remarkable air and water stability characteristics of ruthenium complex [Ru]-**IV**. In 1999, the effectiveness of the ruthenium-based metathesis catalysts was even further enhanced by replacing one of the phosphine ligands with a heterocyclic carbene, the 1,3-bis(2,4,6-trimethylphenyl)-2-imidazolidinylidene ($IMesH_2$) ligand, to give

Metathesis in Natural Product Synthesis: Strategies, Substrates and Catalysts. Edited by Janine Cossy, Stellios Arseniyadis, and Christophe Meyer

ISBN: 978-3-527-32440-8

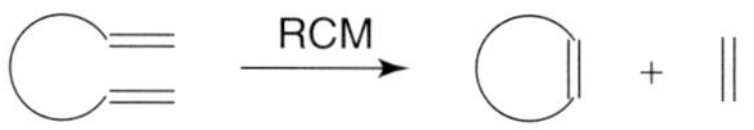

Scheme 2.1 Schematic illustration of the ring-closing metathesis reaction.

catalyst [Ru]-**II** [10]. This carbene complex exhibits high olefin metathesis activity in RCM reactions due to, amongst others, a higher thermal stability and wider functional group tolerance, and is nowadays commercially available as the Grubbs-II or second-generation Grubbs catalyst. Many (commercially available) modifications of these catalysts exist [11]; in particular, ones in which the tricyclohexylphosphine and benzylidene groups have been replaced by an *ortho*-isopropoxy-substituted benzylidene, the so-called Grubbs–Hoveyda catalysts [Ru]-**III** and [Ru]-**IX** [12].

Formation of heterocycles by RCM was first reported by Grubbs and Fu, who found that a di-N-allylated trifluoroacetamide underwent cyclization in the presence of Schrock's catalyst [Mo]-**I** to give the corresponding 3-pyrroline [13] (Figure 2.1). At that time it was shown that the ruthenium–alkylidene complex [Ru]-**IV** gave also excellent yields for similar heterocyclic RCM reactions, even when performed in undistilled solvents and in the presence of air. Although it was well established that free amines are generally incompatible with the metathesis reaction [14], later work demonstrated that this problem could be overcome by the protection of the basic nitrogen atom as an amide or carbamate. In addition, protonation of the amine to give the corresponding ammonium salts and thus avoiding protecting groups is tolerated by [Ru]-**I** and [Ru]-**II**.

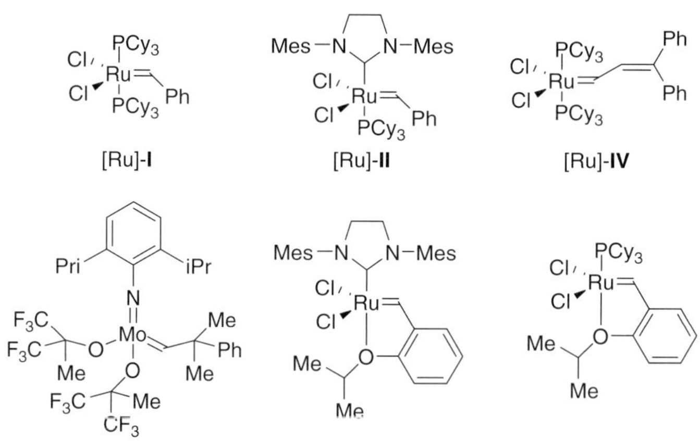

Figure 2.1 Metal–alkylidene complexes as metathesis catalysts: Schrock's catalyst [Mo]-**I**, the first benzylidene ruthenium complex [Ru]-**IV**, Grubbs-I catalyst ($Cl_2(Cy_3P)_2Ru{=}CHPh$) [Ru]-**I**, Grubbs-**II** catalyst ($Cl_2(IMesH_2)(Cy_3P)Ru{=}CHPh$) [Ru]-**II**, Grubbs–Hoveyda-I catalyst [Ru]-**IX**, Grubbs–Hoveyda-II catalyst [Ru]-**III**.

As a result of the compatibility of the Ru catalysts with nitrogen-based functional groups, RCM became a widely applied synthetic tool in the total synthesis of nitrogen-containing natural products. In particular, the formation of medium-sized (five- to eight-membered) heterocyclic systems, which frequently occurs in naturally occurring alkaloids, became a key target for metathesis-mediated synthesis. Hence, the aim of this account is to provide an overview of the majority of syntheses of alkaloid natural products in which a medium-sized nitrogen-containing ring was constructed using RCM. The examples are restricted to pathways that really have led to a natural product and cover the literature until August 2008.

2.2 Five-membered Nitrogen Heterocycles

2.2.1 Dihydropyrroles

Interesting candidates for the application of RCM are the linear marine peptides (aeruginosins) that share a common 1-aza[4.3.0]-bicyclic core unit. Hanessian *et al.* recently described the total syntheses of dysinosin A (**1**) [15] that was isolated from a new genus of sponge of the family Dysideidae and oscillarin (**2**) [16] that was found in the algal cultures of *Oscillatoria agardhii*. Both alkaloids were constructed via two peptide couplings between three common subunits: an indolizidine carboxylic acid (**4** or **6**) derived from L-glutamic acid, an acyclic peptide chain (**5** or **7**) and pyrrolino-ethylamine **3** (Scheme 2.2).

The synthesis of the Δ-3 pyrroline unit **3** is shown in Scheme 2.3 and started with readily available butyrolactone. Six steps involving ring opening to the hydroxy ester and substitution of the reduced ester with allylamine were required to form the metathesis precursor **8**. Cyclization of *N*-Boc-protected **8** into pyrroline **9** employing [Ru]-**I** (10 mol%) went smoothly in a high yield of 90%. The six-membered carbocycle of dysinosin A fragment dihydroxyindolizidine **4** was prepared via RCM as well (99% yield) using the same catalyst (1 mol%) as for **8**.

2.2.2 Pyrrolidine Alkaloids

2.2.2.1 Pyrrolidines

(−)-(*S*)-Nicotine (**13**) is one of the many piperidine and pyrrolidine alkaloids isolated from the leaves of *Nicotaiana tabacum*. Welter and coworkers developed a synthesis route where **13** was obtained in only four steps from pyridinylallyl carbonate **10** involving RCM (Scheme 2.4) [17]. The synthesis commenced with an Ir-catalyzed asymmetric allylic amination of **10** to produce bis-alkene **11** in excellent enantiomeric purity when using phosphoramidate ligand **L1**. Subsequently, secondary amine **11** was N-protected to produce **12** prior to RCM in order to prevent catalyst

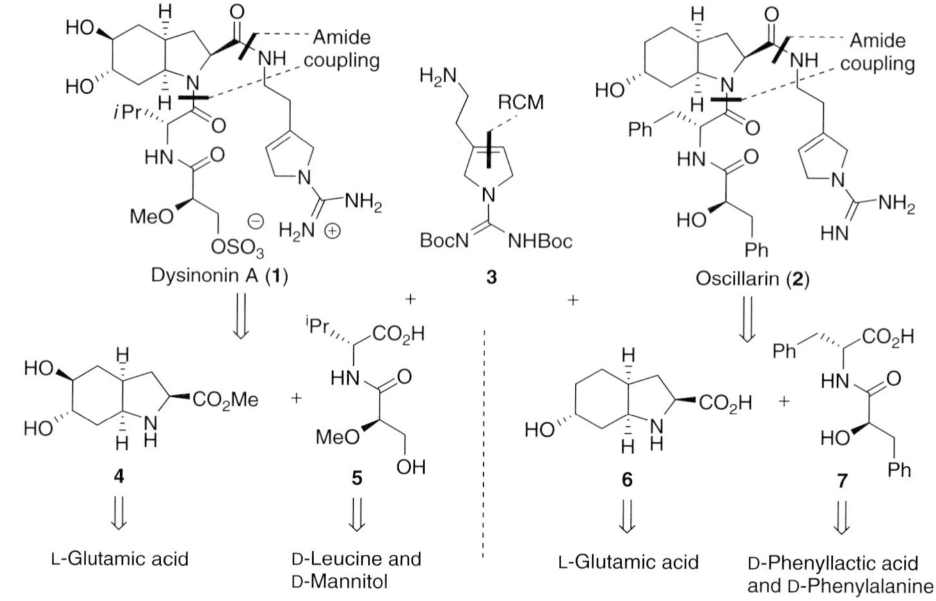

Scheme 2.2 Retrosynthesis of the natural products dysinosin A (**1**) and oscillarin (**2**) sharing the RCM product **3**.

deactivation. Best RCM results were achieved with [Ru]-**II** yielding the corresponding dihydropyrrole derivative in >90% yield. The synthesis of (*S*)-nicotine was completed through a deprotection step and finally reduction of the double bond.

Over the past few decades, the biosyntheses of tropane alkaloids such as (−)-hyoscyamine, (−)-scopolamine, and (−)-cocaine have been studied extensively [18]. The skeletons of all these alkaloids feature the tropinone moiety, produced by oxidation and subsequent cyclization of (+)-hygrine (**17**) [19]. It was envisioned that RCM could be invoked in order to develop an enantioselective synthesis of **17** [20]. As depicted in Scheme 2.5, a phase-transfer catalytic allylation reaction with methallyl bromide converted *t*-butyl ester **13** into product **14** with 97% ee. Metathesis precursor **15** was then prepared in six more steps, including transformation of the *t*-butyl ester of **14** into a terminal olefin. Dihydropyrrole **17** was formed in quantitative yield by RCM employing [Ru]-**II** (8 mol%). Subsequent hydrogenation and deprotection of **16** led to (+)-hygrine **17** with overall yield of 29% and an ee of 97%.

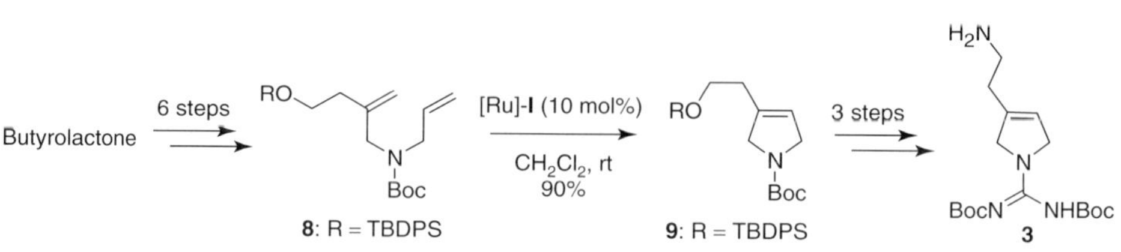

Scheme 2.3

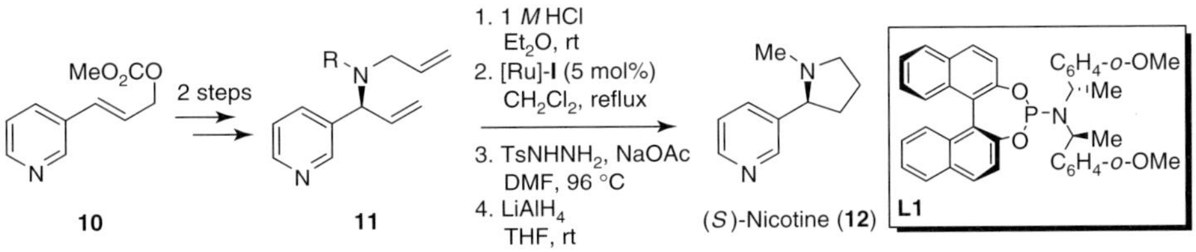

Scheme 2.4

2.2.2.2 Dipyrrolidines

A total synthesis of the dipyrrolidine alkaloid (−)-*trans*-dendrochrysine (**22**), isolated from the orchid endemic *Dendrobium chrysanthum*, was achieved by Blechert and coworkers utilizing RRM [21]. The asymmetric synthesis commenced with preparation of diolefin **20** from commercially available tropone (**19**, Scheme 2.6) in 11 steps involving a solid-phase supported *Candida antarctica* lipase B (Cal-B)-mediated desymmetrization.

Different ruthenium catalysts were evaluated for the RRM process and it appeared that [Ru]-**III** (5 mol%) in refluxing toluene under an ethylene atmosphere [22] gave the best results providing bicycle **21** in 91% yield. Thus, (−)-*trans*-dendrochrysine (**22**) was synthesized with a yield of 5.9% over 18 linear steps.

2.2.2.3 Polyhydroxypyrrolidines

Naturally occurring iminosugars (azasugars) have been prominent target molecules for the utilization of RCM, involving formation of a cyclic olefin, followed by dihydroxylation. The mulberry tree (*Morus alba*) alkaloid 1,4-dideoxy-1,4-imino-D-ribitol [(+)-DRB, **27**] is a potent inhibitor of glucosidases and of eukaryotic DNA polymerases. The Riera synthesis of polyhydroxypyrrolidine **27** featured a five-membered ring formation via RCM as illustrated in Scheme 2.7 [23].

Divinylcarbinol **23** was converted in a number of steps into the enantiomerically pure RCM precursor **25** (ee = 99%) via a Sharpless epoxidation and subsequent Payne rearrangement as the key steps. RCM under the influence of the [Ru]-**II** catalyst (8 mol%) proceeded uneventfully, despite the two unprotected alcohols, to give intermediate **26** in a surprisingly high yield of 90%. Synthesis of (+)-DRB (**27**)

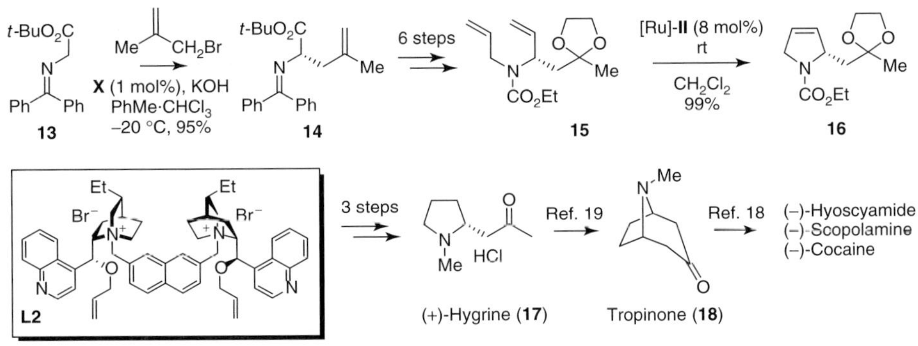

Scheme 2.5

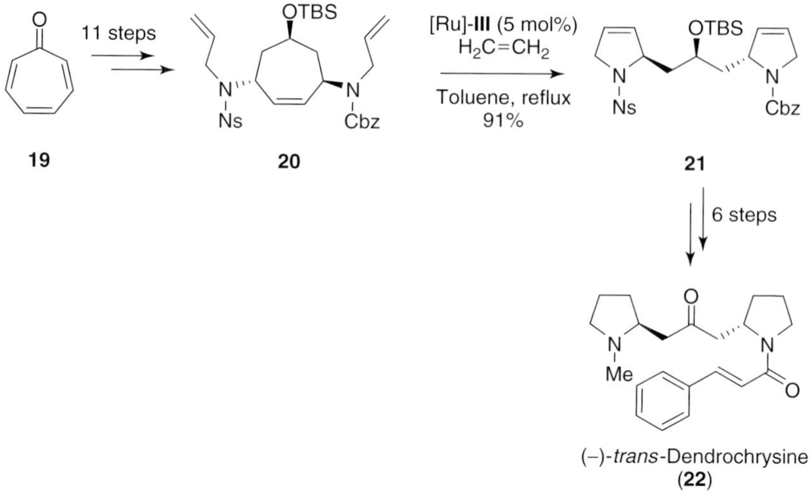

Scheme 2.6

was completed in several steps including a side chain–directed syn-dihydroxylation with OsO_4 of the ring olefin.

A similar metathesis strategy was utilized by Trost *et al.* for the total syntheses of the azasugars 2,5-dideoxy-2,5-imino-D-mannitol [(+)-DMDP, **28**], (−)-bulgecinine (**29**), and the somewhat more complex (+)-broussonetine G (**30**) [24, 25]. (+)-DMDP (**28**) can be isolated from several sources, but was initially discovered in the leaves of *Derris eliptica*. (−)-Bulgecinine was found in bacterial cultures of *Pseudomonas acidophila* and *Pseudomonas mesoacidophila* and the broussonetine family including broussonetine G in branches of the deciduous tree *Broussontia kazinoki*. All three natural products were obtained from the same metathesis product **31** (Scheme 2.8),

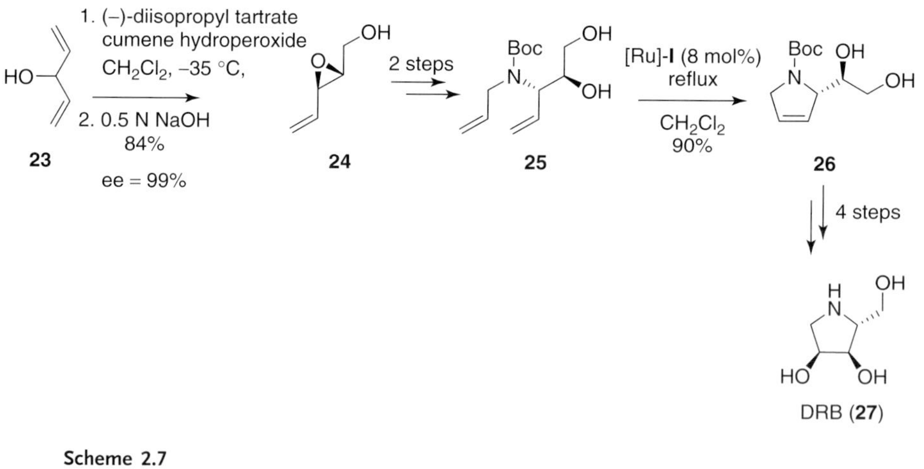

Scheme 2.7

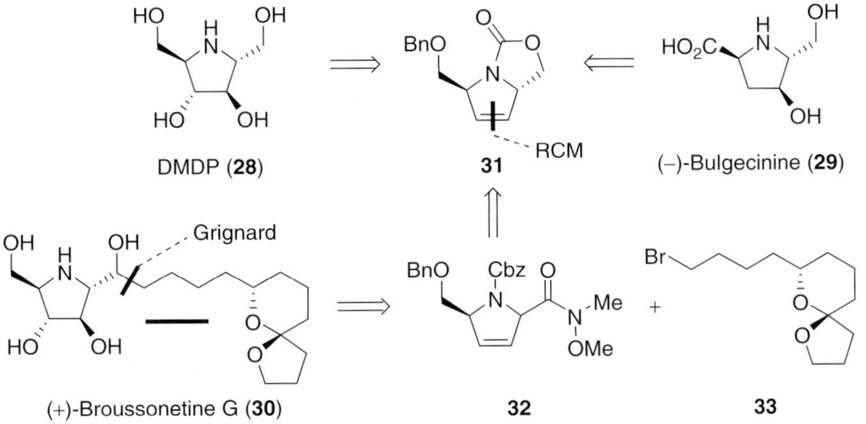

Scheme 2.8 Natural products (+)-DMDP (**28**), (−)-bulgecinine (**29**), and (+)-broussonetine G (**30**) sharing the same RCM-made building block **31**.

from which broussonetine G was derived via a Grignard addition of **33** onto **32**. Key steps in the synthesis of the central intermediate **31** were a palladium-catalyzed dynamic kinetic asymmetric transformation (DYKAT) and RCM.

The synthesis of **31** started with a double dynamic asymmetric allylic amination for the enantioselective construction of diallylamine **36** from allylic epoxide (**34**, Scheme 2.9) involving ligand (*R,R*)-**L3**. While RCM with a free hydroxyl (**36**) appeared troublesome, the ring closure proceeded smoothly in high yield using [Ru]-**II** (1.2 mol%) after benzyl protection (**36**), which presumably prevents undesired coordination to ruthenium. Finally, seven steps including a Grignard reaction between alkyl bromide **33** and the corresponding Weinreb amide of **32** were required to afford broussonetine G (**30**) in 8.2% overall yield.

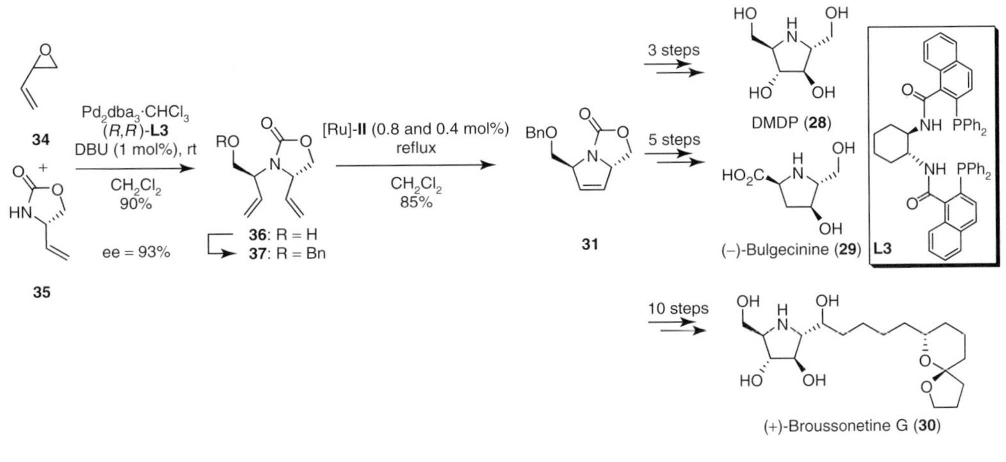

Scheme 2.9

2.2.3
Indolizidine Alkaloids

2.2.3.1 Polycyclic Indolizidines

Alkaloids with polycyclic skeletal frameworks are, when it comes to their synthesis, excellent candidates for RCM. Illustrative are the indolizidines rhynchophylline (**43**) and its C(7)-epimer *iso*-rhynchophylline (**44**), both isolated from the plant *Uncaria rhynchophylla* (Rubiaceae) [26]. Deiters' total synthesis of **43** and **44** started with the efficient construction of diallylamine **39** via amide formation between indole-3-acetic acid (**38**) and diallylamine (Scheme 2.10). One-pot RCM–carbomagnesation of **39** was smoothly achieved with only 1 mol% of [Ru]-**I** catalyst and 4 equiv. of EtMgCl to afford the 2-ethyl-3-butene-amine derivative in 71% yield. It appeared that the electron-withdrawing carbonyl moiety was critical to the success of the RCM–carbomagnesation steps. Amide reduction and subsequent treatment with acryloyl chloride delivered the second metathesis precursor **40**. Cyclization with [Ru]-**I** (5 mol%) then furnished the α,β-unsaturated lactam **41** in a high yield of 91%. Continuation of the total synthesis of alkaloids **43** and **44** included a Bischler–Napieralski cyclization (**42**) and subsequent rearrangement into the oxindole framework.

The erythrina alkaloids form a widely distributed family that lately has received considerable attention as a result of their unique tetracyclic skeleton and biological activities. Erythrocarine (**46**) [27, 28], erythravine (**50**) [29], and (+)-β-erythroidine (**59**) [30] are three members of this family that have been synthesized with metathesis as a key step. The molecular structures of erythrocarine, found in the seeds of *Erythrina caribaea*, and erythravine, isolated from *Erythrina cochleata*, are related. As a result, the total syntheses independently proposed by, respectively, Mori *et al.* and Hatakeyama *et al.* proceeded via the comparable intermediates **45** and **49**, respectively, as shown in Scheme 2.11. Ene–yne **45** was synthesized in a number of steps. This substrate was subjected to HCl, followed by [Ru]-**I** catalyst (10 mol%) to induce a tandem enyne-RCM process to yield the acetylated natural product in a quantitative conversion as a 1 : 1 mixture of diastereoisomers. After

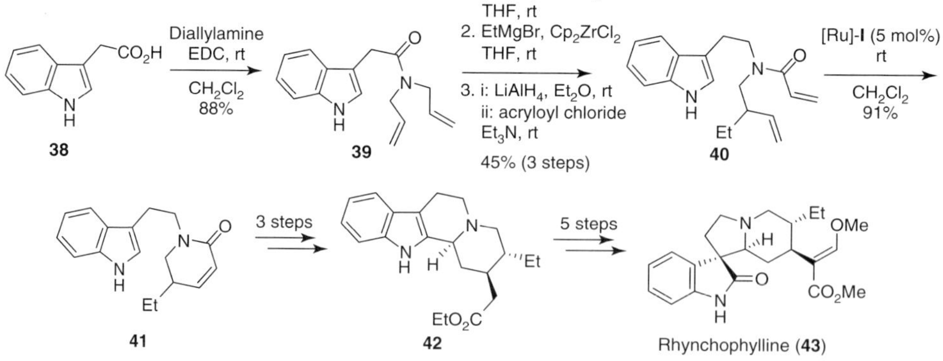

Scheme 2.10

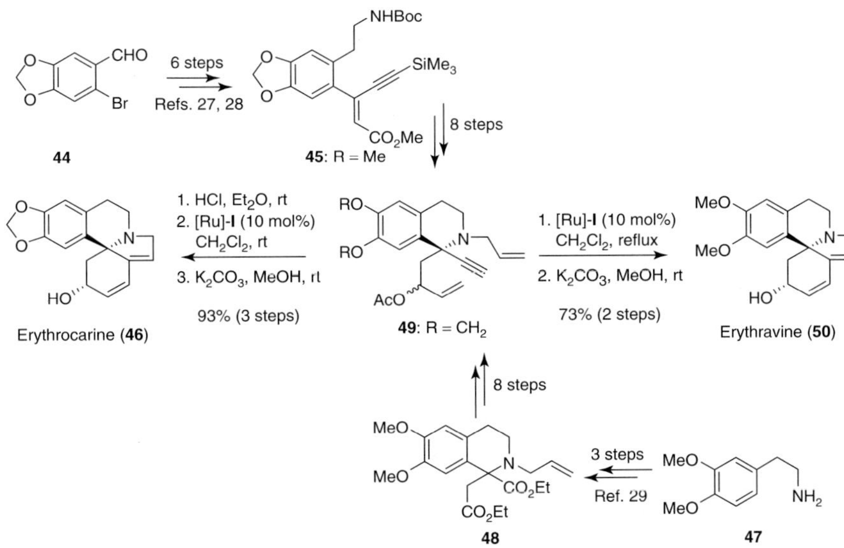

Scheme 2.11

separation of the two isomers, acetyl cleavage provided erythocarine (**46**) in pure form.

Intermediate **49** was prepared from **48** in a large number of steps. The crucial tandem ene–yne-RCM process was also performed with [Ru]-**I** catalyst (10 mol%), which afforded the tetracyclic product in 78% in a diastereomeric ratio of α-OAc-**50**: β-OAc-**50** as 63 : 37. Once more, separation of the products, followed by deacetylation of α-OAc-**50** afforded erythravine (**50**) in an overall yield of 7.0%.

The efficient approach to erythravine paved the way for a second erythrina alkaloid [(+)-β-erythroidine (**56**)] synthesis by Hatakeyama *et al.* This alkaloid was found in several species of the *Erythrina* genus and was synthesized in 26 linear steps as depicted in Scheme 2.12. Starting from **54**, tandem ene–yne RCM of the isomeric mixture **55** underwent cyclization with [Ru]-**I** catalyst (10 mol%) to

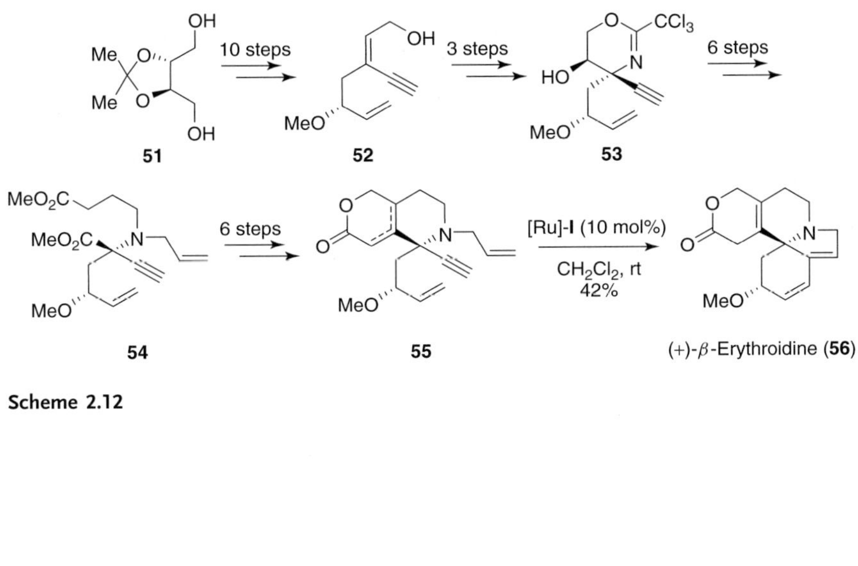

Scheme 2.12

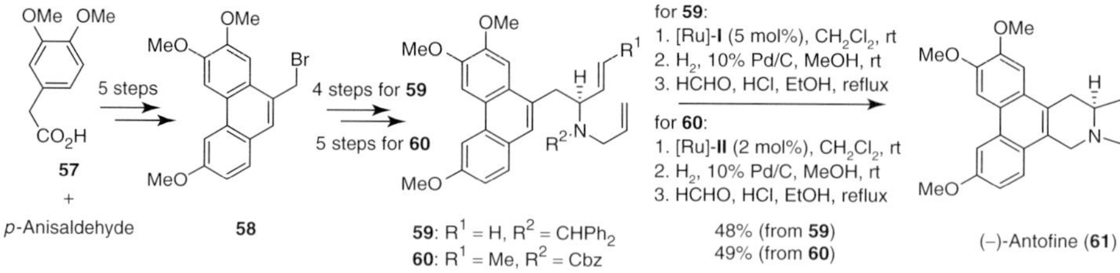

Scheme 2.13

furnish (+)-β-erythroidine (**56**) in a moderate yield together with a small amount of a mono-ring-closed by-product.

(−)-Antofine **61** is a member of the small group of phenanthroindolizidine alkaloids and was initially found in *Vincetoxicum nigrum*, belonging to the Asclepiadaeceae family. Recently, Kim *et al.* described two routes to enantiopure (−)-antofine [31, 32]. Both routes began with the coupling of homoveratric acid (**57**) and *p*-anisaldehyde in order to form phenanthryl alcohol, the precursor of bromide **58** (Scheme 2.13). In the first article, the synthesis was continued via enantioselective catalytic phase-transfer alkylation of a protected glycine derivative with **58**. The glycine fragment was further transformed into the diallylamine of **59** in three steps. In the more recent article, enantiopure **60** was obtained via a Stille coupling with an enantiopure fragment and subsequent Overman rearrangement. Both routes converged to **61** after an RCM step that was high yielding in both cases, followed by hydrogenation and Pictet–Spengler cyclization. Interestingly, RCM of diolefin **60** took place 10 times faster and was somewhat higher yielding in comparison with **59**, probably as a result of different catalysts. Overall, it can be concluded that the second approach to (−)-antofine was, though one step longer, more efficient and higher yielding.

The tricyclic tetraponerines form an unusual class of alkaloids representing the major constituents of contact poison of the New Guinean ant *Tetraponera* sp. A flexible synthesis of tetraponerines T1–T8 is challenging as they differ in side chain, stereochemistry at C-9, and size of one of the rings. Stragies *et al.* proposed a combined ROM–RCM strategy that led to four of these alkaloids [33]. One of them was tetraponerine T6 (**65**), synthesized in only six steps. The synthesis commenced with a palladium-catalyzed domino allylic alkylation on dicarbonate **62** to deliver the disulfonamide **63**. [Ru]-**I**-induced (5 mol%) ROM–RCM yielded an equilibrium between precursor **63** and product **64** in a 1 : 10 ratio. Subsequent elaboration of the side chain, followed by cyclization, finally provided tetraponerine T6 (**65**), in a good 30% overall yield (Scheme 2.14).

One of the simple representatives of the indolizidine series isolated as a trace compound from the skin of the Neotropical frog *Dendrobates speciosus* is indolizidine 167B (**70**). Although neither the relative nor the absolute configuration of **70** was known, Blechert and coworkers synthesized the alkaloid in accordance with the literature data via RRM (Scheme 2.15) [34]. The two olefinic side chains of product **67** required for RRM process were introduced in four steps from cycloheptenediol-monoacetate (**66**). RRM under the influence of [Ru]-**I** (5 mol%)

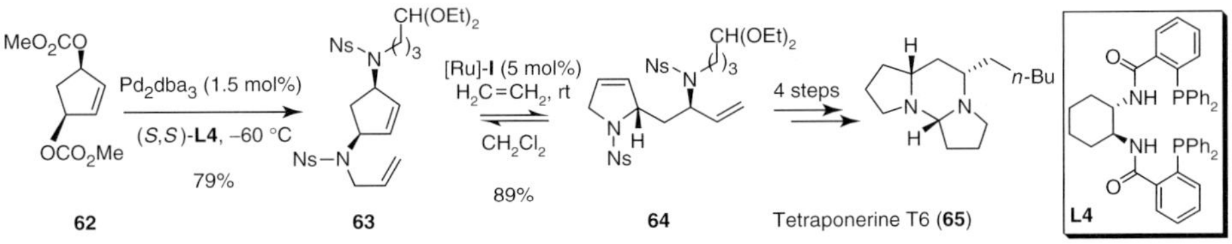

Scheme 2.14

was *in situ* followed by TBAF-mediated silyl ether cleavage into alcohol **69** in an outstanding 92% yield. Final ring closure via reductive amination and hydrogenation delivered natural product **70** in a stereocontrolled way in an overall yield of 35%.

Another member of the indolizidine family derived from the frog genera *Dendrobatid* is (−)-indolizidine 223A (**73**). Davis *et al.* discovered that an intramolecular Mannich reaction combined with metathesis provided a high yield and enantioselective solution for its synthesis [35]. Reaction of 4-heptanone with enantiopure sulfinimine **71**, followed by treatment with crotonaldehyde, gave the tetrasubstituted piperidinone **72** via an intramolecular Mannich reaction (Scheme 2.16). Subsequent RCM with [Ru]-**I** (5 mol%), followed by hydrogenation furnished as the key steps, afforded (−)-indolizidine 223A **73** in an overall yield of 9.3%.

2.2.3.2 **Polyhydroxyindolizidines**

Despite the availability of various existing synthetic procedures, application of RCM in the synthesis of polyhydroxyindolizidines recently became increasingly popular. The skeletons of all alkaloids described below were produced according to the same strategy involving N-alkenylation and RCM for the formation of five- and six-membered nitrogen heterocycles (Scheme 2.17).

The total synthesis of lengtinosine, isolated from *Australahus lentiginosus*, was performed by the group of Spino. Uniflorine A, recently discovered in the tree *Eugenia uniflora* L., and initially proposed as structure **75**, was independently synthesized by various groups including those of White, Pyne, and Dhavale. However, NMR data of **75** did not correspond with the data of natural uniflorine A.

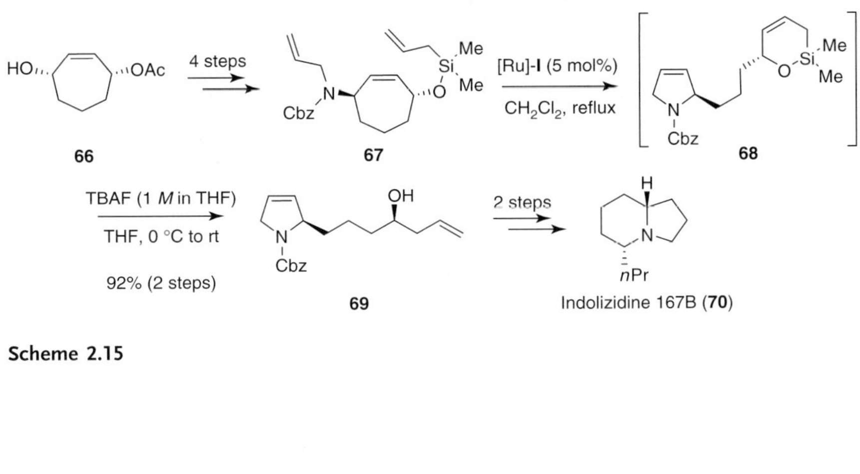

Scheme 2.15

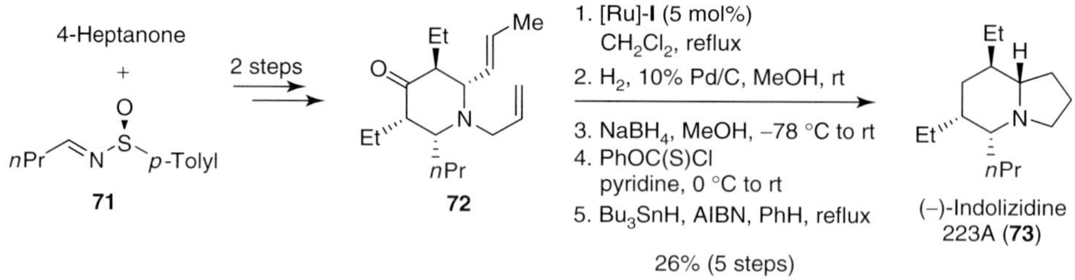

Scheme 2.16

Synthesis of (+)-lentiginosine's metathesis precursor **80** commenced with the stereoselective transformation of commercially available L-menthone into compound **79** (Scheme 2.18). RCM precursor **80** was efficiently cyclized with [Ru]-**I** (5 mol%) to dihydropyrrolidine **76**, and the L-menthone auxiliary was recovered. From there, lentiginosine **74** was eventually synthesized in five steps.

In 2004, White *et al.* were the first to report a synthesis for putative uniflorine **75** in which RCM was used as the key step (Scheme 2.19). A three-component boronic acid–Mannich reaction (Petasis reaction) of L-xylose, allylamine, and (*E*)-styrene boronic acid provided the amino-tetraol **81** as a single diastereoisomer in good yield. Application of [Ru]-**I** (10 mol%) on *N*-Boc- and *O*-trityl-protected **81** afforded metathesis product **77** in 86% yield. The route to tetrahydroxyindolizidine **75** was completed by catalytic syn-dihydroxylation of **74** and six-membered ring formation. Further investigations involving five-membered RCM in the quest to find the correct structure of uniflorine A provided 2-*epi*-**75** and 1,2-di*epi*-**75** [36] and 8a-*epi*-**75** and 2,8a-di*epi*-**75** [37]. Yet, all these putative structures delivered mismatching NMR dataas well. In 2008, White *et al.* suggested, therefore, that uniflorine A is 1,2,6,7-tetrahydroxy-3-hydroxymethylpyrrolizidine [36].

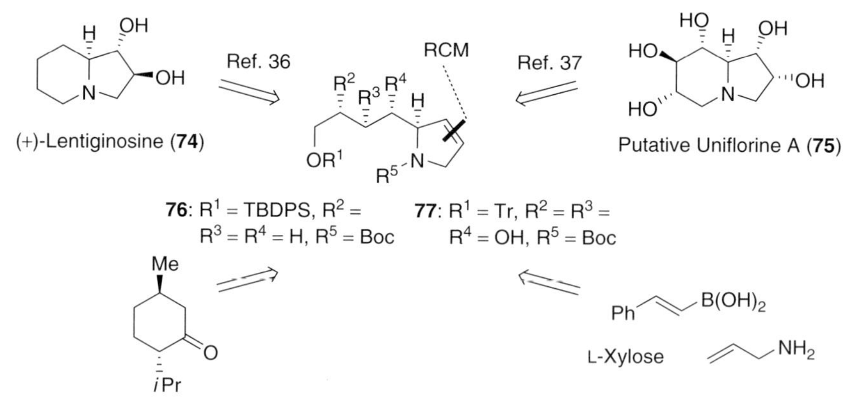

Scheme 2.17 Retrosyntheses of (+)-lentiginosine (**74**) and the putative structure of uniflorine A (**75**) proceed via similar RCM precursors (**76** for **74**, **77** for **75**).

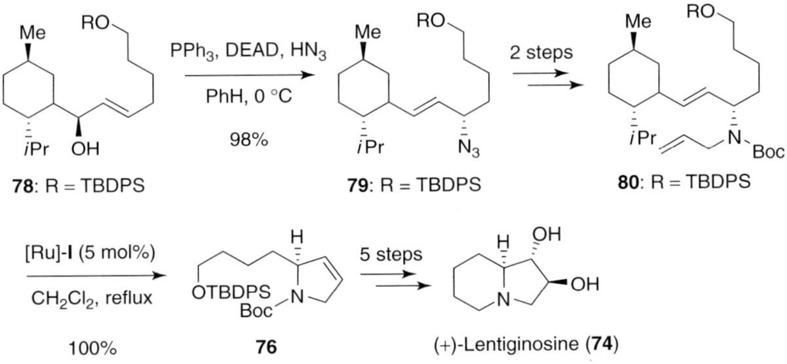

Scheme 2.18

Next, Pyne *et al.* successfully utilized the synthetic methodology created for putative uniflorine **75** to construct the analogous castanospermine (**84**, Scheme 2.19) [38]. In this case, amino-tetraol **81** had to be protected with three different protecting groups resulting in compound **82**. Treatment of **82** with [Ru]-**II** (10 mol%) yielded pyrrolo-oxazolone **83** in 88% yield. Diastereoselective syn-dihydroxylation of the olefin, followed by several steps, led then to castanospermine (**84**).

Trihydroxyindolizidine (−)-swainsonine **89** was first isolated from the fungus *Rhizoctonia leguminicolain* and has since then attracted great attention from chemists and biologists. In 2002, both Blechert *et al.* [39] and Pyne *et al.* [40] reported a metathesis-based approach to alkaloid **89** via the creation of a five-membered heterocycle. The route by Blechert *et al.* began with the enantioselective formation of oxazolidinone **86** via palladium-catalyzed desymmetrization of the *meso*-bis-carbamate derivative of diol **85** (Scheme 2.20). From there, the metathesis precursor **84** was delivered in excellent yield. Tandem RRM employing [Ru]-**I** catalyst (5 mol%) under an ethylene atmosphere converted compound **87** into the desired 3-pyrroline **88** in yields up to 98%, while the stereochemical integrity of the two chiral centers was completely preserved during the reaction. Next, elaboration

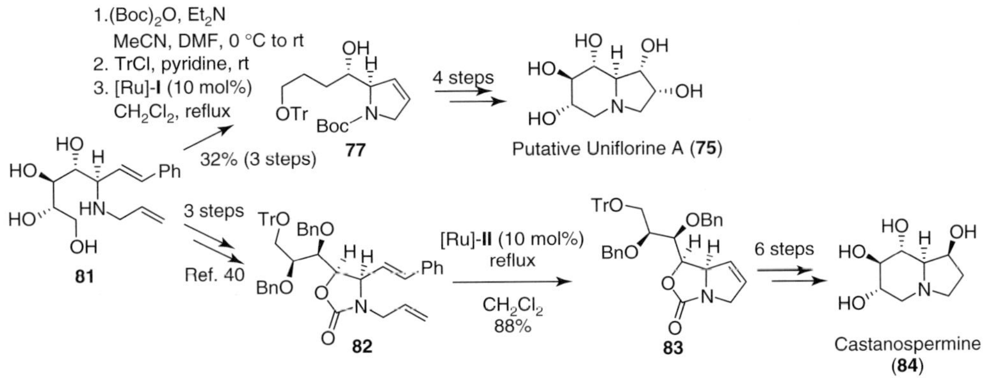

Scheme 2.19

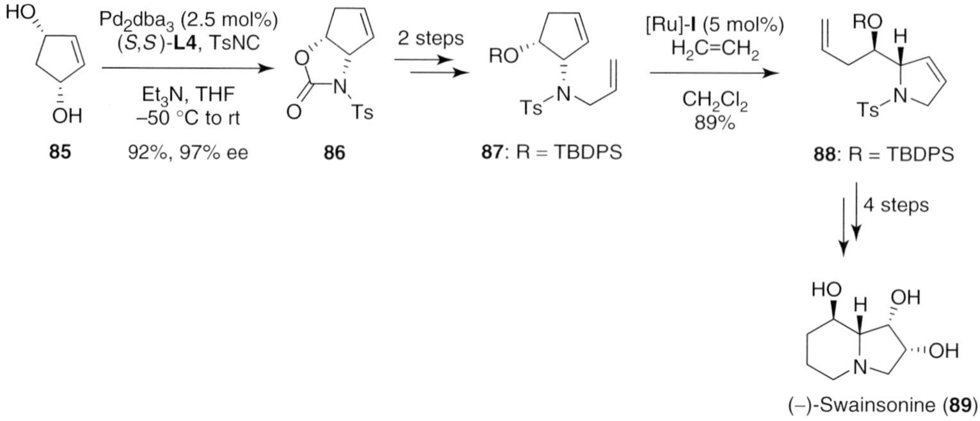

Scheme 2.20

of the side chain, ring closure, and olefin dihydroxylation using commercially available AD-mix-α afforded swainsonine **89** in an impressive overall yield of 40%.

Pyne and coworkers reported a completely different synthetic approach to (−)-swainsonine (**89**) starting from readily available 4-pentynol (Scheme 2.21). This was elaborated via *cis*-diol **90** into the enantiopure metathesis precursor **91** in a number of steps. Next, RCM with [Ru]-**II** (6.5 mol%) converted **91** into the cyclic product **92** in 95% yield. Intramolecular N-alkylation and stereoselective cis-dihydroxylation via the same methodology as applied by Blechert *et al.* completed the synthesis of swainsonine (**89**).

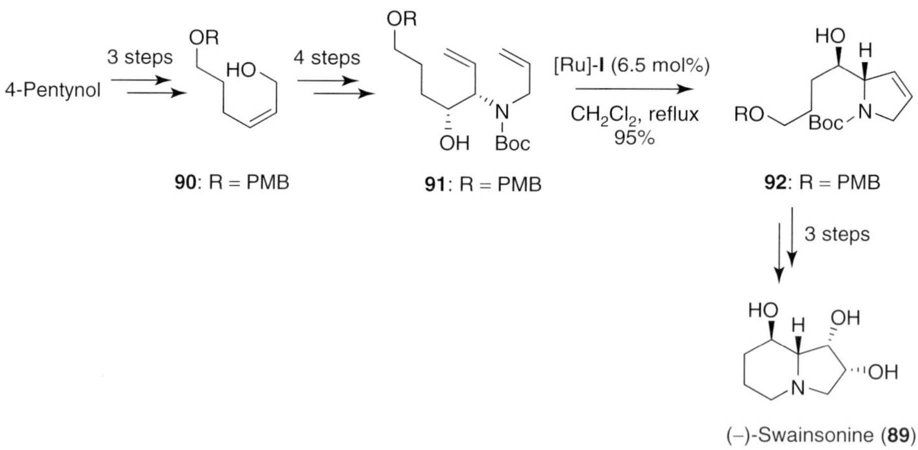

Scheme 2.21

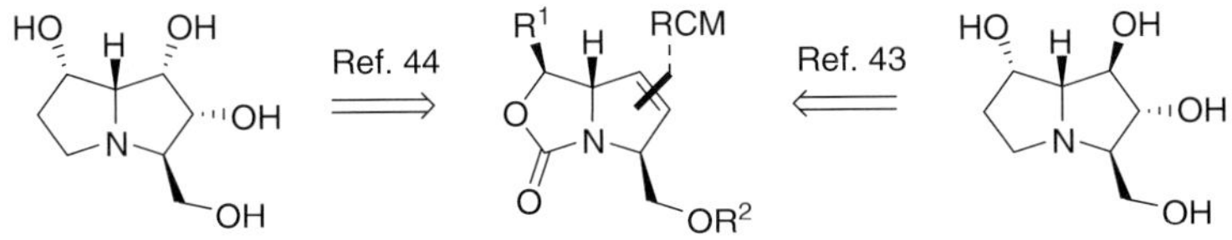

1-Epiaustraline (**93**)   **94**: $R^1 = CH_2CH_2OPMB$, $R^2 = Bn$; **95**: $R^1 = CH{=}CH_2$, $R^2 = H$ (+)-Australine (**96**)

Scheme 2.22 (+)-1-*epi*-Australine and (+)-australine, which were derived from a similar RCM precursor.

2.2.4
Pyrrolizidine Alkaloids

The family of polyhydroxyindolizidine alkaloids is not the only target of interest in the field of natural product synthesis when it comes to application of RCM. Several approaches for the formation of the hindered 5,5-fused bicyclic (poly)hydroxypyrrolidines have been reported as well. The RCM methodology is, in some of these cases, applied to construct the second pyrrole, while on the other hand a larger number of cases exist for both natural and nonnatural products where the first cycle is created this way.

(+)-Australine (**96**) [41] and its epimer (+)-1-*epi*-australine (**93**) [42] constitute two examples in which RCM was used in the construction of the second pyrrole (Scheme 2.22). Both alkaloids were, together with other australine epimers, isolated from the seeds of *Castanospermum australe*.

Pyne *et al.* were the first to use RCM in the synthesis of a tetrahydropyrrolizidine, namely, 1-*epi*-australine **93**. Ring opening of epoxide **97**, readily obtained from 3-butynol, with an enantiopure allylic amine provided the desired diallylamine in moderate selectivity (Scheme 2.23). Conversion of the diallylamine into oxazolidinone **98** opened the door to the first RCM cyclization. Portionwise addition of [Ru]-**I** catalyst (30 mol% in total) gave product **95** in 97% yield. When less catalyst was used, significantly lower yields of **95** were obtained. syn-Dihydroxylation of the 3-pyrroline **95** and an additional five more steps were required to obtain (+)-1-*epi*-australine (**93**) as a colorless oil.

Unfortunately, an identical approach failed for the synthesis of (+)-australine (**96**) since a Mitsunobu ring closure of the second ring did not take place. A few years later, however, Trost and coworkers published a different route that made an RCM-mediated synthesis of (+)-australine (**96**) possible. In a similar procedure as

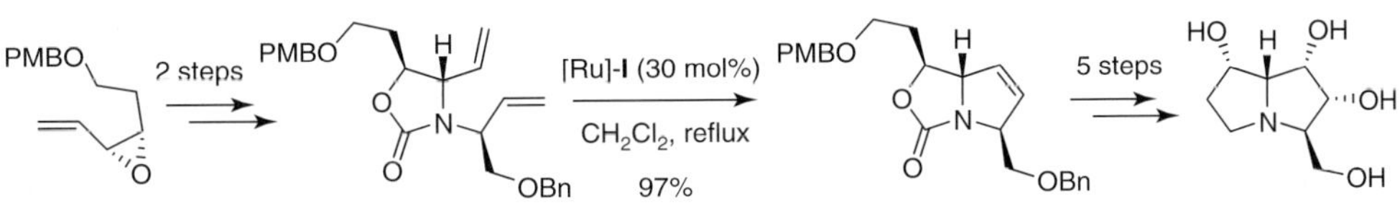

97   **98**   **95**   (+)-Epi-australine (**93**)

Scheme 2.23

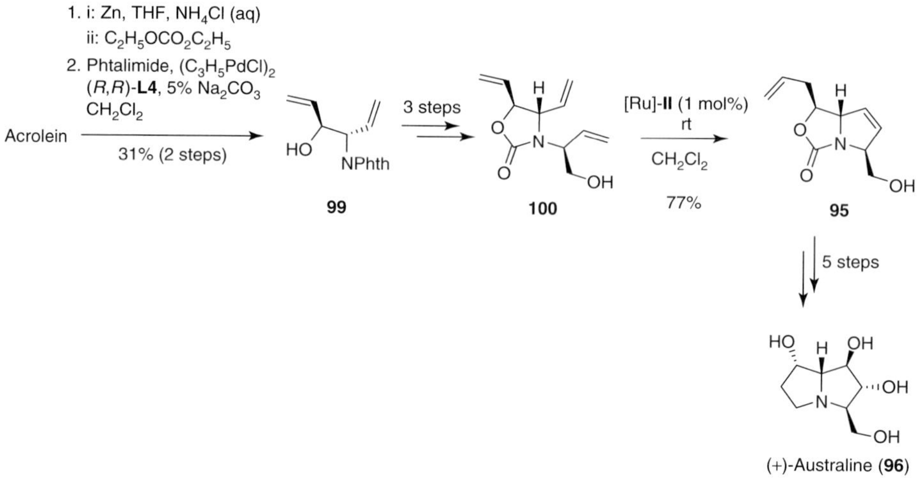

Scheme 2.24

described for the synthesis of compound **31** [24, 25], triolefin **99** was prepared in an enantiomerically pure form. Interestingly, application of RCM on unprotected **100** with only 1 mol% of [Ru]-**II** afforded the corresponding product **95** in 77% yield. Next, benzylation of the primary alcohol, followed by selective epoxidation and subsequent opening with benzyl alcohol, provided an improved precursor for the second cyclization. Thus, (+)-australine (**96**) was eventually successfully prepared and isolated as its hydrochloride salt (Scheme 2.24).

In Martin's synthesis of (+)-hyacinthacine A_2 (**103**), one of the alkaloids occurring in the bulbs of *Muscari armeniacum* (Hyacintheceae), RCM was used for the second ring formation (Scheme 2.25) [43]. Starting from D-arabinofuranose, only four steps were required to prepare the metathesis precursor **102**, including reductive amination with allylamine – an intramolecular displacement of the allylic benzoate ester. RCM was performed on the corresponding hydrochloride salt **102** in the presence of [Ru]-**I** (16 mol%) in toluene to provide the corresponding trihydroxypyrrolizine in 30% yield (75% based on recovered starting material). (+)-Hyacinthacine A_2 (**103**) was eventually delivered by concomitant debenzylation and double bond hydrogenation.

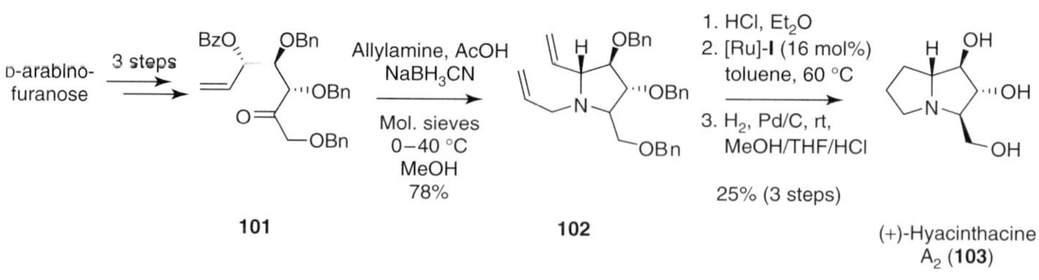

Scheme 2.25

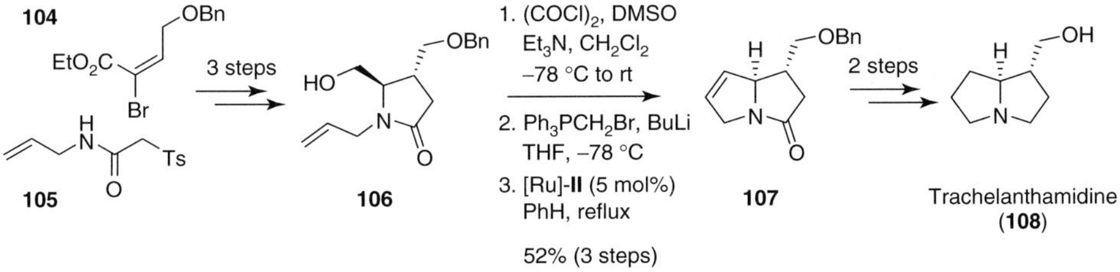

Scheme 2.26

Chang *et al.* reported a synthesis of the monohydroxypyrrolidine (−)-trachelanthamidine (**108**) also utilizing RCM to create the second cycle (Scheme 2.26) [44]. The synthesis of this alkaloid, isolated from the plant *Trachelanthus korolkovi*, commenced with a stereo- and regioselective stepwise [3+2]-annulation reaction of bromoacrylate **104** with sulfonylacetamide **105** leading in two more steps to compound **106**. Preparation of the diene for the metathesis reaction was achieved by subsequent Swern oxidation and Wittig olefination. The pyrrolidizine skeleton **107** was formed upon ring closure, employing [Ru]-**II** (5 mol%), in 81%. Through hydrogenation and subsequent lactam reduction, the authors achieved a synthesis of (−)-trachelanthamidine (**108**) in a total of seven steps.

2.3 Six-membered Nitrogen Heterocycles

Alkaloids incorporating six-membered nitrogen heterocycles are abundantly present in nature and occur in a multitude of alkaloid natural product families. Many of these alkaloids exhibit a wide range of diverse biological activities. As a result, extensive research has been devoted to the synthesis of many of the natural products and derivatives. This has led to an enormous variety of synthetic methodology that has been efficiently applied to synthesize many of these alkaloids. Among these methods, metathesis-based processes have played an increasingly important role as is illustrated in the sequel.

2.3.1 Piperidine Alkaloids

2.3.1.1 Piperidines

(−)-Allosedamine (**113**) and (+)-sedamine (**119**), two piperidine alkaloids isolated from *Lobelia inflate*, have been used for the treatment of respiratory disorders such as asthma, bronchitis, and pneumonia. While several racemic syntheses have been reported, there are much less asymmetric ones. An enantioselective route to (−)-allosedamine (**113**) was developed by Raghaven *et al.*, which commenced with addition of lithium compound **110** to the imine **109** providing the separable allylic amine **111** as a 3 : 1 mixture of diastereoisomers (Scheme 2.27) [45, 46]. Six

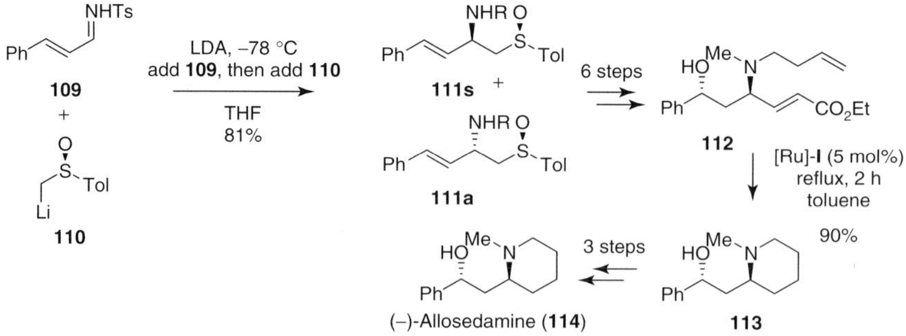

Scheme 2.27

additional steps were required to form metathesis precursor **112** from the major isomer **111a**. Straightforward RCM ([Ru]-**I**, 5 mol%) provided the corresponding heterocycle **113**, which upon hydrogenation, deprotection, and methylation afforded (−)-allosedamine (**114**).

Yadav *et al.* developed via a similar strategy stereoselective syntheses of both (−)-allosedamine (**114**) and (+)-sedamine (**119**), starting from the Sharpless epoxidation product of cinnamyl alcohol (**114**, Scheme 2.28) [47]. The latter compound was converted in a number of steps into a separable diastereomeric mixture of amino alcohols **116a** and **116b**. N-Allylation gave in both cases the corresponding RCM precursors, which on exposure to [Ru]-**I** catalyst (20 mol%, benzene, 50 °C) yielded the cyclic compounds **117** and **118** in 90 and 92% yield, respectively. Finally, deprotection and methylation led to (−)-allosedamine (**114**) and (+)-sedamine (**119**) via identical procedures.

A different approach to synthesize (+)-sedamine (**119**), reported by Cossy *et al.* [48, 49], was based on a double enantioselective allyltitanation of the aldehydes **120** and **122** (Scheme 2.29). Next, transformation of the resulting homoallylic alcohol **123** into amine **124** was accomplished via a Mitsunobu reaction, which was in a two-step procedure converted into the RCM precursor **124**. Completion of the synthesis was achieved through RCM (catalyst [Ru]-**I**, benzene, reflux), followed by hydrogenation of the alkene and deprotection, in 52% over four steps.

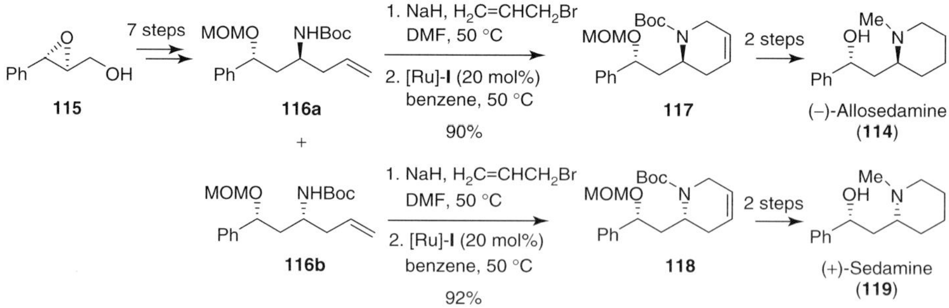

Scheme 2.28

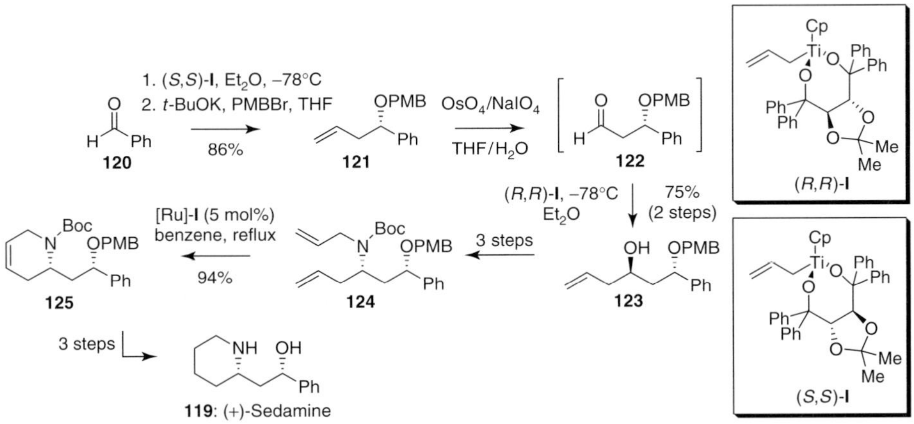

Scheme 2.29

A synthesis of (+)-allosedamine, described by Chang *et al.* in 2004, started with a hydrolytic kinetic resolution of **126**, which in turn was prepared in three steps from (+)-styrene oxide (Scheme 2.30) [50]. Subsequent formation of the diallylamine **129** as shown above, followed by [Ru]-**II**-catalyzed RCM (10 mol%, benzene, 70 °C), afforded (+)-allosedamine (**131**).

α-Conhydrine, one of the alkaloids occurring in hemlock, was isolated from the seeds and leaves of the poisonous plant *Conium maculatum*. A synthesis of α-conhydrine (**135**) [51] started with transformation of prolinol derivative **132** into amino alcohol **133** via a regioselective Baeyer–Villiger reaction of ketone **132**

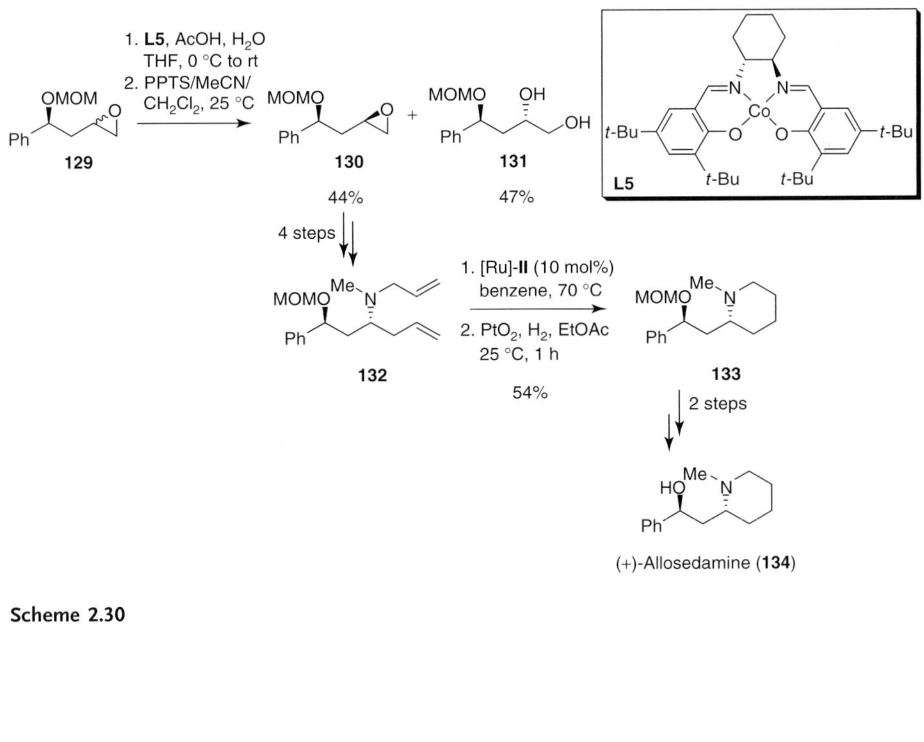

Scheme 2.30

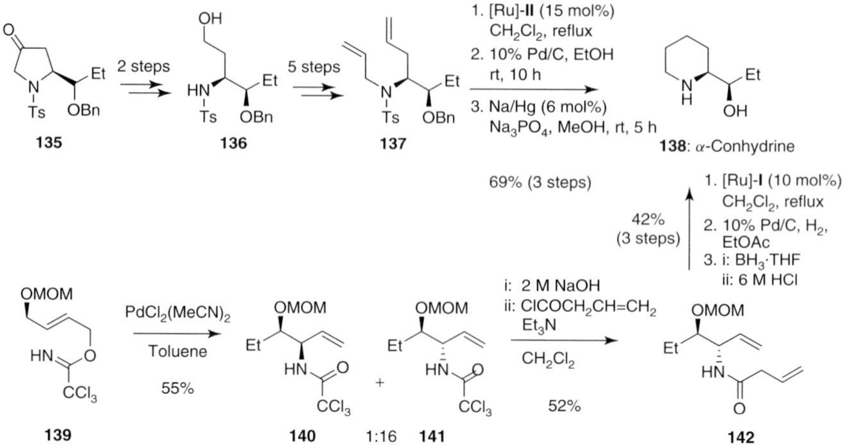

Scheme 2.31

(Scheme 2.31). Compound **133** was then subjected to oxidation and subsequent Wittig olefination to afford the RCM precursor **134**. Smooth RCM under the influence of catalyst [Ru]-**II** (15 mol%, CH_2Cl_2, reflux) gave the corresponding unsaturated heterocycle in 92% yield, which upon hydrogenation and desulfonylation provided the target compound **135**.

Another approach to the synthesis of α-conhydrine (**138**) involved an Overman rearrangement of **136**, which was prepared from (*S*)-glycidol in seven consecutive steps [52]. Deprotection of the amine **139**, followed by acylation, provided the RCM precursor **139**. Ring closure under the influence of catalyst [Ru]-**I** (CH_2Cl_2, reflux) afforded the corresponding unsaturated piperidine in quantitative yield. Finally, hydrogenation and subsequent MOM deprotection afforded **135** in 42% yield.

An auxiliary-based synthesis of the related alkaloid (+)-β-conhydrine (**144**) commenced with conversion of enantiopure benzyl-protected 2-hydroxypropionaldehyde into hydrazone **140b** (Scheme 2.32). Diastereoselective alkylation and subsequent acylation gave the RCM precursor **141b**, which upon treatment with [Ru]-**II** (5 mol%, CH_2Cl_2, rt) led to the corresponding product in 75% yield. Subsequent hydrogenation and deprotection led to (+)-β-conhydrine (**144**) in a yield of 23% over six steps [53]. Fairly similar procedures were applied in the synthesis of (*S*)-(+)-coniine (**143**) [54].

A formal synthesis of (*S*)-(+)-coniine (**143**) was developed by Chang *et al.* in 1999 starting from L-norvaline (**144**), which was converted into diolefin **145** in several steps [55]. Finally, RCM ([Ru]-**I**, CH_2Cl_2, rt), followed by hydrogenation, led to the known intermediate **146** [56] thus completing the formal synthesis (Scheme 2.33).

Another approach in the synthesis of functionalized piperidine alkaloids (−)-coniine (**149a**) and (−)-pipecoline (**149b**) started with the imines **147a** and **147b** using (*R*)-α-methylbenzylamine as a chiral auxiliary (Scheme 2.34) [57]. Consecutive diastereoselective vinylation and N-allylation afforded compounds **148a** and **148b** as two diastereoisomers (71% de, major isomer shown). RCM

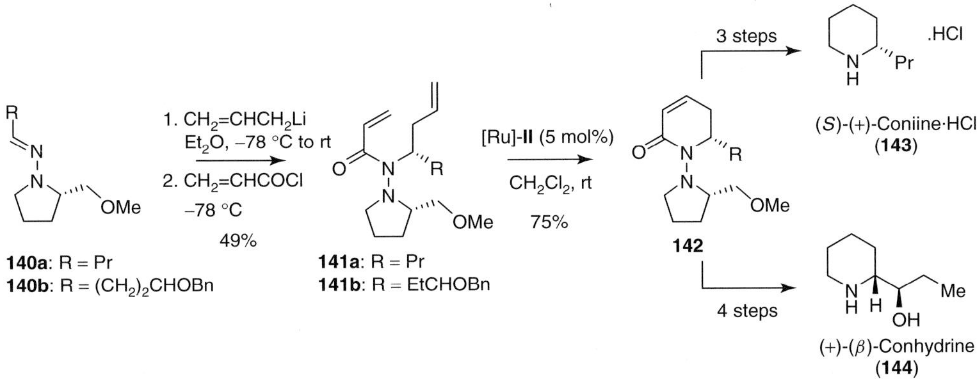

Scheme 2.32

([Ru]-**I**, CH_2Cl_2, rt) gave rise to the corresponding cyclic products in good yields, of which the diastereoisomers were separated via silica gel chromatography. Eventually, straightforward hydrogenation led to the natural products **149a** and **149b**.

An auxiliary-mediated approach reported by Agami *et al.* also led to a synthesis of (−)-β-conhydrine (**156**, Scheme 2.35) [58]. This time (*S*)-phenylglycinol (**150**) was used as the starting material, which via Weinreb amide **151** and *N*,*O*-acetal **152** was converted into oxazolidinone **153**. The latter compound was then diastereoselectively allylated using the corresponding *N*-acyliminium ion, deprotected, and allylated at the nitrogen atom. The resulting diolefin **154** was then subjected to RCM [catalyst [Ru]-**I** (4 mol%), CH_2Cl_2, reflux, 79%] and another four consecutive steps to give (−)-β-conhydrine (**156**).

Probably, the most well-known alkaloid in *N. tabacum* is (*S*)-nicotine (Section 2.2.2), which plays an important role in the modulation of nicotinic acetylcholine receptors (nAChRs). However, related alkaloids have also been isolated from the same source, such as the homologous analogs (*S*)-anabasine (**160a**) and (*S*)-anatabine (**161a**). In 2000, Felpin *et al.* published a synthesis of the latter alkaloids starting from an asymmetric allylboration of 3-pyridinecarboxaldehyde into compound **157** [59, 60]. Azide substitution of the alcohol, followed by reduction and functionalization, provided the RCM precursor **158**. Subsequent protonation of the pyridine ring, followed by RCM, proceeded uneventfully ([Ru]-**I** catalyst, CH_2Cl_2) to give the ring-closed product **159** that via hydrogenation was converted into both (*S*)-anabasine (**160a**) and its methylated counterpart **160b**.

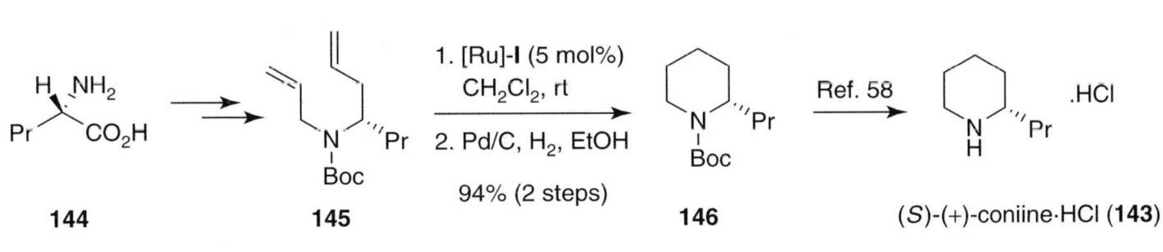

Scheme 2.33

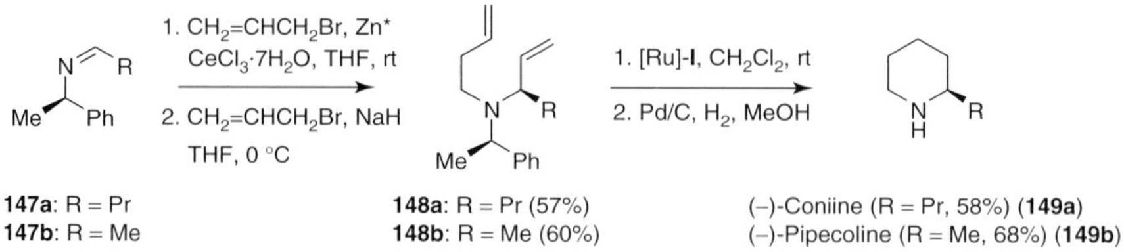

Scheme 2.34

Alternatively, (*S*)-anatabine (**161a**) and its methylated derivative **161b** could be directly obtained from the metathesis intermediate **159** (Scheme 2.36).

En route to the natural products *ent*-CP-999,994 (**165**) and *ent*-L-733,060 (**166**), in 2005 Nakano *et al.* reported a highly enantioselective palladium-catalyzed asymmetric allylic amination (>99% ee) of the allylic acetate **162** using the ligand **L6** [61]. Conversion of the resulting allylic amine **163** with 3-butenoic acid in the presence of DCC followed by RCM (catalyst [Ru]-**II**, CH_2Cl_2, reflux, 94%) yielded the piperidinone core structure **164**, which was readily converted into the target molecules **165** and **166** (Scheme 2.37).

2.3.1.2 Piperidine Carboxylic Acids

A significant number of alkaloids contain the 2-piperidinecarboxylic acid (pipecolic acid) unit as a core structure. Being a proline homolog, the pipecolic acid moiety has been used in many modified peptides and synthetic drugs. For example, naturally occurring 4-hydroxy-2-pipecolic acids have been isolated from various green plants and constitute important chiral building blocks of biologically active molecules such as palinavir. A synthesis of 4-hydroxy-2-pipecolic acid derivatives described by Johnson *et al.* involved conversion of (*S*)-**167** into the corresponding oxazolidinone, which was N-allylated using allyl bromide and then subjected to [Ru]-**I** catalyst

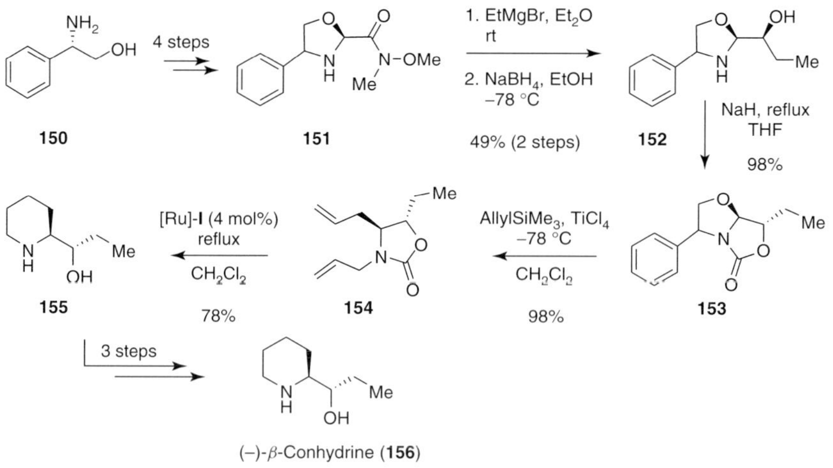

Scheme 2.35

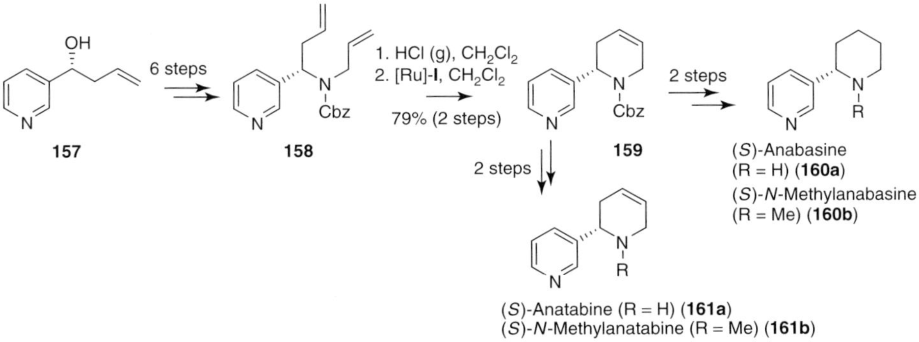

Scheme 2.36

(CH_2Cl_2, rt) to give the cyclic product **168** [62]. The metathesis product was either readily converted into the protected *trans*-4-hydroxypipecolic acid derivative **169**, but upon elaboration of the side chain also to the natural product (2*S*,4*S*)-(−)-SS20846A (**171**) (Scheme 2.38).

Baikiain (**175**), another example of a natural product containing the pipecolic acid core structure, was synthesized in 2000 by Rutjes *et al.* via a metathesis route [63]. Amidopalladation of methyl propadienyl ether with the enantiopure L-allylglycine derivative **172** provided the RCM precursor **173** in good yield. Cyclization ([Ru]-**I** catalyst, 10 mol%, CH_2Cl_2, rt) afforded the corresponding cyclic *N*,*O*-acetal **174**, which was readily transformed into the natural product via *N*-acyliminium-mediated reduction with Et_3SiH, followed by deprotection (Scheme 2.39).

A different approach to baikiain (**175**) was also reported two years later by Riera *et al.* [64]. Their synthesis commenced with a Sharpless epoxidation of 2,5-hexadienol to give the enantiopure epoxide **176**. Subsequent epoxide ring opening with allylamine and *N*-Boc protection afforded the RCM precursor **177**, which was readily ring closed ([Ru]-**I**, CH_2Cl_2) in 72% to the tetrahydropyridine derivative. Finally, oxidative cleavage of the diol, followed by Pinnick oxidation, led to the *N*-Boc-protected natural product **178** (Scheme 2.40).

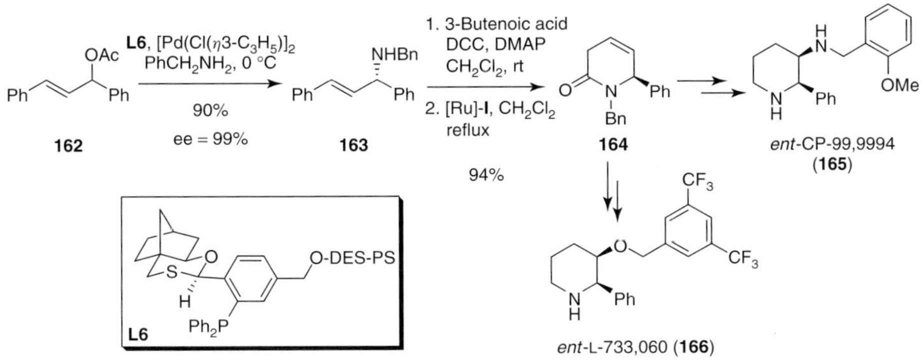

Scheme 2.37

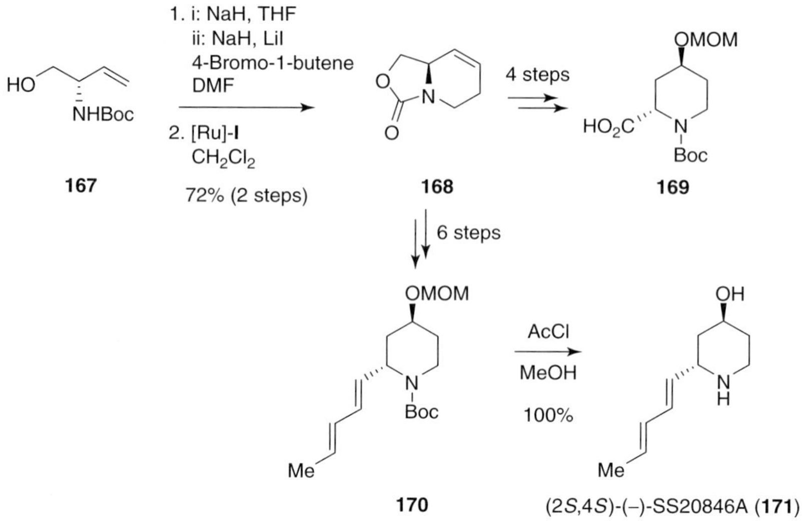

Scheme 2.38

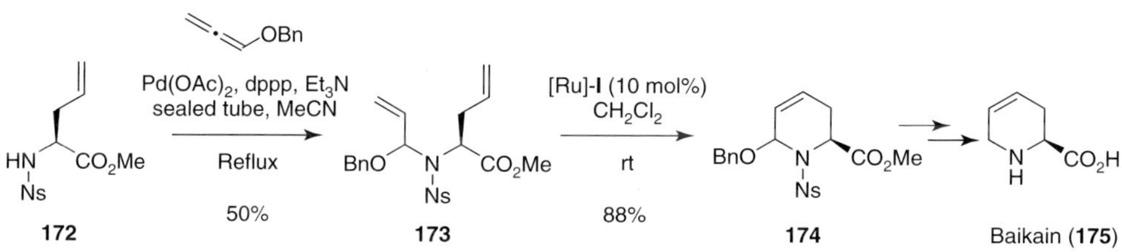

Scheme 2.39

2.3.1.3 Piperidones

The 2-(1*H*)-pyridone ring system is abundantly encountered not only in a wide variety of naturally occurring alkaloids, but also in the synthetic intermediates of piperidine, (iso)quinoline, indolizidone, and quinolizidine alkaloids. Plants belonging to genus *Lycopodium* are known to contain alkaloids possessing unique skeletal characteristics and biological activities, such as acetylcholine esterase (AChE) inhibition. One of them is the Phlegmarine-type alkaloid lycoposerramine-V (**183**) [65]. Its synthesis featured the initial formation of cyclohexene **179** from (*R*)-3-methylcyclohexanone (Scheme 2.41). *t*-Butyldiphenylsilyl removal, subsequent Swern oxidation and elaboration of the double bond provided aldehyde **180**,

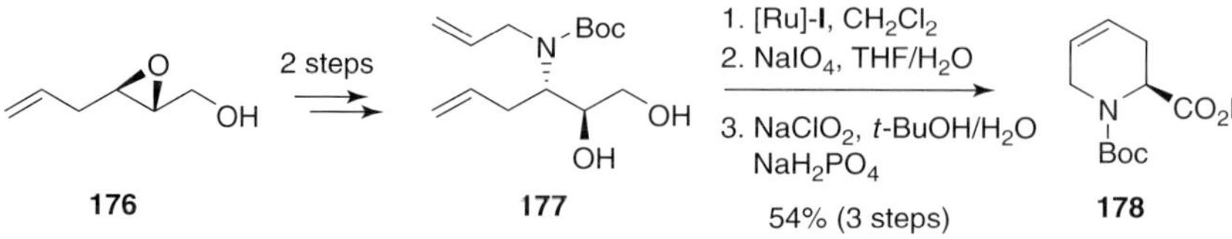

Scheme 2.40

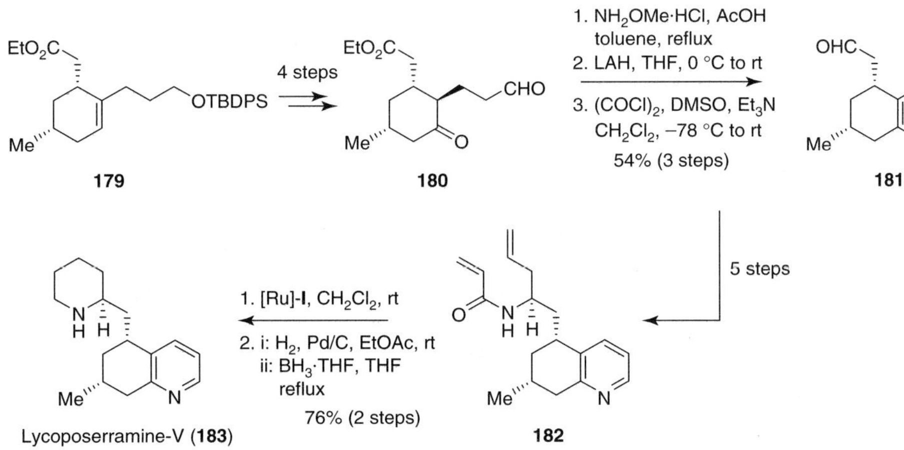

Scheme 2.41

which was then converted into quinoline derivative **181**. The aldehyde function was in a number of steps transformed into diolefin **182**, proceeding via an asymmetric allylation with *B*-allyldiisopinocampheylborane. RCM ([Ru]-**II**, CH_2Cl_2, rt) proceeded in near quantitative yield and was followed by hydrogenation and lactam reduction to afford the targeted lycoposerramine-V (**183**).

Pipermethystine is a major constituent of the leaves of *Piper methysticum*, a large shrub indigenous to the Islands of the South Pacific [66]. Its synthesis commenced with butenamide **184**, which upon subjection to [Ru]-**I** catalyst (4 mol%, CH_2Cl_2, reflux) gave the pyridinone **185** in good yield. Epoxidation of **185**, followed by treatment with *t*-BuOK, afforded the allylic alcohol **186**, which was enzymatically resolved with a lipase to give the fully protected enantiopure (*S*)-pipermethystine (**187**). The nonreacted alcohol (*R*)-**187** was afterwards also converted into (*R*)-pipermethystine (**188**) upon esterification with acetic anhydride (Scheme 2.42).

2.3.1.4 Polyhydroxypiperidines

In addition to five-membered ring azasugars (Section 2.2.2.3), a large number of different six-membered ring azasugars exist in nature. Since these compounds resemble the glycopyranosyl cation after protonation of the nitrogen atom, their structures show promising potential in the treatment of several diseases. Two syntheses of six-membered ring azasugars are described in Scheme 2.43, both starting from the protected diacetone glucose derivative **189**. After inversion of the free hydroxyl group, compound **190** was converted into the RCM precursor **191**. [Ru]-**I**-mediated ring closure (CH_2Cl_2, rt) led to the unsaturated bicyclic intermediate **192** in 90% yield. From there, D-1-deoxyallonojirimycin (**193**) was prepared via face-selective dihydroxylation of the cyclic olefin [67]. In addition, L-1-deoxyallonojirimycin (**194**) was directly prepared from diacetone glucose derivative **189** according to identical procedures.

A similar approach to synthesize six-membered ring azasugars was published by Takahata *et al.*, which commenced with a Wittig reaction of the D-serine-derived Garner aldehyde **195**. Facile elaboration into the metathesis precursor **196**, followed by treatment with [Ru]-**I** (CH_2Cl_2, rt), afforded tetrahydropyridine (**197**) in virtually quantitative yield. The latter structure was shown to be a good starting point for the preparation of fagomine (**200**) and its congeners **199** and **201** (Scheme 2.44) [68].

2.3.2 Indolizidine Alkaloids

Coniceine (**205**), a bicyclic alkaloid containing the simplest indolizidine skeleton, has attracted much attention as a building block of indolizidine alkaloids. An early RCM-based synthesis was described by Chang *et al.* [69]. It started with the ethyl ester of proline (**202**), which was converted in four steps into the RCM precursor **203**. Subsequent metathesis under the influence of catalyst [Ru]-**II** (CH_2Cl_2, rt) provided the corresponding cyclic olefin (**204**), which is a known intermediate en route to coniceine (**205**) (Scheme 2.45) [70].

Another synthesis was reported in 2005 by Génisson *et al.* who constructed the six-membered indolizidine skeleton via RCM starting from the epoxide **206** (Scheme 2.46). Selective epoxide hydrolysis, followed by alkylation with the triflate of 3-butenol, led to the diolefin **207** in 72%. Subsequent RCM ([Ru]-**II** [5–10 mol%], toluene, 70 °C) afforded the tetrahydropyridine **208** in 66% yield. Completion of the synthesis enclosed hydrogenolysis followed by ring closure to afford the targeted alkaloid (−)-lentiginosine (**209**) [71].

In 2006, Jung *et al.* reported another synthesis of (−)-lentiginosine (**209**) starting from the D-xylose-derived olefin **211** [72]. Subjection to chlorosulfonyl isocyanate (CSI) provided the allylic amine **212** with a high diastereoselectivity (*syn*/*anti* =

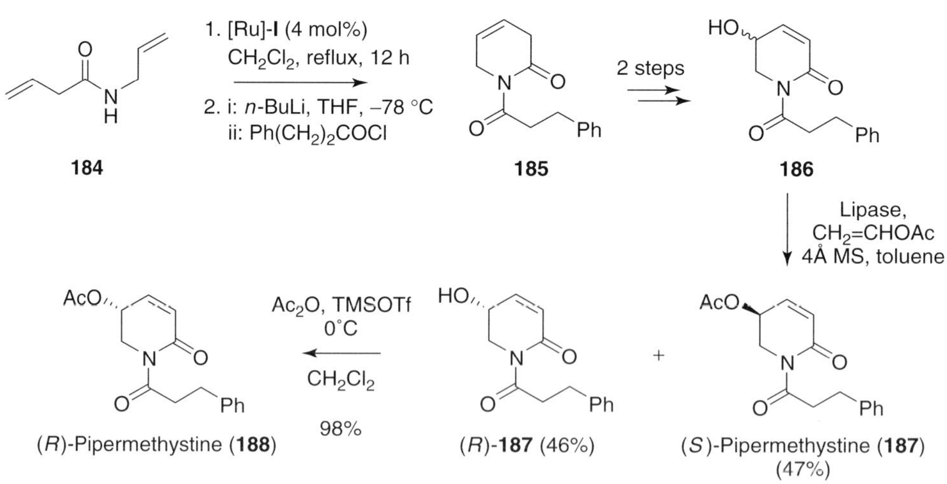

Scheme 2.42

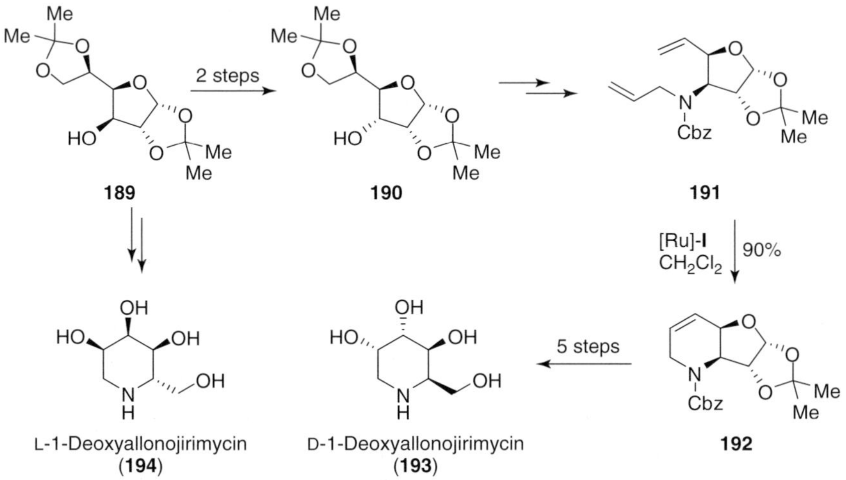

Scheme 2.43

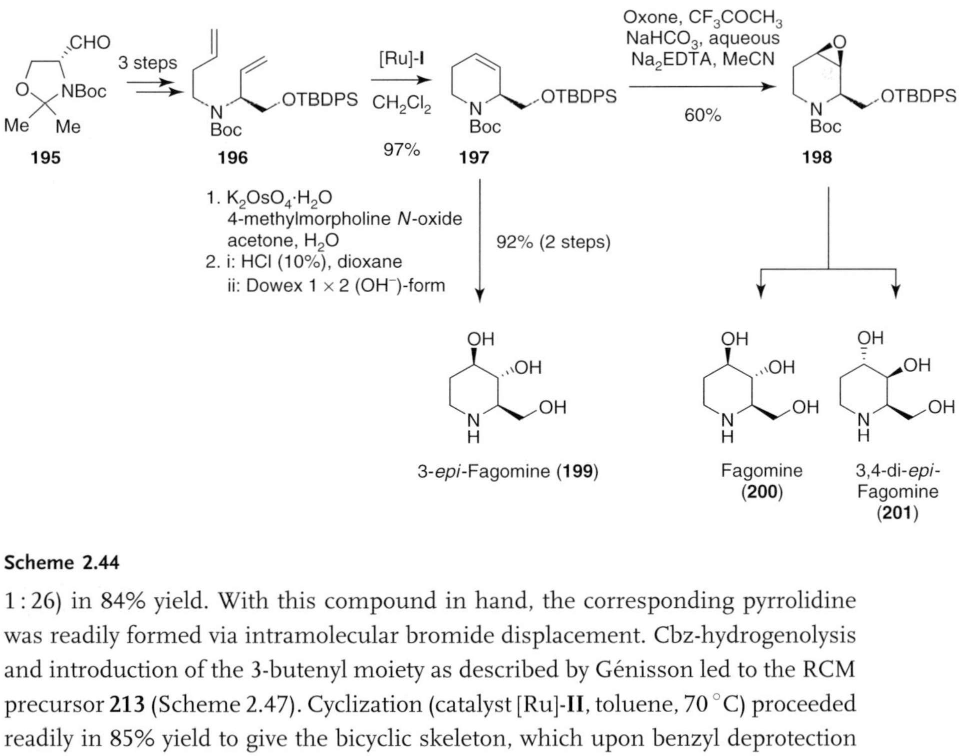

Scheme 2.44

1 : 26) in 84% yield. With this compound in hand, the corresponding pyrrolidine was readily formed via intramolecular bromide displacement. Cbz-hydrogenolysis and introduction of the 3-butenyl moiety as described by Génisson led to the RCM precursor **213** (Scheme 2.47). Cyclization (catalyst [Ru]-**II**, toluene, 70 °C) proceeded readily in 85% yield to give the bicyclic skeleton, which upon benzyl deprotection and dihydroxylation of the olefin gave (−)-lentiginosine (**209**).

A different approach leading to *ent*-lentiginosine (**77**) was adopted by Schmidt *et al.* [73] who prepared pyrrolidine **214** in three steps from L-tartrate. *N*-Acyliminium ion–mediated allylation provided **215** in quantitative yield as a diastereomeric mixture (1 : 1), which was separated via column chromatography. RCM ([Ru]-**I** [0.4 mol%], CH_2Cl_2, reflux, 12 hours) afforded the corresponding indolizidine in 78% yield, which was converted into the natural product via deacetylation and hydrogenation (Scheme 2.48).

In 1994, the tricyclic perhydropyrrolo[2,1-*j*]quinolone lepadiformine (**222**) [74] was isolated from the tunicate *Clavelina lepadiformis* by Biard *et al.* and exhibits moderate cytotoxic activity against various tumor cell lines *in vitro*, as well as high cardiovascular effects *in vitro* and *in vivo*. Unfortunately, many synthetic efforts were initially directed to a wrongly assigned structure. However, since 2001, several elegant syntheses have been reported. One of them included RCM of **219** (Scheme 2.49), which was prepared via an Ireland–Claisen rearrangement of amino acid ester **217**. Further elaboration provided diolefin **219**, which upon RCM ([Ru]-**II**, CH_2Cl_2, 40 °C, 1 hour) afforded the bicyclic skeleton **220** in 98% yield. Via five more additional steps, the tricyclic cyanide **221** was prepared from which (−)-lepadiformine (**222**) could be readily synthesized following known procedures [75].

Melodinus alkaloids, also known as *meloquinolines*, represent a group of monoterpenoid indole alkaloids isolated from certain Apocynaceae species, amongst which is (+)-meloscine (**229**). Although in 1989 Overman *et al.* described the first racemic synthesis of meloscine, it was not until recently that Bach *et al.* succeeded in synthesizing (+)-meloscine (**229**) in enantiopure form [76]. Its synthesis started with a [2+2]-photocycloaddition of enone **223** in the presence of the chiral auxiliary **A** (Scheme 2.50). Ring expansion of cyclobutane **224** into the α-hydroxycyclopentenone (**225**) was then accomplished by a retro-benzylic ester rearrangement. Further conversion of **225** via **226** provided diolefin **227**, which after RCM (catalyst [Ru]-**II** [15 mol%], toluene, 65 °C) gave the corresponding product **228** in an excellent yield (95%). The latter compound was then transformed in five steps into the targeted natural product (+)-meloscine (**229**).

Daly *et al.* isolated numerous alkaloids such as steroidal batrachotoxins, histrionicotoxins, gephyrotoxins, and pumiliotoxins from the skin of Neotropical poison frogs. One of these alkaloids, (−)-205B (**236**), possessed an unusual aza-acenaphthylene ring system [77]. In 2005, Smith *et al.* published a first total synthesis of alkaloid 205B (**236**) based on a three-component Linchpin coupling of compounds **230–232**, followed by a one-pot sequential cyclization of intermediate **233** to complete the indolizidine core structure **234**. Upon treatment with Me_3SiCl in the presence of LHMDS, the corresponding kinetic silyl enol ether was formed, which was then directly subjected to metathesis catalyst [Ru]-**II** (benzene,

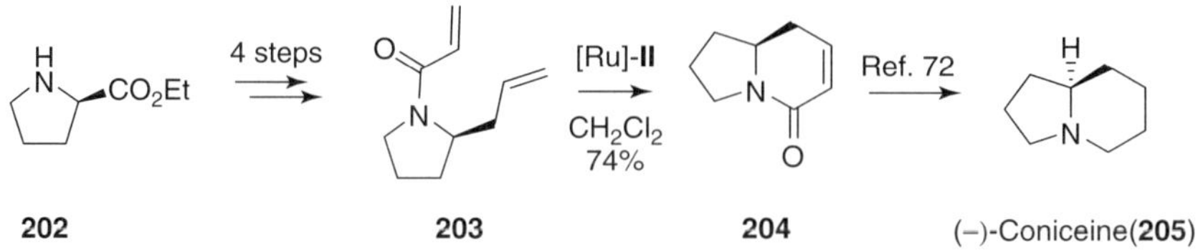

Scheme 2.45

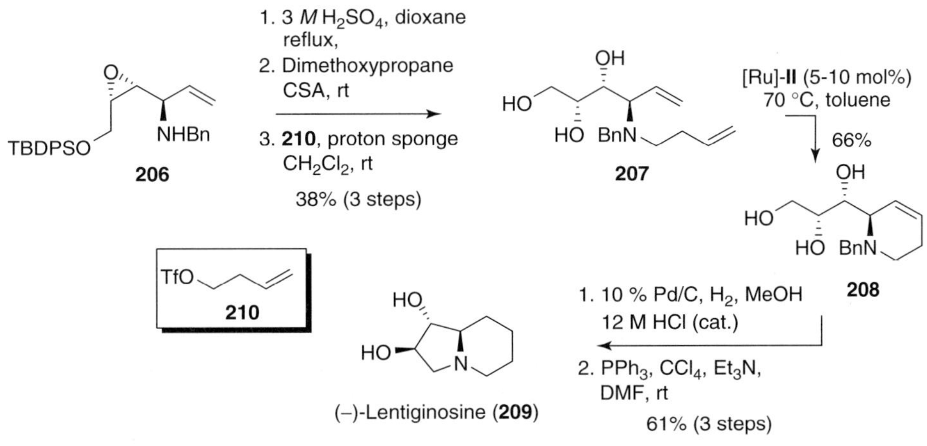

Scheme 2.46

65 °C) to provide after an aqueous workup the cyclic ketone **235** in a satisfactory yield (81%). From this aza-acenaphthylene skeleton, (−)-205B (**236**) was readily synthesized in six steps (Scheme 2.51).

2.3.3
Quinolizidine Alkaloids

The structure of quinolizidine 233A (**242**), obtained from skin extracts of poisonous frogs of the genera *Dendrobates* and *Mantella*, was published in 1993. Following up on racemic syntheses, a formal synthesis of enantiopure **242** was reported by Rutjes *et al.* based on a cationic cyclization of allylic *N,O*-acetals (Scheme 2.52) [78].

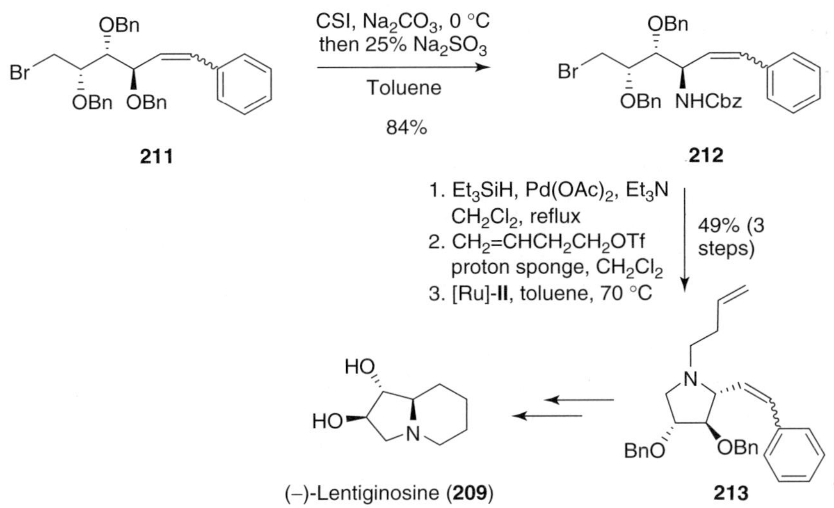

Scheme 2.47

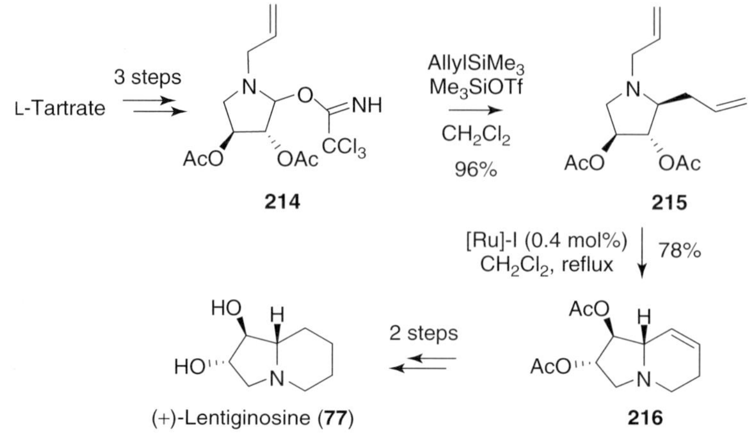

Scheme 2.48

The approach commenced with CM ([Ru]-**IX** catalyst [5 mol%, CH_2Cl_2, reflux) of the enantiopure protected 2-amino-6-heptenoic ester **237** with allyltrimethylsilane to afford the modified side chain via unexpected double bond isomerization. Subsequent amidopalladation with benzyl propadienyl ether gave the allylic *N,O*-acetal **238**, which smoothly cyclized ($Sn(OTf)_2$ [2 mol%], CH_2Cl_2) to give **239** as a 3 : 1

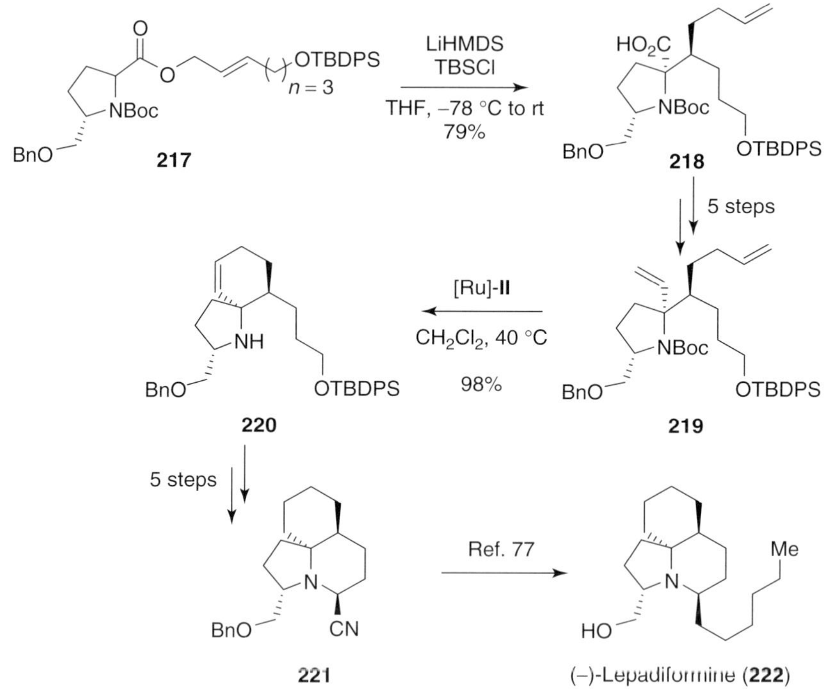

Scheme 2.49

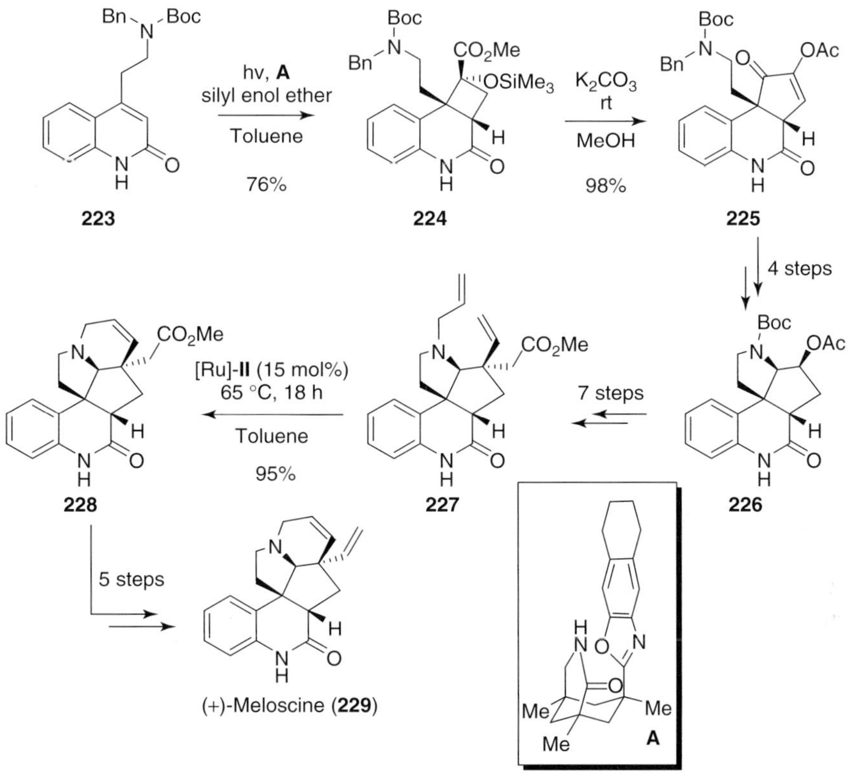

Scheme 2.50

mixture of diastereoisomers (major one shown). Deprotection, introduction of the second alkene, and subsequent RCM (catalyst [Ru]-**II** [5 mol%], CH_2Cl_2, reflux) afforded **240** in 83% yield. Finally, hydrogenation and reduction of both carbonyl groups provided **241**, thereby constituting a formal synthesis of quinolizidine 233A (**242**) [79].

In 2005, Honda *et al.* published a first total synthesis of a quinolizidine alkaloid using RCM as one of the key steps, namely, (−)-deoxynupharidine (**246**) [80]. The alkaloid was isolated from plants of the genus *Nuphar* and exhibited amongst others immunosuppressive activity. Its synthesis commenced with a stereoselective construction of the enantiopure 5,6-disubstituted piperidinone building block **244** by modification and ring expansion of the protected D-pyroglutamic acid derivative **243** (Scheme 2.53). Subsequent RCM of **244** (catalyst [Ru]-**II**, benzene, 60 °C) afforded quinolizidinone **245** in 96% yield, whereafter three consecutive steps led to the natural product itself.

Blaauw *et al.* described a first synthesis of the quinolizidine alkaloid (+)-epiquinamide (**250**) [81]. This compound was isolated from the skin of the Ecuadorian frog *Epipedobates tricolor* and was claimed to show activity on the nicotinic

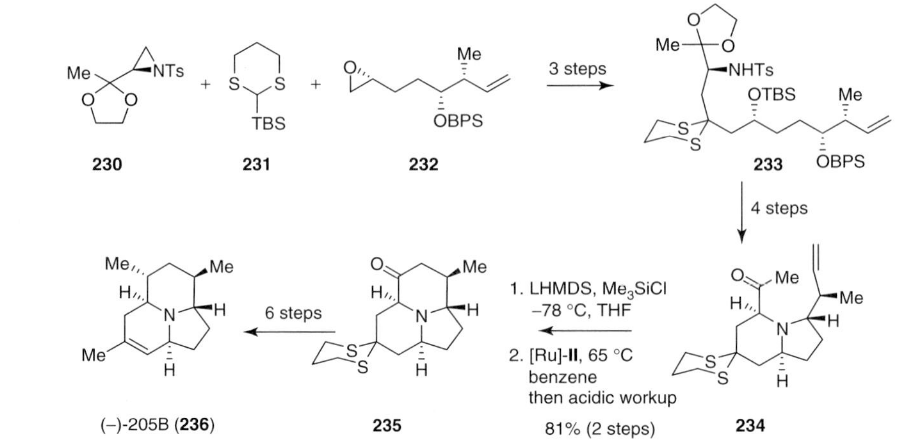

Scheme 2.51

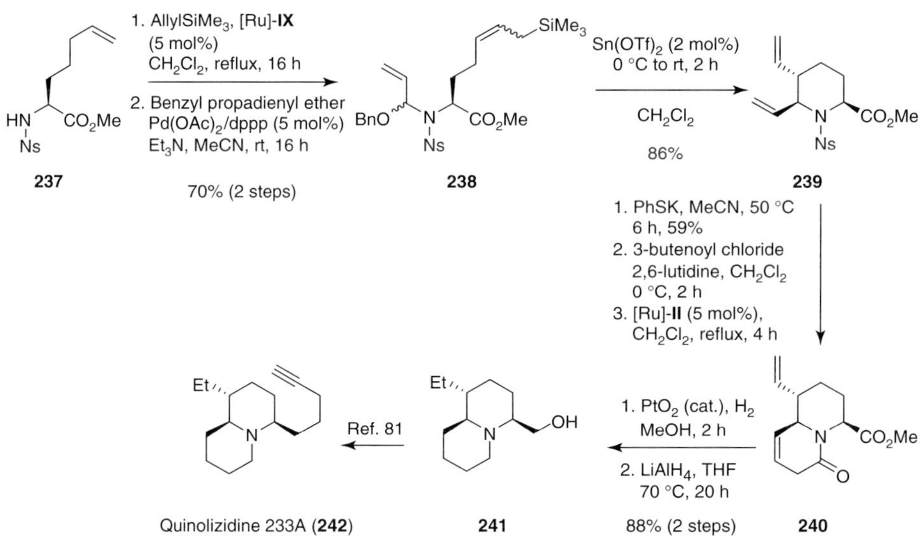

Scheme 2.52

receptor. The approach started with the efficient conversion of commercially available L-allysine ethylene acetal (**247**) in the functionalized pipecolic acid derivative **248**. RCM under the influence of catalyst [Ru]-**II** (toluene, rt), followed by hydrogenation, provided the bicyclic skeleton **249** in excellent yield. Further functional group manipulation then provided (+)-epiquinamide (**250**) in an overall yield of 22% from the RCM precursor **248** (Scheme 2.54).

The Lythraceae plants produce a large variety of bioactive alkaloids. A number of them comprise the quinolizidine scaffold, such as (−)-lasubine I and II. Liao *et al.*

described in 2006 a synthesis of (−)-lasubine I (**254**) [82] involving RCM to construct one of the two rings (Scheme 2.55). Most effort were directed to an enantioselective synthesis of the functionalized piperidine and the RCM precursor **253**. This was achieved via S_N2-type cyclization of amino alcohol **252**, which in turn was prepared via two sequential Roush asymmetric allylboration steps from the protected amino alcohol **251**. RCM under the influence of catalyst [Ru]-**I** (CH_2Cl_2, reflux) gave the corresponding ring-closed product in 92% yield, which upon catalytic hydrogenation in the presence of HF·Et_3N was efficiently converted into (−)-lasubine I (**254**).

A synthesis of the highly conjugated pentacyclic quinolizidine alkaloid mitralactonine (**257**), isolated in 1999 by Takayama *et al.*, shows the usefulness of RCM in constructing polycyclic frameworks [83]. It commenced with a protected tryptamine derivative, which amongst other steps via a Pictet–Spengler cyclization was converted into the RCM precursor **255**. RCM (catalyst [Ru]-**II** [10 mol%], toluene, 80 °C) proceeded smoothly both on the *N*-Ts-protected (87%) as well as on the unprotected compound. The tosyl protection, however, appeared crucial in the next steps involving osmium-catalyzed dihydroxylation and oxidation into the corresponding hydroxyl enone intermediate. TBAF-mediated Ts deprotection then gave rise to the free indole **256**, which is a known precursor for preparing mitralactonine (**257**) (Scheme 2.56) [84].

The sparteine group of alkaloids, abundant in common papilionaceaous plants, is characterized by an asymmetrical exo–endo arrangement of bridgehead hydrogen atoms at C6 and C11. (−)-α-Sparteine, obtained from the commercially available (−)-sparteine by isomerization, has found widespread use as chiral ligand in asymmetric catalysis and asymmetric deprotonation reactions. Therefore, a practical stereocontrolled entry into the sparteine core structure would contribute to exo-ending the versatility of existing asymmetric methods. An elegant approach was detailed in 2005 by Blakemore *et al.* who identified tetraoxobispidine (**260**) as a synthetically versatile scaffold, which could be readily prepared starting from a Knoevenagel condensation of dimethyl malonate (**258**) with paraformaldehyde (Scheme 2.57) [85]. The intermediate was directly treated with ammonia (**259**), followed by cyclization and subsequent double allylation, to provide the key intermediate **260**. Subsequent Grignard addition resulted in a remarkably regioselective diallylation affording the tetraene and the RCM precursor **261**. Subjection to [Ru]-**I** catalyst (4 mol%, CH_2Cl_2, rt) provided the tetracyclic structure **262** in 82% yield.

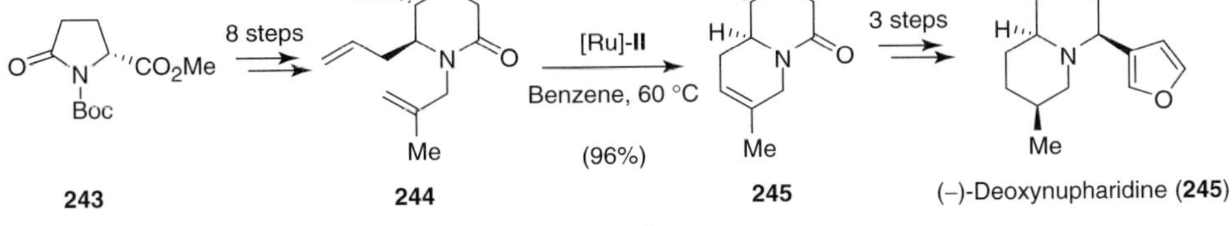

Scheme 2.53

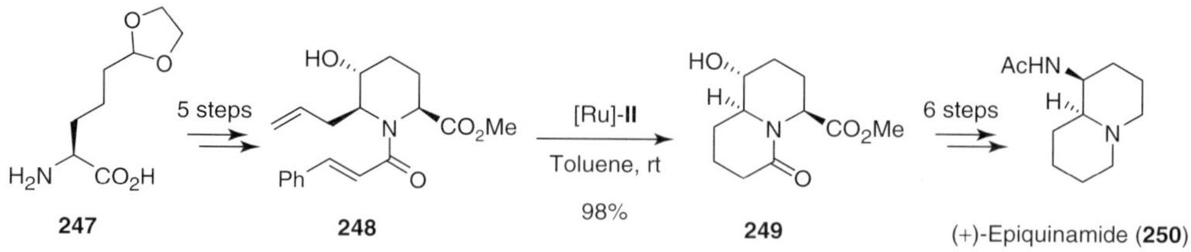

Scheme 2.54

Finally, racemic α-sparteine (266) was obtained upon global reduction of all olefin, aminal and carbonyl functions.

Alkaloids isolated by Rakotoson *et al.* in 1998 from *Galipea officinalis* consisted of several 2-substituted quinolines among which is (+)-(*S*)-angustureine (**266**). An RCM-mediated synthesis of this compound started with a Mitsunobu alkylation of sulfonamide **265** with the enantiopure secondary alcohol **262** to give the desired α,ω-diene **264** in 78% yield. With the RCM precursor in hand, subjection to [Ru]-**II** catalyst (CH_2Cl_2, 50 °C) gave quinoline **265** in 92% yield. Finally, hydrogenation, deprotection, and subsequent methylation provided (+)-(*S*)-angustureine (**266**) in 68% yield (Scheme 2.58) [86].

2.4
Seven-membered Nitrogen Heterocycles

In contrast to the large abundance of RCM examples of five- and six-membered nitrogen heterocycles en route to natural products, RCM to prepare seven-membered rings occurs considerably less frequently. One of the examples concerns a formal synthesis of (−)-balanol (**271**, Scheme 2.59). Balanol was first isolated in 1993 from the fungus *Verticillium balanoids* and is a potent inhibitor of human protein kinase

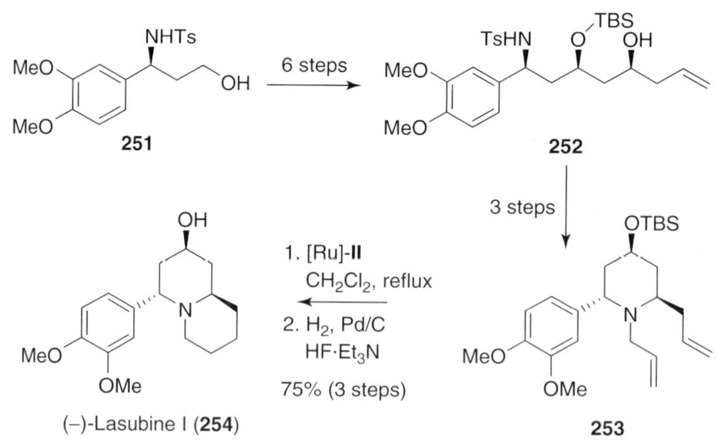

Scheme 2.55

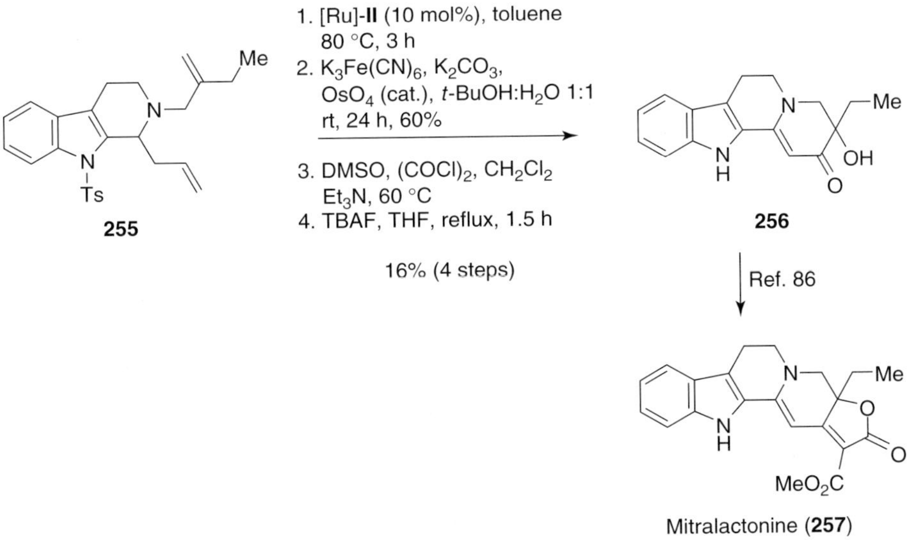

Scheme 2.56

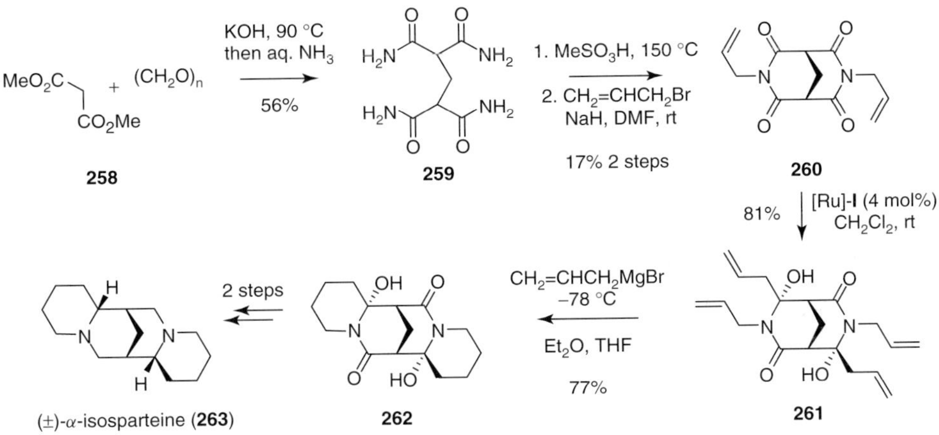

Scheme 2.57

C. Because of its high inhibitory activity, balanol has been a target for a number of synthetic efforts. In 1999, Cook and coworkers published a formal synthesis of balanol (**271**) based on RCM to form a seven-membered nitrogen heterocycle as a key step [87]. The balanol synthesis started with D-serine derivative **267**, which via oxazoline **268** was transformed into the metathesis precursor **269** (Scheme 2.59). RCM using [Ru]-**I** catalyst (2 × 5 mol%, CH_2Cl_2, reflux) provided the corresponding unsaturated heterocycle in 77% yield, which upon acidic hydrolysis of the oxazoline and hydrogenation with Wilkinson's catalyst afforded **270**, thereby completing the formal synthesis [88].

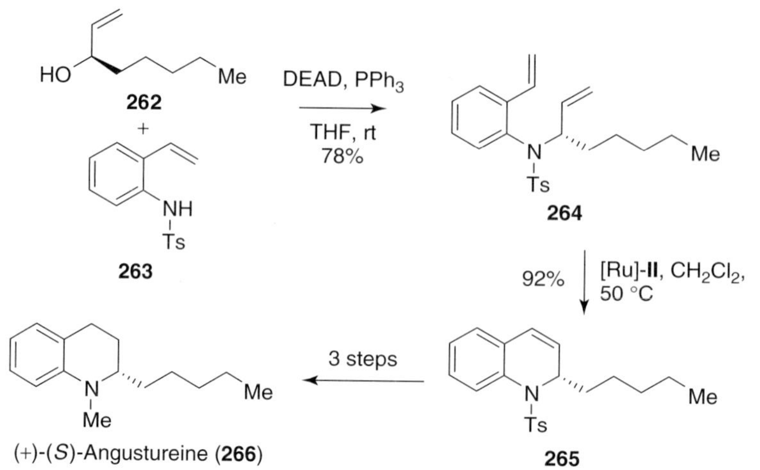

Scheme 2.58

Two years later, Fürstner *et al.* developed a synthesis of azepine **277** based on a selective Sharpless asymmetric epoxidation of divinylcarbinol (Scheme 2.60) [89]. The resulting epoxide **272** was then regioselectively opened with allylamine to give the corresponding diene. Protection of the secondary amine with an *N*-Boc group provided the precursor **275** for RCM. Cyclization proceeded in 94% yield using [Mo]-**I** catalyst (CH_2Cl_2, 30 min, reflux, 94%). Subsequent conversion of the

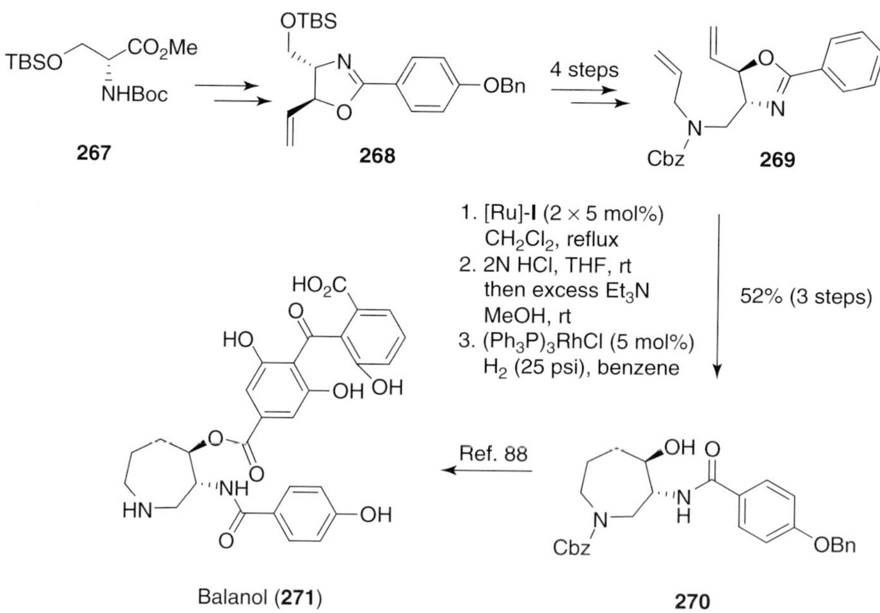

Scheme 2.59

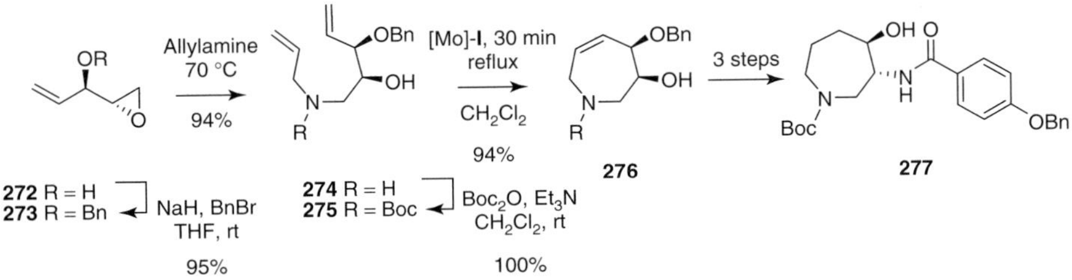

Scheme 2.60

secondary alcohol into the azide, reduction, and acylation gave the Boc-protected derivative of **277**, which also represents a formal synthesis of balanol [88].

An additional example involves the synthesis of (−)-stemoamide **282** (Scheme 2.61) [90]. In a five-step procedure, **278** was synthesized starting from the commercially available (*S*)-pyroglutaminol. Next, a vinyl iodide was installed via a chemoselective iodoboration of the alkyne substituent and subsequently subjected to the Reformatsky reagent derived from ethyl α-bromoacetate to yield the RCM precursor **279**. By using [Ru]-**II** catalyst under high dilution conditions ($c = 0.005$ M), azepine **280** was formed in an excellent yield of 92%. Since installation of the C8-hydroxy group via hydroboration was unsuccessful, an alternative was sought in hydrolyzing the ester and exposure to $CuBr_2$ on alumina, which gave the desired lactone **281**. Conjugate hydride addition from the sterically most accessible β-face using $NiCl_2 \cdot 6H_2O$ and $NaBH_4$, followed by methylation, completed the synthesis.

2.5
Eight-membered Nitrogen Heterocycles

Manzamine A (**288**) has attracted much attention over the past years because of its potent antitumor activity and since it was the first member of a group of

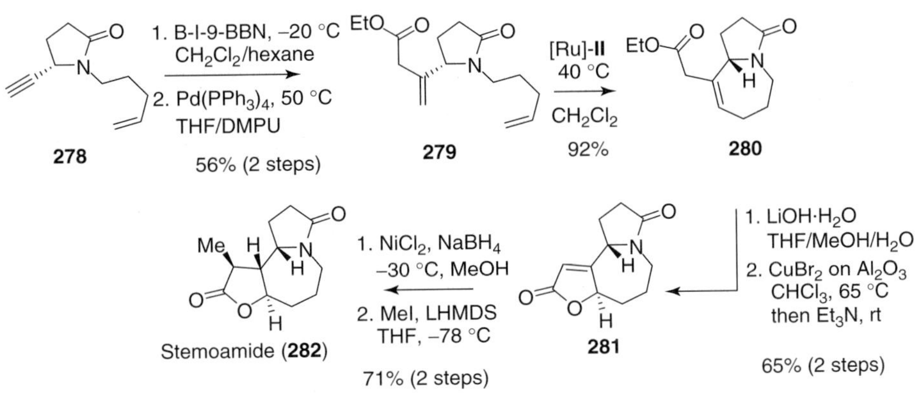

Scheme 2.61

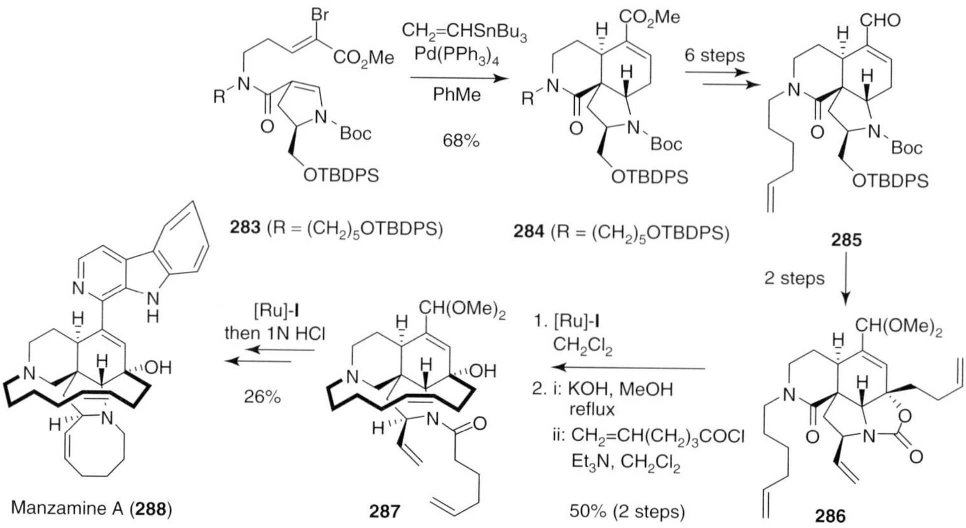

Scheme 2.62

similar alkaloids to be isolated in 1986. The combination of the complex and unusual structure of manzamine A together with its promising biological activity has led to numerous synthetic efforts. One of these comes from Martin *et al.* involving two cyclizations via RCM (Scheme 2.62) [91]. The advanced intermediate **283** was converted by a domino Stille/Diels–Alder reaction into the tricyclic system **284**. From there, eight sequential steps were required for installing the two olefinic moieties required for the macrocyclic ring formation (**286**). This very early RCM macrocyclization was carried out using [Ru]-**I** catalyst giving rise to the 13-membered ring containing predominantly the *Z*-olefin ($Z/E = 8:1$) in a respectable yield (67%). Then, hydrolysis of the cyclic carbamate followed by N-acylation provided diolefin **287**, thereby setting the stage for the formation of the eight-membered ring. RCM with [Ru]-**I** catalyst took place in a moderate yield of 26%. Promotion of RCM with the Schrock catalyst ([Mo]-**I**) did not improve the yield. Nevertheless, in a few more steps the synthesis of manzamine A (**293**) could be successfully completed.

2.6 Conclusion

Only 16 years have elapsed since Grubbs and Fu reported that Schrock's molybdenum catalyst ([Mo-**I**]) could be used to induce efficient cyclization of functionalized α,ω-diene-amines. Since then, a wide range of RCM reactions to form the nitrogen heterocyclic products have been reported, including applications in the synthesis of natural products. These applications have been greatly facilitated by the advent of new and more reactive RCM catalysts, thereby continuously widening the range of substrates and increasing the scope of the metathesis processes. This overview shows that, especially, five- and six-membered heterocyclic rings are ideal targets for RCM and that applications in larger rings are lagging behind. On the one hand, this is due to the lower occurrence of the larger rings in natural products; on the other hand, especially, eight-membered rings are significantly more difficult to form using the metathesis reaction. However, it may be clear from the spectacular developments over the past few years that further progress in the years to come will also provide solutions for problems that we are currently dealing with.

References

1 (a) Chauvin, Y. (2006) *Angew. Chem. Int. Ed.*, **45**, 3740–3747; (b) Grubbs, R.H. (2006) *Angew. Chem. Int. Ed.*, **45**, 3748–3759; (c) Schrock, R.R. (2006) *Angew. Chem. Int. Ed.*, **45**, 3760–3765.

2 Grubbs, R.H. (ed.) (2003) *Handbook of Metathesis*, Wiley-VCH Verlag GmbH, Weinheim.

3 For recent reviews on RCM, see e.g.: (a) Brown, R.C.D. and Satcharoen, V. (2006) *Heterocycles*, **70**, 705–736; (b) Arisawa, M., Nishida, A., and Nakagawa, M. (2006) *J. Organomet. Chem.*, **691**, 5109–5121; (c) Martin, S.F. (2005) *Pure Appl. Chem.*, **77**, 1207–1212; (d) Deiters, A. and Martin, S.F. (2004) *Chem. Rev.*, **104**, 2199–2238.

4 Ivin, K.J. (1983) *Olefin Metathesis*, Academic Press, London.

5 Schrock, R.R., Murdzek, J.S., Bazan, G.C., Robbins, J., DiMare, M., and O'Regan, M. (1990) *J. Am. Chem. Soc.*, **112**, 3875–3886.

6 Nguyen, S.T., Johnson, L.K., and Grubbs, R.H. (1992) *J. Am. Chem. Soc.*, **114**, 3874–3875.

7 Fu, G.C. and Grubbs, R.H. (1992) *J. Am. Chem. Soc.*, **114**, 5426–5427.

8 Nguyen, S.T., Grubbs, R.H., and Ziller, J.W. (1993) *J. Am. Chem. Soc.*, **115**, 9858–9859.

9 Schwab, P., France, M.B., Ziller, J.W., and Grubbs, R.H. (1995) *Angew. Chem. Int. Ed.*, **34**, 2039–2041.

10 Scholl, M., Ding, S., Lee, C.W., and Grubbs, R.H. (1999) *Org. Lett.*, **1**, 953–956.

11 Clavier, H., Grela, K., Kirschning, A., Mauduit, M., and Nolan, S.P. (2007) *Angew. Chem. Int. Ed.*, **46**, 6786–6801.

12 (a) Harrity, J.P.A., Visser, M.S., Gleason, J.D., and Hoveyda, A.H. (1997) *J. Am. Chem. Soc.*, **119**, 1488–1489; (b) Harrity, J.P.A., La, D.S., Cefalo, D.R., Visser, M.S., and Hoveyda, A.H. (1998) *J. Am. Chem. Soc.*, **120**, 2343–2351.

13 Fu, G.C. and Grubbs, R.H. (1992) *J. Am. Chem. Soc.*, **114**, 7324–7325.

14 Philips, A.J. and Abell, A.D. (1999) *Aldrichim. Acta*, **32**, 75–89.

15 Hanessian, S., Margarita, R., Hall, A., Johnstone, S., Tremblay, M., and Parlanti, L. (2002) *J. Am. Chem. Soc.*, **124**, 13342–13343.

16 Hanessian, S., Tremblay, M., and Petersen, J.F.W. (2004) *J. Am. Chem. Soc.*, **126**, 6064–6071.

17 Welter, C., Moreno, R.M., Streiff, S., and Helmchen, G. (2005) *Org. Biomol. Chem.*, **3**, 3266–3268.

18 Leete, E., Bjorklund, J.A., Couladis, M.M., and Kim, S.K. (1991) *J. Am. Chem. Soc.*, **113**, 9286–9292.

19 Leete, E. and Kim, S.H. (1989) *J. Chem. Soc., Chem. Commun.*, 1899–1900.

20 Lee, J.-H., Jeong, B.-S., Ku, J.-M., Jew, S.-S., and Park, H.-G. (2006) *J. Org. Chem.*, **71**, 6690–6692.

21 Dochnahl, M., Schulz, S.R., and Blechert, S. (2007) *Synlett*, 2599–2601.

22 Mori, M., Sakakibara, N., and Kinoshita, A. (1998) *J. Org. Chem.*, **63**, 6082–6083.

23 Murruzzu, C. and Riera, A. (2007) *Tetrahedron: Asymmetry*, **18**, 149–154.

24 Trost, B.M., Horne, D.B., and Woltering, M.J. (2006) *Chem. Eur. J.*, **12**, 6607–6620.
25 Trost, B.M., Horne, D.B., and Woltering, M.J. (2003) *Angew. Chem. Int. Ed.*, **42**, 5987–5990.
26 Deiters, A., Pettersson, M., and Martin, S.F. (2006) *J. Org. Chem.*, **71**, 6547–6561.
27 Shimizu, K., Takimoto, M., Sato, Y., and Mori, M. (2006) *J. Organomet. Chem.*, **691**, 5466–5475.
28 Shimizu, K., Takimoto, M., and Mori, M. (2003) *J. Org. Chem.*, **5**, 2323–2325.
29 Fukumoto, H., Esumi, T., Ishihara, J., and Hatakeyama, S. (2003) *Tetrahedron Lett.*, **44**, 8047–8049.
30 Fukumoto, H., Takahashi, K., Ishihara, J., and Hatakeyama, S. (2006) *Angew. Chem. Int. Ed.*, **45**, 2731–2734.
31 Kim, S., Lee, J., Lee, T., Park, H.-G., and Kim, D. (2003) *Org. Lett.*, **5**, 2703–2707.
32 Kim, S., Lee, T., Lee, E., Lee, J., Fan, G.-J., Lee, S.K., and Kim, D. (2004) *J. Org. Chem.*, **69**, 3144–3149.
33 Stragies, R. and Blechert, S. (2000) *J. Am. Chem. Soc.*, **122**, 9584–9591.
34 Zaminer, J., Stapper, C., and Blechert, S. (2002) *Tetrahedron Lett.*, **43**, 6739–6741.
35 Davis, F.A. and Yang, B. (2005) *J. Am. Chem. Soc.*, **127**, 8398–8407.
36 Davis, A.S., Ritthiwigrom, T., and Pyne, S.G. (2008) *Tetrahedron*, **64**, 4868–4879.
37 Karanjule, N.S., Markad, S.D., and Dhavale, D.P. (2006) *J. Org. Chem.*, **71**, 6273–6276.
38 Machan, T., Davis, A.S., Liawruangrath, B., and Pyne, S.G. (2008) *Tetrahedron*, **64**, 2725–2732.
39 Buschmann, N., Rückert, A., and Blechert, S. (2002) *J. Org. Chem.*, **67**, 4325–4329.
40 Lindsay, K.B. and Pyne, S.G. (2002) *J. Org. Chem.*, **67**, 7774–7780.
41 Trost, B.M., Aponick, A., and Stanzl, B.N. (2007) *Chem. Eur. J.*, **13**, 9547–9560.
42 Pyne, S.G. and Tang, M. (2004) *Tetrahedron*, **60**, 5759–5767.
43 Rambaud, L., Compain, P., and Martin, O.R. (2001) *Tetrahedron: Asymmetry*, **12**, 1807–1809.
44 Chang, M.-Y., Wu, D.-C., and Chang, N.-C. (2005) *Heterocycles*, **65**, 2965–2971.
45 Raghaven, S. and Rajender, A. (2004) *Tetrahedron Lett.*, **45**, 1919–1922.
46 Raghaven, S. and Rajender, A. (2004) *Tetrahedron*, **60**, 5059–5067.
47 Yadav, J.S. (2006) *Synthesis*, **23**, 4005–4012.
48 Cossy, J., Willis, C., Bellosta, V., and Bouzbouz, S. (2000) *Synlett*, 1461–1463.
49 Cossy, J., Willis, C., Bellosta, V., and Bouzbouz, S. (2002) *J. Org. Chem.*, **67**, 1982–1992.
50 Kang, B. and Chang, S. (2004) *Tetrahedron*, **60**, 7353–7359.
51 Chang, M., Kung, Y., and Chen, S. (2006) *Tetrahedron*, **62**, 10843–10848.
52 Jamieson, A.G. and Sutherland, A. (2007) *Org. Lett.*, **9**, 1609–1611.
53 Lebrun, S., Couture, A., Deniau, E., and Grandclaudon, P. (2008) *Tetrahedron: Asymmetry*, **19**, 1245–1249.
54 Lebrun, S., Couture, A., Deniau, E., and Grandclaudon, P. (2007) *Org. Lett.*, **9**, 2473–2476.
55 Jo, E., Na, Y. and Chang, S. (1999) *Tetrahedron Lett.*, **40**, 5581–5582.
56 (a) Enders, D. and Tiebes, J. (1993) *Liebigs Ann. Chem.*, **106**, 173–177; (b) Reding, M.T. and Buchwald, S.L. (1998) *J. Org. Chem.*, **63**, 6344–6347; (c) Davies, S.B. and McKervey, M.A. (1999) *Tetrahedron Lett.*, **40**, 1229–1232 and references cited therein.
57 Pachamuthu, K. and Vankar, Y. (2001) *J. Organomet. Chem.*, **1-2**, 359–363.
58 Agami, C., Couty, F., and Rabasso, N. (2000) *Tetrahedron Lett.*, **41**, 4113–4116.
59 Felpin, F., Vo-Thanh, G., Robins, R., Villieras, J., and Lebreton, J. (2000) *Synlett*, 1646–1648.
60 Felpin, F., Girard, S., Vo-Thanh, G., Robins, R., Villieras, J., and Lebreton, J. (2001) *J. Org. Chem.*, **66**, 6305–6312.
61 Takahashi, K., Nakano, H., and Fujita, R. (2005) *Tetrahedron Lett.*, **46**, 8927–8930.
62 Sabat, M. and Johnson, C. (2001) *Tetrahedron Lett.*, **42**, 1209–1212.
63 Tjen, K.C.M.F., Kinderman, S.S., Schoemaker, H.E., Hiemstra, H., and Rutjes, F.P.J.T. (2000) *Chem. Commun.*, 699–700.

64 Ginesta, X., Pericas, M.A., and Riera, A. (2002) *Tetrahedron Lett.*, **43**, 779–782.
65 Shigeyama, T., Katakawa, K., Kogure, N., Kitajima, M., and Takayama, H. (2007) *Org. Lett.*, **9**, 4069–4072.
66 Arrayas, R.G., Alcudia, A., and Liebeskind, L.S. (2001) *Org. Lett.*, **3**, 3381–3383.
67 Ghosh, S., Shashidhar, J., and Dutta, S.K. (2006) *Tetrahedron Lett.*, **47**, 6041–6044.
68 Yasunori, B., Chiemi, A., Hideo, N., Atsushi, K., Isao, A., and Hiroki, T. (2001) *Tetrahedron: Asymmetry*, **12**, 817–820.
69 Park, S.H., Kang, H.J., Ko, S., Park, S., and Chang, S. (2001) *Tetrahedron: Asymmetry*, **12**, 2621–2624.
70 Nukui, S., Sodeoka, M., and Shibasaki, M. (1993) *Tetrahedron Lett.*, **34**, 4965.
71 Ayad, T., Génisson, Y., and Baltas, M. (2005) *Org. Biomol. Chem.*, **3**, 2626–2631.
72 Kim, I.S., Zee, O.P., and Jung, Y.H. (2006) *Org. Lett.*, **8**, 4101–4104.
73 El-Nezhawy, A.O.H., El-Diwani, H.I., and Schmidt, R.R. (2002) *Eur. J. Org. Chem.*, **24**, 4137–4142.
74 Marsden, S.P. (2005) *Synlett*, 2528–2530.
75 (a) Sun, P., Sun, C., and Weinreb, S.M. (2002) *J. Org. Chem.*, **67**, 4337–4345; (b) Sun, P., Sun, C., and Weinreb, S.M. (2001) *Org. Lett.*, **3**, 3507–3510.
76 Selig, P. and Bach, T. (2008) *Angew. Chem. Int. Ed.*, **47**, 5082–5084.
77 Smith, A.B. and Kim, D. (2005) *Org. Lett.*, **7**, 3247–3250.
78 Kinderman, S.S., de Gelder, R., van Maarseveen, J., Schoemaker, H.E., Hiemstra, H., and Rutjes, F.P.J.T. (2004) *J. Am. Chem. Soc.*, **126**, 4100–4101.
79 Michel, P. and Rassat, A. (1999) *Chem. Commun.*, 2281.
80 Katoh, M., Mizutani, H., and Honda, T. (2005) *Tetrahedron Lett.*, **46**, 5161–5163.
81 Wijdeven, M.A., Botman, P.N.M., Wijtmans, R., Schoemaker, H.E., Rutjes, F.P.J.T., and Blaauw, R.H. (2005) *Org. Lett.*, **7**, 4005–4007.
82 Liu, S., Fan, Y., Peng, X., Wang, W., Hua, W., Akber, H., and Liao, L. (2006) *Tetrahedron Lett.*, **47**, 7681–7684.
83 Chavan, S.P., Sharma, P., Sivappa, R., and Kalkote, R.T. (2006) *Tetrahedron Lett.*, **47**, 9301–9303.
84 Takayama, H., Kurihara, M., Kitajima, M., Said, I.M., and Aimi, N. (1999) *J. Org. Chem.*, **64**, 1772–1773.
85 Blakemore, P.R., Kilner, C., Norcross, N.R., and Astles, P.C. (2005) *Org. Lett.*, **7**, 4721–4724.
86 Theeraladanon, C., Ariswa, M., Nakagawa, M., and Nishida, A. (2005) *Tetrahedron: Asymmetry*, **16**, 827–831.
87 Cook, G.R., Shanker, P.S., and Peterson, S.L. (1999) *Org. Lett.*, **1**, 615–618.
88 Nicolaou, K.C., Bunnage, M.E., and Koide, K. (1994) *J. Am. Chem. Soc.*, **116**, 8402–8403.
89 Fürstner, A. and Thiel, O.R. (2000) *J. Org. Chem.*, **65**, 1738–1742.
90 Torssell, S., Wanngren, E., and Somfai, P. (2007) *J. Org. Chem.*, **72**, 4246–4249.
91 Humphrey, J.M., Liao, Y., Ali, A., Rein, T., Wong, Y., Chen, H., Courtney, A.K., and Martin, S.F. (2002) *J. Am. Chem. Soc.*, **124**, 8584–8592.

3
Synthesis of Natural Products Containing Medium-size Oxygen Heterocycles by Ring-closing Alkene Metathesis

Jon D. Rainier

3.1
Introduction

Medium-sized cyclic ethers are present in a number of interesting and important natural products. As illustrated in Figure 3.1, this list includes members of the polycyclic ether family (brevetoxins, ciguatoxins, etc.), the laurencins, heliannuol A, along with members of the eunicellin family. In addition to this are medium ring lactones, including octalactin A, and the 10-membered lactones, microcarpalide and herbarumin III.

Recognizing the general difficulty associated with the synthesis of medium rings along with the specific synthetic challenges associated with the generation of the targets listed above, synthetic chemists have devised a number of creative methodologies to medium ring substrates [1]. As the title suggests, the focus of this review is on those methods that utilize alkene metathesis [2]. It was not long after the development of ring-closing metathesis (RCM) that scientists interested in organic synthesis became intrigued with the notion of medium ring generation [3]. However, these chemists also quickly came to realize that the medium rings are thermodynamically unstable, when coupled with their slower rates of formation, that RCM is reversible, and that undesired side products are often formed, for example, from isomerization and intermolecular reactions, and so meant that one needed to pay close attention to the product stability, starting material conformation, and reactivity. A daunting task indeed! This chapter attempts to define what is known about the scope and limitations of oxygenated medium ring synthesis via RCM [4]. As the focus here is on natural product synthesis and the utility of the reaction, the chemistry of the cyclization precursors and their cyclization products is also highlighted. As it is of particular interest to the author, examples of carbonyl-olefination cyclizations using a stoichiometric titanium reagent are also included.

Obvious to even the casual follower of the metathesis literature is that a large number of catalysts have been developed and employed in RCM [5] (Figure 3.2). Highlighted here are cyclizations that employ both the Grubbs I ([Ru]-**I**) and II

Metathesis in Natural Product Synthesis: Strategies, Substrates and Catalysts. Edited by Janine Cossy, Stellios Arseniyadis, and Christophe Meyer

ISBN: 978-3-527-32440-8

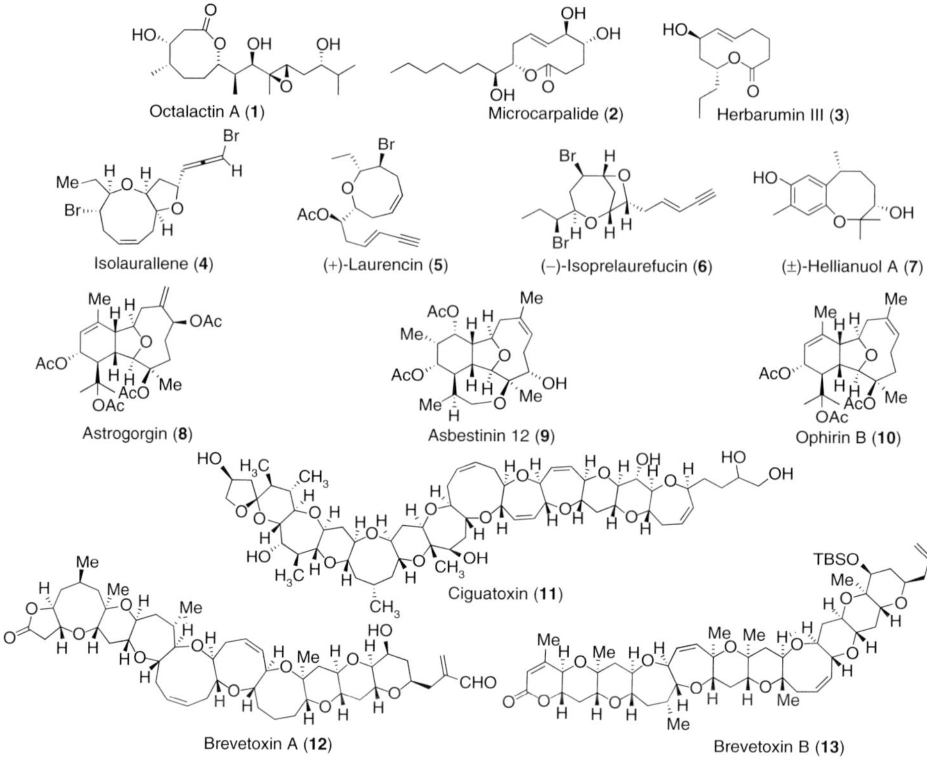

Figure 3.1 Natural products containing medium ring oxygen heterocycles.

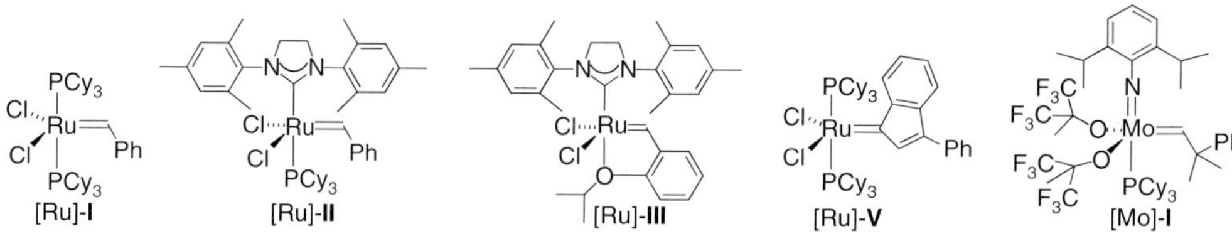

Figure 3.2 Catalysts/reagents that have been employed in RCM to medium oxygen rings.

([Ru]-**II**) catalysts [6, 7], along with reactions using the Hoveyda–Grubbs ([Ru]-**III**) [8], Fürstner ([Ru]-**V**) [9], or Schrock ([Mo]-**I**) [10] catalyst.

For the purposes of this chapter, *medium ring heterocycles* are defined as ring systems that are 8-(oxocine)- to 9-(oxonene)- and 10-(oxecin)-membered-rings. Because their synthesis is often problematic, representative examples of seven-(oxepine)-membered rings are also included. While this chapter has mostly been divided into natural product class, also included are more general RCM reactions that lead to medium ring ethers. The reader will find that it is not hard to imagine these methods being used in total synthesis studies.

3.2 General RCM Approaches to Medium Rings

Clark *et al.* was one of the very first to recognize that RCM could be used to generate medium rings. In model studies aimed at the generation of the medium rings present in brevetoxin A, Clark and Kettle showed that allyl ethers having pendant olefins undergo RCM cyclizations to give oxocines and oxonenes (Table 3.1) [11]. Interestingly, in the generation of oxonenes, the stereochemistry of the ether dictated the efficiency of the reactions (Table 3.1, entries 3 and 4). Diastereomers **18** and **20** gave dramatically different yields of the cyclic products **19** and **21**, respectively. This was one of the first studies that demonstrated the significant role of substituents and conformation on RCM cyclizations to medium rings.

Clark and Kettle also demonstrated that [Mo]-**I** catalyst can be used to give dihydropyrans and oxepines from enol ether–olefin RCM (Table 3.2, entries 1–4) [12].

Table 3.1 Generation of oxocines and oxonenes.

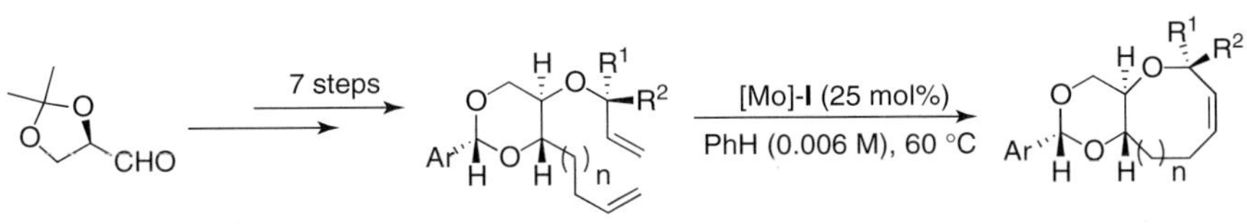

Entry	Diene	Cyclic ether	Yield (%)
1	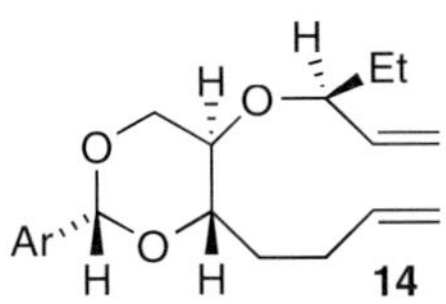 **14**	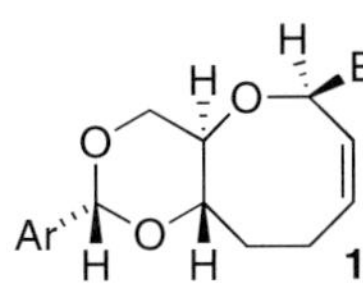 **15**	97
2	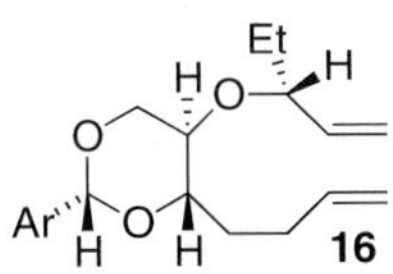 **16**	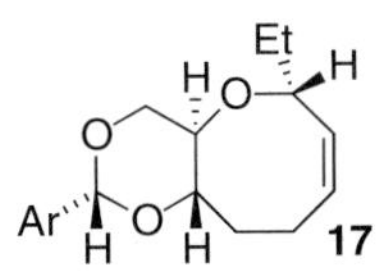 **17**	86
3	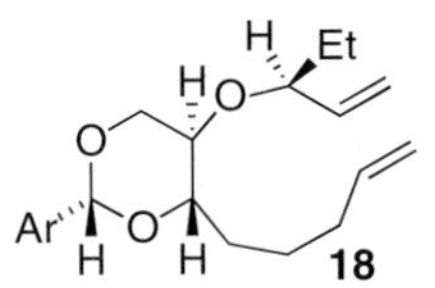 **18**	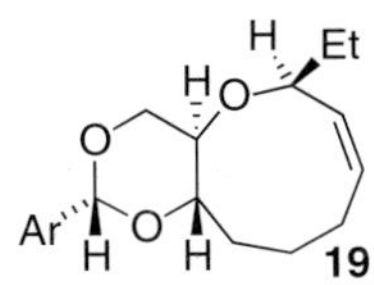 **19**	86
4	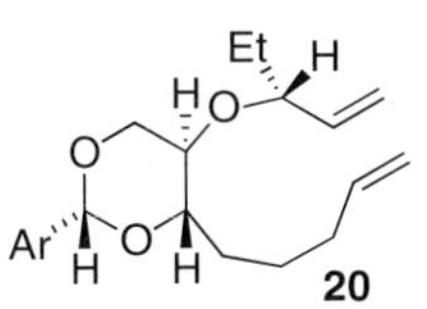 **20**	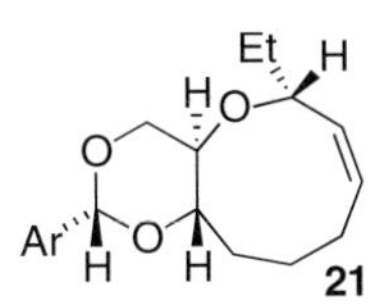 **21**	14

Table 3.2 Enol ether–olefin RCM to cyclic enol ethers.

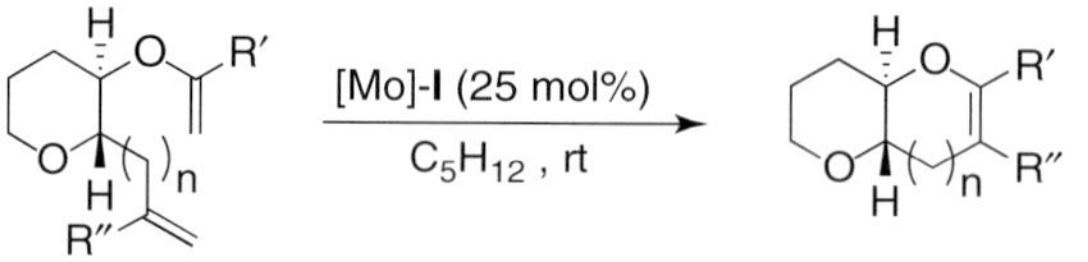

Entry	Acyclic enol ether	Product(s)	Yield (%)
1	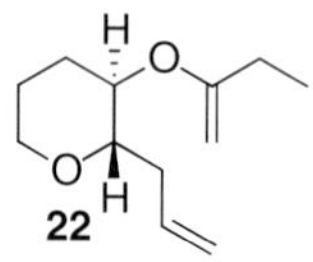 **22**	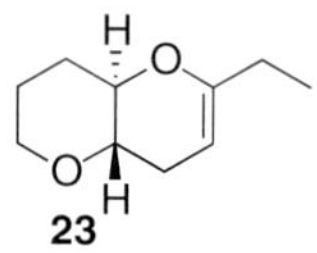 **23**	72
2	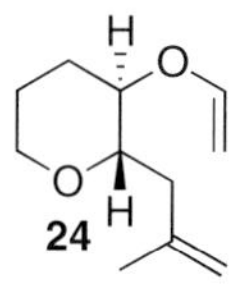 **24**	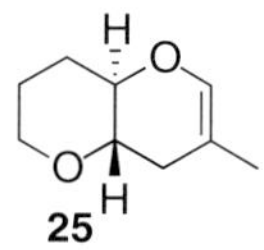 **25**	93
3	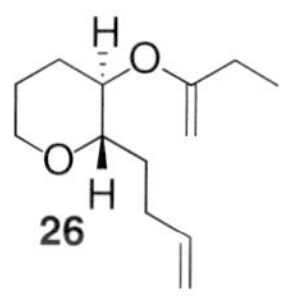 **26**	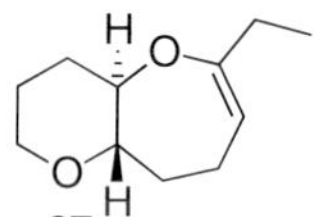 **27**	42
4	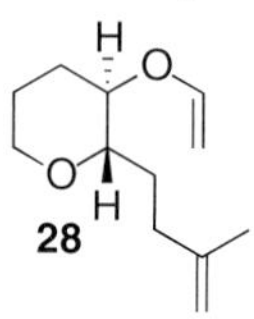 **28**	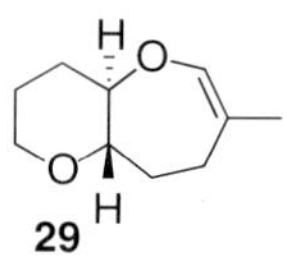 **29**	94
5[a]	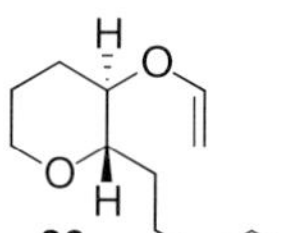 **30**	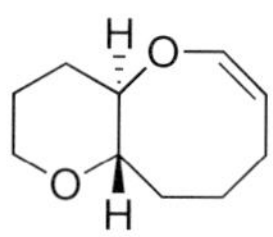 **31** (27%) 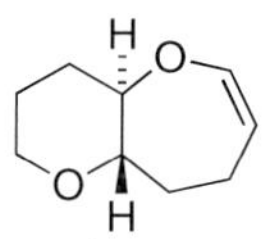 **32** (13%) 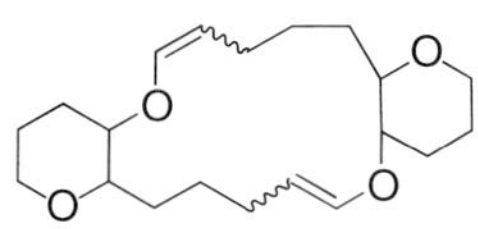 **33** (40%)	
6[a]	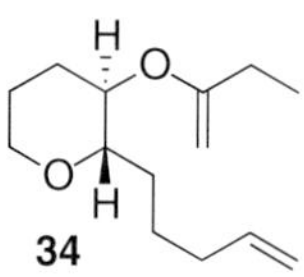 **34**	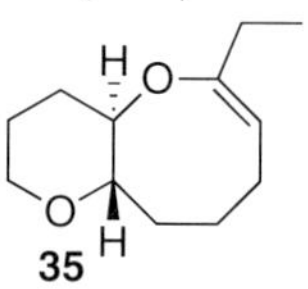 **35**	0

[a] Reaction was run at 60 °C in C_6H_6.

Cyclization reactions to the corresponding oxocines were plagued by competitive isomerization and/or dimerization (Table 3.2, entries 5 and 6). Electron-rich olefins, such as those studied by Clark and Kettle, continue to be problematic in RCM due to their slow cyclizations, thereby allowing competing reactions (isomerization

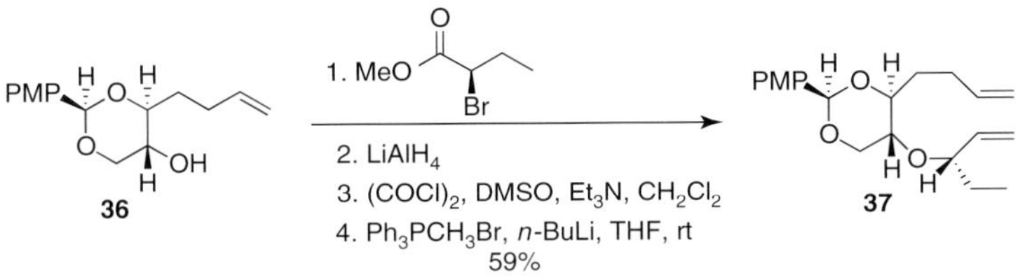

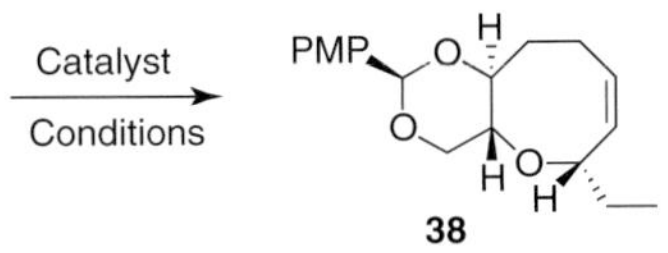

Catalyst	Conditions	Yield
[Mo]-**I** (25–30 mol%)	PhH (<0.02 M), 60 °C	97(%)
[Ru]-**I** (10–20 mol%)	CH_2Cl_2 (<0.01 M), rt	86(%)

Scheme 3.1 Comparison of the Schrock and Grubbs I catalysts in diene RCM.

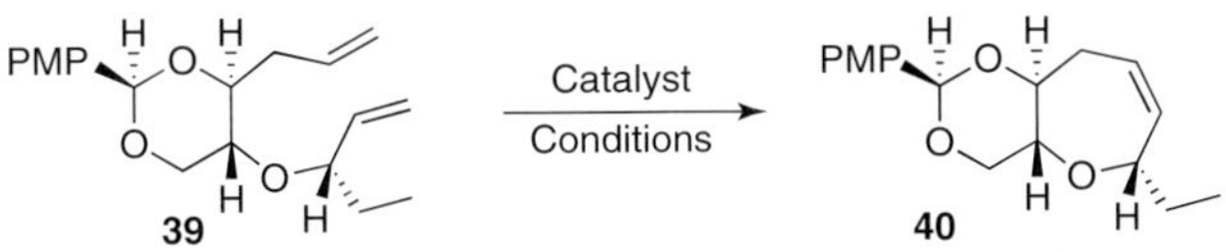

Catalyst	Conditions	Yield
[Mo]-I (25–30 mol%)	PhH (<0.02 M), rt	86(%)
[Ru]-I (10–20 mol%)	CH_2Cl_2 (<0.01 M), rt	94(%)

Scheme 3.2 Catalyst comparisons to oxepines.

and dimerization) to take place. Advantageous is that these reactions are generally irreversible.

Clark *et al.* have compared the ability of [Ru]-**I** and [Mo]-**I** catalysts to generate oxocines (Scheme 3.1) and oxepines (Scheme 3.2) from the neutral diene RCM reactions [13]. The precursors for these studies were prepared from the addition of the appropriate carbon nucleophile to D-glyceraldehyde, followed by acetonide removal and reprotection of the resulting triol as the cyclic acetal, to give **36**. Ether and diene formation gave the metathesis precursor **37**. A comparison of [Mo]-**I** and [Ru]-**I** catalysts revealed a similar efficiency for the synthesis of oxocine **38**.

In a similar fashion to the cyclization to **37**, the cyclization to oxepine **40** was highly successful with both [Mo]-**I** and [Ru]-**I** catalysts (Scheme 3.2) [13].

In contrast to the cyclization to the corresponding dihydropyrans, Clark *et al.* found the cyclization to oxepines **42a** and **42b** using ene–yne RCM and [Ru]-**I** catalyst to be sluggish and low yielding (Scheme 3.3) [14].

In the course of synthesizing model substrates of *trans*-fused polyether natural products, Delgado and Martín utilized the Grubbs catalyst [Ru]-**I** in the synthesis of medium and larger cyclic ether rings (Table 3.3) [15]. Cyclization using [Ru]-**I** (10 mol%) gave the corresponding oxepines and oxocines in excellent yield (Table 3.3, entries 1–5). In contrast, the generation of oxonene **54** and oxecine **56** was less efficient despite the use of higher temperatures and more catalyst (Table 3.3, entries

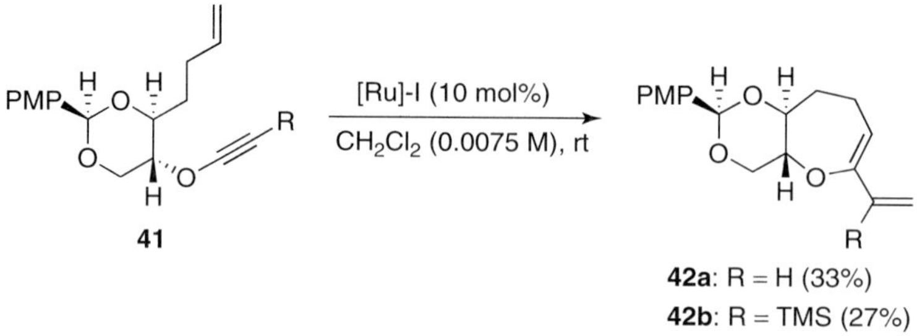

Scheme 3.3 Ene–yne cyclizations to oxepines.

6 and 7). Interestingly, the authors found that the yield of cyclic products increased with increasing oxygen incorporation in the tether between the alkenes (Table 3.3, entry 8).

Fujiwara *et al.* met the challenge of stereoselectively synthesizing oxocines by coupling an Ireland–Claisen rearrangement reaction with RCM (Scheme 3.4) [16]. As outlined throughout this chapter, glycolates have been popular precursors to dienes and medium ring ethers. Fujiwara's approach began with the use of glycolate ester **59** as a precursor to an Ireland–Claisen rearrangement to give **60** after the formation of methyl ester. A subsequent RCM cyclization using [Ru]-**I** catalyst gave oxocine **61** in 87% yield.

Illustrating again the important role of conformation on the medium ring RCM process and in contrast to the results in Scheme 3.4, the product from the rearrangement of the *Z*-enol ether **62** did not undergo cyclization with [Ru]-**I** catalyst (Scheme 3.5) [16]. The use of the more reactive [Ru]-**II** catalyst did work, however, giving **64** in 53% yield.

Fujiwara *et al.* extended their methodology to the synthesis of optically active medium rings [17]. As outlined in Scheme 3.6, the use of L-serine-derived oxazolidinones in Ireland–Claisen rearrangement resulted in the generation of the optically active dienes **66a–c**. Exposure of **66a–c** to [Ru]-**II** catalyst resulted in the formation of dihydropyrans, oxepines, and oxocines in reasonable yields. A secondary benefit of the cyclization reaction was that the auxiliary was removed in the course of the cyclization.

Gais *et al.* have described a useful approach to medium ring lactones, which utilizes sulfoximines as both cyclization precursors and as alkene protecting groups (Scheme 3.7) [18]. They demonstrated that sulfoximine **68** was a precursor to homoallyl ether **69**. The subsequent allylation of **69** was followed by deprotection and esterification to give the triene cyclization precursors **71a** and **71b**. RCM using [Ru]-**II** catalyst gave the 9- and 11-membered lactones **72a** and **72b**, respectively, in high yield.

Illustrating the importance of the sulfoximine, Waldmann and Basu found that related conjugated dienes lacking sulfoximine substitution gave the smaller of the two possible rings when exposed to [Ru]-**I**, [Ru]-**II**, or [Ru]-**III** catalyst (Scheme 3.8) [19]. For example, when triene **73** was subjected to any of these catalysts, only oxepine **74** was isolated.

Table 3.3 Synthesis of medium ring ethers.

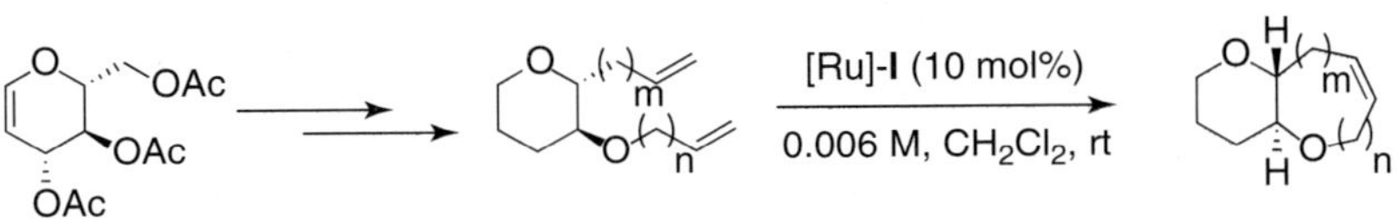

Entry	Diene	Cyclic ether	Yield (%)
1	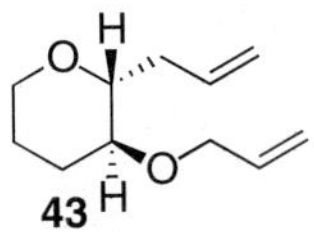 **43**	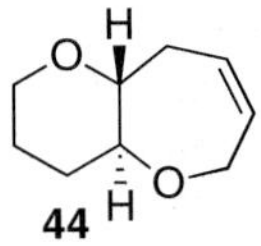 **44**	>95
2	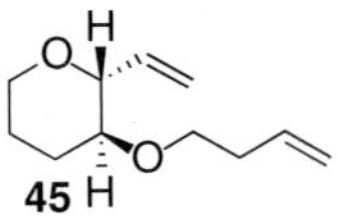 **45**	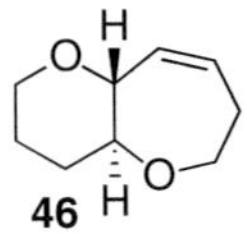 **46**	>95
3	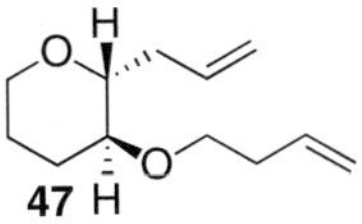 **47**	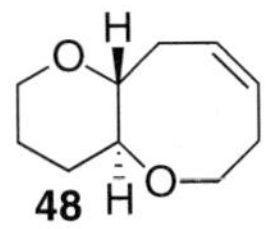 **48**	>95
4	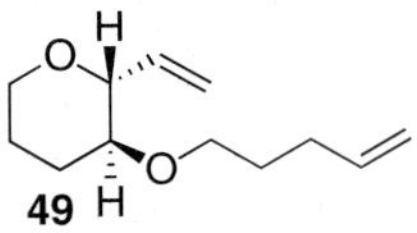 **49**	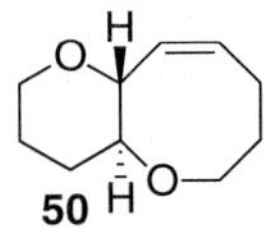 **50**	>95
5	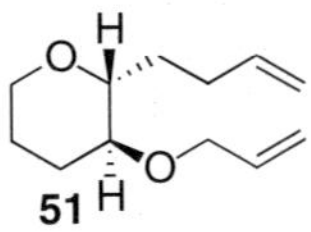 **51**	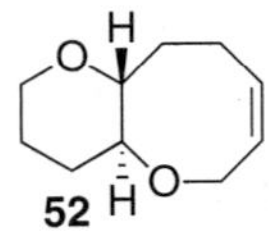 **52**	>95
6	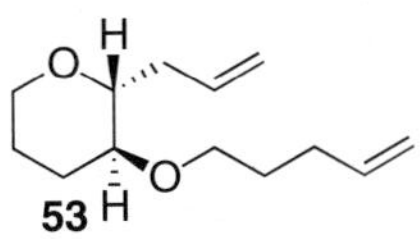 **53**	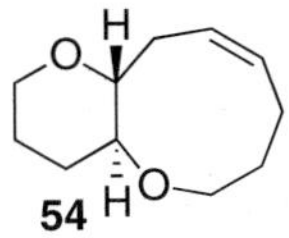 **54**	45[a]
7	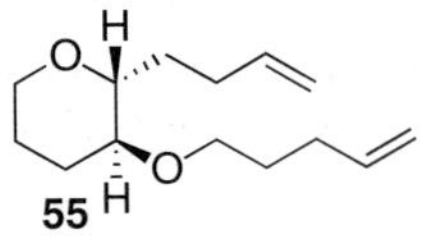 **55**	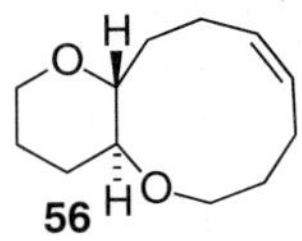 **56**	42[a]
8	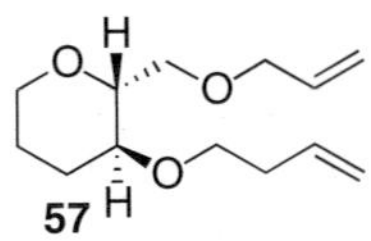 **57**	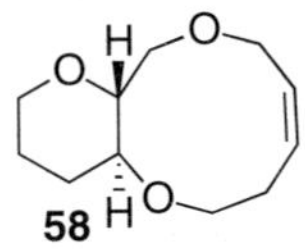 **58**	63

[a]Conditions: [Ru]-**I** (30 mol%), 0.005 M, CH_2Cl_2, 45 °C.

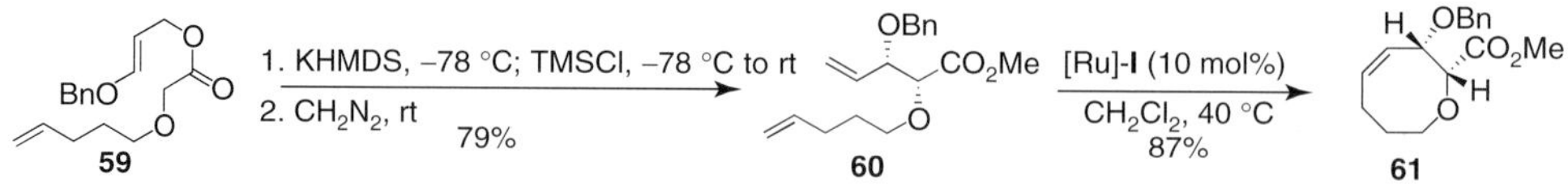

Scheme 3.4 Ireland–Claisen, RCM sequence to oxocines.

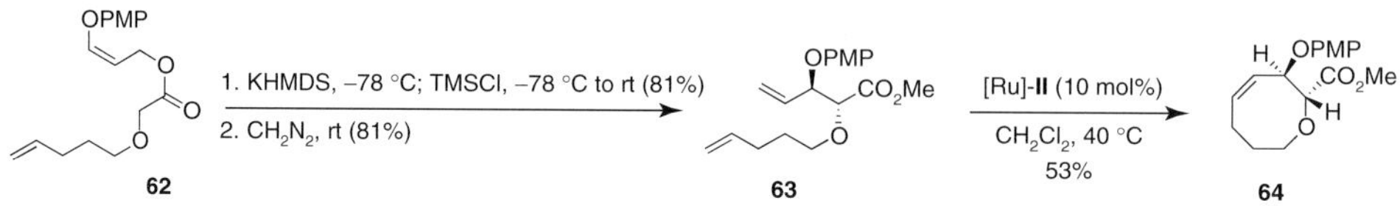

Scheme 3.5 Ireland–Claisen, RCM sequence: Z-enol ethers.

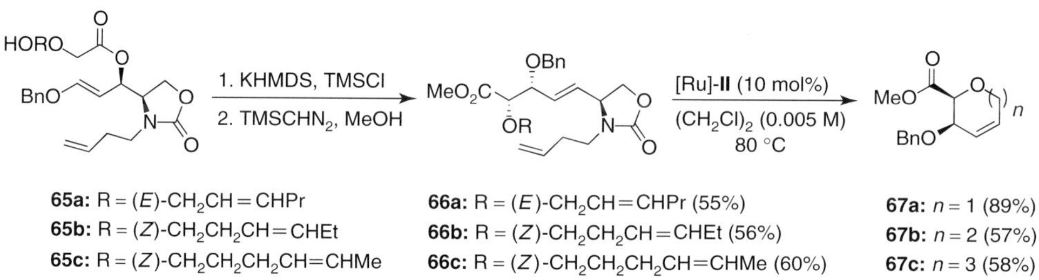

Scheme 3.6 The synthesis of optically active medium rings using sequential Ireland–Claisen RCM technology.

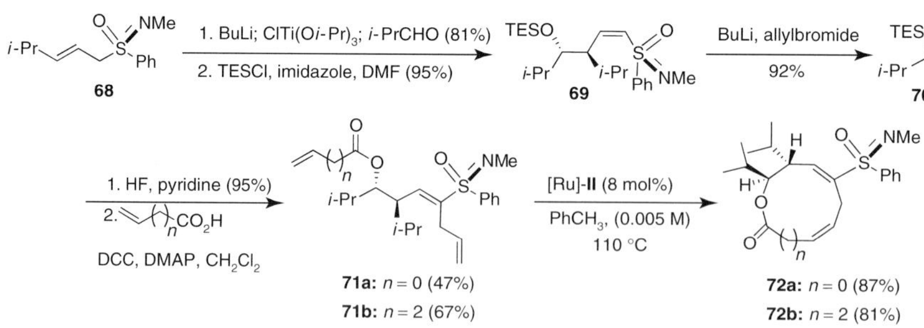

Scheme 3.7 The synthesis of medium ring lactones.

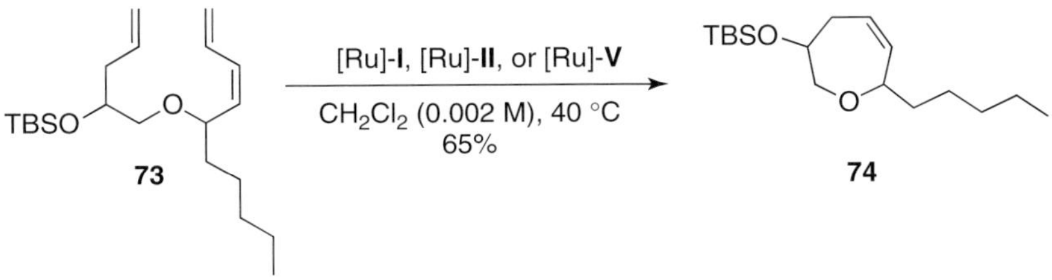

Scheme 3.8 Triene cyclizations.

Linderman *et al.* have found that oxepines and oxocines can be efficiently generated from the RCM of acyclic tributylstannyl-substituted dienes using [Ru]-**I** catalyst [20]. As illustrated in Table 3.4, stannylated dienes gave very high yields of the corresponding oxocines regardless of the relative position of the olefins. In contrast to these results, attempts at employing *t*-butyl groups in the place of the stannane gave only polymeric material indicating that the stannane is not affecting the reaction through a steric effect.

That the α-alkoxy stannane is synthetically useful is also beneficial (Scheme 3.9). Linderman *et al.* converted oxocene **83** into ketone **84** by first transmetallating with *n*-BuLi and then quenching the resulting organolithium with *N,N*-dimethyl benzamide to give **84** in 75% yield (+5% *trans*-isomer) [20].

3.3 Laurencin

In the first of a series of studies employing asymmetric enolate methodology to RCM precursors, Crimmins *et al.* employed an aldol-RCM strategy to synthesize medium ring heterocycles that are found in the red algae *Laurencia* sp. Central to the success of the strategy was the recognition that the vicinal diol, required for the synthesis of the natural product, would facilitate ring closure through a gauche effect placing the substrate in a favorable conformation for cyclization [21]. As outlined in Schemes 3.10–3.12, this strategy was used to effect the generation of oxepines, oxocines, and oxonenes. The synthesis of oxepine **88** is illustrated in Scheme 3.10 [22]. Using glyoxylate oxazolidinone **85** as a staging point, aldol condensation with acrolein provided **86** (de > 99%), and reductive removal of the auxiliary and RCM using [Ru]-**I** catalyst gave **88** in 95% yield.

The use of this strategy in the synthesis of oxocines and a formal synthesis of laurencin is illustrated in Scheme 3.11 [22]. From acyloxazolidinethione **89**, aldol condensation with 3-butenal provided **91** after reductive removal of the auxiliary. As with the synthesis of oxepine **88**, the generation of oxocine **92** was successfully carried out using [Ru]-**I** catalyst. Subsequent manipulation of the protecting group enabled the interception of **93**, an intermediate used in the synthesis of (+)-laurencin by Holmes *et al.* [23]. As an illustration of the importance of substitution on the ring-closing reaction, earlier attempts to carry out a related

Table 3.4 Stannyl ketals in RCM.

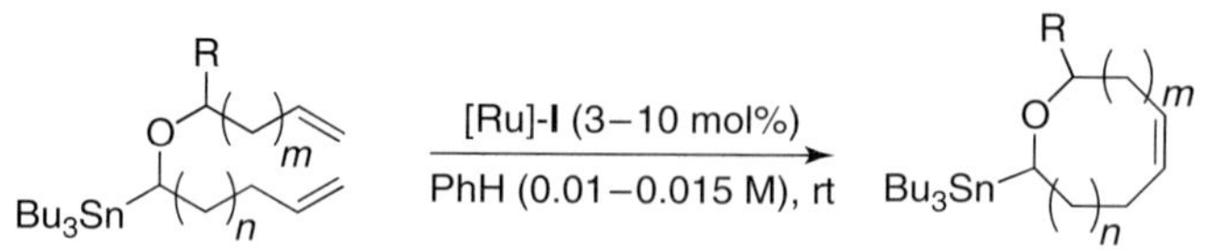

Entry	Diene	Product	Yield (%)
1	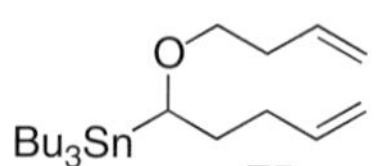 **75**	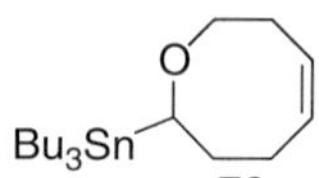 **76**	74
2	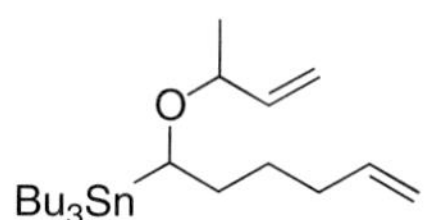 **77**	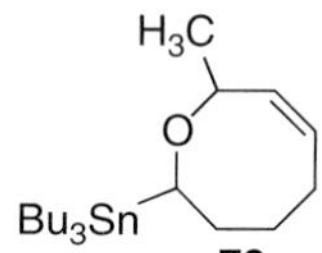 **78**	38
3	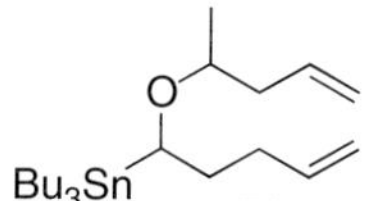 **79**	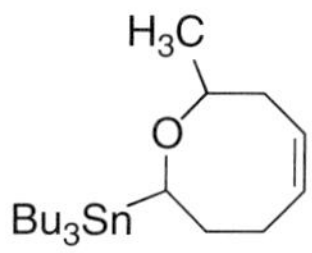 **80**	84
4	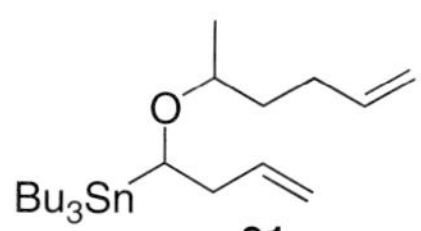 **81**	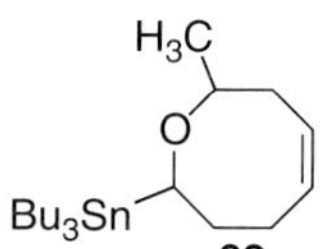 **82**	96

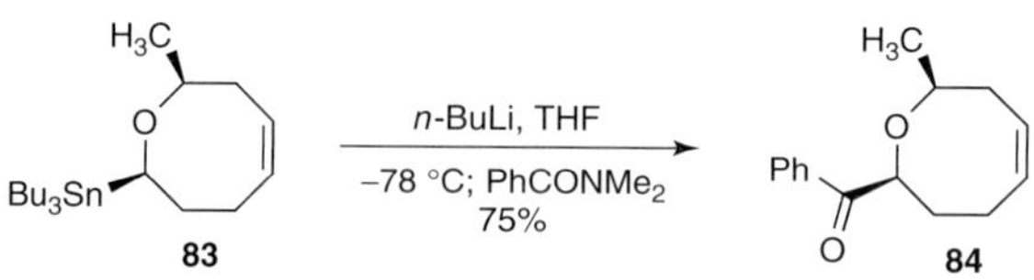

83 → **84**

Scheme 3.9 Stannyl ketal manipulations.

cyclization on a substrate lacking the substitution in **77** gave only dimers and other oligomers [24].

The aldol-RCM strategy was also used to generate oxonene **97** (Scheme 3.12) [22].

Crimmins *et al.* also used the strategy outlined above in the synthesis of isolaurallene [25]. The synthesis of the diene cyclization precursor is shown in

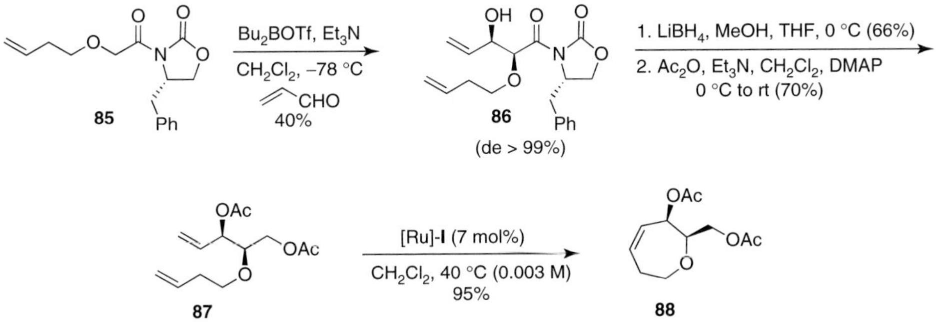

Scheme 3.10 Asymmetric enolate-RCM sequence: oxepines.

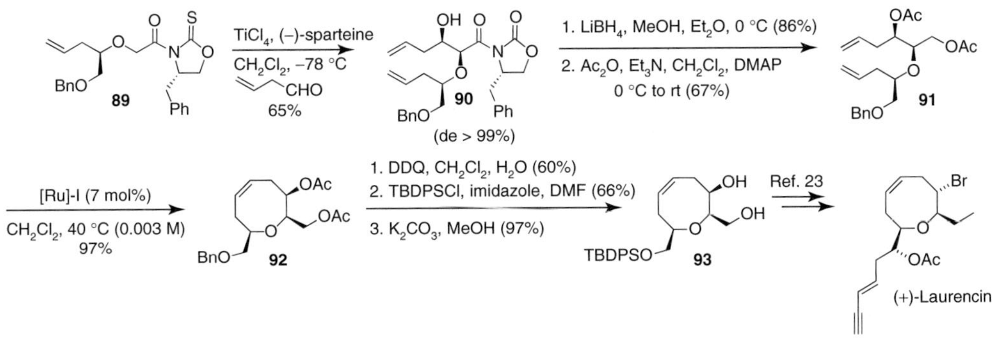

Scheme 3.11 Formal synthesis of (+)-laurencin.

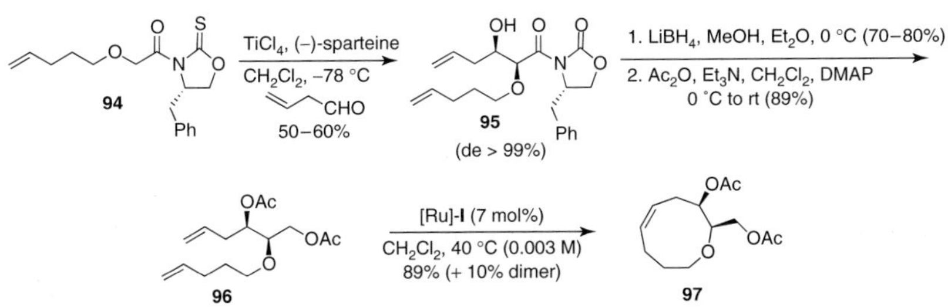

Scheme 3.12 Asymmetric aldol, RCM reactions: oxonenes.

Scheme 3.13. From benzyl glycidyl ether **98**, ring opening and esterification gave **100** after attachment of oxazolidinone **99**. Aldol condensation gave **101** in 58% yield along with 20% of a mixture of two other isomers. Following the manipulatoin of the protecting group and oxidation, the second alkene was inserted into the substrate using Brown's allyl borane chemistry to give the cyclization precursor **103**.

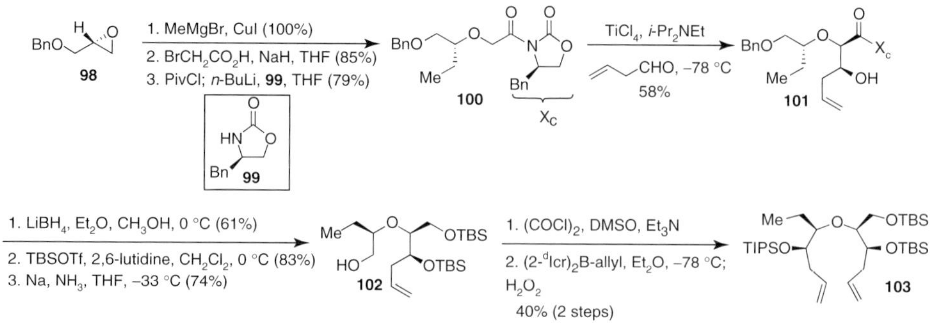

Scheme 3.13 Synthesis of isolaurallene.

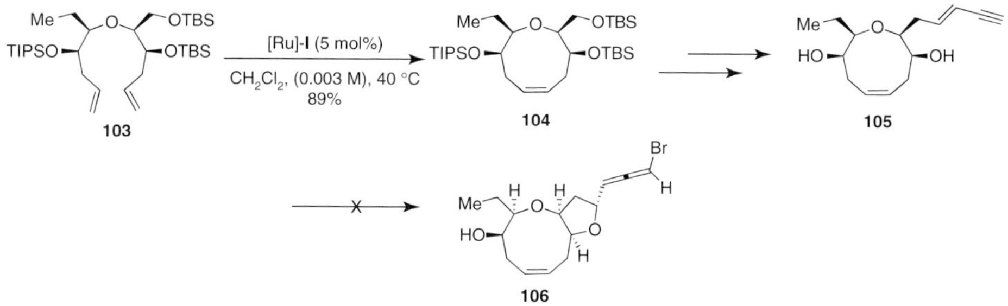

Scheme 3.14 RCM cyclization to the isolaurallene oxonene.

RCM to isolaurallene's oxonene was successful using [Ru]-**I** catalyst (Scheme 3.14) [25]. Unfortunately, the subsequent conversion of **105** into the requisite furan **106** for the completion of isolaurallene using bromonium ion activation was unsuccessful.

Having been unsuccessful in the cyclization to **106**, Crimmins *et al.* took an alternate approach to isolaurallene involving a hydroxy-epoxide cyclization to the furan [25]. To this goal, triene **107** was synthesized using an approach that was similar to that used to generate **103**. Attempted cyclization of **107** to oxonene **108** again illustrated some of the difficulties associated with medium ring formation (Scheme 3.15). When **107** was subjected to [Ru]-**I** catalyst, a mixture of the desired **108** and dihydropyran **109** from competitive cyclization onto the disubstituted olefin was isolated. Attempts at optimization involving the examination of the effect of alcohol protecting group (R^1, R^2, R^3) on the cyclization reaction were only moderately successful.

Taking advantage of the mild nature of the RCM conditions and the ruthenium catalysts, Crimmins *et al.* were able to successfully complete the synthesis of isolaurallene by incorporating an epoxide rather than an alkene into the RCM [25]. This resulted in the generation of oxonene **114** in 94% yield (Scheme 3.16). Epoxide **114** was subsequently converted into isolaurallene.

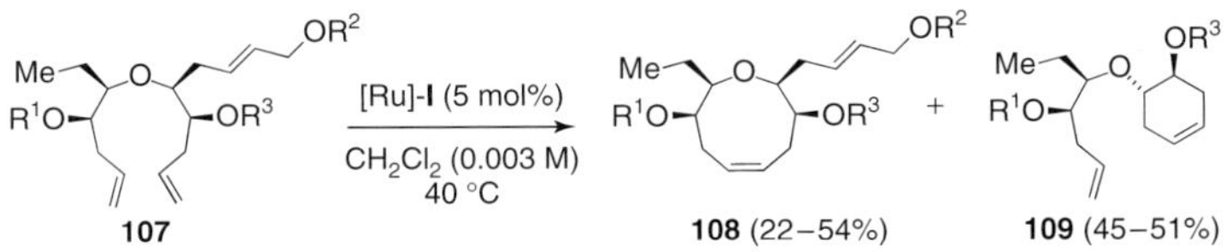

Scheme 3.15 Oxonenes from trienes.

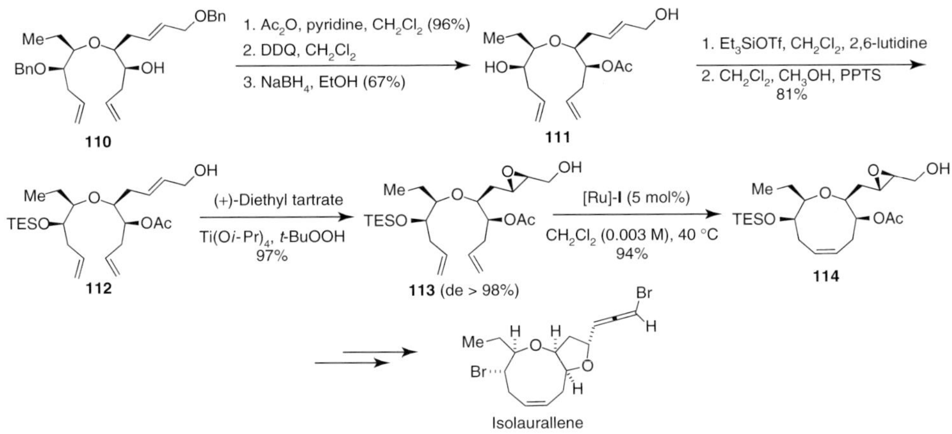

Scheme 3.16 Epoxy diene RCM to isolaurallene.

In addition to utilizing aldol-RCM sequences to access medium rings, Crimmins *et al.* have also generated medium rings by coupling glycolate oxazolidinone alkylations with RCM reactions [26]. As illustrated in Scheme 3.17, Crimmins and Emmitte's synthesis of (+)-laurencin began with the allylation of **118** to give the cyclization precursor **119**. Cyclization using [Ru]-**I** catalyst gave oxocine **120** in 94% yield. Subsequently, **120** was transformed into laurencin in an additional 10 steps.

Other research groups have also employed asymmetric glycolate alkylation/RCM strategies to synthesize medium rings that are relevant to the laurencin family. For example, Lee *et al.* have utilized such a strategy in the synthesis of (−)-isoprelaurefucin [27]. As outlined in Scheme 3.18, the allylation or prenylation of the enolate derived from glycolate amide **121** provided the corresponding alkylated RCM precursor **122** with high diastereoselectivity. Of interest was that the nature of the secondary hydroxyl-protecting group on **121** was important in this reaction. With TIPS (triisopropyl silyl) rather than PMB (para-methoxy benfyl), the other diastereomer predominated. Subsequent to the alkylation, RCM using [Ru]-**I** catalyst gave the desired oxepine **123** in 65% overall yield for the two steps. Further manipulation of **123** resulted in the generation of (−)-isoprelaurefucin.

Clark *et al.* have also employed a glycolate alkylation–RCM strategy to synthesize medium ring ethers (Scheme 3.19) [28]. Specifically, their work to (+)-laurenyne's

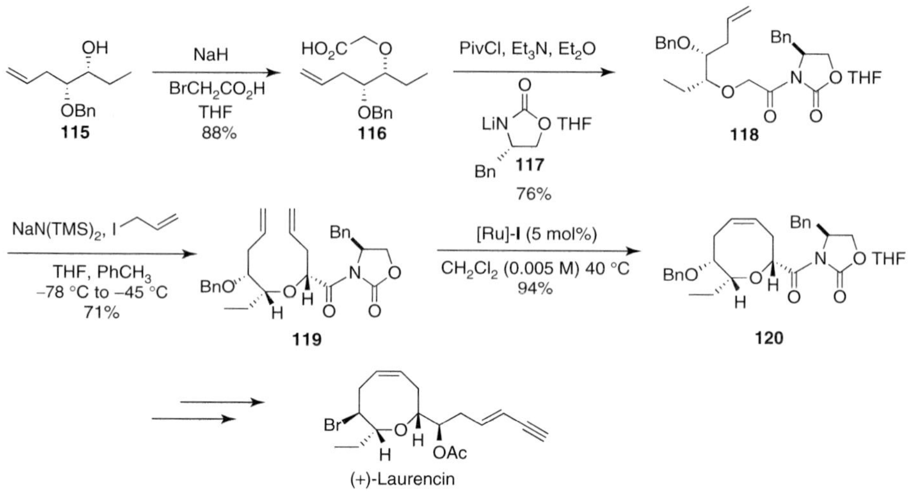

Scheme 3.17 Synthesis of (+)-laurencin.

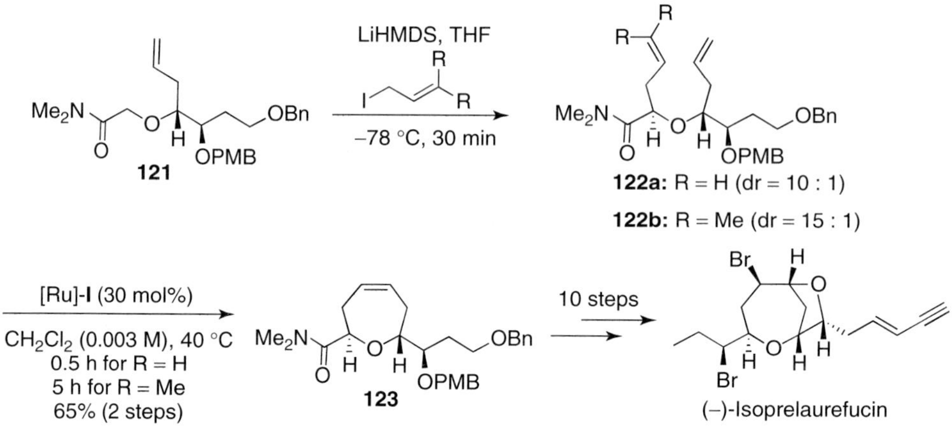

Scheme 3.18 Synthesis of (−)-isoprelaurefucin.

oxocine ring involved the coupling of a diastereoselective enolate allylation to **125** with RCM to give **126** using either the [Ru]-**I** or [Ru]-**II** catalyst. With [Ru]-**II** catalyst, the yield for the RCM transformation was enhanced when the alcohol was protected as a TBDPS (tert-butyl diphenyl silyl) ether. For the cyclization using [Ru]-**I** catalyst, the yield was enhanced when the catalyst was added to the reaction in several small portions.

Adsool and Pansare coupled diastereoselective morpholine dione addition chemistry with diene RCM using [Ru]-**II** catalyst to generate oxocine **133** (Scheme 3.20)

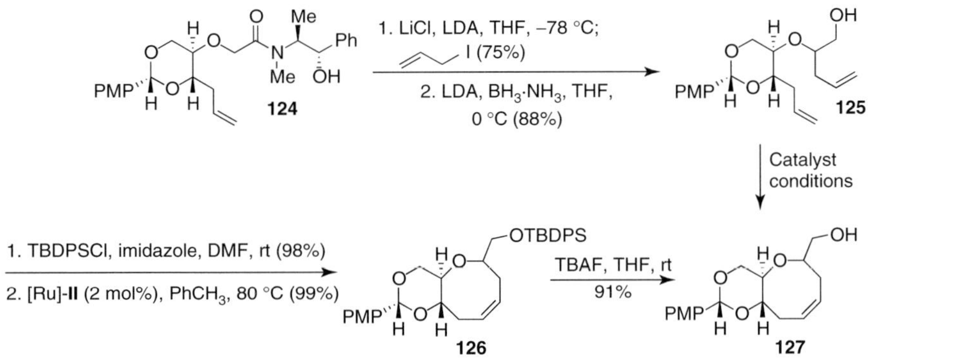

Entry	Catalyst	Conditions	Yield
1	[Ru]-I (5 mol%)	CH_2Cl_2, rt	48(%)
2	[Ru]-I (10 mol%)	CH_2Cl_2, rt, catalyst added (5 × 2 mol%)	63(%)
3	[Ru]-II (5 mol%)	$PhCH_3$, 80 °C	75(%)

Scheme 3.19 Alkylation–RCM strategy to (+)-laurenyne.

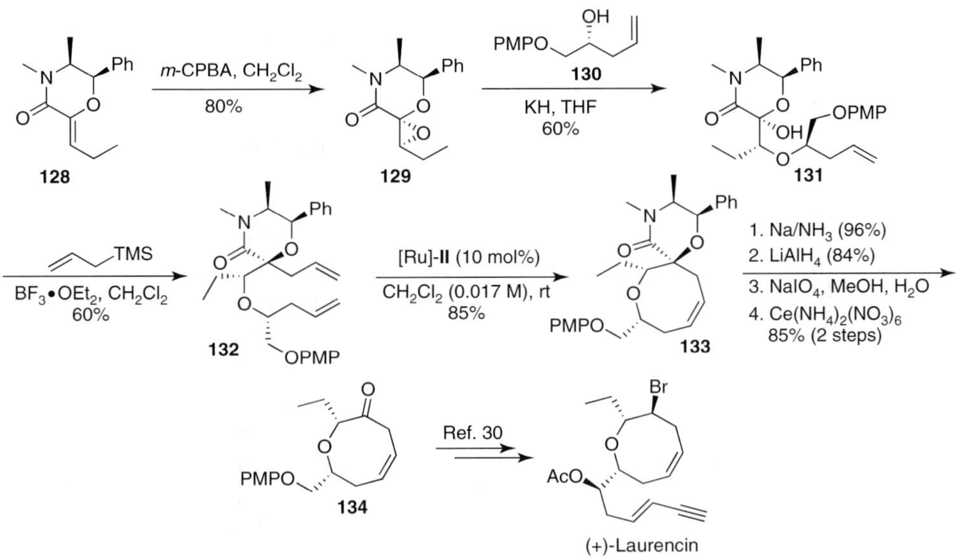

Scheme 3.20 Formal synthesis of laurencin.

[29]. Oxocine **133** was converted into ketone **134**, thus intercepting an intermediate that had previously been synthesized by Overman *et al.* on their route to laurencin [30]. In addition to the key RCM reaction, Adsool and Pansare also effectively utilized hemiketal **131** in a coupling reaction with allylsilane to give the RCM precursor.

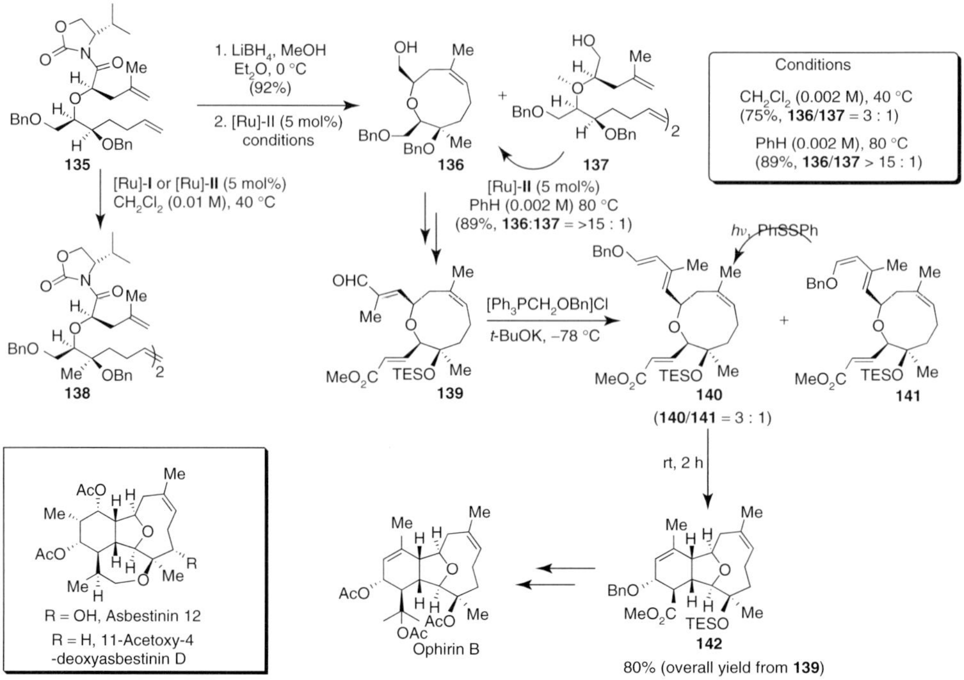

Scheme 3.21 Synthesis of ophirin B.

3.4 Eunicellins/Eleutherobin

In another beautiful illustration of the power of their RCM methodology, Crimmins *et al.* have coupled asymmetric glycolate alkylation/RCM with intramolecular Diels–Alder cycloadditions to synthesize members of the eunicellins and asbestinins [31]. As outlined in Scheme 3.21, the synthesis of ophirin B began with diene **135** that was obtained from Crimmins *et al.* previously discussed glycolate alkylation protocol [32]. When diene **135** was exposed to either [Ru]-**I** or [Ru]-**II** catalyst at 40 °C, exclusive generation of dimer **138** was observed. Reasoning that an unfavorable conformation about the oxazolidinone might be hindering cyclization, Crimmins *et al.* reduced the oxazolidinone and studied the cyclization chemistry of the resulting alcohol. Pleasingly, the alcohol from **135** underwent the desired cyclization to oxonene **136** when [Ru]-**II** catalyst was used at 40 °C but a large quantity of dimer **137** as it was also formed. The authors subsequently showed that the amount of dimer could be minimized by carrying out the reaction at 80°C in benzene. That compound **137** was the kinetic product was demonstrated through its conversion into the desired product **136** by resubjecting **137** to [Ru]-**II** catalyst. Having solved the oxonene generation, the synthesis of the eunicellin

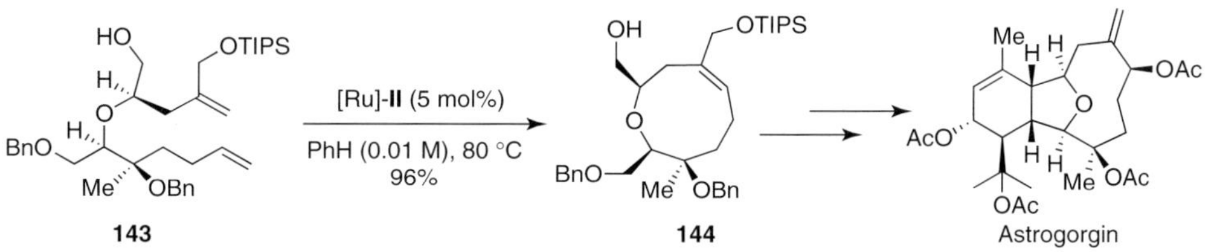

Scheme 3.22 Synthesis of astrogorgin.

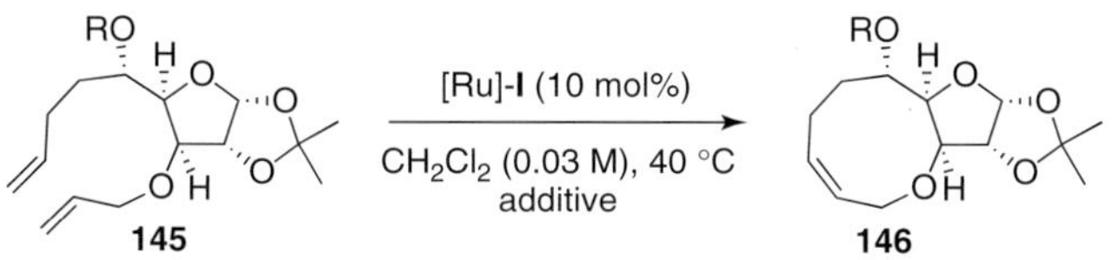

Without additive
R = TBS (30%)
R = Me (33%)

Additive: $Ti(Oi\text{-}Pr)_4$
R = TBS (83%)
R = Me (63%)

Scheme 3.23 Generation of Eleutherobin's oxonene ring.

six-membered ring from a Diels–Alder strategy was examined next. To this goal, **136** was converted into enal **139**. When **139** was subjected to benzyloxymethylenetriphenylphosphorane, a 3 : 1 mixture of alkene isomers **140/141** was formed from which only the *E/E*-stereoisomer **140** underwent the desired intramolecular Diels–Alder cycloaddition to give **142** upon standing at room temperature for 2 hours. Subsequently, it was found that the undesired *Z/E*-isomer **141** could be equilibrated into a 1 : 1 mixture of dienes and ultimately into the cycloadduct by exposing it to light and PhSSPh. Utilizing this protocol on the mixture until all the diene had reacted resulted in an 80% overall yield of adduct **142**. Tricycle **142** was converted into ophirin B in an additional six steps. Crimmins and Ellis employed a very closely related approach in their generation of 11-acetoxy-4-deoxyasbestinin D and asbestinin-12 [33].

With the exception that an allylic ether was employed in the RCM reaction, Crimmins *et al.* employed a nearly identical strategy to generate astrogorgin (Scheme 3.22) [34].

Using the tubulin stabilizing natural product eleutherobin as an inspiration, Kaliappan and Kumar have investigated the generation of medium rings that are fused to ribose subunits using RCM [35]. Initially targeting eleutherobin's oxonene ring, they exposed **145** to [Ru]-**I** catalyst and realized low yield of cyclic product (Scheme 3.23). Following the discovery that the alcohol protecting group had little impact on the overall efficiency of the reaction as substrates with R = TBS and Me performed similarly, the effect of additives on the reaction was examined and

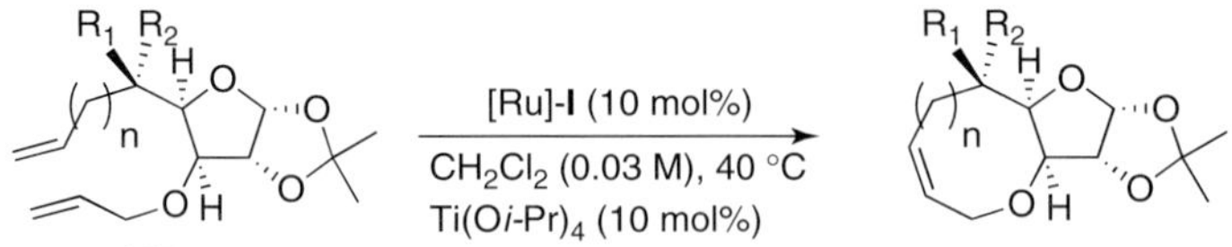

147

148a: $n = 0$, $R_1 = H$, $R_2 = OMe$ (0%, no reaction)
148b: $n = 0$, $R_1 = H$, $R_2 = OH$ (84%)
148c: $n = 1$, $R_1 = OMe$, $R_2 = H$ (68%)
148d: $n = 3$, $R_1 = H$, $R_2 = OMe$ (82%)

Scheme 3.24 The generation of bicyclic fused ethers using RCM.

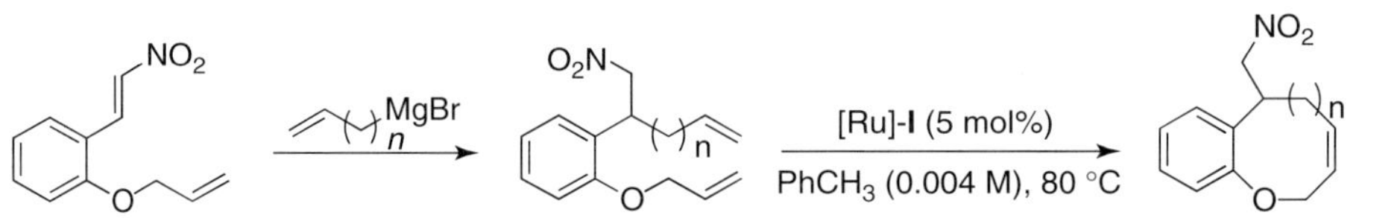

149

150a: $n = 1$ (78%)
150b: $n = 2$ (75%)

151a: $n = 1$ (60%)
151b: $n = 2$ (65%)

Scheme 3.25 Synthesis of benzo-fused medium rings using sequential Michael-RCM technology.

it was found that catalytic amounts of Ti(O*i*-Pr)$_4$ dramatically increased the yield of **146**. Presumably, Ti(O*i*-Pr)$_4$ affects the conformation of the substrate through chelation [36].

Kaliappan and Kumar were also able to successfully synthesize oxocines and oxecines using this strategy (Scheme 3.24) [35]. Interestingly, the synthesis of the corresponding oxepine required that the alcohol not be protected, presumably a result of the influence of allylic substitution on RCM. As with the oxonenes, the use of Ti(O*i*-Pr)$_4$ as an additive was crucial to the success of these reactions.

3.5
Helianane

Medium-sized oxygenated rings that are fused to aromatic systems have also been targeted using RCM. Deb *et al.* have developed an approach to the oxocines and oxonenes that are present in the helianane family, which couple β-nitrostyrene Michael addition chemistry with RCM (Scheme 3.25) [37]. For example, the addition of Grignard reagents containing a terminal olefin to the nitrostyrene **149** resulted in the generation of the corresponding Michael adducts **150**. RCM cyclization gave the corresponding medium rings **151** in reasonable yields.

Macias *et al.* have developed a stereoselective route to heliannuol A using diene metathesis to generate the requisite oxocine (Scheme 3.26) [38]. The cyclization precursor was obtained from the sequential propargyl ether formation and allyl Grignard addition to the acetophenone derivative **152**. Reductive removal of the

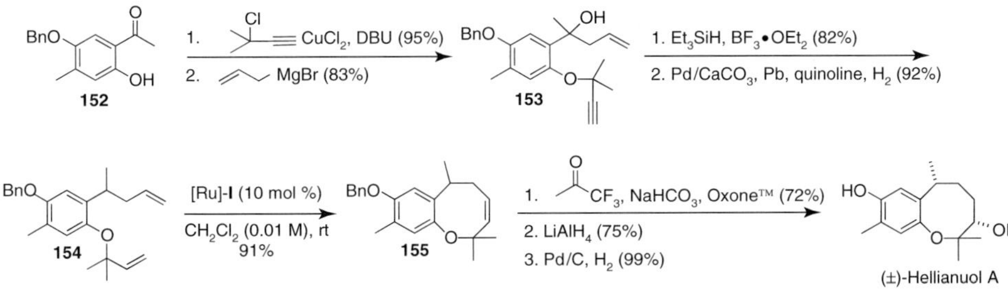

Scheme 3.26 Synthesis of heliannuol A.

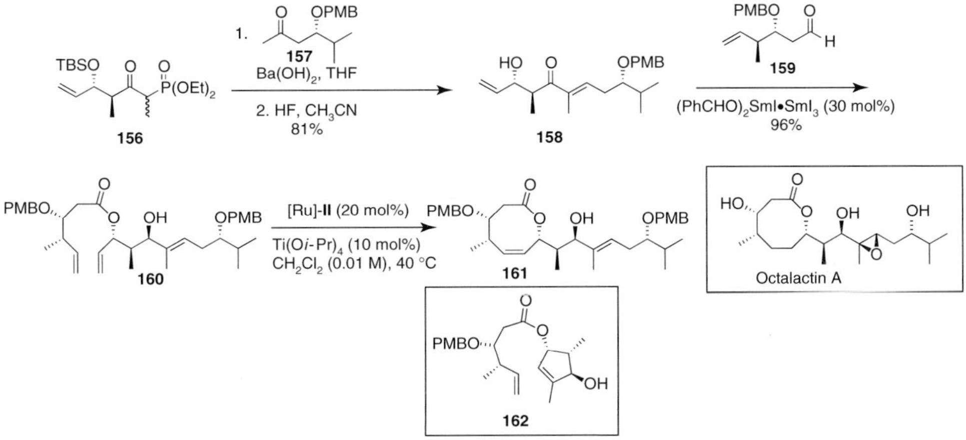

Scheme 3.27 Synthesis of the octalactin A core using a sequential Evans–Tishchenko/RCM strategy.

benzylic alcohol and Lindlar reduction of the alkyne resulted in the formation of the cyclization precursor **154**. RCM using [Ru]-**I** catalyst provided heliannuol A after epoxidation, reduction, and removal of the benzyl protecting group.

3.6
Octalactin A

Medium ring lactone containing natural products have also been the target of RCM reactions. In their work targeting octalactin A, Hulme *et al.* utilized a sequential Evans–Tishchenko ester formation/RCM sequence to generate octalactin's oxocine ring (Scheme 3.27) [39]. From phosphonate **156**, Horner–Wadsworth–Emmons coupling with methyl ketone **157** gave enone **158**. The use of the Evans–Tishchenko reaction between **158** and **159** gave the ester cyclization precursor **160** as a single diastereomer in 96% yield. RCM using [Ru]-**II** catalyst resulted in the generation of lactone **161** and cyclopentene **162** (1 : 1 mixture). Although the yield of the desired cyclization product was not improved by using other RCM catalysts, the authors

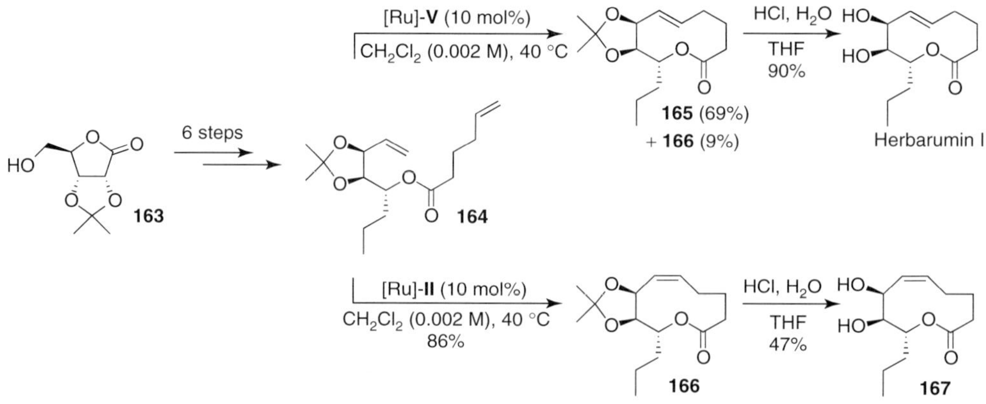

Scheme 3.28 Synthesis of herbarumin I.

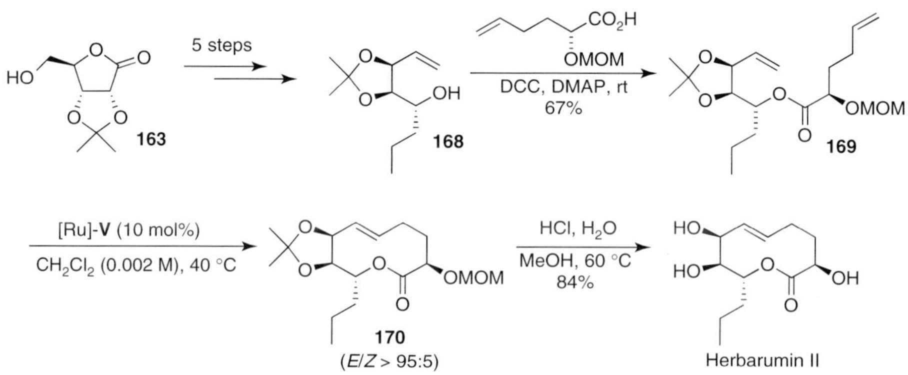

Scheme 3.29 Synthesis of herbarumin II.

were able to overcome this problem by utilizing Ti(O*i*-Pr)$_4$ as a cocatalyst giving **161** in 70% yield. This additive presumably affects the conformation of the cyclization precursor through chelation [36].

3.7
Microcarpalide and the Herbarums

Concerned with the generation of the thermodynamically less stable *E*-isomer of herbarumin I, Fürstner *et al.* examined the ability of both [Ru]-**V** and [Ru]-**II** catalysts to effect the cyclization of diene **164** (Scheme 3.28) [40]. Impressively, [Ru]-**V** catalyst gave *E*-alkene isomer **165** as the major product in 69% yield along with 9% of *Z*-isomer **166**. The authors found that the amount of **166** did not increase with time indicating that the observed *E* : *Z* ratio was the result of a thermodynamic control. Interestingly, the more reactive [Ru]-**II** catalyst resulted in the exclusive generation of *Z*-isomer **166**.

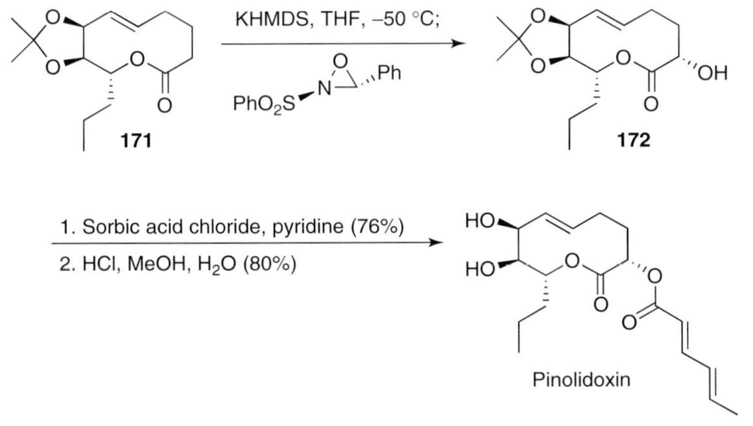

Scheme 3.30 Conversion of the herbarumin I precursor into pinolidoxin.

Fürstner *et al.* attempted the generation of herbarumin II from herbarumin I via the oxidation of a herbarumin I enolate, which resulted in the generation of the C-2 herbarumin II epimer. For this reason, the hydroxyl group was incorporated into the RCM precursor and herbarumin II was generated using [Ru]-**V** catalyst as illustrated in Scheme 3.29 [40]. Of note was the formation of less than 5% of the corresponding *Z*-alkene isomer during the cyclization event.

In addition to forming the herbarumins, Fürstner *et al.* were able to synthesize pinolidoxin by carrying out the aforementioned hydroxylation reaction on herbarumin I precursor **71** (Scheme 3.30) [40]. This work effectively corrected previously misassigned stereochemistry of pinolidoxin [41].

Davoli *et al.* have combined a diastereoselective homologation reaction that was initially developed by Matteson *et al.* with diene RCM to generate (−)-microcarpalide [42]. As illustrated in Scheme 3.31, the use of Matteson *et al.* chemistry resulted in the generation of alcohol **177** after oxidation of **176**. Esterification with acid **178** and RCM using [Ru]-**I** catalyst gave a 2 : 1 mixture of alkene isomers favoring the *E*-isomer. Similar results were observed with [Ru]-**V** catalyst.

Gurjar *et al.* have also utilized diene RCM in the generation of the medium ring lactones microcarpalide and herbarumin III (Scheme 3.32) [43]. To microcarpalide, esterification of **180** with **181** led to the cyclization precursor **182** in 76% yield. RCM using [Ru]-**I** catalyst gave the corresponding 10-membered lactone and microcarpalide after the removal of the benzyl and MEM (methoxy ethoxy methyl) protecting groups. The obvious difference between the work of Gurjar *et al.* and Davoli *et al.* lies in the choice of alcohol protecting groups, MEM and Bn, respectively, again demonstrating the dramatic effect of substrate substitution on the cyclization reaction.

Although a related sequence of reactions was utilized to generate herbarumin III, of note was the presence of the free allylic hydroxyl group in the

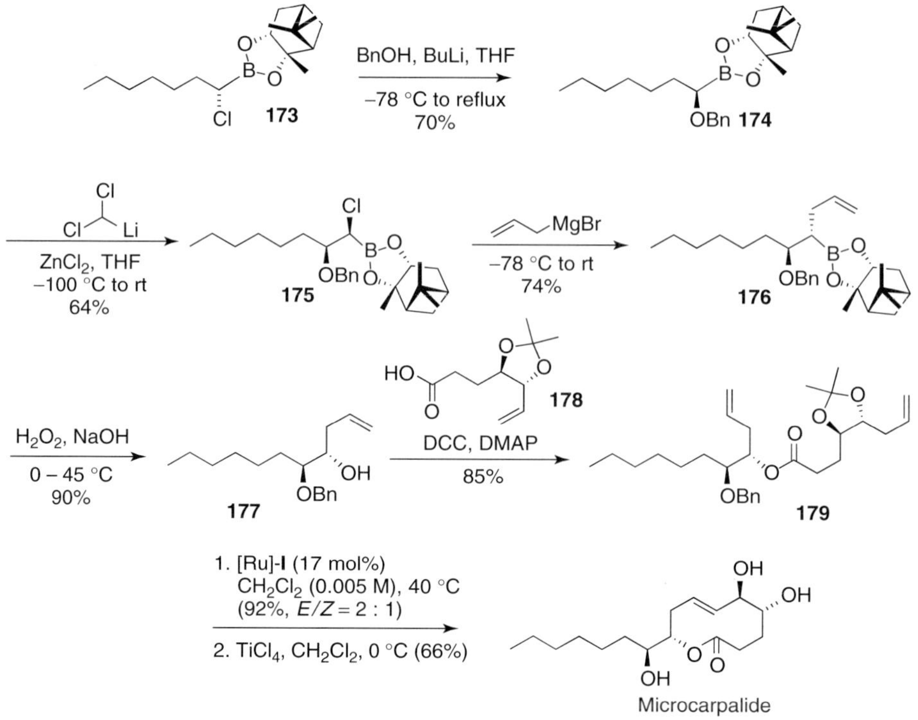

Scheme 3.31 Synthesis of (−)-microcarpalide.

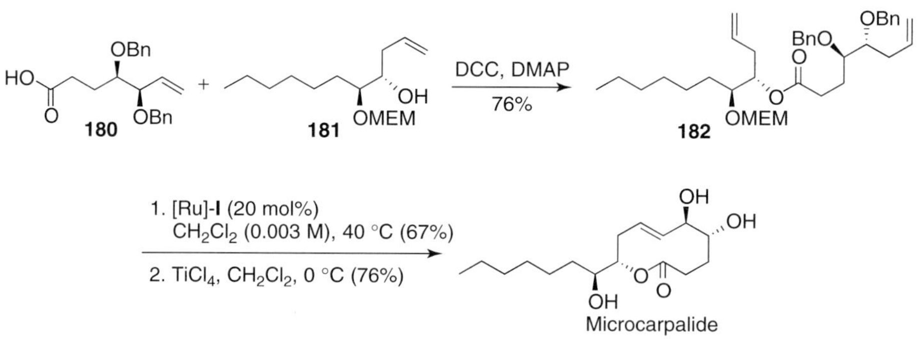

Scheme 3.32 Diene cyclization to microcarpalide.

cyclization reaction and that [Ru]-**II** catalyst was much more successful than [Ru]-**I** catalyst (Scheme 3.33) [43]. When compared with Fürstner *et al.* work (Scheme 3.28), it was interesting that [Ru]-**II** catalyst gave the desired *E*-olefin in this cyclization.

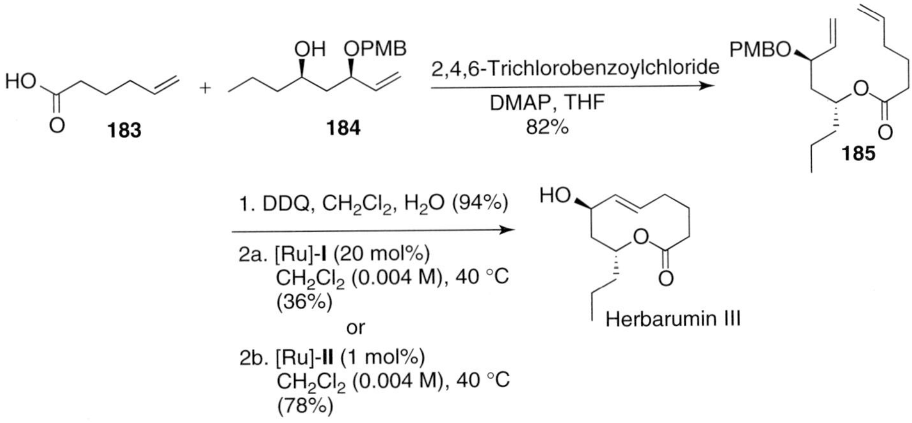

Scheme 3.33 Diene cyclization to herbarumin III.

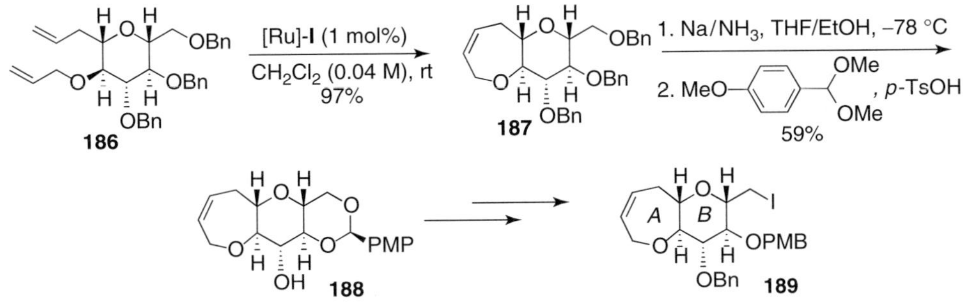

Scheme 3.34 Synthesis of the ciguatoxin A-ring.

3.8 Marine Ladder Toxins

3.8.1 Ciguatoxin

RCM has played an important role in the synthesis of members of the ladder toxin family of natural products. Arguably, the most impressive use of RCM to generate medium rings in the total synthesis of ladder toxins has come from the Hirama laboratory and their synthesis of ciguatoxin. Their efforts to synthesize ciguatoxin are outlined in Schemes 3.34–3.40, which began with the synthesis of the A-ring from C-glucoside **186** (Scheme 3.34) [44]. Cyclization using 1 mol% of the Grubbs catalyst [Ru]-**I** gave oxepine **187** in 97% yield. Reductive removal of the benzyl ethers was accomplished in 59% yield after the generation of PMP (para-methoxyphenyl) acetal **188**. The allylic ether in **187** is sensitive and was partially cleaved (21%) when exposed to Na/NH_3. Additional protecting group manipulation gave the A,B-precursor **189**.

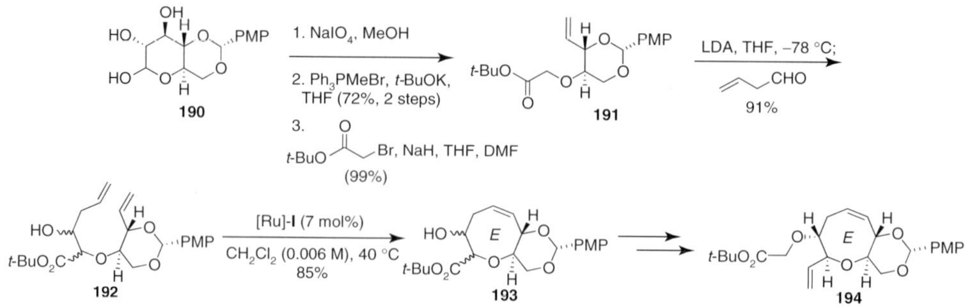

Scheme 3.35 Synthesis of the ciguatoxin E-ring.

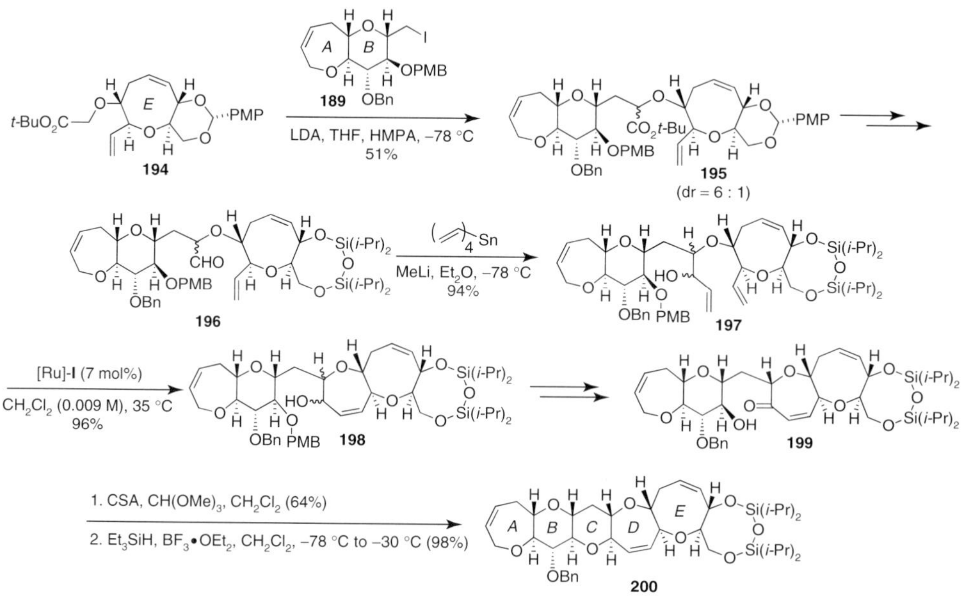

Scheme 3.36 Synthesis of the ciguatoxin C,D-ring.

The generation of the eight-membered ciguatoxin E-ring coupled glycolate aldol chemistry with RCM and began with protected D-glucose (Scheme 3.35) [44]. Oxidative cleavage and Wittig reaction gave **191** after glycolate ester formation. Aldol condensation gave **191** as a mixture of three diastereomers. RCM was carried out on the mixture using [Ru]-**I** catalyst to give the eight-membered ring substrate **193** in 85% yield. Following their separation, each of the diastereomers was independently processed to the E-ring coupling precursor **194**.

Using RCM, Hirama, Inoue, *et al.* generated the seven-membered ciguatoxin D-ring after coupling the A,B- and E-precursors (Scheme 3.36) [44]. Alkylation of the E-ring glyoxylate enolate from **194** with A,B-ring iodide **189** resulted in **195**

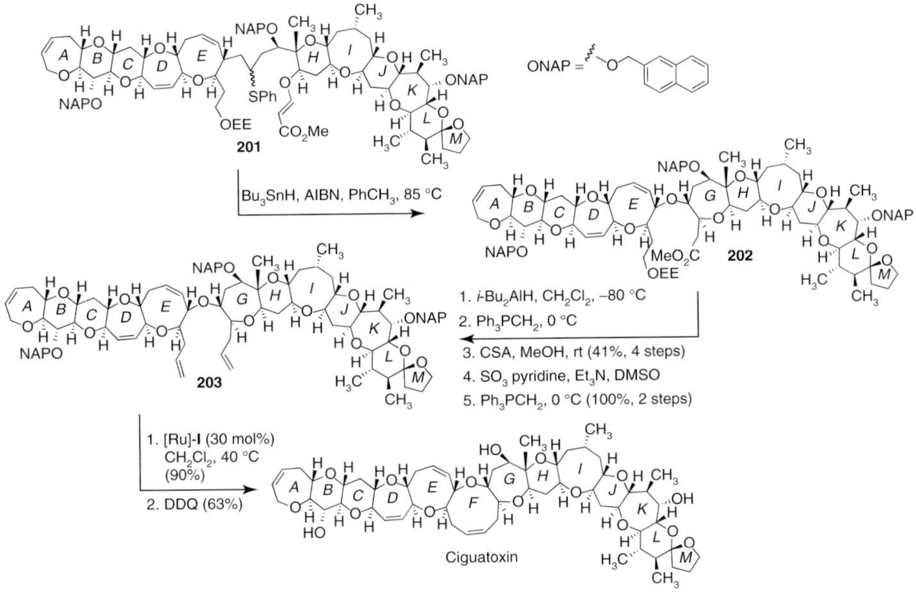

Scheme 3.37 Radical cyclization–RCM sequence to ciguatoxin.

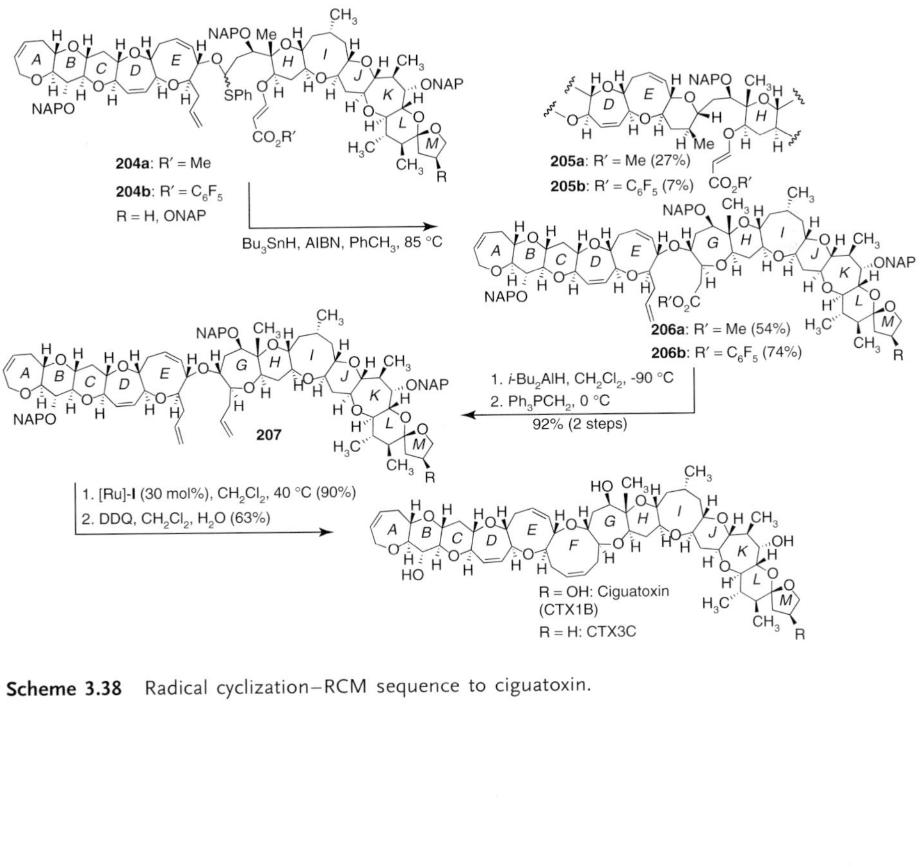

Scheme 3.38 Radical cyclization–RCM sequence to ciguatoxin.

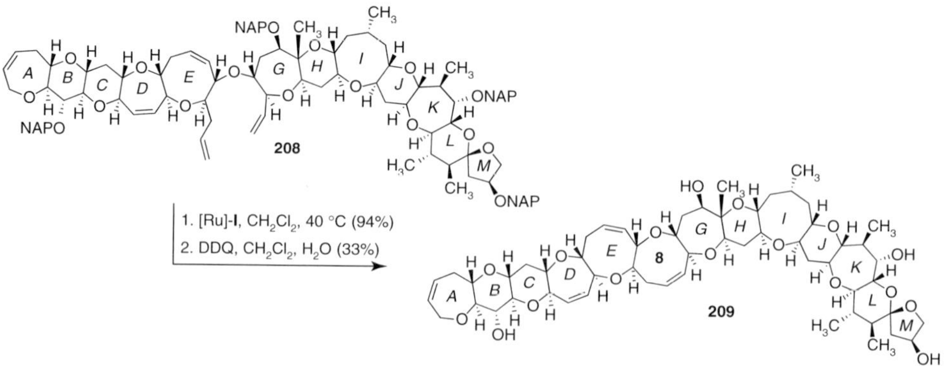

Scheme 3.39 Radical cyclization–RCM sequence to ciguatoxin analogs.

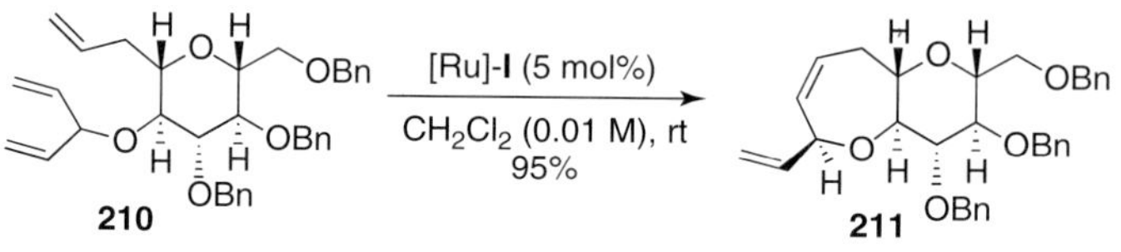

Scheme 3.40 Diastereoselective RCM to the ciguatoxin A-ring.

as a 6 : 1 mixture of diastereomers. Following the manipulation of protecting group and oxidation state, vinyllithium addition to **196** gave the allylic alcohol RCM precursor **197** as a mixture of four diastereomers. Cyclization using [Ru]-**I** catalyst (7 mol%) provided a 96% yield of the D-ring oxepine as **198**. All four diastereomers were converted into enone **199** that was then transformed into the A–E ciguatoxin subunit **200** using an acid-catalyzed cyclization and reduction sequence.

Hirama *et al.* were able to combine the A–E and H–J precursors by using a thioacetal/free-radical cyclization strategy to give the G-ring oxepine **202** (Scheme 3.37) [45]. Additional manipulation resulted in the diene cyclization precursor **203**. In what is arguably the most impressive demonstration of RCM to medium rings (four alkenes, nine-membered ring formation), the cyclization of **203** using [Ru]-**I** catalyst gave the F-ring and the entire ciguatoxin skeleton in 90% yield. In contrast to this result, the attempted cyclization of **203** with [Ru]-**II** catalyst resulted in a lower yield of the oxonene and the generation of significant quantities of a by-product came from a competitive RCM–ROM (ring-opening metathesis) reaction onto the E-ring alkene [46]. The synthesis of ciguatoxin was completed after the global removal of the napthyl protecting groups.

Hirama *et al.* also utilized dienyl substrate **204a** as a precursor to the F,G-ciguatoxin ring system (Scheme 3.38) [47]. Although the use of **204a** enabled a more convergent coupling sequence, it suffered from a competitive free-radical cyclization reaction to generate the pyranyl substrate **205a**. Subsequently,

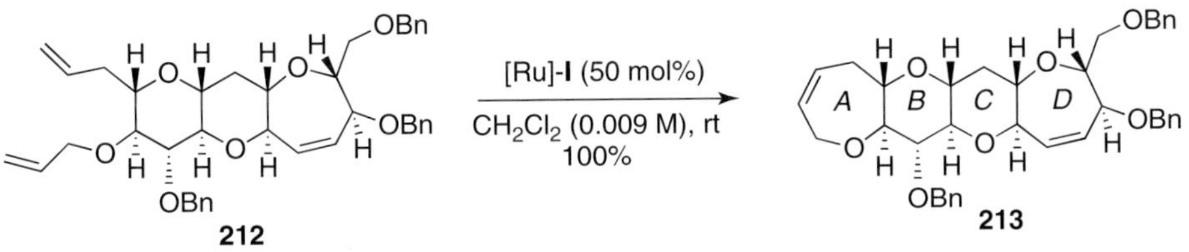

Scheme 3.41 Generation of the ciguatoxin A-ring.

α,β-unsaturated pentafluorophenyl ester **205b** was used as the radicalphile resulting in a 74% yield of the desired product **206b** along with 7% of pyran **205b**. Compounds **206a** and **206b** were converted into the cyclization precursor **207** as illustrated. RCM using [Ru]-**I** catalyst as before gave 51-hydroxy CTX3C and CTX1B (ciguatoxin) after the removal of the protecting group.

Their late stage RCM reaction to the F-ring enabled Inoue, Hirama, *et al.* to examine the effect of the size of the F-ring on ciguatoxin's toxicity and ability to bind to voltage-gated sodium channels (VGSCs) [48]. As illustrated in Scheme 3.39, the synthesis of an oxocine analog of the F-ring **209** was accomplished in 93% yield employing [Ru]-**I** catalyst. Interestingly, **209** showed a 173-fold lower affinity for VGSC binding than did CTX3C and was not toxic to mice. The importance of the nine-membered F-ring was further demonstrated when an acyclic variant of CTX3C also showed a much lower affinity for VGSC binding.

In addition to their other accomplishments, Hirama *et al.* utilized an interesting diastereoselective RCM reaction in their synthesis of the ciguatoxin (CTX1B) A-ring [49]. When triene **210** (available from triacetyl-D-glucose) was subjected to [Ru]-**I** catalyst, oxepine **211** was isolated in 95% yield as a single diastereoisomer. In the presence of an external olefin and [Ru]-**I** catalyst, compound **211** subsequently underwent a competitive ROM reaction, illustrating the sensitivity of medium rings having olefins imbedded in them to RCM conditions.

Sasaki *et al.* have utilized RCM to generate the seven-membered ciguatoxin A-ring (Scheme 3.41) [50]. From the intact B–D subunit **212** that was synthesized using a novel enol phosphonate alkyl Suzuki coupling sequence, RCM using [Ru]-**I** catalyst gave tetracycle **213**.

In the course of synthesizing a model of ciguatoxin F-ring, Sasaki *et al.* developed the radical cyclization, RCM sequence that was used by Hirama *et al.* in their ciguatoxin synthesis (Scheme 3.42) [51]. The conversion of acetal **214** into seleno- and thio-mixed acetals **215a** and **215b** was accomplished using i-$Bu_2AlSePh$ and i-Bu_2AlSPh, respectively. Interestingly, the bicyclic acetal in **214** was necessary for conversion into the mixed acetal using the Al reagent; when monocyclic acetals were used, mixtures of products were isolated. Both **216a** and **216b** were converted into the corresponding oxepine **217** via 7-*exo*-trig radical cyclizations. Elaboration of **217** gave the cyclization precursor **218**, which when treated with [Ru]-**I** catalyst provided the ciguatoxin F-ring as **219** in 61% yield.

By coupling a reductive isomerization sequence with RCM, Fujiwara *et al.* were also able to generate ciguatoxin's F-ring oxonene (Scheme 3.43) [52]. From oxepine

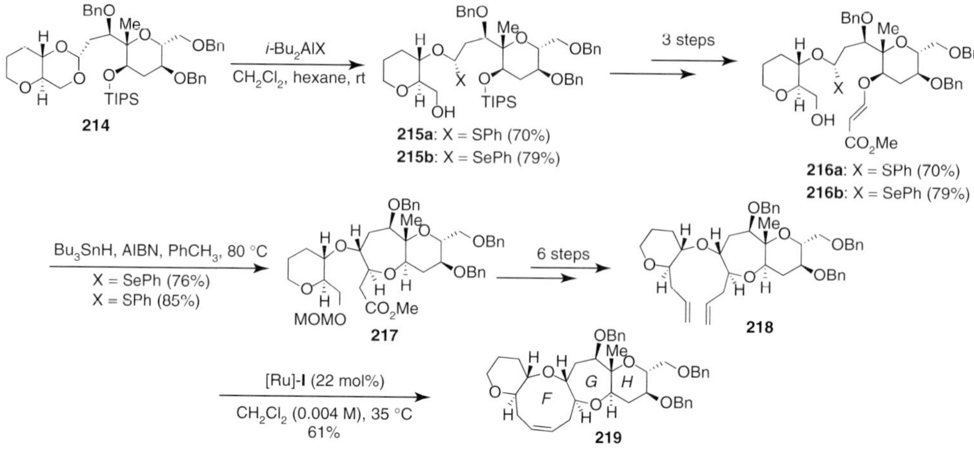

Scheme 3.42 Synthesis of the ciguatoxin F–H subunit.

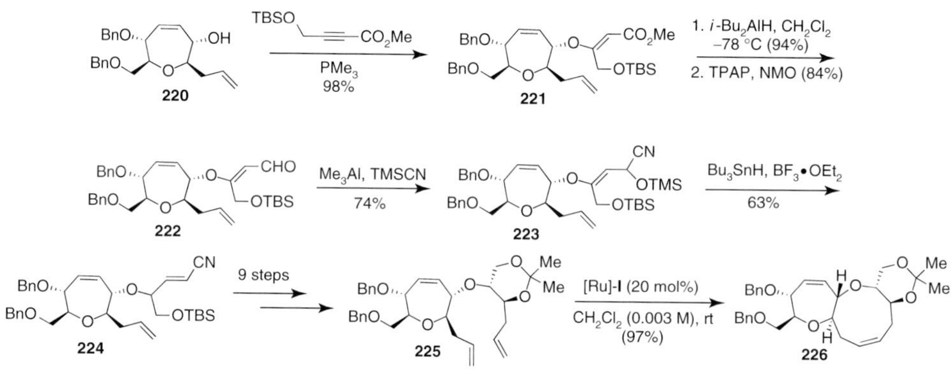

Scheme 3.43 Synthesis of ciguatoxin F-ring.

220, conjugate addition and conversion of the resulting ester to the corresponding aldehyde gave **222**. Cyanohydrin formation was followed by reduction with Bu_3SnH to give acrylonitrile derivative **224** as a 1 : 1 mixture of diastereomers. Conversion of **224** into the F-ring cyclization precursor **225** involved an epoxidation, Wittig sequence, and the reduction of an intermediate π-allyl complex. The diastereomers resulting from the reduction of **223** were separated after the formation of a 1,3-diol acetonide. Diastereomer **225** was cyclized and subjected to [Ru]-**I** catalyst (20 mol%) to provide the F-ring oxonene **226** (97%).

Another example of some of the problems associated with the use of RCM in the formation of medium rings comes from the Fujiwara *et al.* work on ciguatoxin [52]. The generation of the E-ring oxepine with the F-ring intact resulted in the generation of not only the desired product (**228**) but also a significant quantity of bis-oxepene **229** from a competitive ROM/RCM tandem sequence (Scheme 3.44).

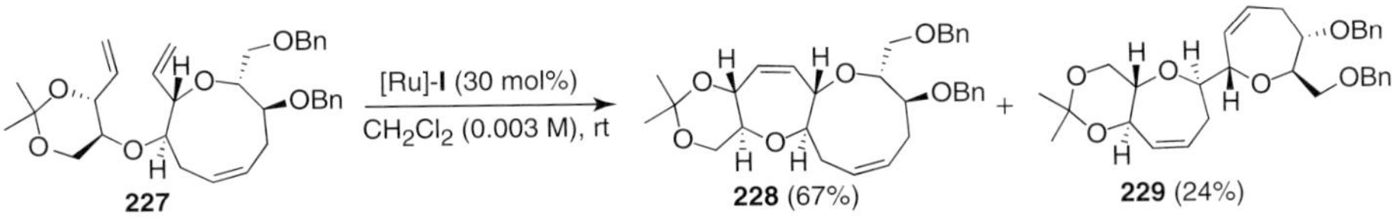

Scheme 3.44 Ciguatoxin's E,F-ring system.

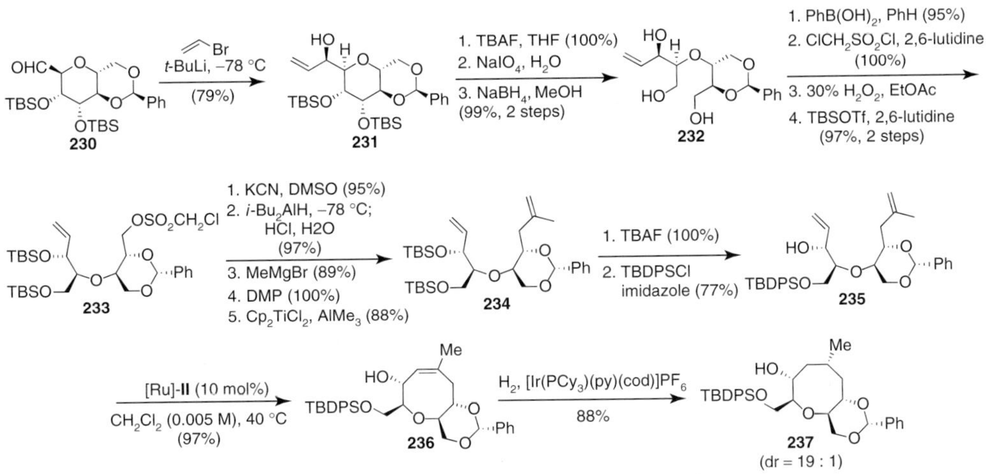

Scheme 3.45 RCM to ciguatoxin I-ring.

Fujiwara and Murai have used tri-O-acetyl-D-glucal as a precursor to both the ciguatoxin F-ring and the I-ring oxocine [53]. The coupling of aldehyde **230** with vinyl lithium gave **231** (Scheme 3.45). Oxidative cleavage of the pyran ring gave **232** after reduction. The manipulation of the protecting group and the formation of the disubstituted olefin gave the cyclization precursor **235**. Impressively, RCM using [Ru]-**II** catalyst resulted in the requisite trisubstituted olefin and oxocine **236** in high yield. Interestingly, hydrogenation of **236** using Crabtree's catalyst showed unusual stereoselectivity. Rather than the anticipated directed reduction, the reaction gave **237**. Also interesting was that the use of PtO_2 as a catalyst for the hydrogenation resulted in the formation of the epimer of **237** as the major product.

A related sequence of reactions was used to generate the ciguatoxin F-ring oxonene as illustrated in Scheme 3.46 [53]. RCM using [Ru]-**I** catalyst gave oxonene **241** in 98% yield.

In a related fashion to their aforementioned Ireland–Claisen/RCM sequence (Schemes 3.4–3.6), Goto *et al.* have coupled a stereoselective [2,3]-Wittig rearrangement with RCM as a mean to synthesize the ciguatoxin E,F-ring system (Scheme 3.47) [54]. From **242**, conjugate addition provided **243** in 99% yield after one recycle. Stereoselective ketone reduction and alkylation with bromo-*t*-butyl acetate gave the rearrangement precursor **244**. The [2,3]-rearrangement of **244** was

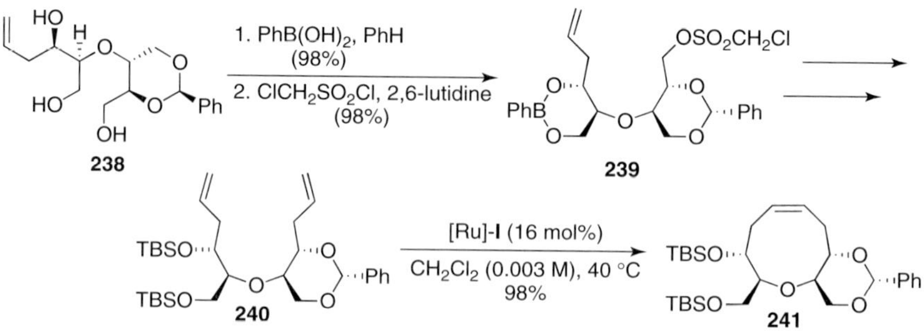

Scheme 3.46 Synthesis of the ciguatoxin F-ring.

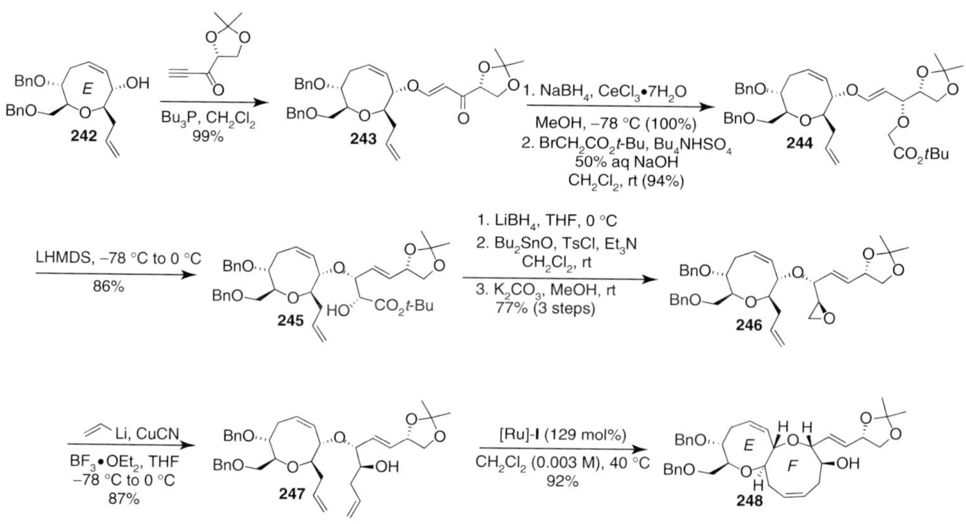

Scheme 3.47 [2,3]-Wittig/RCM sequence to ciguatoxin F-ring.

induced with LHMDS (lithium hexamethyldisilazide) to give **245** in 86% yield with almost complete diastereoselectivity. Reduction of the ester, epoxide formation, and coupling with vinyl cuprate provided the F-ring precursor **247**. Despite the presence of four olefins, [Ru]-**I** catalyst reacted chemoselectively with the terminal olefins to give the F-ring oxonene in 92% yield as **248**.

As part of their efforts targeting the A–E subunit of ciguatoxin CTX3C, Clark *et al.* utilized a two-directional RCM sequence to synthesize the A- and E-rings in a single flask (Scheme 3.48) [55, 56]. From glucal derivative **249**, the functional group manipulation resulted in triene–yne cyclization precursor **250**. Two-directional diene, ene–yne RCM using [Ru]-**II** catalyst gave A,B,C-tricycle **251** in 58% yield. Selective oxidation of the enol ether and enone formation gave the D-ring precursor **252**. RCM using [Ru]-**II** catalyst gave oxepine **253** in 70% yield. Functional group

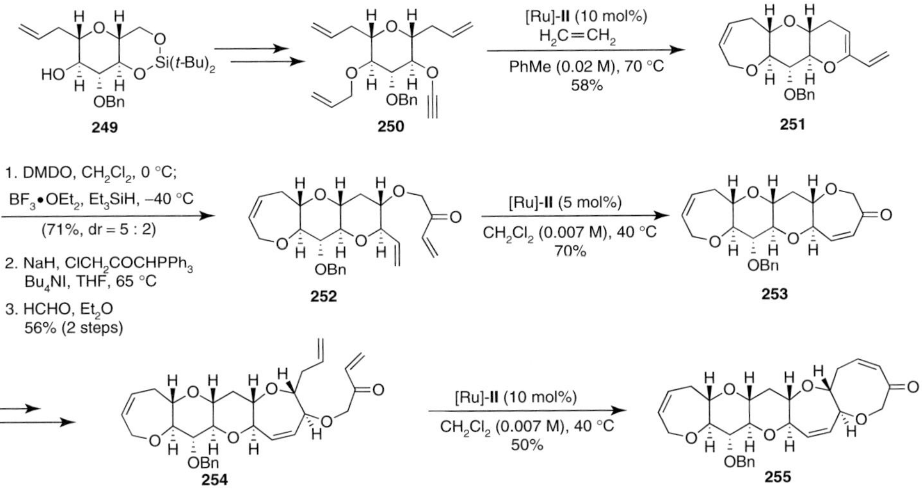

Scheme 3.48 Two-directional synthesis of the ciguatoxin A–E precursor.

manipulation and alkylation provided dienyl ketone **254** as the E-ring precursor. A third RCM reaction gave the E-ring oxocine and completed the synthesis of the A–E ciguatoxin precursor **255**.

3.8.2
Brevetoxin

Kadota, Yamamoto *et al.* have developed a novel subunit coupling strategy that combines an intramolecular allylstannane/acetal cyclization with diene RCM. This sequence of reactions was used to link the B–C- and F–G-rings of brevetoxin B (Scheme 3.49) [57]. Beginning with the B–C rings containing substrate **256**, *N*-chlorosuccinimide (NCS)-induced oxidation and AgOTf-catalyzed coupling with F,G-rings containing alcohol **258** gave **259** in 81% yield for the two steps. Removal of the TBS group and incorporation of an allylstannane moiety using allylstannyl ether **260** gave the cyclization precursor **261** following the elimination of MeOH. Intramolecular allylation catalyzed by AgOTf gave the D-ring as **262** in 84% yield as a 78 : 22 mixture with epimer **263**, respectively. Diene **262** was subjected to [Ru]-**I** catalyst to give oxepene **264** in 72% yield. The use of [Ru]-**II** catalyst in the cyclization of diene **263** gave a quantitative yield of oxepine **265**. Interestingly, **262** did not undergo RCM under the conditions used to cyclize **263**.

Kadota, Yamamoto *et al.* took a related approach to the generation of the H–I brevetoxin B rings and the completion of the synthesis (Scheme 3.50) [57]. Esterification of A–G subunit **267** with J,K subunit **266** gave ester **268**. Allylstannane generation and reduction of the ester gave the acetal cyclization precursor **269** after the formation of chloroacetate. As anticipated, the cyclization to the brevetoxin B I-ring was highly diastereoselective giving pyran **270** as a single diastereomer in

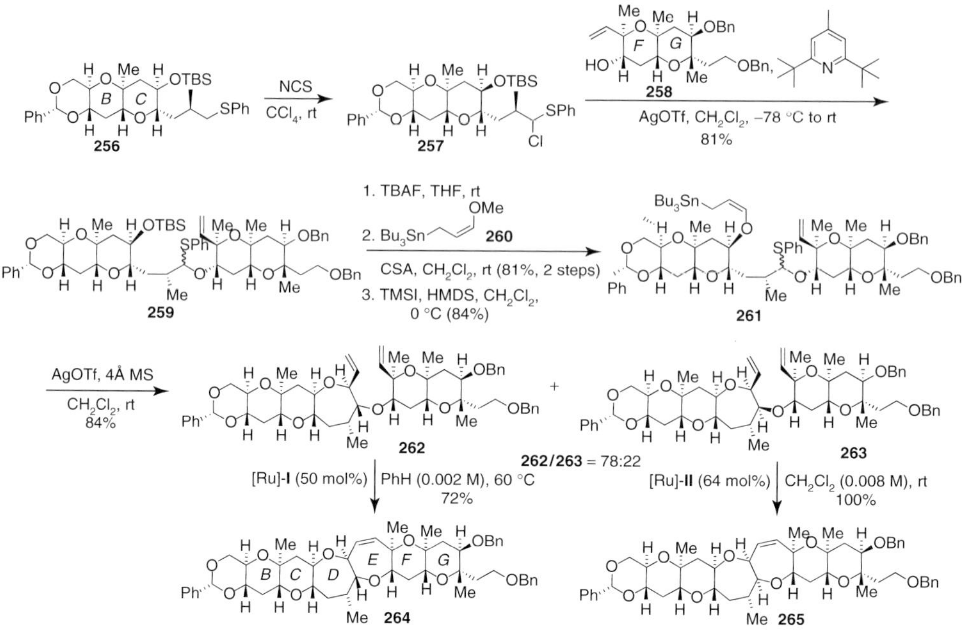

Scheme 3.49 The use of RCM in convergent subunit couplings to the brevetoxin B B–G subunit.

84% yield. RCM using [Ru]-I catalyst provided an 81% yield of the H-ring oxocine and the fully intact brevetoxin B skeleton as **271**. Oxidation of the A-ring and the allylic ether side chain completed the synthesis of brevetoxin B.

Crimmins *et al.* have used their sequential glycolate alkylation/RCM and glycolate aldol/RCM protocols to synthesize the marine natural product brevetoxin A. The synthesis of the B-ring began with the glycolate coupling of **272** with methallyl iodide to give adduct **273** (Scheme 3.51) [58]. Protecting group manipulation then led to the glycolate aldol precursor **274**. Coupling of **274** with benzyl iodomethyl ether gave **275** after the reductive removal of the auxiliary. Oxidation and vinyl Grignard addition gave the cyclization precursor **276** (major diastereomer shown) as a 3 : 1 mixture of diastereomers. Cyclization using [Ru]-**II** catalyst delivered the B-ring oxocine as tricyclic alkene **277** in 61% yield for the three steps. The diastereomers arising from the Grignard addition were separated after cyclization and the minor diastereomer was recycled to **277** using an oxidation–reduction sequence. The alcohol in **277** was then used as a handle to direct the hydrogenation of the alkene to give the desired C6 brevetoxin A precursor **278** and ultimately **279** after modification.

The E-ring was synthesized as outlined in Scheme 3.52 [58]. Beginning from glycolate aldol–derived adduct **280**, alkylation with bromoacetonitrile gave **281** following the reductive removal of the auxiliary. Oxidation of the primary alcohol, diastereoselective addition of allyl stannane to the resulting aldehyde, and hydrolysis gave lactone **282**. Protecting group and oxidation state manipulation gave **283** as

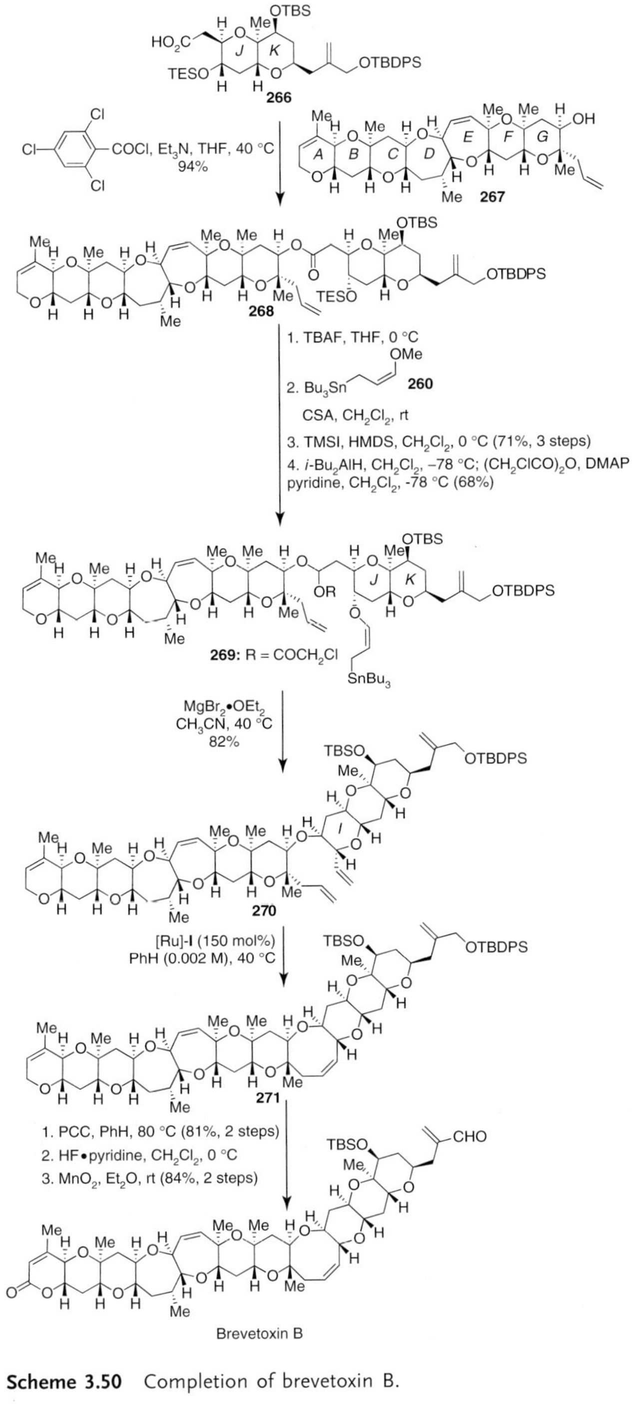

Scheme 3.50 Completion of brevetoxin B.

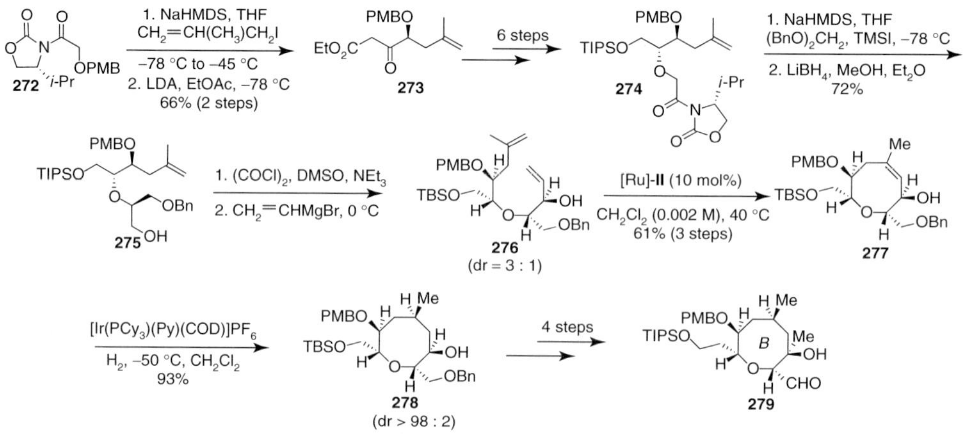

Scheme 3.51 The alkylation/RCM approach to the brevetoxin A B-ring.

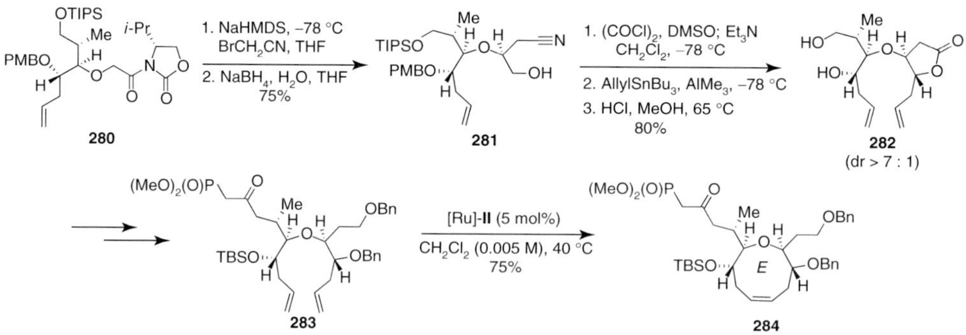

Scheme 3.52 Synthesis of the brevetoxin A E-ring.

the E-ring cyclization precursor. Exposure of **283** to [Ru]-**I** catalyst resulted in the smooth conversion to the corresponding oxonene **284**.

A related strategy was used to build the brevetoxin A G-ring oxocine **288** (Scheme 3.53) [59]. In an exceptionally convergent fashion, the A-, E-, and G-ring fragments (**279**, **284**, and **288**, respectively) were coupled and carried forward to brevetoxin A [60].

3.8.3
Gambierol, Gambieric Acid, Olefinic-ester Cyclizations

Johnson *et al.* synthesis of gambierol employed enol ether–olefin RCM to generate the seven-membered E- and H-rings [61]. The use of RCM to generate the H-ring is illustrated in Scheme 3.54. Olefinic ester **289** containing the intact the F- and G-rings was subjected to the two-step metathesis protocol that was initially described by Clark *et al.* (Table 3.2) where the ester was first converted into the corresponding

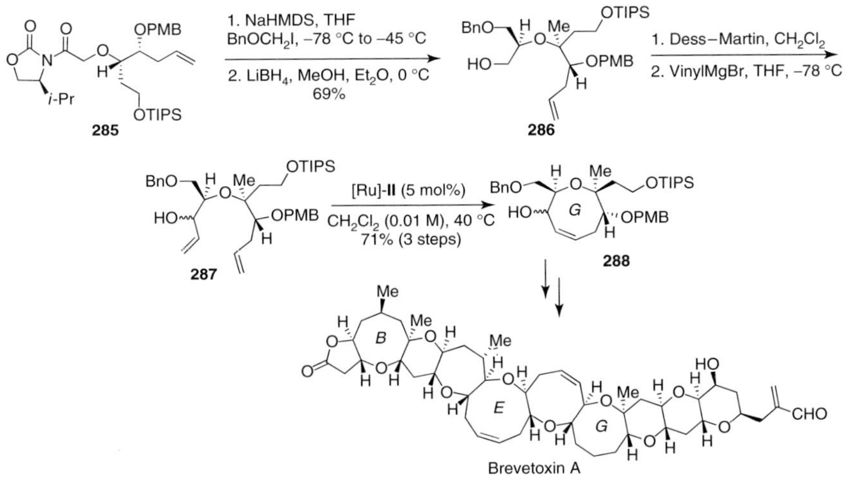

Scheme 3.53 The synthesis of the brevetoxin A G-ring and completion.

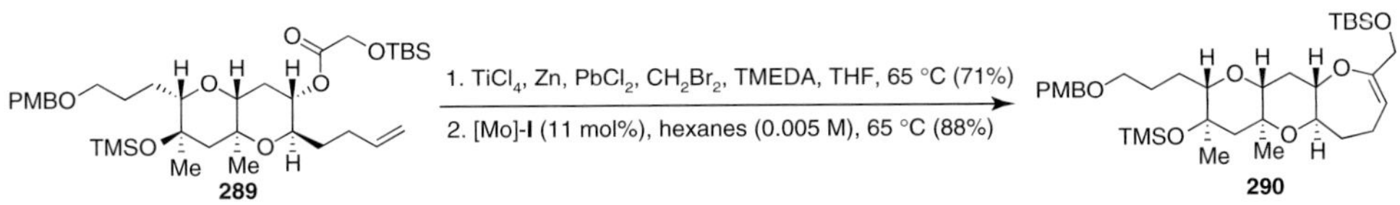

Scheme 3.54 The synthesis of the gambierol H-ring.

acyclic enol ether and this substrate was then subjected to either [Mo]-**I** catalyst or [Ru]-**II** catalyst to give **290**. For **289**, [Mo]-**I** catalyst was more reactive.

The significant advantage to enol ether–olefin RCM is that the precursor to the enol ether is an ester thus enabling the coupling of fairly elaborate substrates. An illustration of this is Johnson *et al.* completion of gambierol where the A–C and F–H precursors **291** and **292** were coupled and then subjected to the same two-step RCM sequence outlined for **290** (Scheme 3.55). In contrast to the protocol from **289**, a titanium ethylidene reagent was used in the initial enol ether forming reaction to avoid the substrate decomposition that was seen with the use of the corresponding titanium methylidene. Surprisingly, this modification led to the generation of a 2 : 1 mixture of oxepine **294** and expected acyclic enol ether **293**. After separation, enol ether **293** was also converted into **294** using [Ru]-**II** catalyst. This latter reaction also resulted in the generation of a small amount of dihydropyran **295**. The overall yield to **294** was 80%. Oxepine **294** was converted into gambierol in an additional 12 steps.

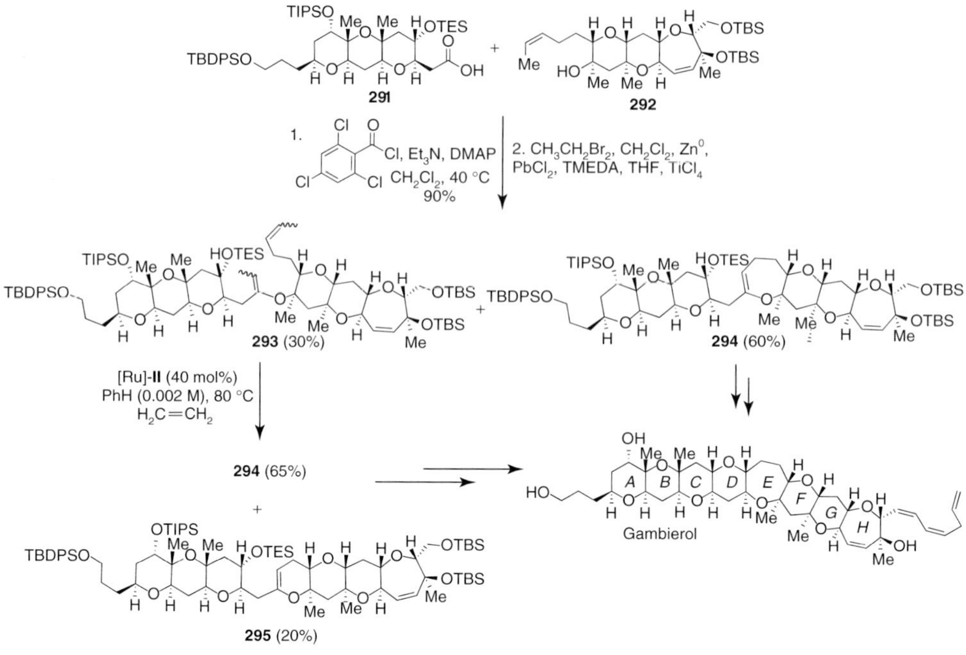

Scheme 3.55 Completion of gambierol.

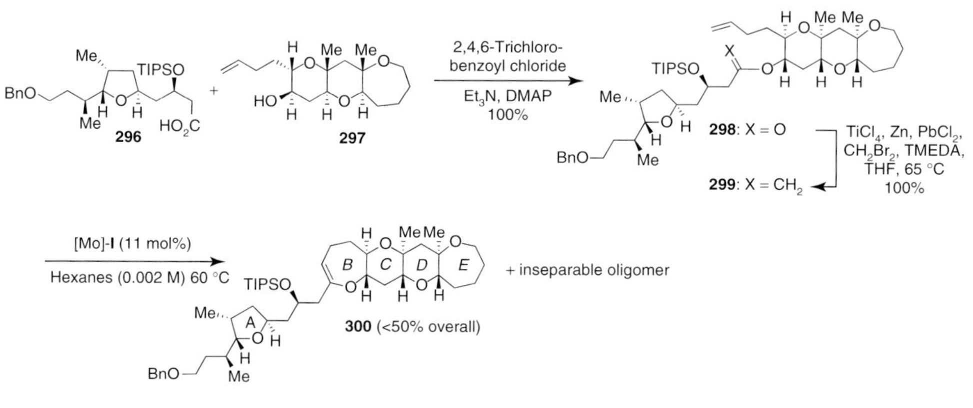

Scheme 3.56 Enol ether–olefin RCM to the gambieric acid B-ring.

As a continuation on the theme of using enol ether–olefin RCM in coupling sequences, Roberts and Rainier targeted the A–H ring system of gambieric acid (Scheme 3.56) [62]. The coupling of C–E tricycle **297** with A-ring tetrahydrofuran **296** gave **298**. Although the formation of acyclic enol ether **299** using the *in situ* generated titanium methylidene reagent was quantitative, the subsequent RCM using either [Mo]-**I** or [Ru]-**II** catalyst was inefficient leading to a mixture of the desired oxepine **300** along with a large amount of inseparable oligomer.

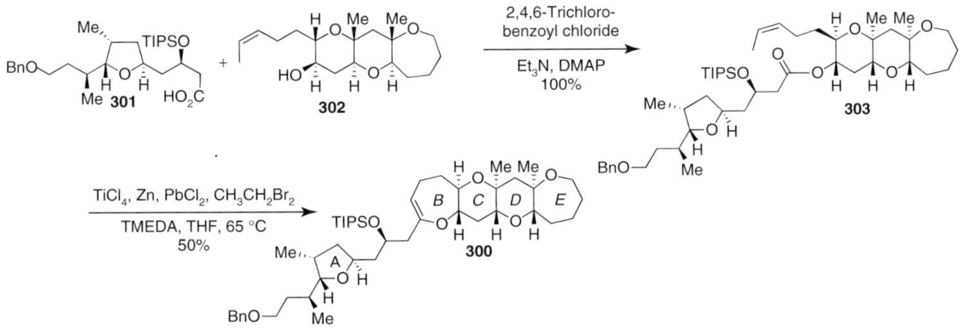

Scheme 3.57 Olefin metathesis, carbonyl olefination to the gambieric acid B-ring.

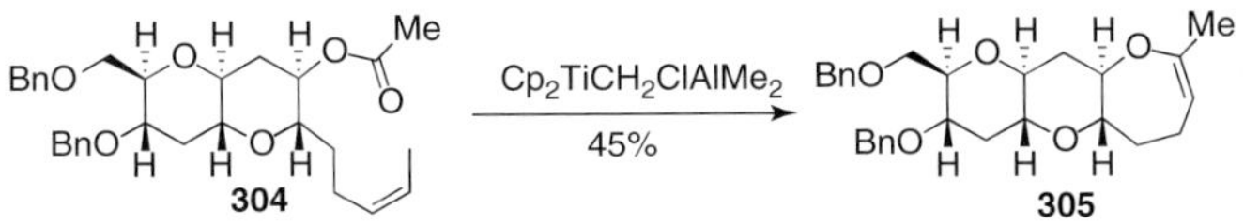

Scheme 3.58 Olefinic-ester cyclizations to oxepenes.

To avoid oligomer formation, Roberts and Rainier utilized an internal rather than a terminal olefin as a precursor to the cyclization reaction (Scheme 3.57) [62]. To this goal, internal olefin **303** was synthesized and subjected to enol ether formation but with the titanium ethylidene rather than the methylidene reagent for the acyclic enol ether forming reaction. Surprisingly, this relatively minor modification resulted in the conversion of **303** into cyclic enol ether **300** [63]. No acyclic enol ether was observed. The authors argued that the relatively moderate yield was due to the instability of the products and was not necessarily a result of an inefficient reaction.

The reaction to **300** is reminiscent of related work from Nicolaou's laboratory where it was demonstrated that dihydropyrans and oxepenes could be generated from olefinic esters using the Tebbe reagent or dimethyl titanocene (Scheme 3.58) [63d,e]. For example, oxepene **305** was generated from olefinic ester **304** in 45% yield when it was exposed to the Tebbe reagent. Evidence suggested that the transformation to **305** proceeded through an intermediate acyclic enol ether [64].

Iyer and Rainier followed up on the gambieric acid studies (Scheme 3.57) to show that oxepines can generally be formed using titanium ethylidenes. One example is given in Scheme 3.59 [65]. In contrast to Nicolaou's Tebbe reagent protocol, evidence is suggestive of an olefin metathesis, carbonyl-olefination cyclization mechanism [63a,b,f], [64d], [66].[1)] Subsequently, this reaction has been used by the Rainier

1) Grubbs has proposed a related mechanism for ring-opening metathesis, carbonyl-olefination reactions.

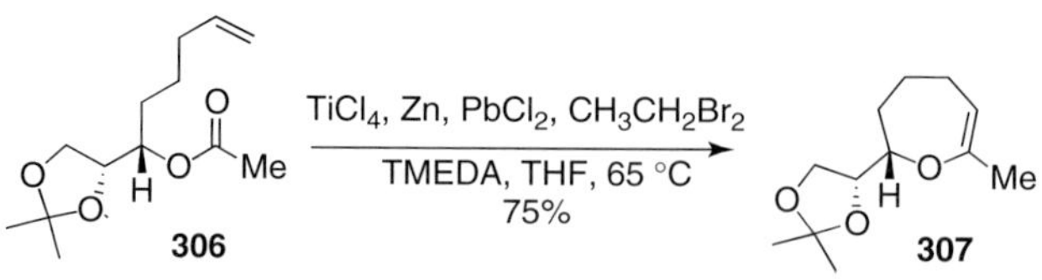

Scheme 3.59 Olefinic-ester cyclizations to oxepenes.

group to generate subunits of adriatoxin, in a two-directional approach to libraries of polycyclic ethers and in olefinic-lactone cyclizations to medium rings [67–69].

3.9 Conclusion

Described here have been RCM reactions to oxygenated 7-, 8-, 9-, and 10-membered rings with a focus on natural product synthesis. Although considerable progress has been made, challenges including the continued application of these reactions to the synthesis of complex molecules and overcoming the relatively high catalyst loadings that are typically required remain.

Acknowledgments

I am deeply indebted to the students and postdocs from my laboratory, who have been involved in the use of RCM and olefinic-ester cyclizations to synthesize oxygen containing heterocycles. In particular, Shawn Allwein, Jason Cox, Henry Johnson, Utpal Majumder, Scott Roberts, Karthik Iyer, Jie Zhou, Yuan Zhang, Jin Liu, and John Rohanna have all made and continue to make seminal contributions. I am also indebted to the scientists from across the globe whose work has been highlighted in this chapter and who have dedicated themselves to the brilliant strategies that have been presented. I look forward to hearing and reading about their continued successes. Finally, I would like to thank the NIH for funding our work (GM56677).

References

1 For a recent review see Chattopadhyay, S.K., Karmakar, S., Biswas, T., Majumdar, K.C., Rahaman, H., and Roy, B. (2007) *Tetrahedron*, **63**, 3919–3952.

2 For recent reviews describing the generation of heterocycles see: (a) Deiters, A. and Martin, S.F. (2004) *Chem. Rev.*, **104**, 2199–2238; (b) Brown, R.C.D. and Satcharoen, V. (2006) *Heterocycles*, **70**, 705–736; For recent reviews on alkene metathesis see: (c) Grubbs, R.H. (2004) *Tetrahedron*, **60**, 7117–7140; (d) Schrock, R.R. and Hoveyda, A.H. (2003) *Angew. Chem. Int. Ed.*, **42**, 4592–4633.

3 For recent reviews of the use of metathesis in medium-ring formation see: reference 1 and Maier, M.E. (2000) *Angew. Chem. Int. Ed.*, **39**, 2073–2077.

4 For reviews of RCM to smaller oxygenated rings see: references 1, 2 and Marco, J.A., Carda, M., Murga, J., and Falomir, E. (2007) *Tetrahedron*, **63**, 2929–2958.

5 For a review see Deshmukh, P.H. and Blechert, S. (2007) *Dalton Trans.*, 2479–2491.

6 (a) Schwab, P., Grubbs, R.H., and Ziller, J.W. (1996) *J. Am. Chem. Soc.*, **118**, 100–110; (b) Schwab, P., France, M.B., Ziller, J.W., and Grubbs, R.H. (1995) *Angew. Chem. Int. Ed.*, **34**, 2039–2041.

7 Scholl, M., Ding, S., Lee, C.W., and Grubbs, R.H. (1999) *Org. Lett.*, **1**, 953–956.

8 (a) Garber, S.B., Kingsbury, J.S., Gray, B.L., and Hoveyda, A.H. (2000) *J. Am. Chem. Soc.*, **122**, 8168–8179; (b) Gessler, S., Randl, S., and Blechert, S. (2000) *Tetrahedron Lett.*, **41**, 9973–9976.

9 Fürstner, A., Guth, O., Düffels, A., Seidel, G., Liebl, M., Gabor, B., and Mynott, R. (2001) *Chem. Eur. J.*, **7**, 4811–4820.

10 Schrock, R.R., Murdzek, J.S., Bazan, G.C., Robbins, J., DiMare, M., and O'Regan, M. (1990) *J. Am. Chem. Soc.*, **112**, 3875–3886.

11 Clark, J.S. and Kettle, J.G. (1997) *Tetrahedron Lett.*, **38**, 127–130.

12 Clark, J.S. and Kettle, J.G. (1997) *Tetrahedron Lett.*, **38**, 123–126.

13 Clark, J.S., Hamelin, O., and Hufton, R. (1998) *Tetrahedron Lett.*, **39**, 8321–8324.

14 Clark, J.S., Trevitt, G.P., Boyall, D., and Stammen, B. (1998) *Chem. Commun.*, 2629–2630.

15 Delgado, M. and Martín, J.D. (1999) *J. Org. Chem.*, **64**, 4798–4816.

16 Fujiwara, K., Goto, A., Sato, D., Kawai, H., and Suzuki, T. (2005) *Tetrahedron Lett.*, **46**, 3465–3468.

17 Sato, D., Fujiwara, K., Kawai, H., and Suzuki, T. (2008) *Tetrahedron Lett.*, **49**, 1514–1517.

18 Lejkowski, M., Gais, H.-J., Banerjee, P., and Vermeeren, C. (2006) *J. Am. Chem. Soc.*, **128**, 15378–15379.

19 Basu, S. and Waldmann, H. (2006) *J. Org. Chem.*, **71**, 3977–3979.

20 Linderman, R.J., Siedlecki, J., O'Neill, S.A., and Sun, H. (1997) *J. Am. Chem. Soc.*, **119**, 6919–6920.

21 For a discussion of the gauche effect see: Eliel, E.L. and Wilen, S.H. (1994) *Stereochemistry of Organic Compounds*. John Wiley & Sons, Inc., New York, pp. 606–610.

22 Crimmins, M.T. and Choy, A.L. (1999) *J. Am. Chem. Soc.*, **121**, 5653–5660.

23 Burton, J.W., Clark, J.S., Derrer, S., Stork, T.C., Bendall, J.G., and Holmes, A.B. (1997) *J. Am. Chem. Soc.*, **119**, 7483–7498.

24 Visser, M.S., Heron, N.M., Didiuk, M.T., Sagal, J.F., and Hoveyda, A.H. (1996) *J. Am. Chem. Soc.*, **118**, 4291–4298.

25 Crimmins, M.T., Emmitte, K.A., and Choy, A.L. (2002) *Tetrahedron*, **58**, 1817–1834.

26 Crimmins, M.T. and Emmitte, K.A. (1999) *Org. Lett.*, **1**, 2029–2032.

27 Lee, H., Kim, H., Yoon, T., Kim, B., Kim, S., Kim, H.-D., and Kim, D. (2005) *J. Org. Chem.*, **70**, 8723–8729.

28 Clark, J.S., Freeman, R.P., Cacho, M., Thomas, A.W., Swallow, S., and Wilson, C. (2004) *Tetrahedron Lett.*, **45**, 8639–8642.

29 Adsool, V.A. and Pansare, S.V. (2008) *Org. Biomol. Chem.*, **6**, 2011–2015.

30 Bratz, M., Bullock, W.H., Overman, L.E., and Takemoto, T. (1995) *J. Am. Chem. Soc.*, **117**, 5958–5966.

31 For a review of approaches to cembranoids see: Ellis, J.M. and Crimmins, M.T. (2008) *Chem. Rev.*, **108**, 5278–5298.

32 Crimmins, M.T. and Brown, B.H. (2004) *J. Am. Chem. Soc.*, **126**, 10264–10266.

33 Crimmins, M.T. and Ellis, J.M. (2008) *J. Org. Chem.*, **73**, 1649–1660.

34 Crimmins, M.T., Brown, B.H., and Plake, H.R. (2006) *J. Am. Chem. Soc.*, **128**, 1371–1378.

35 Kaliappan, K.P. and Kumar, N. (2005) *Tetrahedron*, **61**, 7461–7469.

36 Fürstner, A. and Langemann, K. (1997) *J. Am. Chem. Soc.*, **119**, 9130–9136.

37 Deb, I., John, S., and Namboothiri, I.N.N. (2007) *Tetrahedron*, **63**, 11991–11997.

38 Macías, F.A., Chinchilla, D., Molinillo, J.M.G., Fronczek, F.R., and Shishido, K. (2008) *Tetrahedron*, **64**, 5502–5508.

39 Aird, J.L., Hulme, A.N., and White, J.W. (2007) *Org. Lett.*, **9**, 631–634.

40 Fürstner, A., Radkowski, K., Wirtz, C., Goddard, R., Lehmann, C.W., and Mynott, R. (2002) *J. Am. Chem. Soc.*, **124**, 7061–7069.

41 De Napoli, L., Messere, A., Palomba, D., Piccialli, V., Evidente, A., and Piccialli, G. (2000) *J. Org. Chem.*, **65**, 3432–3442.

42 Davoli, P., Spaggiari, A., Castagnetti, L., and Prati, F. (2004) *Org. Biomol. Chem.*, **2**, 38–47.

43 Gurjar, M.K., Nagaprasad, R., Ramana, C.V., Karmakar, S., and Mohapatra, D.K. (2005) *Arkivoc*, **iii**, 237–257.

44 Maruyama, M., Inoue, M., Oishi, T., Oguri, H., Ogasawara, Y., Shindo, Y., and Hirama, M. (2002) *Tetrahedron*, **58**, 1835–1851.

45 (a) Hirama, M., Oishi, T., Uehara, H., Inoue, M., Maruyama, M., Oguri, H., and Satake, M. (2001) *Science*, **294**, 1904–1907; (b) Inoue, M., Miyazaki, K., Uehara, H., Maruyama, M., and Hirama, M. (2004) *Proc. Natl. Acad. Sci.*, **101**, 12013–12018.

46 Imai, H., Uehara, H., Inoue, M., Oguri, H., Oishi, T., and Hirama, M. (2001) *Tetrahedron Lett.*, **42**, 6219–6222.

47 Inoue, M., Miyazaki, K., Ishihara, T., Tatami, A., Ohnuma, Y., Kawada, Y., Komano, K., Yamashita, S., Lee, N., and Hirama, M. (2006) *J. Am. Chem. Soc.*, **128**, 9352–9354.

48 Inoue, M., Lee, N., Miyazaki, K., Usuki, T., Matsuoka, S., and Hirama, M. (2008) *Angew. Chem. Int. Ed.*, **47**, 8611–8614.

49 Oguri, H., Sasaki, S.-Y., Oishi, T., and Hirama, M. (1999) *Tetrahedron Lett.*, **40**, 5405–5408.

50 Sasaki, M., Ishikawa, M., Fuwa, H., and Tachibana, K. (2002) *Tetrahedron*, **58**, 1889–1911.

51 (a) Sasaki, M., Noguchi, T., and Tachibana, K. (1999) *Tetrahedron Lett.*, **40**, 1337–1340; (b) Sasaki, M., Noguchi, T., and Tachibana, K. (2002) *J. Org. Chem.*, **67**, 3301–3310.

52 (a) Takemura, A., Fujiwara, K., Shimawaki, K., Murai, A., Kawai, H., and Suzuki, T. (2005) *Tetrahedron*, **61**, 7392–7419; (b) Takemura, A., Fujiwara, K., Murai, A., Kawai, H., and Suzuki, T. (2004) *Tetrahedron Lett.*, **45**, 7567–7571.

53 Fujiwara, K., Koyama, Y., Doi, E., Shimawaki, K., Ohtaniuchi, Y., Takemura, A., Souma, S.-I., and Murai, A. (2002) *Synlett*, 1496–1499.

54 Goto, A., Fujiwara, K., Kawai, A., Kawai, H., and Suzuki, T. (2007) *Org. Lett.*, **9**, 5373–5376.

55 Clark, J.S., Conroy, J., and Blakc, A.J. (2007) *Org. Lett.*, **9**, 2091–2094.

56 (a) Clark, J.S. (2006) *Chem. Commun.*, 3571–3581; (b) Clark, J.S., Kimber, M.C., Robertson, J., McErlean, C.S.P., and Wilson, C. (2005) *Angew. Chem. Int. Ed.*, **44**, 6157–6162; (c) Clark, J.S. and Hamelin, O. (2000) *Angew. Chem. Int. Ed.*, **39**, 372–374.

57 (a) Kadota, I., Takamura, H., Nishii, H., and Yamamoto, Y. (2005) *J. Am. Chem. Soc.*, **127**, 9246–9250; (b) Kadota, I., Nishii, H., Ishioka, H., Takamura, H., and Yamamoto, Y. (2006) *J. Org. Chem.*, **71**, 4183–4187.

58 (a) Crimmins, M.T., McDougall, P.J., and Ellis, J.M. (2006) *Org. Lett.*, **8**, 4079–4082; (b) Crimmins, M.T., McDougall, P.J., and Emmitte, K.A. (2005) *Org. Lett.*, **7**, 4033–4036.

59 Crimmins, M.T., Zuccarello, J.L., Cleary, P.A., and Parrish, J.D. (2006) *Org. Lett.*, **8**, 159–162.

60 Crimmins, M.T., Zuccarello, J.L., Ellis, J.M., McDougall, P.J., Haile, P.A., Parrish, J.D., and Emmitte, K.A. (2009) *Org. Lett.*, **11**, 489–492.

61 (a) Majumder, U., Cox, J.M., Johnson, H.W.B., and Rainier, J.D. (2006) *Chem. Eur. J.*, **12**, 1736–1746; (b) Johnson, H.W.B., Majumder, U., and Rainier, J.D. (2006) *Chem. Eur. J.*, **12**, 1747–1753; (c) Johnson, H.W.B., Majumder, U., and Rainier, J.D. (2005) *J. Am. Chem. Soc.*, **127**, 848–849.

62 Roberts, S.W. and Rainier, J.D. (2007) *Org. Lett.*, **9**, 2227–2230.

63 Grubbs, Nicolaou, and Rainier had previously reported related reactions. See (a) Stille, J.R. and Grubbs, R.H. (1986) *J. Am. Chem. Soc.*, **108**, 855–856; (b) Stille, J.R., Santarsiero, B.D., and Grubbs, R.H. (1990) *J. Org. Chem.*, **55**, 843–862; (c) Fu, G.C. and Grubbs, R.H. (1993) *J. Am. Chem. Soc.*, **115**, 3800–3801; (d) Nicolaou, K.C., Postema, M.H.D., and Claiborne, C.F. (1996) *J. Am. Chem. Soc.*, **118**, 1565–1566; (e) Nicolaou,

K.C., Postema, M.H.D., Yue, E.W., and Nadin, A. (1996) *J. Am. Chem. Soc.*, **118**, 10335–10336; (f) Majumder, U. and Rainier, J.D. (2005) *Tetrahedron Lett.*, **46**, 7209–7211.

64 It has been reported that the Tebbe reagent induced cyclization has a limited scope. See: (a) Oishi, T., Uehara, H., Nagumo, Y., Shoji, M., Le Brazidec, J.-Y., Kosaka, M., and Hirama, M. (2001) *Chem. Commun.*, 381–382; (b) Uehara, H., Oishi, T., Inoue, M., Shoji, M., Nagumo, Y., Kosaka, M., Le Brazidec, J.-Y., and Hirama, M. (2002) *Tetrahedron*, **58**, 6493–6512; (c) Kadota, I., Kadowaki, C., Park, C.-H., Takamura, H., Sato, K., Chan, P.W.H., Thorand, S., and Yamamoto, Y. (2002) *Tetrahedron*, **58**, 1799–1816; (d) Allwein, S.P., Cox, J.M., Howard, B.E., Johnson, H.W.B., and Rainier, J.D. (2002) *Tetrahedron*, **58**, 1997–2009.

65 Iyer, K. and Rainier, J.D. (2007) *J. Am. Chem. Soc.*, **129**, 12604–12605.

66 Rainier, J.D., Allwein, S.P., and Cox, J.M. (2001) *J. Org. Chem.*, **66**, 1380–1386.

67 Osei Akoto, C. and Rainier, J.D. (2008) *Angew. Chem. Int. Ed.*, **47**, 8055–8088.

68 Zhang, Y. and Rainier, J.D. (2009) *Org. Lett.*, **11**, 237–239.

69 Rohanna, J.C. and Rainier, J.D. (2009) *Org. Lett.*, **11**, 493–495.

4 Phosphorus and Sulfur Heterocycles via Ring-closing Metathesis: Application in Natural Product Synthesis

Christopher D. Thomas and Paul R. Hanson

4.1 Introduction

The development of new synthetic strategies allowing for efficient asymmetric syntheses of complex biologically active targets, with minimal protecting group manipulations and chemical steps, is an enormous challenge in natural product synthesis. An effective way of addressing this challenge is through the use of convergent methodologies employing the temporary tethering of two advanced intermediates. Historically, silicon has been the most widely used temporary tether due to its facile installation/cleavage attributes as well as its innate protecting group properties [1]. Moreover, the ability of silicon tethers to undergo myriad functional group transformations positions them as ideal tethers in the realm of total synthesis [2]. Phosphorus and sulfur, namely, phosphates and sultones, possess alternative properties as tethers in that they can serve as leaving groups in mediating a variety of nucleophilic substitution reactions while still retaining orthogonal stability. Both tethers can also serve as the central linchpin in the formation of the corresponding heterocycle via ring-closing metathesis (RCM) catalyzed by the Grubbs catalysts [3] [Ru]-**I** and [Ru]-**II** [4, 5]. In addition, their facile cleavage result in by-products that are easily removed through aqueous workup (phosphine oxide) [4] or extrusion (SO_2) [5]. Rapid access to complex, functionalized intermediates have allowed these methodologies to emerge in the realm of natural product synthesis [6, 7].

4.2 Synthesis and Reactivity of Sultones Derived from RCM

In 2002, Metz *et al.* reported the first synthesis of an unsaturated sultone, of type **4**, resulting in a method producing a variety of ring sizes [8]. These heterocycles were achieved in a two-step protocol starting from a variety of olefinic sulfonyl chlorides **1** and alkenols **2** (Scheme 4.1) [8a].

Metathesis in Natural Product Synthesis: Strategies, Substrates and Catalysts. Edited by Janine Cossy, Stellios Arseniyadis, and Christophe Meyer

ISBN: 978-3-527-32440-8

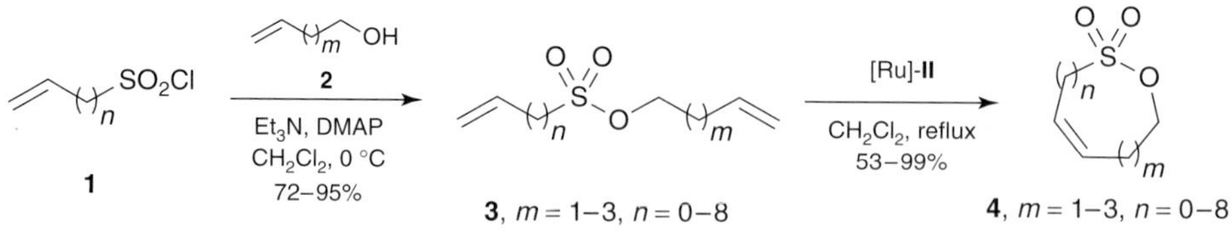

Scheme 4.1

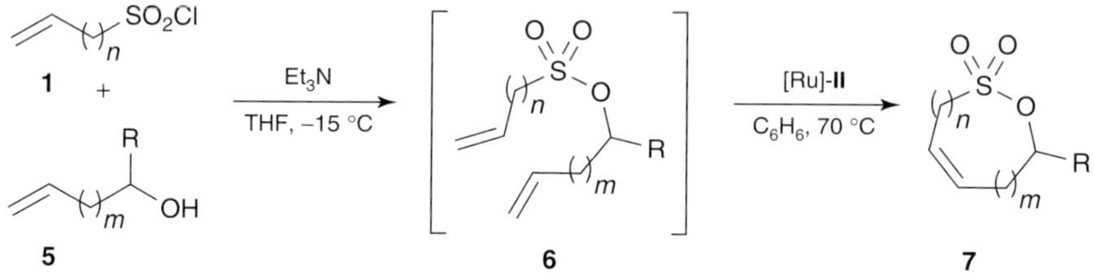

m	n	R	Yield of **7** (%)
0	1	$(CH_2)_3OBn$	76
0	1	CO_2Et	54
1	0	CH_2OPiv	74
1	0	$CH_2OTBDPS$	65
1	1	$(CH_2)_3OBn$	65

Scheme 4.2

In 2003, Cossy *et al.* demonstrated the ability to construct sultones from secondary alkenols, thus augmenting previous work by Metz with primary alkenols [5]. The unsaturated sulfonates derived from secondary alkenols were more difficult to construct and exhibited reduced stability. This problem was circumvented by subjecting the crude sulfonates **6** to catalyst [Ru]-**II** (C_6H_6, 70 °C) thus affording a variety of substituted unsaturated sultones **7** in good yields (Scheme 4.2).

The aforementioned heterocyclic building blocks possess the potential for exploring numerous reaction pathways [9]. Hence, Cossy *et al.* highlighted α-alkylation of various sultones **8** in the presence of *n*-butyllithium followed by the addition of different alkyl halides (Scheme 4.3) [5]. The choice of base was also found to be crucial, as the use of LDA resulted in the formation of a sulfene intermediate, which rapidly dimerized to product **10** through a [2+2]-cycloaddition.

The leaving group ability of the sultone was exploited in an innovative alkylation/elimination sequence. The removal of the tether is facilitated by metallation of **11** (*n*-BuLi) and treatment with iodomethylmagnesium chloride (generated from CH_2I_2, *i*-PrMgCl, THF, −78 °C) to produce the reactive Grignard species **12**. This species undergoes subsequent β-elimination resulting in loss of SO_2 to afford

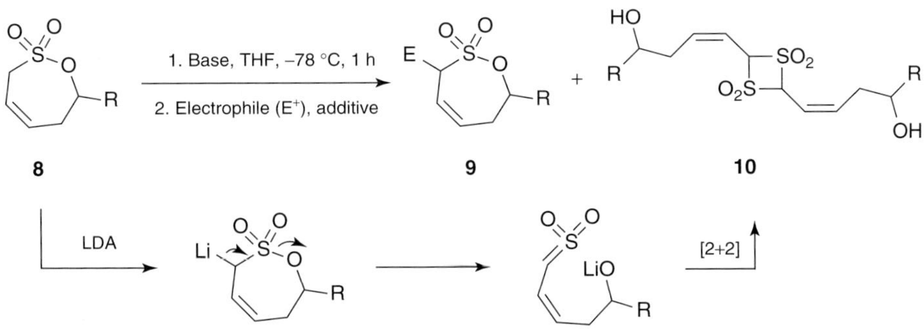

R	E+	Base	Additive	Yield of **9** (%)	Yield of **10** (%)
H	MeI	*n*-BuLi (1 equiv.)	None	99	0
$(CH_2)_3OBn$	MeI	*n*-BuLi (1 equiv.)	None	85	0
$(CH_2)_3OBn$	$Br(CH_2)_3OBn$	*n*-BuLi (1 equiv.)	HMPA	84	0
$(CH_2)_3OBn$	None	LDA (3 equiv.)	None	0	55[a]

[a] 60 : 40 mixture of *trans*/*cis* diastereomers.

Scheme 4.3

homoallylic alcohols bearing 1,3-dienes with an internal *Z*-disubstituted olefin and a terminal or α,α-disubstituted alkene, **13a/13b**, respectively (Scheme 4.4) [5].

A number of bioactive natural products possess a similar homoallylic conjugated *Z*-dienol, which can be produced from a sultone heterocycle (Figure 4.1) [10]. The approach of Cossy *et al.* to this synthon is unique compared to previous methodologies that do not utilize olefin metathesis to afford the internal *Z*-olefin of the 1,3-dienol [11].

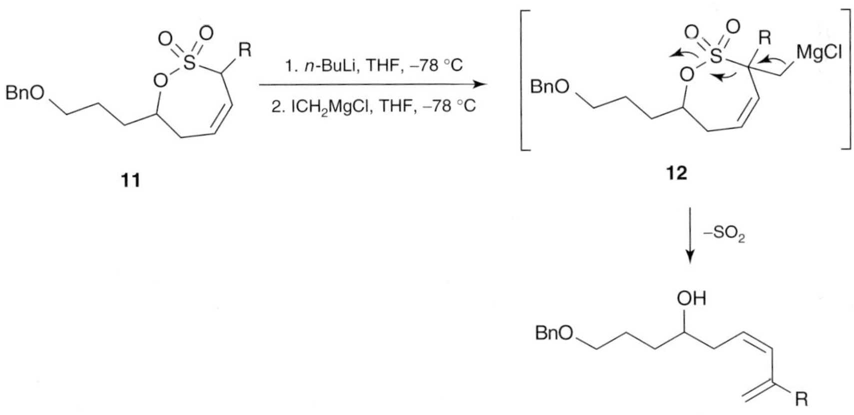

13a: R = H (60%)
13b: R = Me (53%)

Scheme 4.4

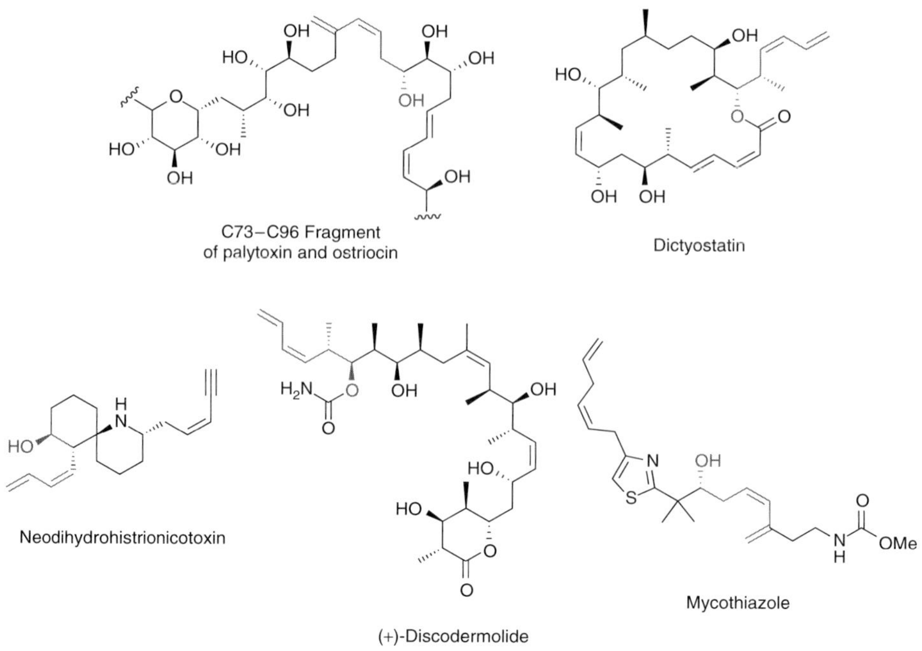

Figure 4.1 Bioactive Z-dienol-containing natural products.

4.3 Total Synthesis of the Originally Proposed Structure of (±)-Mycothiazole

The development of a method toward constructing conjugated *Z*-dienols led Cossy *et al.* to pursue the total synthesis of (±)-mycothiazole [5, 6], with the originally proposed structure outlined in Scheme 4.5 [12]. The synthesis was designed for the C7–C11/C20 subunit to be accessed through formation of a sultone tether from homoallylic alcohol **15** through sulfonylation and RCM (Scheme 4.5). Subsequent alkylation of the α-position to the sulfonyl would afford requisite functionalization to append the C12–C13 side chain. Overall, the retrosynthetic approach started from readily available 2,4-dibromothiazole (**17**) [6], which established the necessary connectivity for the uniquely substituted thiazole ring.

The key sequence began with the formation of the sultone derived from the advanced homoallylic alcohol **15**, where sulfonylation and subsequent RCM with [Ru]-**II** afforded sultone **18**, bearing the necessary C9–C10 *Z*-olefin, as a single diastereomer in 70% yield over two steps (Scheme 4.6). Subsequent alkylation of the sultone with 1,1-dimethoxy-3-iodopropane, furnished sultone **19** as a 1 : 1 mixture of diastereomers at C11. Construction of the dienol commenced via deprotonation of **19** with *n*-BuLi, followed by addition of ICH_2MgCl (supplying the C20 terminal methylene group), yielding the desired *Z*-dienol **20** in 60% yield [5, 6].

Following assembly of the key C9–C11/C20 subunit, conjugated *Z*-dienol, the dimethyl acetal **20** was converted to the required carbamate in 33% yield over

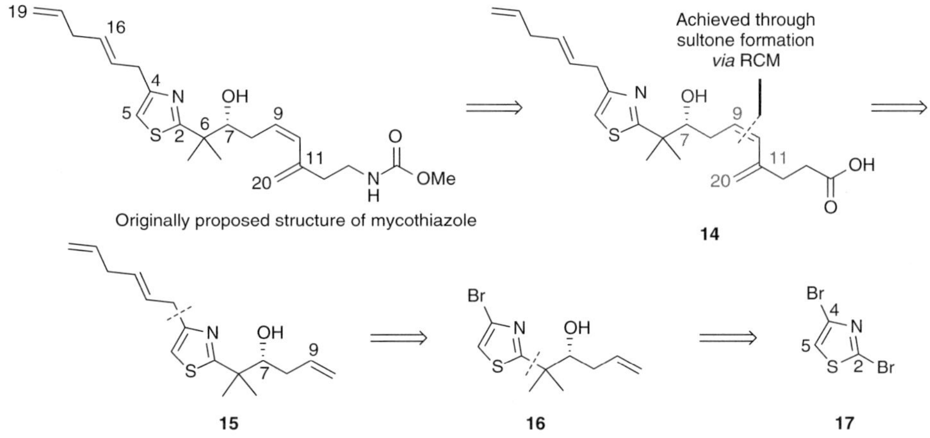

Scheme 4.5

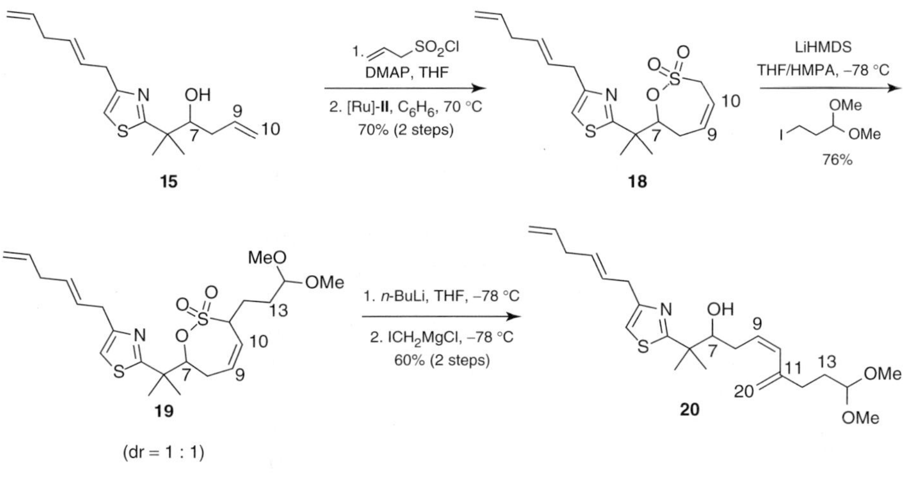

Scheme 4.6

four steps, achieving a racemic synthesis of the originally proposed structure of mycothiazole in 5% overall yield and 18 steps from 2,4-dibromothiazole **17** (Scheme 4.7). This was the first example of utilization of a sulfur-mediated tether in the context of the total synthesis of a biologically active natural product [5b, 12].

Overall, the construction of sultones via sulfonylation/RCM from the corresponding allylic alcohol fulfills the requirements of an effective temporary tether in synthesis. The sultone functionality provides sufficient reactivity under basic conditions to afford complex *Z*-substituted 1,3-dienols after the facile removal of the sulfonyl through the extrusion of SO_2. The prospect of utilizing additional selective reaction pathways has the potential to generate new synthons furnishing a variety of pathways to biologically active natural products [8, 10].

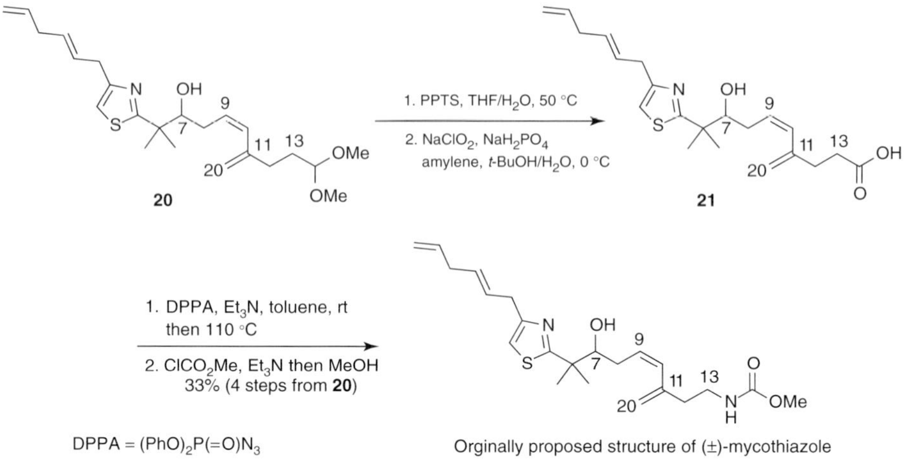

Scheme 4.7

4.4 Synthesis and Reactivity of Phosphates from RCM

Despite the enormous research on organophosphorus compounds, the general use of phosphate triesters in synthesis has been relatively limited until recently. Phosphorus has inherent properties that make it an attractive tether candidate, including the ability to mediate di- and tripodal coupling, provide orthogonal protection, and serve as a leaving group. Irrespective of these attributes, the application of phosphates in synthesis is largely centered on monovalent activation of a single phosphate ester appendage. Such classical use of phosphates in complex synthesis have focused on nucleophilic displacement reactions of allylic phosphates [13], cross-coupling/reduction reactions with enol phosphates [14], and more recent applications employing direct displacements of phosphates in cyclization protocols [15]. Additional uses of phosphate triesters in iodophosphonylation procedures [16] and their role in oligonucleotide synthesis [17] further highlight both nucleophilic properties and facile coupling characteristics innate to phosphates. These reaction processes utilize the leaving group ability of a phosphate monoanion, where a single ester moiety is activated (monovalent activation) and the remaining ester positions serve as ancillary substituents.

In 2005, Hanson *et al.* investigated the possibility of exploiting both multipodal coupling and multivalent activation attributes within phosphate triesters, which by their nature presented opportunities to investigate selective processes. This study resulted in a new method that embodied several underdeveloped areas of phosphate chemistry, namely, their use as removable, functionally active tethers capable of multipodal coupling and multivalent activation and their subsequent role as latent leaving groups in a number of unprecedented selective cleavage reactions. In addition, stereoelectronic effects within the bicyclic framework lend orthogonal protecting group stability. In 2005, Hanson *et al.* implemented this protocol for constructing cyclic phosphates from allylic alcohols [4c] and

1,5-diene-2,4-diols [4] to yield unique mono- and bicyclic heterocycles that undergo several selective transformations while imparting orthogonal stability. The utilization of a tether to desymmetrize a C_2-symmetric diol was inspired by the work of Burke *et al.*, who demonstrated that ketalization of 1,5-diene-3,4-diol **21** with 5-chloropentan-2-one and subsequent elimination of the primary chloride with *tert*-butoxide could generate triene **22** (Scheme 4.8) [18]. After undergoing cyclization in the presence of [Ru]-**I**, RCM afforded chiral, nonracemic bicycle **23** in 86% yield. Diastereotopic differentiation of the pseudo-C_2-symmetric ketal **22** using RCM was facilitated by the proximity of the newly installed allyl group, which allowed the differentiation of the previously homotopic terminal vinyl groups.

In 2005, Hanson *et al.* utilized this concept with a phosphate tether in the desymmetrization of C_2-symmetric (3*S*,5*S*)-1,6-diene-diol (*S*,*S*)-**24** (Scheme 4.9) [4a]. After coupling with $POCl_3$, allyl alcohol was added to the phosphoryl chloride species, yielding phosphate triene (*S*,*S*)-**25**. In the presence of [Ru]-**II**, the proximity of the two 1,3-diaxial terminal olefins in ring conformer **25a** allowed for facile RCM affording the *P*-chiral, nonracemic bicyclo[4.3.1]phosphate scaffold (*S*,*S*,P_R)-**25**.

The nature of bicyclo[4.3.1]phosphate (*S*,*S*,P_R)-**26** showed promise in pursuing selective transformations, whereby the tripodal phosphate tether imparts multivalent activation of the three carbinols en route to differentiated polyol fragments [4a]. Consistent with previous reports of cyclic phosphines and phosphates, it was shown that in acidic media, **26** exhibited excellent stability due to the decrease in proton affinity (basicity) of the phosphonyl oxygen (Equation 1, Scheme 4.10), thus hindering activation pathways [19]. Different nucleophiles were shown to yield selective reaction pathways through their hard/soft characteristics [20]. First, addition of $LiAlH_4$ resulted in hydride addition to phosphorus and facile removal of the phosphate thus affording triol **27** (Equation 2, Scheme 4.10). This result supports the ability of phosphate to be utilized as a temporary tether in coupling diol **24**

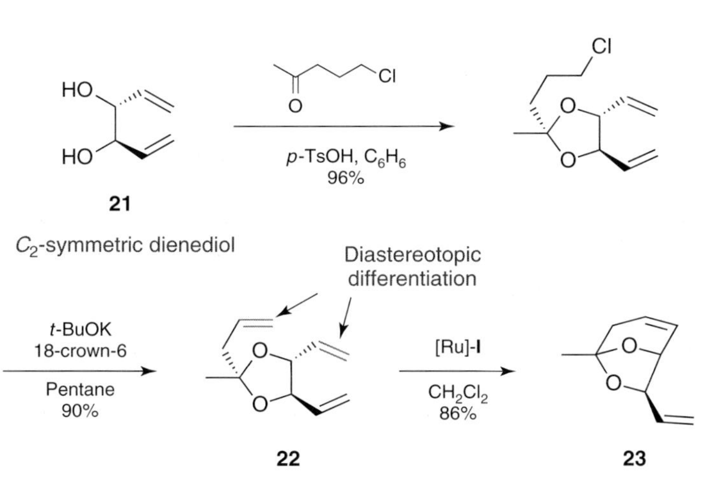

Scheme 4.8

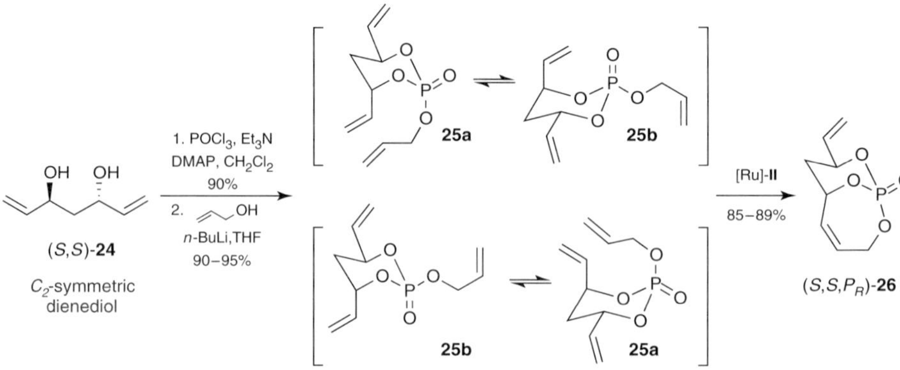

Scheme 4.9

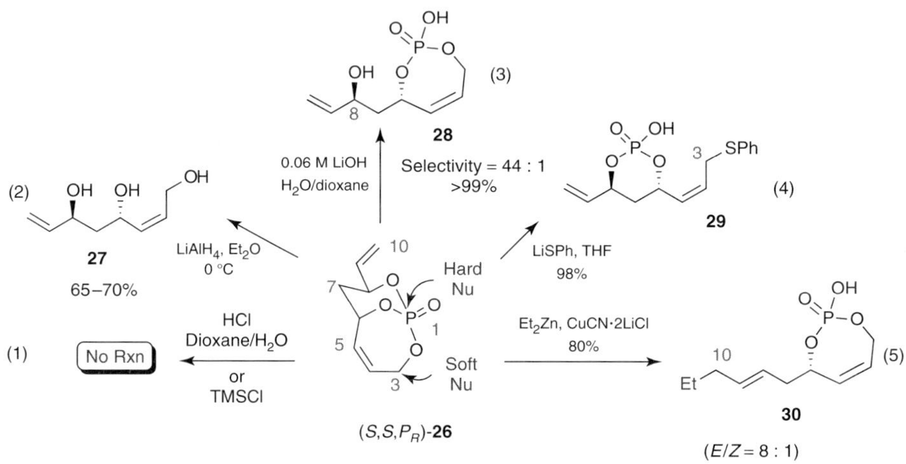

Scheme 4.10

with allyl alcohol through a phosphate-mediated sequence [4a]. Basic hydrolysis (Equation 3, Scheme 4.10) also resulted in attack at phosphorus, however, due to the characteristics of the phosphate monoanion [21], an excess of LiOH only displaced one of the three P–O bonds, yielding phosphate acid **28**. Variation from an oxygen nucleophile to a sulfur nucleophile such as LiSPh (Equation 4, Scheme 4.10) demonstrated the preference for addition onto the most sterically accessible C3 carbinol position, furnishing phosphate acid **29**. Lastly, cuprate addition resulted in allylic displacement of the more sterically accessible olefin (Equation 5, Scheme 4.10), resulting in phosphate acid **30**, along with preferential formation of the *E*-isomer, which is consistent with previous reports in the field of cuprate additions to allylic phosphates [22].

The preference of cuprate nucleophiles to add to the more sterically accessible olefin of **26** led into examining the possible functionalization of the endocyclic olefin [4a] After regioselective hydrogenation of the exocyclic olefin in the presence of Wilkinson's catalyst, cuprate addition to phosphate **31** preferentially occurred at

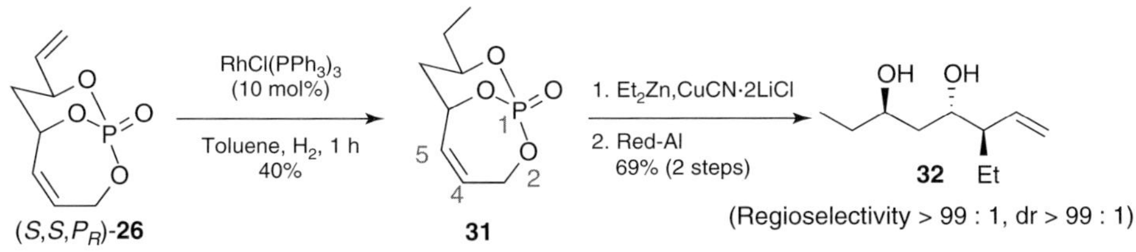

Scheme 4.11

the C5 position with high regioselectivity (Scheme 4.11). Removal of the phosphate acid with Red-Al produced diol **29** bearing an 1,3,4-*anti,anti*-stereotriad, as a single diastereomer over the three-step sequence from (S,S,P_R)-**26**.

The remarkable regioselectivity in this transformation is attributed to inherent stereoelectronic effects of the C4–C5 olefin and P–O bond of the corresponding allylic phosphate. For addition into the C5 position (path A, Figure 4.2, X-ray of structure **26**), orthogonal alignment of the π^* orbital of the C=C bond and σ^* orbital of the P–O provide proper electronics for the allylic displacement to occur. However, in a potential C4 addition (path B, Figure 4.2), S_N2' addition does not take place due to coplanar alignment of the aforementioned π^* orbital of the C=C bond to the P–O σ^* orbital [4a].

Facial selectivity was attributed to the overall geometry of the bicyclic skeleton of **26**, whereby the bicyclo[4.3.1]phosphate system hinders approach of the cuprate from the concave face (Figure 4.3, X-ray of structure **26**) [4a].

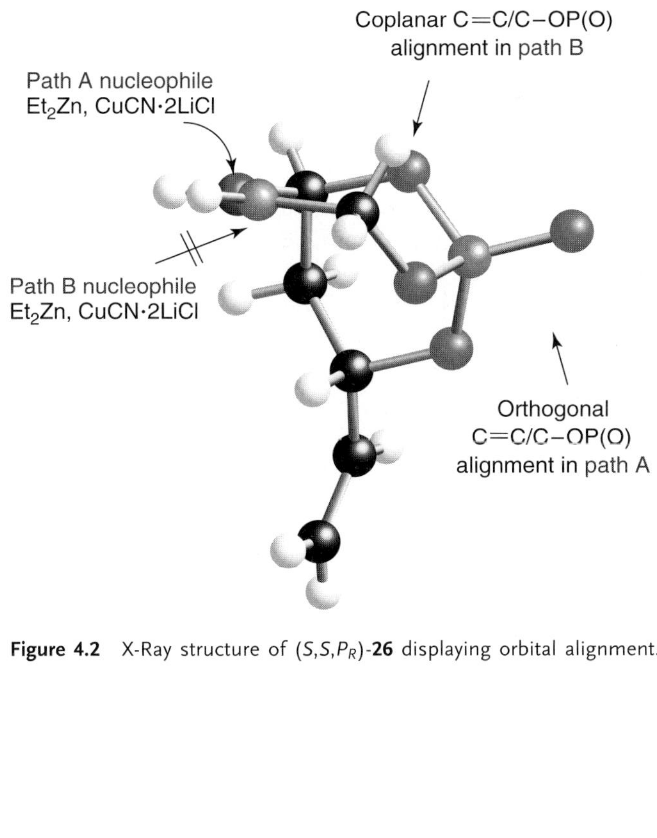

Figure 4.2 X-Ray structure of (S,S,P_R)-**26** displaying orbital alignment.

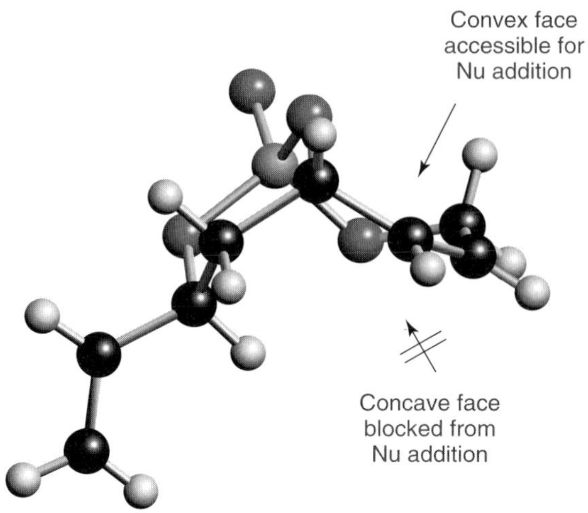

Figure 4.3 X-Ray structure of (S,S,P_R)-**26** displaying facial bias.

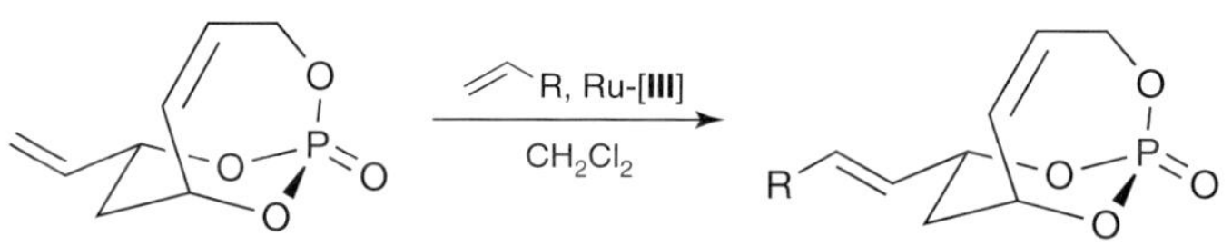

(R,R,P_S)-**26**

R	Yield (%)	*E/Z*	Type
CH_2OH	86	>99 : 1	I
CH_2OTBS	87	>99 : 1	I
CH_2NHBoc	69	>99 : 1	I
$CH_2OP(O)(OMe)_2$	80	2 : 1	I
CO_2Me	78	8 : 1	II
CO_2t-Bu	60	5 : 1	II
CHO	78	>99 : 1	II
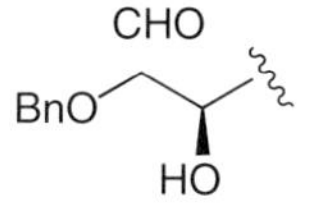	72	>99 : 1	II

Scheme 4.12

Another strong attribute of the bicyclic phosphate **26** is the ability to mediate cross-metathesis (CM) with various type I and II olefins [23] as shown in Scheme 4.12 [4b]. Studies performed with (R,R,P_S)-**26** (synthesized from (R,R)-1,6-diene-3,5-diol) provided stereoselective formation of the *E*-isomer in the presence of the catalyst [Ru]-**III**.

When attempting to undergo CM with type III partners [23], very poor yields were observed in the presence of [Ru]-**III** (Table 4.1). In accordance with Grubbs

Table 4.1

Cross partner	Conc.	Yield
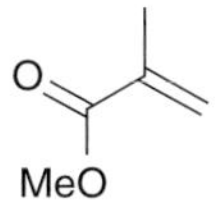	0.05 M	< 5%
	0.05 M	< 5%
	neat	< 5%
	neat	< 5%
none	0.05 M	no rxn.

classification of CM partners, type III olefins do not undergo homodimerization, and are therefore ideal heteropartners with substrates bearing type I and II olefins and unreactive with substrates possessing type III olefins [23]. On the basis of the observations seen and the fact that **26** did not undergo homodimerization, the exocyclic olefin in **26** was deemed to be type III [4b].

By combining the CM and cuprate methodologies, an advanced intermediate **37** containing the key aspects of the dolabelide family was constructed (Scheme 4.13) [24]. Hence, phosphate-mediated coupling of (*R,R*)-**24** with 1,1-dimethylallyl alcohol afforded the bicyclic phosphate **33** over a three-step sequence. CM of the latter with but-3-en-1-ol in the presence of [Ru]-**III** and subsequent hydrogenation of the exocyclic olefin of **34** furnished an extended side chain bearing the C15–C19 structure of dolabelide. After benzylation of the primary alcohol in **35** with *p*-methoxybenzyl trichloroacetimidate, addition of a methyl cuprate, methylation of the phosphate acid ($TMSCHN_2$), and removal of the phosphate ester (Red-Al) yielded advanced intermediate **37**, a key synthon toward the synthesis of dolabelide C [4b].

The differentiated 1,3,4-*anti,anti*-stereotriad seen in **37** is also present in a number of biologically active natural products, including dolabelide C, bitungolide E, (+)-discodermolide, dictyostatin, and salicylihalamide A (Figure 4.4) [10, 21–24].

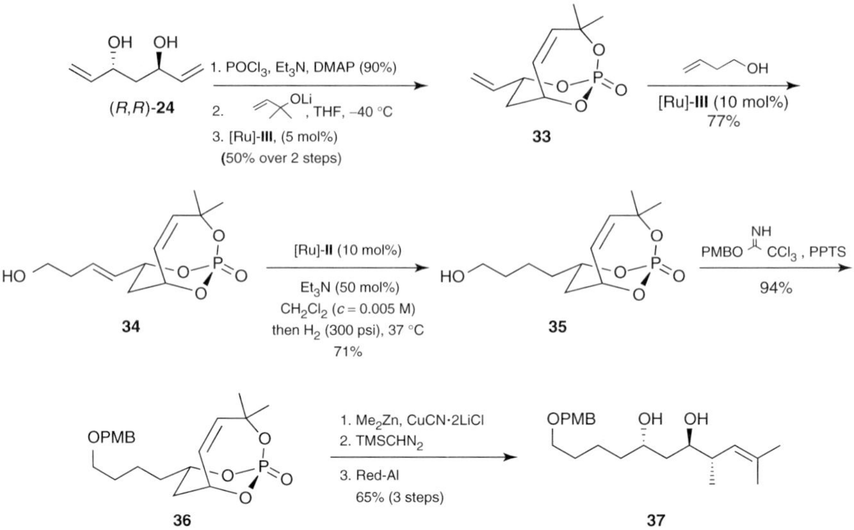

Scheme 4.13

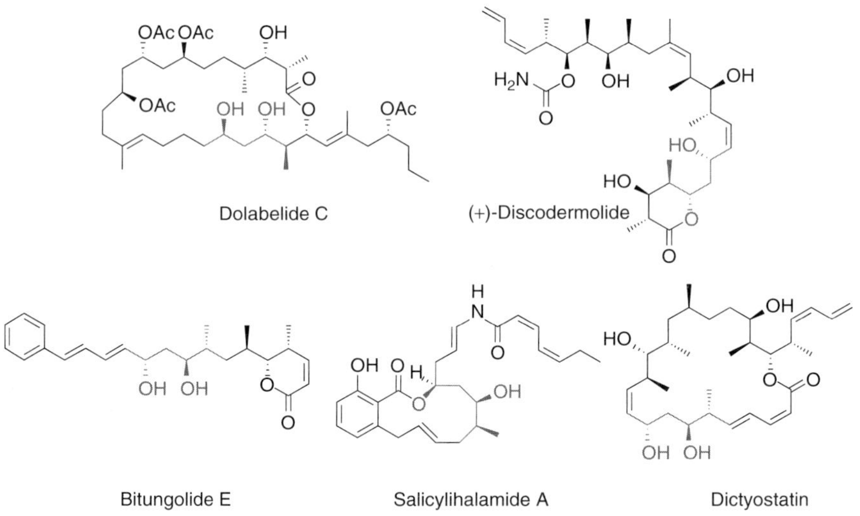

Figure 4.4 Cross metathesis of (S,S,P_R)-**26** with type III olefins.

4.5 Applications of Phosphate Tethers in the Synthesis of Dolabelide C

In 2008, Hanson *et al.* reported the synthesis of two key subunits of dolabelide C, which possess the *anti*-1,3-diol motif observed at the C7/C9 positions and the 1,3,4-*anti,anti*-stereotriad at the C19/C21–C22 positions. Retrosynthetic analysis segmented the 24-membered macrolide into a C1–C14 subunit **38** and

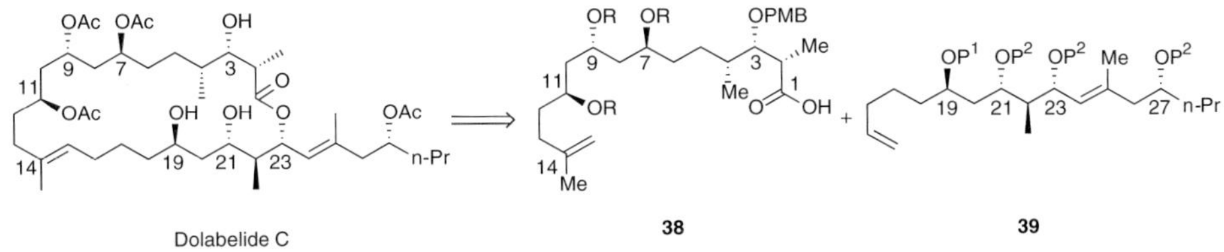

Scheme 4.14

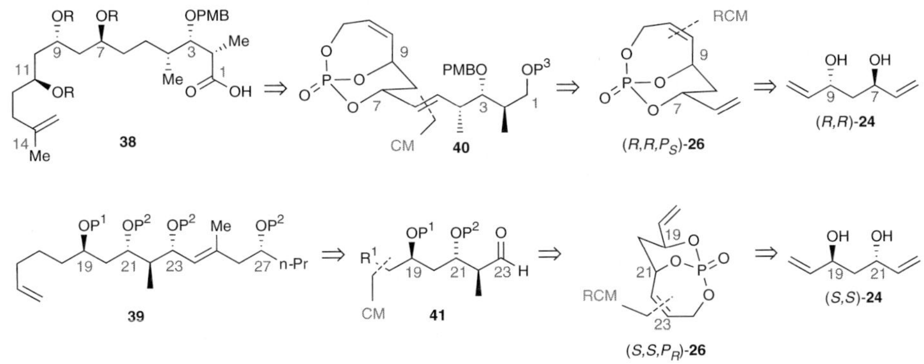

Scheme 4.15

C15–C30 subunit **39** from disconnecting the lactone bond and C14–C15 olefin (Scheme 4.14) [7].

Each subunit was constructed by a phosphate-mediated desymmetrization of **24**, followed by the aforementioned CM and allylic phosphate displacements. The C1–C14 subunit was envisioned to have the C1–C6 side chain appended via CM to yield advanced intermediate **40** from (R,R,P_S)-**26**. Subsequent regioselective hydrogenation of the C5–C6 olefin and regioselective Pd(0)-mediated allylic hydride addition (to C10) would afford a terminal olefin armed for oxidation and subsequent installation of the C11 carbinol center (Scheme 4.15) [7a]. The C15–C30 subunit largely depended on the CM/cuprate approach (Scheme 4.13) and the advanced intermediate **41** would be synthesized from (S,S,P_R)-**26** [7b].

The synthesis of the C1–C14 subunit was initiated from readily available (R,R,P_S)-**26**, which was coupled with olefin **42**, possessing the necessary C2–C4 stereochemistry, by CM in the presence of [Ru]-**III** (Scheme 4.16). Regioselective hydrogenation of the external olefin in **43** was accomplished using *in situ* formation of diimide from *o*-nitrobenzenesulfonyl hydrazine [25] and Et_3N to provide partially reduced phosphate **44**. This was followed by another regioselective ring opening, and a Pd(0)-mediated reduction of allylic phosphate **44** via formation of hydride from formic acid and Et_3N [26]. Overall, this transformation resulted in an allylic displacement of the endocyclic phosphate with hydride, which after methylation with $TMSCHN_2$ yielded phosphate ester **45**. Removal of the phosphate ester with $LiAlH_4$ produced advanced intermediate **46**, bearing the C1–C10 framework [7a].

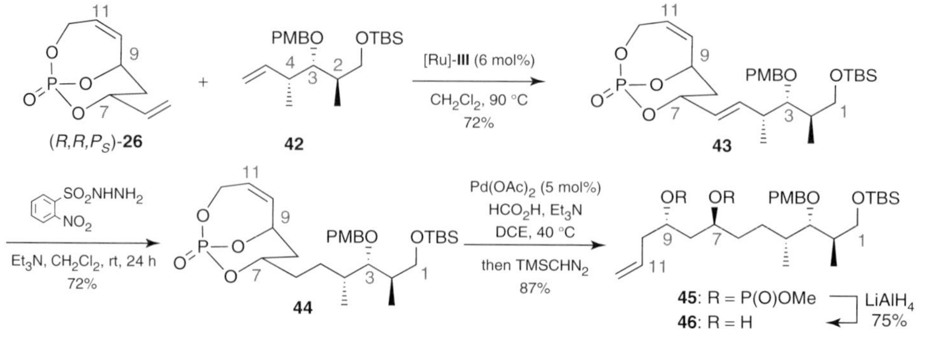

Scheme 4.16

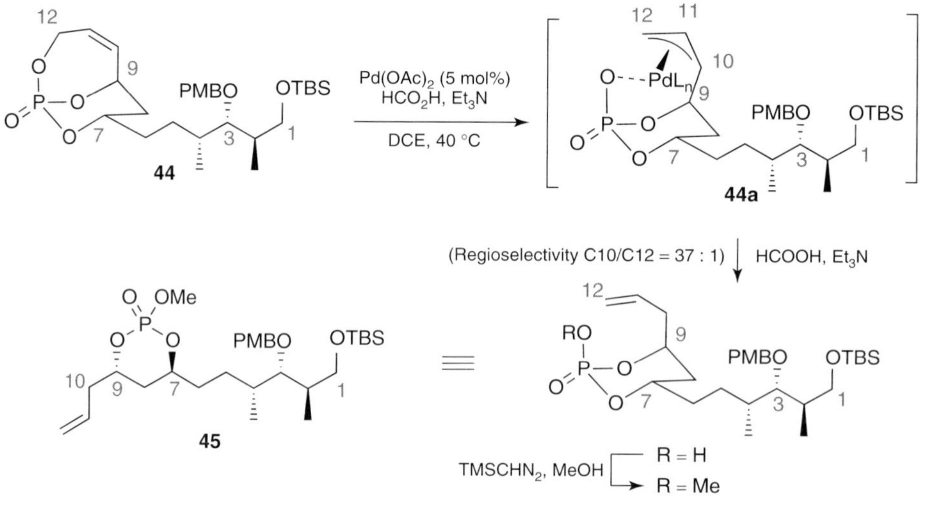

Scheme 4.17

Introducing Pd(0) to phosphate **44** gave rise to π-allyl complex **44a** (Scheme 4.17). The Pd(0)-mediated allylic hydride addition [26] to phosphate **44** gave the desired regioselectivity of hydride addition at the internal position (C10) of **44a** (regioselectivity C10/C12 = 37 : 1), similar to the previous cuprate additions to **26**. Both examples are illustrative of the preference in regioselectivity of allylic phosphate displacements with the endocyclic π bond.

The C15–C30 subunit was synthesized by means of two approaches. Initially, regioselective hydroboration [27] of the exocyclic olefin in (S,S,P_R)-**26**, followed by PMB protection of the primary alcohol, provided **47** (Scheme 4.18). Allylic displacement using a methyl cuprate derived from Me_2Zn and CuCN·2LiCl furnished the aforementioned 1,3,4-*anti,anti*-stereotriad of the C19/C21–C22 subunit as a single diastereomer. Methylation of the phosphate acid and cleavage of the resulting phosphate ester **48** with $LiAlH_4$ afforded 1,3-diol **49** over the five-step sequence from bicyclic phosphate **26** [7b].

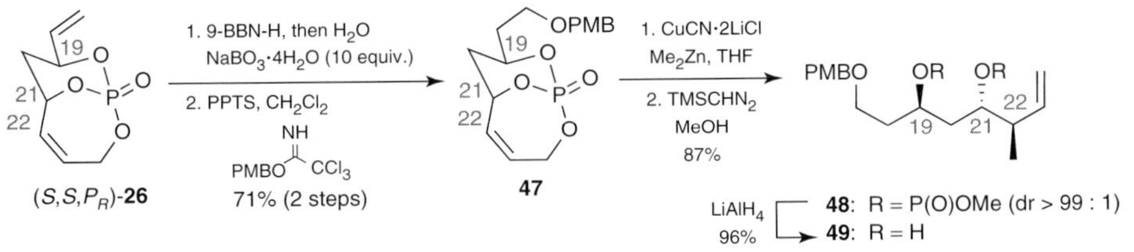

Scheme 4.18

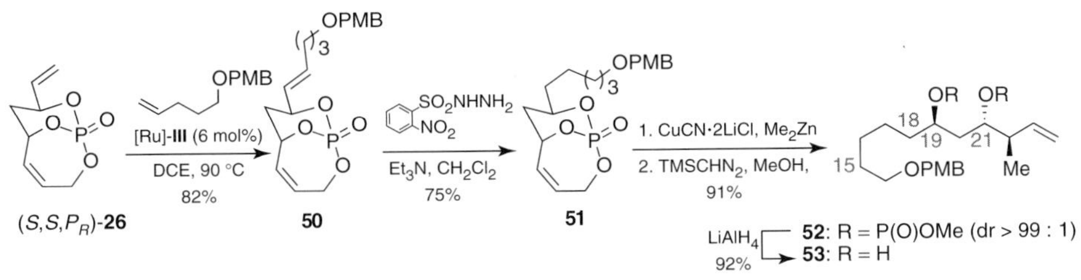

Scheme 4.19

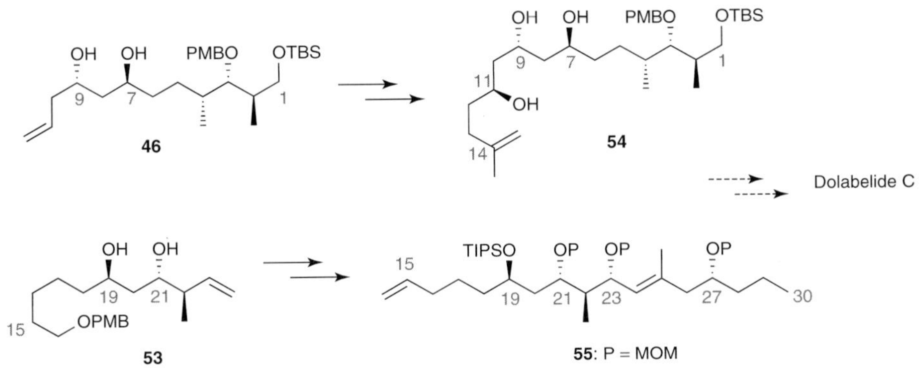

Scheme 4.20

An alternative route utilized CM of (S,S,P_R)-**26** and PMB-protected pent-4-en-1-ol to give phosphate **50** (Scheme 4.19). Selective hydrogenation of the exocyclic olefin gave **51**, which possesses the C15–C18 side chain of the C15–C30 subunit. Again, allylic displacement with a methyl cuprate derived from Me_2Zn and CuCN·2LiCl occurred regioselectively and gave phosphate **52** (after methylation with $TMSCHN_2$). Phosphate cleavage with $LiAlH_4$ afforded 1,3-diol **53** in a five-step protocol from **26** in higher overall yields than the previous route, with the C15–C18 side chain [7b].

With the carbon framework in hand, advanced intermediates **46** and **53** were converted to the C1–C14 (**54**) [7a] and C15–C30 (**55**) [7b] subunits of dolabelide C respectively, setting the stage for final efforts en route to the total synthesis of the natural product (Scheme 4.20).

The synthesis of bicyclic phosphate triester **26** has been accomplished via a desymmetrization of C_2-symmetric diol **24** using a phosphate-tether mediated coupling/RCM sequence with allyl alcohol. The facile assembly of complex polyol synthons **54** and **55** can be realized by exploiting selective reactivity patterns within **26**, namely, CM, cuprate additions, hydroboration, and hydrogenation. Overall, this work represents the first report of using a phosphate-mediated approach to constructing complex intermediates contained in a natural product. Additional biologically active natural products bearing the 1,3,4-*anti,anti*-stereotriad can be accessed through these phosphate-mediated processes [24].

4.6 Conclusion

The use of temporary tethers in synthesis represents a powerful approach to the coupling of two advanced structures en route to more advanced synthons. RCM plays a key role in the construction of sultone and phosphate heterocycles providing effective methods where both sulfur and phosphorus serve as temporary tethers mediating several selective transformations. The use of sultones to construct 1,3-dienols possessing an internal *Z*-olefin has been utilized in the synthesis of the originally proposed structure of (±)-mycothiazole and has the potential of expanding to other biologically active natural products. Phosphate tethers have served a central role in the construction of asymmetric 1,3-*anti*-diol subunits, which have been demonstrated in the construction of both the C1–C14 and C15–C30 subunits of dolabelide C [7]. Additional uses of these versatile heterocycles are in order and will be reported in future studies.

Acknowledgment

The authors would like to thank the National Science Foundation (NSF CHE-0503875) and the National Institute of General Medical Sciences (NIH RO1 GM077309) for their kind and generous support of their program. The authors also kindly acknowledge Materia, Inc., for supplying metathesis catalyst and helpful suggestions.

References

1 (a) Fensterbank, L., Malacria, M., and Sieburth, S. (1997) *Synthesis*, **8**, 813–854; (b) Gauthier, D.R., Zandi, K.S., and Shea, K.J. (1998) *Tetrahedron*, **54**, 2289–2338.

2 (a) White, J.D. and Carter, R.G. (2001) in *Science of Synthesis: Houben-Weyl Methods of Molecular Transformations*, vol. 4 (ed. I. Fleming), Georg Thieme Verlag, New York, pp. 371–412 and references cited therein; (b) Evans, P.A., Cui, J., Gharpure, S.J., Polosukhin, A., and Zhang, H.R. (2003) *J. Am. Chem. Soc.*, **125**, 14702–14703; (c) Cox, L.R. and Ley, S.V. (2000) in *Templated Organic Synthesis* (eds F. Diederich and P.J. Stang), Wiley-VCH Verlag GMBH, Weinheim, pp. 275–395.

3 Nosse, B., Schall, A., Jeong, W.B., and Reiser, O. (2005) *Adv. Synth. Catal.*, **347**, 1869–1874.

4 (a) Whitehead, A., McReynolds, M.D., Moore, J.D., and Hanson, P.R. (2005) *Org. Lett.*, **7**, 3375–3378; (b) Waetzig, J.D. and Hanson, P.R. (2006) *Org. Lett.*, **8**, 1673–1676; (c) Whitehead, A., McParland, J.P., and Hanson, P.R. (2006) *Org. Lett.*, **8**, 5025–5028.

5 (a) Le Flohic, A., Meyer, C., Cossy, J., Desmurs, J.-R., and Galland, J.-C. (2003) *Synlett*, **5**, 667–670; (b) Le Flohic, A., Meyer, C., and Cossy, J. (2006) *Tetrahedron*, **62**, 9017–9037.

6 Le Flohic, A., Meyer, C., and Cossy, J. (2005) *Org. Lett.*, **7**, 339–342.

7 (a) Waetzig, J.D. and Hanson, P.R. (2008) *Org. Lett.*, **10**, 109–112; (b) Whitehead, A., Waetzig, J.D., Thomas, C.D., and Hanson, P.R. (2008) *Org. Lett.*, **10**, 1421–1424.

8 (a) Karsch, S., Schwab, P., and Metz, P. (2002) *Synlett*, **12**, 2019–2022; (b) Karsch, S., Freitag, D., Schwab, P., and Metz, P. (2004) *Synthesis*, **10**, 1696–1712.

9 (a) Roberts, D.W. and Williams, D.L. (1987) *Tetrahedron*, **43**, 1027–1062; (b) Burglass, A.J. and Tillett, J.G. (1991) in *The Chemistry of Sulfonic Acids, Esters and their Derivatives* (eds S. Pataï and Z. Rappoport), John Wiley & Sons, Inc., New York, pp. 789–878; (c) Metz, P. (1998) *J. Prakt. Chem.*, **340**, 1–10.

10 Palytoxin: (a) Uemura, D., Ueda, K., Hirata, Y., Naoki, H., and Iwashita, T. (1981) *Tetrahedron Lett.*, **22**, 2781–2784; Ostreocin: (b) Ukena, T., Satake, M., Usami, M., Oshima, Y., Naoki, H., Fujita, T., Kan, Y., and Yasumoto, T. (2001) *Biosci. Biotechnol. Biochem.*, **65**, 2585–2588; (c) (+)-Discodermolide: Gunasekera, S.P., Gunasekara, M., and Longley, R.E. (1990) *J. Org. Chem.*, **55**, 4912–4915; additions and corrections (d) Gunasekera, S.P., Gunasekera, M., and Longley, R.E. (1991) *J. Org. Chem.*, **56**, 1346; Dictyostatin: (e) Pettit, G.R., Chichacz, Z.A., Gao, F., Boyd, M.R., and Schmidt, J.M. (1994) *J. Chem. Soc., Chem. Commun.*, 1111–1112; (f) Paterson, I., Britton, R., Delgado, O., Meyer, A., and Poullennec, K.G. (2004) *Angew. Chem. Int. Ed.*, **43**, 4629–4633; (g) Shin, Y., Fournier, J.-H., Fukui, Y., Brückner, A.M., and Curran, D.P. (2004) *Angew. Chem. Int. Ed.*, **43**, 4634–4637; Neodihydrohistrionicotoxin: (h) Tokuyama, T., Yamamoto, J., Daly, J.W., and Highet, R.J. (1983) *Tetrahedron*, **39**, 49–53; Mycothiazole: (i) Crews, P., Kakou, Y., and Quiñoà, E.J. (1988) *J. Am. Chem. Soc.*, **110**, 4365–4368.

11 Palladium/Nickel Cross Coupling: (a) Diederich, F. and Stang., P.J. (eds) (1998) *Metal-Catalyzed Cross-Coupling Reactions*, Wiley-VCH Verlag GmbH, Weinheim; (b) Nerenburg, J.B., Hung, D.T., Somers, P.K., and Schreiber, S.L. (1993) *J. Am. Chem. Soc.*, **115**, 12621–12622; (c) Sugiyama, H., Yokokawa, F., and Shioiri, T. (2003) *Tetrahedron*, **59**, 6579–6593; (d) Porée, F.-H., Barbion, J., Dhulut, S., Betzer, J.-F., Pancrazi, A., and Ardisson, J. (2004) *Synthesis*, **18**, 3017–3022; Nozaki-Hiyama-Kishii Coupling: (e) Armstrong, R.W., Beau, J.-M., Cheon, S.H., Christ, W.J., Fujioka, H., Ham, W.-H., Hawkins, L.D., Jin, H., Kang, S.H., Kishi, Y., Martinelli, M.J., McWhorter, W.W. Jr., Mizuno, M., Nakata, M., Stutz, A.E., Talamas, F.X., Taniguchi, M., Tino, J.A., Ueda, K., Uenishi, J., White, J.B., and Yonaga, M. (1989) *J. Am. Chem. Soc.*, **111**, 7525–7530; Wittig Olefination: (f) Stork, G. and Zhao, K. (1989) *Tetrahedron Lett.*, **30**, 2173–2174; Allylic Organometallic Species: (g) Ikeda, Y., Ukai, J., Ikeda, N., and Yamamoto, H. (1987) *Tetrahedron*, **43**, 723–730; (h) Hodgson, D.M. and Wells, C. (1992) *Tetrahedron Lett.*, **33**, 4761–4762; (i) Tsai, D.J.S. and Matteson, D.S. (1981) *Tetrahedron Lett.*, **22**, 2751–2752; (j) Roush, W.R. and Grover, P.T. (1992) *Tetrahedron*, **48**, 1981–1998; Cu-Promoted Coupling with Alkynyl Silane: (k) Marshall, J.A., Chobanian, H.R., and Yanik, M.M. (2001) *Org. Lett.*, **3**, 4107–4110.

12 Structure of mycothiazole was later determined that the C14-C15 was incorrectly assigned E instead of Z: Sonnenschein, R.N., Johnson, T.A.,

Tenney, K., Valeriote, F.A., and Crews, P. (2006) *J. Nat. Prod.*, **69**, 145–147.

13 (a) Corey, E.J. and Boaz, N.W. (1984) *Tetrahedron Lett.*, **25**, 3063–3066; (b) Yanagisawa, A., Noritake, Y., Nomura, N., and Yamamoto, H. (1991) *Synlett*, 251–253; (c) Lipshutz, B. (1990) *Synlett*, 119–128; (d) Magid, R.M. (1980) *Tetrahedron*, **36**, 1901–1930; (e) Piarulli, U., Daubos, P., Claverie, C., Roux, M., and Gennari, C. (2003) *Angew. Chem. Int. Ed.*, **42**, 234–235.

14 (a) Takai, K., Oshima, K., and Nozaki, H. (1980) *Tetrahedron Lett.*, **21**, 2531–2534; (b) Sato, M., Takai, K., Oshima, K., and Nozaki, H. (1981) *Tetrahedron Lett.*, **22**, 1609–1612; (c) Fugami, K., Oshima, K., and Utimoto, K. (1987) *Chem. Lett.*, 2203–2206; (d) Nicolaou, K.C. and Namoto, K. (1998) *Chem. Commun.*, 1757–1758; (e) Lepifre, F., Clavier, S., Bouyssou, P., and Coudert, G. (2001) *Tetrahedron*, **57**, 6969–6975; (f) Kadota, I., Takamura, H., Sato, K., and Yamamoto, Y. (2001) *Tetrahedron Lett.*, **42**, 4729–4731.

15 Morin, M.D. and Rychnovsky, S.D. (2006) *Org. Lett.*, **7**, 2051–2053.

16 (a) Bartlett, P.A. and Jernstedt, K.K. (1977) *J. Am. Chem. Soc.*, **99**, 4829–4830; (b) Bartlett, P.A. and Jernstedt, K.K. (1980) *Tetrahedron Lett.*, **21**, 1607–1610.

17 (a) Letsinger, R.L., Finnan, J.L., Heavner, G.A., and Lunsford, W.B. (1975) *J. Am. Chem. Soc.*, **97**, 3278–3279; (b) Beaucage, S.L. and Caruthers, M.H. (1996) in *Bioorganic Chemistry: Nucleic Acids* (ed. S.M. Hecht), Chapter 2, Oxford University Press, New York, pp. 36–74.

18 (a) Burke, S.D., Muller, N., and Beaudry, C.M. (1999) *Org. Lett.*, **1**, 1827–1829; (b) Burke, S.D. and Voight, E.A. (2001) *Org. Lett.*, **3**, 237–240; (c) Lambert, W.T. and Burke, S.D. (2003) *Org. Lett.*, **5**, 515–518.

19 (a) Deslongchamps, P. (1983) *Stereoelectronic Effects in Organic Chemistry*, Pergamon Press, Oxford; (b) Vande Griend, L.J., Verkade, J.G., Pennings, J.F.M., and Buck, H.M. (1977) *J. Am. Chem. Soc.*, **99**, 2459–2463.

20 Pearson, R.G. (1963) *J. Am. Chem. Soc.*, **85**, 3533–3539.

21 (a) Westheimer, F.H. (1987) *Science*, **235**, 1173–1178; (b) Thatcher, G.R.J. and Kluger, R. (1989) Mechanism and catalysis of nucleophilic substitution in phosphate esters. *Adv. Phys. Org. Chem.*, **25**, 99–265; (c) Miyashita, K., Ikejiri, M., Kawasaki, H., Maemura, S., and Imanishi, T. (2003) *J. Am. Chem. Soc.*, **125**, 8238–8243; (d) Cox, J.R. Jr. and Ramsay, J.O.B. (1964) *Chem. Rev.*, **64**, 317–352; (e) Bunton, C.A. (1970) *Acc. Chem Res.*, **3**, 257–265; (f) Westheimer, F. (1968) *Acc. Chem. Res.*, **1**, 70–78; (g) Gorenstein, D.G., Rowell, R., Taira, K. (1981) *Stereoelectronic Effects in Phosphate Esters*, ACS Symposium Series, vol. 171, pp. 69–75; (h) Fanni, T., Taira, K., Gorenstein, D.G., Vaidyanathaswamy, R., and Verkade, J.G. (1986) *J. Am. Chem. Soc.*, **108**, 6311–6314 and references cited therein.

22 (a) Yanagisawa, A., Noritake, Y., Nomura, N., and Yamamoto, H. (1991) *Synlett*, **4**, 251–253; (b) Belelie, J.L. and Chong, M. (2001) *J. Org. Chem.*, **66**, 5552–5555; (c) Calaza, M.I., Hupe, E., and Knochel, P. (2003) *Org. Lett.*, **5**, 1059–1061; (d) Piarulli, U., Daubos, P., Claverie, C., Roux, M., and Gennari, C. (2003) *Angew Chem. Int. Ed. Engl.*, **42**, 234–236; (e) Piarulli, U., Claverie, C., Daubos, P., Gennari, C., Minnaard, A.J., and Feringa, B.L. (2003) *Org. Lett.*, **5**, 4493–4496; (f) Kacprzynski, M.A. and Hoyveda, A.H. (2004) *J. Am. Chem. Soc.*, **126**, 10676–10681; (g) Larsen, A.O., Leu, W., Oberhuber, C.N., Campbell, J.E., and Hoyveda, A.H. (2004) *J. Am. Chem. Soc.*, **126**, 11130–11131.

23 Chatterjee, A.K., Choi, T.-L., Sanders, D.P., and Grubbs, R.H. (2003) *J. Am. Chem. Soc.*, **125**, 11360–11370.

24 Dolabelides A-D: (a) Ojika, M., Nagoya, T., and Yamada, K. (1995) *Tetrahedron Lett.*, **36**, 7491–7494; (b) Suenaga, K., Nagoya, T., Shibata, T., Kigoshi, H., and Yamada, K. (1997) *J. Nat. Prod.*, **60**, 155–157; Salicylihalamides: (c) Wu, Y., Seguil, O.R., and

De Brabander, J.K. (2000) *Org. Lett.*, **2**, 4241–4244; (d) Holloway, G.A., Hugel, H.M., and Rizzacasa, M.A. (2003) *J. Org. Chem.*, **68**, 2200–2204; (e) Snider, B. and Song, F. (2001) *Org. Lett.*, **3**, 1817–1820; Bitungolides A-F: (f) Sirirath, S., Tanaka, J., Ohtani, I.I., Ichiba, T., Rachmat, R., Ueda, K., Usui, T., Osada, H., and Higa, T. (2002) *J. Nat. Prod.*, **65**, 1820–1823; Rhizoxin D: (g) Lafontaine, J.A., Provencal, D.P., Gardelli, C., and Leahy, J.W. (1999) *Tetrahedron Lett.*, **40**, 4145–4148; (+)-Discodermolide: (h) Smith, A.B.III, Kaufman, M.D., Beauchamp, T.J., LaMarche, M.J., and Arimoto, H. (1999) *Org. Lett.*, **1**, 1823–1826.

25 Myers, A.G., Zheng, B., and Movassaghi, M. (1997) *J. Org. Chem.*, **62**, 7507.

26 (a) Hughes, G., Lautens, M., and Wen, C. (2000) *Org. Lett.*, **2**, 107–110; (b) Chau, A., Paquin, J.-F., and Lautens, M. (2006) *J. Org. Chem.*, **71**, 1924–1933.

27 Lucas, B.S., Luther, L.M., and Burke, S.D. (2004) *Org. Lett.*, **6**, 2965–2968.

5
Synthesis of Natural Products Containing Macrocycles by Alkene Ring-closing Metathesis

Ana Gradillas and Javier Pérez-Castells

5.1
Introduction

The impact of metathesis in modern synthetic chemistry is evidenced by the number of publications and reviews that have appeared in the last decade, some of them covering the specific field of macrocyclization in natural product synthesis [1]. It is therefore difficult not to repeat the reports done by other authors or duplicate their critical views on the chemistry. This chapter is exclusively devoted to the total syntheses of natural bioactive macrocycles involving diene and the diene–ene metathesis, and focuses on the reaction conditions used in the macrocyclization key step and on the stereochemical outcome of the reaction. As we shall see, the choice of the catalyst, solvent, temperature, concentration, and reaction time is crucial in these transformations. In addition, we show how subtle variations on the structure of the substrate lead to modifications in the conformational organization of the molecule, which are critical for the success of ring-closing metathesis (RCM). In particular, the substitution pattern, the steric hindrance of the substituents, the ring size to be formed, and the presence of coordinating heteroatoms are particularly important as they have an important influence on the outcome of the reaction. This chapter has been organized around the various families of natural products in order to compare the reaction conditions for compounds bearing structural similarities. We have deliberately decided to reveal only the information regarding the metathesis step and thus not disclose other details of the syntheses.

Synthetic efforts toward macrocycles are often laborious and unpractical. The classical strategies involved are macrolactonizations, macrolactamizations, and macroaldolizations. The macrocyclization approach encounters several problems including the competition between the desired intramolecular reaction and intermolecular processes, which result in the formation of polymers. RCM has the advantage of being compatible with a wide variety of functional groups and forming double bonds that can eventually be converted into other functionalities. The main disadvantages are the control of the stereochemistry of the newly formed double bond and the optimization of the yield. On the other hand, the concomitant loss of ethylene makes this reaction highly atom economical and entropically

Metathesis in Natural Product Synthesis: Strategies, Substrates and Catalysts. Edited by Janine Cossy, Stellios Arseniyadis, and Christophe Meyer

ISBN: 978-3-527-32440-8

driven. Regarding the prediction of the configuration of the newly formed double bond, cyclization of 1,6-, 1,7-, 1,8-, and 1,9-dienes to, respectively, 5-, 6-, 7-, and 8-membered rings generally gives the thermodynamic (*Z*)-alkenes, while larger rings are obtained as *E*/*Z* mixtures where the (*E*)-isomer is dominant. For cycle going from 10- to 14-membered rings, however, the outcome is not as trivial and the factors that establish the *E*/*Z* selectivity still remain unclear. In addition, the *E*/*Z* ratio of the product may be influenced by secondary isomerizations mediated by the metathesis complexes.

In our study, *macrocycles* are defined as compounds containing at least one ring with 12 or more covalently connected atoms. It is worth noting that they constitute more than 3% of the total amount of natural products described to date.

Approximately, 35–40% of all current trade drugs are natural products, or analogs [2]. The presence of macrocycles is particularly common in antitumorals, antibiotics, and antifungals. Nature has done a selection along millions of years to reach bioavailable compounds with good ability to interact with natural macromolecules. The balance between specificity and affinity allows these products, which come generally from bacteria and fungi, to interact efficiently not only with their natural molecules but also with certain receptors of mammals, thus becoming useful for humans. Probably there are not so huge differences between the structure of the natural targets and those present in humans, but, obviously, there is a great functional difference [3].

Using natural products to access leads is therefore highly desirable, in particular, naturally occurring macrocycles that have potent biological activities. As they usually have complex structures, with a variety of chemical motifs and ring sizes, their total synthesis plays a crucial role in order to confirm their structure and to develop derivatives that can modulate the biological activity or the pharmacokinetic properties. Macrocycles have equilibrium between flexibility and conformational preorganization to achieve optimal binding properties with respect to their biological target. However, they are many times toxic and cause undesired effects, which make necessary their derivatization.

With regard to the structures, the most common ring size goes from 11- to 16-membered rings, with a special incidence for 14-membered rings. This group is dominated either by cycles containing only carbon atoms or by macrocyclic lactones and ethers. The first comes biogenetically from the mevalonate pathway and comprises compounds with around 20 carbons that correspond to diterpenoids. One particularly interesting group of macrocycles are those bearing aromatic rings in their structure including phenyl ethers, *ansa* cycles, and biphenyl structures. They generally have cycles going from 14- to 20-membered rings. The other abundant group is formed by oxygen containing compounds, including macrolactones with a possible polyketide origin. These compounds are generally named macrolides, and some of them have two lactone groups and are known as *dienolides*. In addition, some 13- to 15-membered oxygen containing macrocycles are not olides but ethers. It is presumed that they come from alkaloid origin. The majority of the nitrogen containing macrocycles are lactams, and generally have larger cycles reaching sometimes 30-membered cycles. Many of these are cyclopeptides, possessing three to six amino acids. The presence of nonnatural amino acids and unnatural stereochemistry is frequent [4].

5.2
Organization of the Chapter

In this chapter, a special emphasis is given on the results of the RCM macrocyclization step. This key step is illustrated in schemes in which the starting material used for the metathesis, the resulting product, the yield and the eventual ratio of stereoisomers, the best reaction conditions used, the structure of the natural product, and the references are showed. The classification of the compounds follows the biogenetic one and the family group.

The variety in natural macrocyclic structures is high but the biosynthetic pathways that give these products are limited to four main origins and several subsequent transformations, including the cyclization step. The biosynthesis of macrocycles can follow the polyketide pathway, that is, the acetate metabolism, which leads to iterative C2 unit couplings and an aldol or Claisen type cyclization. The second pathway is the mevalonic pathway, that is, isoprenoid metabolism, involving the coupling of C5 units and cyclization, which gives mostly diterpenes. A third group of macrocycles come from the metabolism of amino acids including cyclopeptides, cyclodepsipeptides, and alkaloids. Finally, the metabolism of sugars leads to cyclic oligosaccharides and glycolipids; the latter is a combination with the first pathway, as they include fatty acid units in their structure.

Following the biogenetic classification, we first address macrolactones, divided into the main family groups that have received synthetic attention via RCM (resorcinylic, salycilates, other antibiotic macrolides, macrocyclic musks, epothilones, amphidinolides, other polyketides, and natural cyclophanes), then the terpenoids, followed by the macrocycles obtained from the amino acid metabolism (lactams, depsipeptides, alkaloids), and finally the glycolipids.

From the historical point of view this story begins in the late 1990s, mainly with the RCM-based epothilone syntheses [5], and the early synthesis of the 18-membered α,β-unsaturated macrolide aspicilin [6]. It is worth noting that previously, some groups had disclosed RCM to form macrocycles without completing the total synthesis of the natural products [7]. The first example of regioselective diene–ene metathesis was disclosed in 1999, when the synthesis of simplified macrolide analogs of the immunosuppressant sanglifehrin was carried out [8]. With regard to the reaction conditions used, three generations of ruthenium complexes and the molybdenum Schrock catalyst have been used. Catalyst [Ru]-**I** is the cheapest, although it is thermally unstable and with low reactivity implying its failure with substituted olefins [9]. However, in some cases, this low reactivity is desirable and in the most recent contributions these complexes are still used. Thermal stability plays a critical role in the lifetime of the species and turnover number (TON), and it is known that [Ru]-**I** decomposes after 1 hour at 100 °C in toluene-d_8 up to 75%. Some derivatives of this first-generation catalyst such as [Ru]-**IV** and

[Ru]-**V** improve the kinetic initiation and have been used, although generally in early contributions. The second-generation catalysts [Ru]-**II** and [Ru]-**VI** give better results with substituted olefins, as does also the Hoveyda–Grubbs catalyst [Ru]-**III**. Because of the incorporation of the nitrogen containing heterocycle (NHC), such as in [Ru]-**III**, the stability of the ruthenium complex is increased (stable after 1 hour at 100 °C); it can be recovered by chromatography and reused. This latter complex has been modified also by substitution of the aromatic ring, giving rise to a new family of complexes that can be named "third-generation" catalysts. It is worth noting that the newly described water compatible ruthenium catalysts [Ru]-**VII** and [Ru]-**VIII** have not yet found use in macrocyclization chemistry. The molybdenum species [Mo]-**I** is highly reactive but sensitive to water and oxygen and not compatible with several functionalities, thus it has been much less used than the other catalysts in macrocyclization [10].

5.3 Macrocyclic Polyketides

Many macrocyclic structures are polyketides. The majority of them have one or more lactone linkages (poliolides) and are generally referred to as *macrolides*. Since they present interesting biological activities, they are ideal targets for synthetic studies. These compounds are isolated from bacteria or fungi and are in general of marine origin [11].

5.3.1 Resorcinylic Macrolides

The resorcinylic macrolides (also known as *resorcinylic acid lactones* or RALs) are a family of naturally occurring homologous macrolides. They are derived from polyketide biogenesis in different fungi and have been isolated from different fungal strains [12]. They have unique chemical structures and potent biological activities (antitumoral, antibiotic, and antimalarial activities) [13].

Among these compounds the most popular are radicicol (**3**) and its derivatives. In addition, aigialomycin D (**7**), zeranol (**25**), zeralenone (**28**), and the pochonins (**16**, **19**, **22**) share the basic skeleton consisting of a 14-membered ring lactone that includes a 1,2-phenylene. The major contribution in this field was achieved by Danishefsky *et al.* who, despite the relative similarity of the target molecules, have developed a wide variety of synthetic strategies and conditions. The disconnections and the studies on the RCM reaction conditions developed by this group were reviewed by Winssinger *et al.* [12]. Among the biological activities of the resorcinylic macrolides, the ability of these compounds to inhibit Hsp90 (heat shock protein 90) is outstanding.

The syntheses of radicicol (**3**) are based on RCM of a diene–ene system, which avoids the problem of regioselectivity, compatibility with labile groups such as a vinyl epoxide, the dithiane protecting group, and the *E*/*Z* selectivity [14]. In these

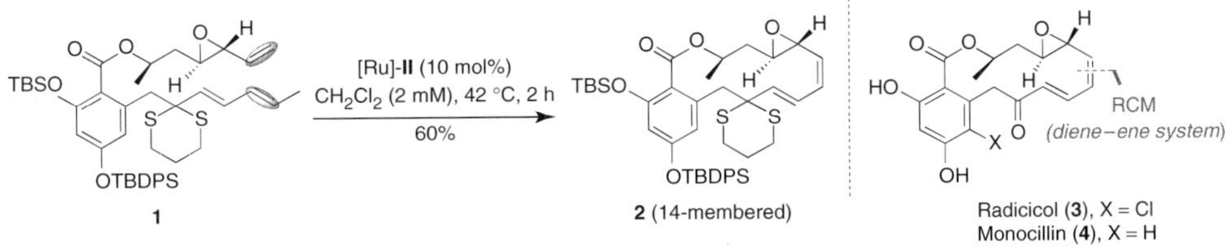

Scheme 5.1

initial syntheses, the role of both the temperature and the concentration appeared to have a tremendous impact on the outcome of RCM [15]. In compound **1**, RCM occurred under mild conditions (2 mM, CH_2Cl_2, 42 °C) to give the monomeric macrocyclic product **2** (Scheme 5.1).

However, when similar conditions were applied to the cyclopropyl precursor ($X = CH_2$), the major isolated product was a dimeric 28-membered ring macrocycle. On the other hand, performing the reaction at 110 °C in toluene and quenching the reaction after a short period resulted in a clean conversion and a 54% yield of the desired macrocycle (Scheme 5.2).

In a more recent and improved approach to cycloproparadicicol (synthetic cyclopropane analog of radicicol) the synthesis was achieved via a stable cobalt complex, which led to the cyclized products as a 2 : 1 mixture of isomers in 57% yield (Scheme 5.3). The complexation of the ethynyl group with $Co_2(CO)_8$ avoids a competitive ene–yne metathesis and brings into proximity the double bonds favoring RCM.

This so-called ynolide methodology was also applied to the synthesis of aigialomycin D (**7**), a natural product with antimalarial and antitumoral activity, and also a CDK_1 and CDK_5 inhibitor, isolated from marine mangrove fungus *Aigialus parvus* (Scheme 5.4) [16]. Oxidative metal decomplexation followed by Diels–Alder reaction using a disiloxydiene was used to elaborate the benzenic

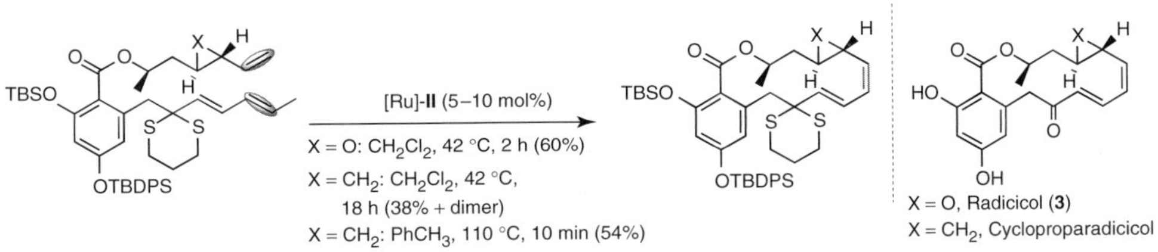

Scheme 5.2

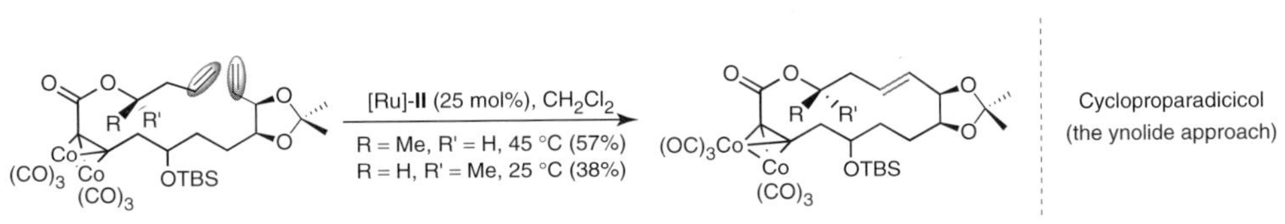

Scheme 5.3

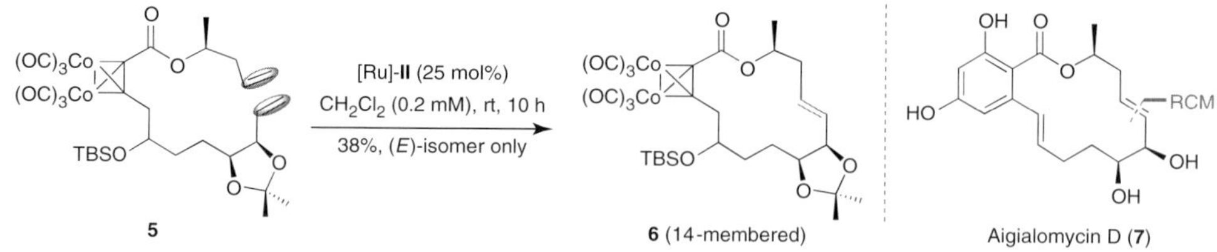

Scheme 5.4

system. The main advantage of this route is the possibility of scaling up the synthesis producing, for example, multigram of cycloproparadicicol. More recently, Winssinger *et al.* reported a flexible polymer-supported synthesis of aigialomycin D (**7**) and a library of analogs thereof [17].

The syntheses of *cis*- and *trans*-resorcilides (**10**, **13**), which are plant growth inhibitors isolated from *Penicillium* sp., were based on RCM of enolizable enones. For the cyclization of diene **8**, a combination of high dilution in refluxing CH_2Cl_2 (0.5 mM) and low catalyst loading (2 mol%) was necessary to reach reasonable yields of **9** (Equation 1, Scheme 5.5). Diene **11** gave *trans*-alkene **12** as the unique stereoisomer in high yield (Equation 2, Scheme 5.5) [18].

Pochonins are a new and interesting group of resorcinylic lactones isolated from *Pochonia chlamydosporia*. Pochonin C (**16**) is active against HSV (Herpes Simple Virus); however, it has less inhibitory activity than radicicol against Hsp90 due to a different conformational behavior of the macrocycle. On the other hand, pochonin D (**22**) can adopt the so-called L-shape bioactive conformation against Hsp90 with a small energy penalty, and is a stable and structurally simple Hsp90 inhibitor. More structurally related to radicicol, pochonin A (**19**) has both Hsp90 and HSV inhibition activity. Winssinger *et al.* have disclosed the syntheses of this family of RALs. Polyene RCM was used for the synthesis of pochocin C (**16**) (Equation 1, Scheme 5.6) [19]. Thus, treatment of intermediate **14** with catalyst [Ru]-**II** in refluxing toluene led exclusively to **15** in 10 minutes. The synthesis of pochonin A

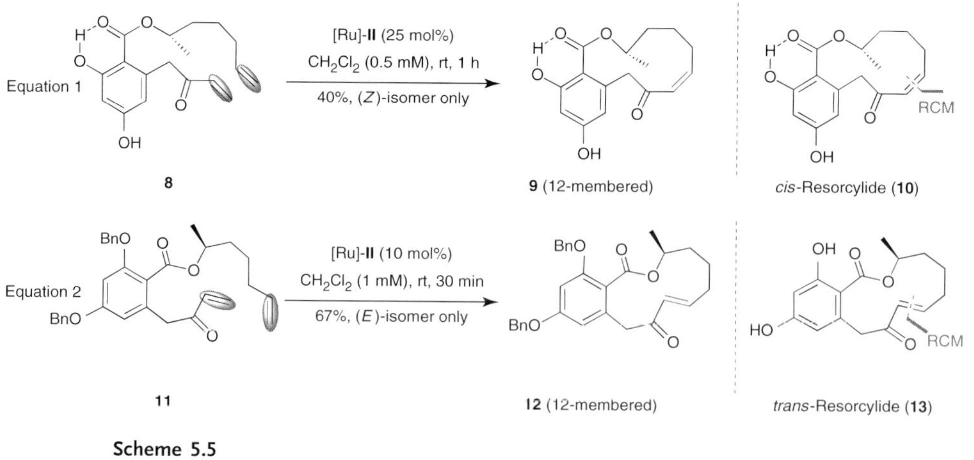

Scheme 5.5

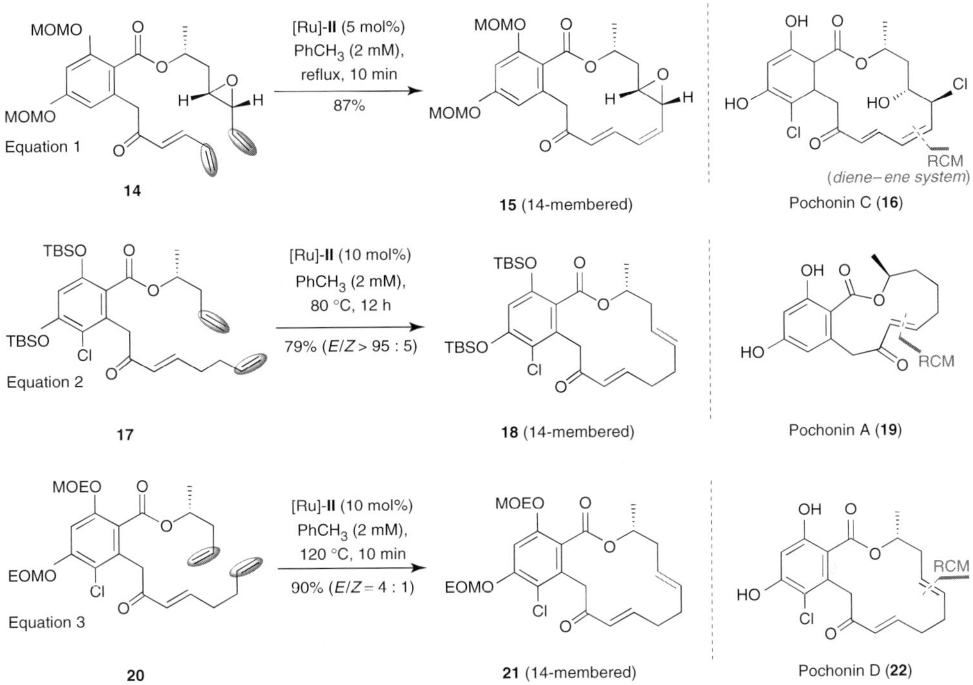

Scheme 5.6

(**19**) [20] and D (**22**) [21] was achieved, respectively, from **17** and **20** using [Ru]-**II** in refluxing toluene (Equations 2 and 3, Scheme 5.6). For the synthesis of **19**, the double bond resulting from RCM was epoxidized.

Included in this resorcinylic group are a number of 12-membered ring lactones. On the basis of the RCM of ketone **23**, Fürstner *et al.* developed a synthetic approach to zeranol (**25**), a nonsteroidal estrogenic growth promoter (Equation 1, Scheme 5.7). The same authors synthesized (*S*)-(−)-zearalenone (**28**) and (*R*)-lasiodiplodin (**31**), two inhibitors of the prostaglandin biosynthesis isolated from the fungus *Botryosphaeria rhodina* [22]. In all these syntheses, a complete lack of reactivity was observed when [Ru]-**I** was utilized. On the contrary, when [Ru]-**II** and [Ru]-**V** were used, the cyclized products were isolated in good to excellent yield (Equations 2 and 3, Scheme 5.7).

5.3.2
Salicylate Macrolides

One closely related family to resorcinylic macrolides are the salicylate enamide macrolides, which are secondary marine metabolites with a 12-membered ring benzolactone and a dienyl enamide side chain. Salicylihalamide A (**34**) is a potent inhibitor of the mammalian vacuolar ATPase and exerts great differential cytotoxicity in the National Cancer Institute (NCI) 60 cell lines [23]. Other members

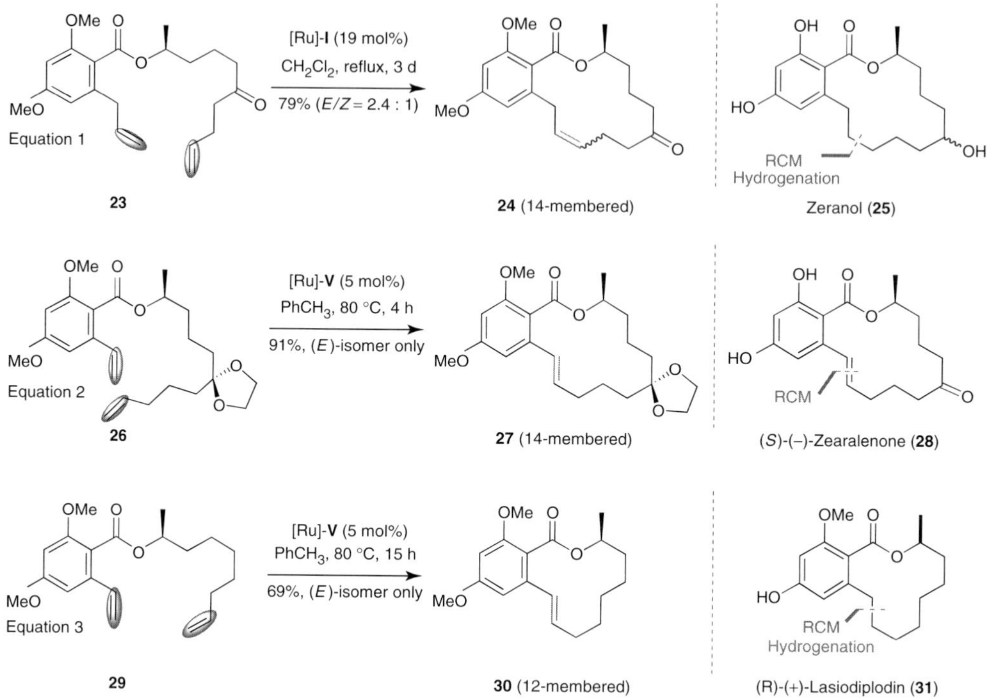

Scheme 5.7

of this family are the apicularens, which bear an identical enamide side chain; the lobatamides A–F; and the oximidines II and III, which possess an enamide side chain terminated by an *O*-methyl oxime [24].

Mitsunobu/RCM strategy, Suzuki cross-coupling, Stille cross-coupling are the three general approaches for the synthesis of salicylihalamides A (**34**) and B (**35**) [25]. After a thorough optimization, the best reaction conditions involved the use of [Ru]-**II** (10 mol%) in toluene at 70 °C for 2 hours (Scheme 5.8).

However, the synthesis of **34** has been the subject of many efforts in order to obtain a good (*E*/*Z*) stereoselectivity and deserve some comments (Table 5.1). The remote phenolic hydroxy group is of importance as it favors the undesired

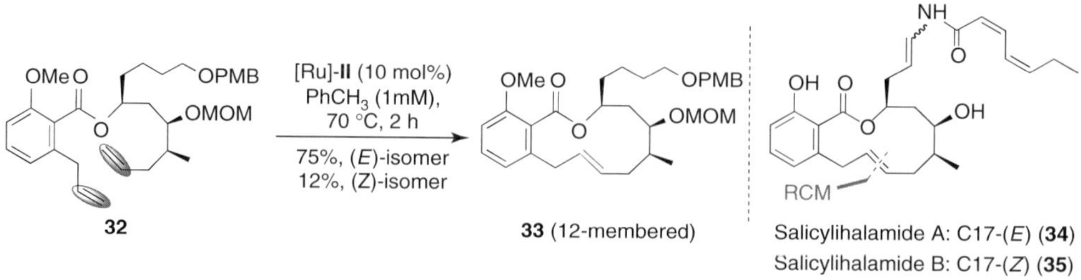

Scheme 5.8

Table 5.1 Conditions for the synthesis of salicylihalamide precursors.

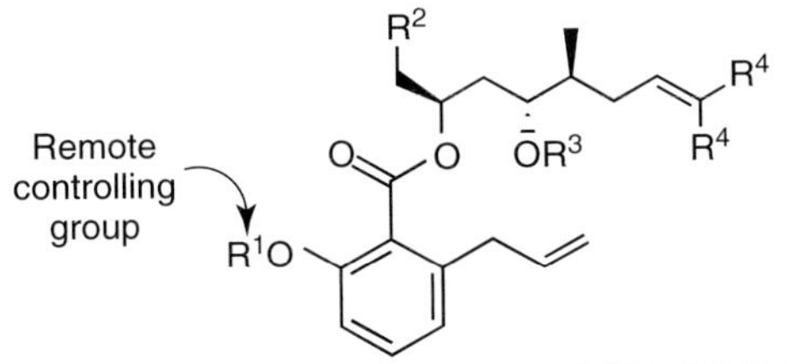

R¹	R²	R³	R⁴	Catalyst	Solvent	T (°C)	Time	Yield (%)	(E/Z)	Reference
Me	CH=CH–CH$_2$OPMB	TBS	H	[Ru]-**I**	CH_2Cl_2	rt	40 min	85	10/1	[26]
TBS	CH=CH–COOMe	H	H	[Ru]-**I**	CH_2Cl_2	Δ	4 d	57	4/1	[27]
H	CH=CH–COOMe	H	H	[Ru]-**I**	CH_2Cl_2	Δ	4 d	Low	*Z*	[27]
Me	CH$_2$–CH$_2$OBn	TBDPS	H	[Ru]-**I**	CH_2Cl_2	Δ	3.0 h	75	9/1	[28]
Me	CH$_2$OPMB	MOM	H	[Ru]-**I**	CH_2Cl_2	Rt	1.3 h	95	9/1	[29]
MOM	CH$_2$OPMB	MOM	H	[Ru]-**I**	CH_2Cl_2	Rt	1.3 h	95	9/1	[29]
MOM	CH$_2$OPMB	MOM	H	[Ru]-**I**	CH_2Cl_2	Rt	18 h	93	5/1	[29]
H	CH$_2$OPMB	MOM	Me	[Ru]-**VI**	$PhCH_3$	80	20 h	69	*Z*	[30]
Me	CH$_2$OPMB	MOM	Me	[Ru]-**VI**	$PhCH_3$	80	1.5 h	93	2/1	[30]
TBS	CH$_2$OPMB	MOM	Me	[Ru]-**VI**	$PhCH_3$	80	1.0 h	91	1/1.5	[30]
MOM	CH$_2$OPMB	MOM	H	[Ru]-**II**	CH_2Cl_2	rt	1.3 h	98	2/1	[29]

(*Z*)-isomer. If the phenolic hydroxy group is protected, the (*E*)-isomer is the major compound. The influence of the catalyst is also crucial, as with catalyst [Ru]-**I**, which was used by Smith *et al.* [26], Snider *et al.* [27], and Labreque *et al.* [28], a high (*E*)-selectivity was observed, especially if the phenolic hydroxy group was protected as a methyl ether (Table 5.1). De Brabander *et al.* [29] showed that this process was under kinetic control. On the other hand, with the second-generation catalysts used by Fürstner *et al.* [30] and de Brabander *et al.*, the equilibrium is reached thus leading to a *E*/*Z* ratio of 2 : 1. An isomerization process can compete during the experiment.

Oximidine II (**38**) and III (**41**), isolated from *Pseudomonas*, were constructed using a relay RCM strategy. Wang *et al.* and Porco *et al.* reported, in the synthesis of oximidine II (**38**), the first example of a successful diene–diene RCM to construct a macrocyclic conjugated triene. A well-defined substrate possessing two differentially functionalized RCM alkene partners was required for the RCM process. The presence of the terminal methyl group on the conjugated diene with an (*E*,*Z*) configuration and the use of [Ru]-**II** catalyst were crucial to avoid oligomeric products or the formation of smaller cycles resulting from the reaction with internal double bonds. Fast addition of the substrate to the catalyst solution resulted in reduced decomposition of the starting material (Scheme 5.9) [31].

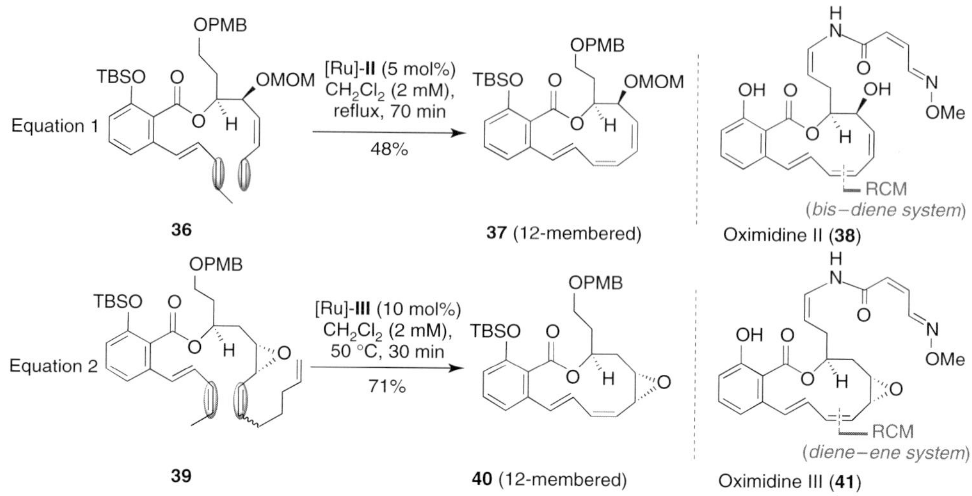

Scheme 5.9

5.3.3 Other Antibiotic Macrolides

Numerous biologically active macrocycles are antibiotics isolated from various microorganisms, with a wide variety of structures and ring sizes ranging from 12- to 28-membered rings.

Fourteen natural antibiotic lactones and diolides synthesized using RCM are shown in Schemes 5.10–5.24. 10-Deoxymethynolide (**44**) is the aglycone of the methymycin family of macrolides and was constructed via an exclusive *E* selective

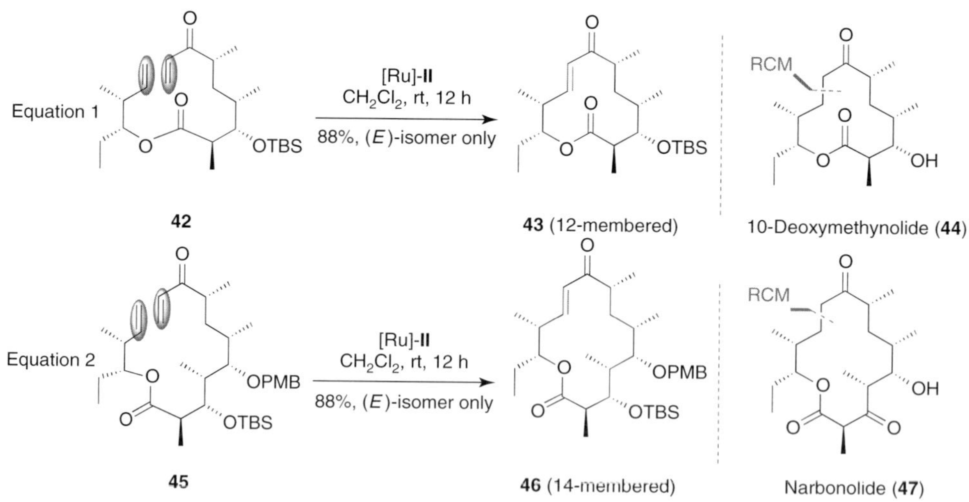

Scheme 5.10

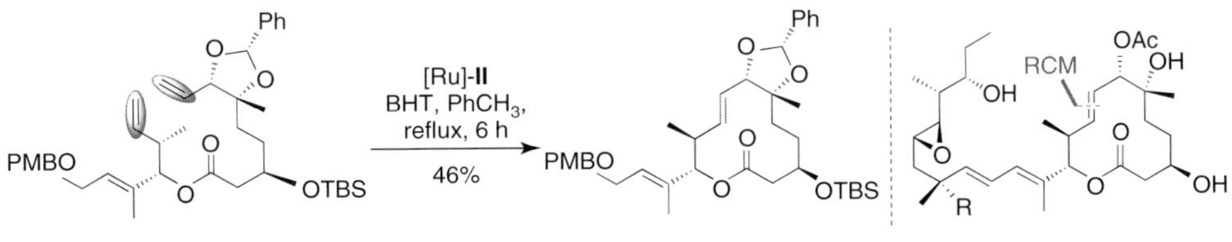

48

49 (12-membered)

Pladienolide B: R = H (50)
Pladienolide D: R = OH (51)

Scheme 5.11

RCM using [Ru]-**II** in CH_2Cl_2 at room temperature (Equation 1, Scheme 5.10). The same conditions were used for the synthesis of narbonolide (**47**), which possesses a 14-membered ring lactone instead of a 12-membered ring. Narbonolide (**47**) is the aglycone of the pikromycin family of antibiotic macrolides (Equation 2, Scheme 5.10) [32].

Recently, the synthesis of two members of methymycins, namely, pladienolides B (**50**) and D (**51**), possessing antineoplasic activity and isolated from *Streptomyces platensis*, was described. These examples used a less efficient RCM but a better catalyst loading was employed (Scheme 5.11) [33].

In addition to these four compounds, other similar 12-membered ring lactones were obtained such as sporiolode B (**54**), isolated from fungal strain *Cladosporium* sp. [34], and (10*S*,12*R*)-10-hydroxy-12-methyl-1-oxacyclododecan-2,5-dione (**57**) [35], which were formed by RCM leading to (*E*/*Z*) mixtures (Equations 1 and 2, Scheme 5.12).

Two early examples disclosed by Fürstner *et al.*, ricinelaidic acid (**60**) and gloeosporone (**63**) (germination self-inhibitors isolated from *Colletotrichum gloeosporioide*), were built by using [Ru]-**I** (Equations 1 and 2, Scheme 5.13). In the synthesis

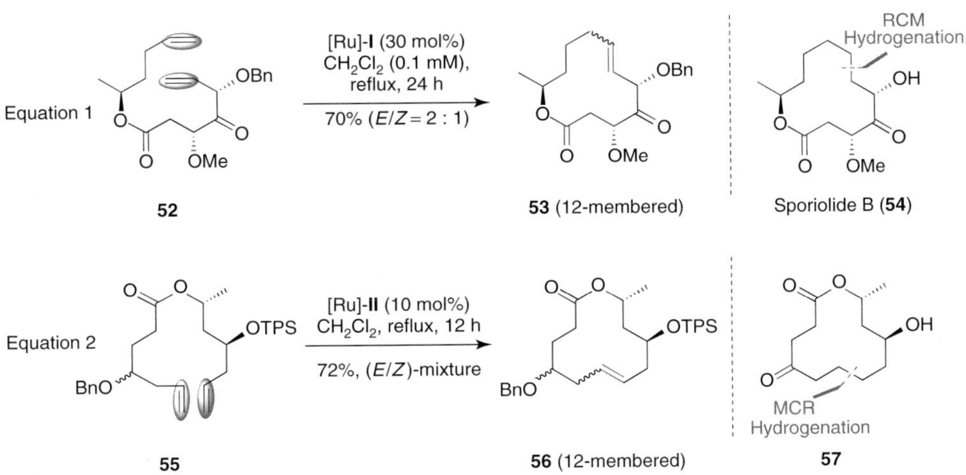

52

53 (12-membered)

Sporiolide B (54)

55

56 (12-membered)

57

Scheme 5.12

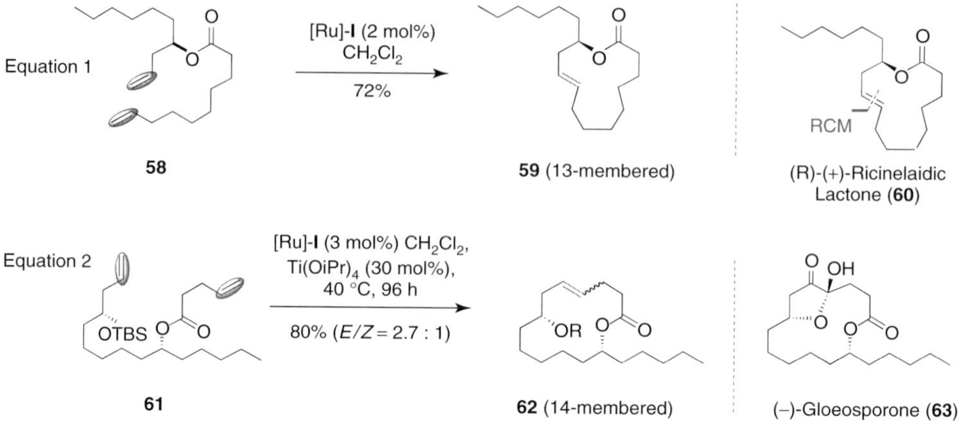

Scheme 5.13

of gloeosporone (**63**), RCM was performed in the presence of a Lewis acid in an attempt to avoid any interaction between the oxygen electron pairs and the catalyst. In both cases, yields were good – although low (*Z*/*E*) selectivities were observed [36].

Migrastatin (**66**), isolated from *S. platensis*, inhibits the migration of human tumor cells. This compound was synthesized via RCM of the fully functionalized tetraene **64** (Scheme 5.14). The reaction was highly chemoselective and proceeded in a highly (*E*)-selective fashion [37].

Among the three RCM-based syntheses of aspicilin (**69**) – a natural product isolated from *Lecanoraceae* lichen – only the most efficient one is reported herein (Scheme 5.15). Important variations in the (*E*/*Z*) selectivity were observed; however, this had no detrimental impact on the synthesis, as hydrogenation of the olefin was required in order to obtain the natural product [38].

Sch 351448 (**72**), isolated from *Micromonospora*, is a novel activator of low-density lipoprotein receptor (LDL-R) promoter with an IC_{50} of 25 nM, which features a 28-membered macrodiolide consisting of two identical hydroxy carboxylic acid units. In its synthesis, intramolecular olefin metathesis of **70** mediated by [Ru]-**II** proceeded smoothly, and the macrodiolide was obtained after a hydrogenation/hydrogenolysis sequence (Scheme 5.16) [39]. Other syntheses of this

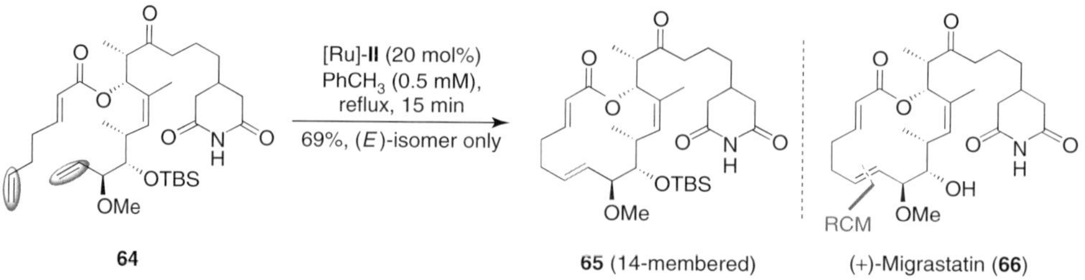

Scheme 5.14

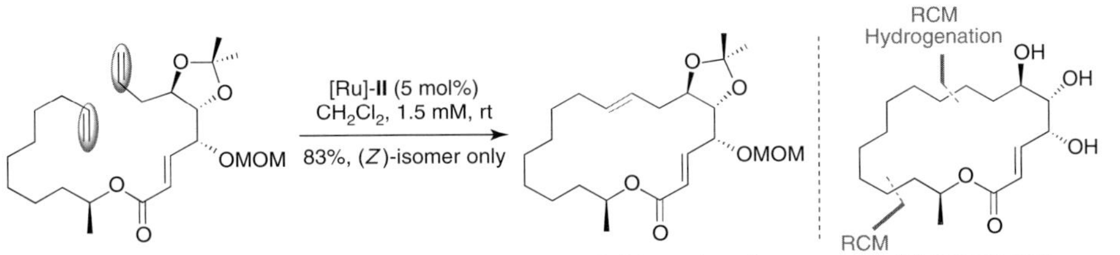

67 68 (18-membered) (+)-Aspicilin (69)

Scheme 5.15

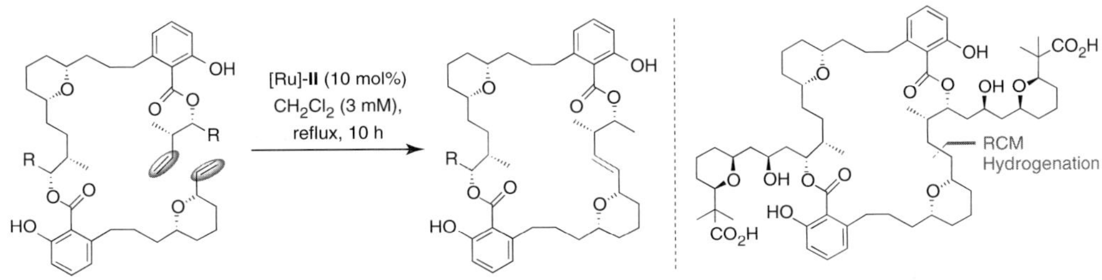

70 71 (28-membered) (+)-Sch 351448 (72)

Scheme 5.16

compound have also appeared in the literature, the most recent one, in 2008, involved a templated olefin metathesis strategy to assemble the C_2-symmetric core of the natural product, and a stepwise fragment coupling to complete the target molecule [40].

Macrosphelides A (**75**) and B (**76**), isolated from *Macrosphaeropsis* sp. FO-5050 and *Periconia byssoides*, have antitumoral properties. For their synthesis, several RCM conditions were used, but 1 equiv. of [Ru]-**II** catalyst and long reaction time were necessary to obtain the cyclized product in reasonable yields (60–65%, Scheme 5.17). The authors pointed out that the addition of a Lewis acid under catalytic conditions was ineffective [41].

Zampanolide (**79**), a macrolide which exhibits significant activity against a variety of tumor cell lines, was recently isolated from the sponge *Dactylospongia* sp. The RCM cyclization step for the synthesis of this compound was done on substrate **77**; however, as with other substrates bearing protected hydroxy groups, the reaction

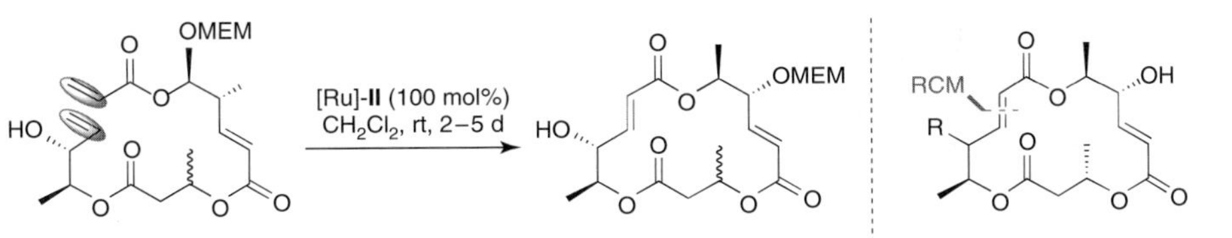

73 74 (16-membered) Macrosphelide A: R = C-OH (**75**)
Macrosphelide B: R = CO (**76**)

Scheme 5.17

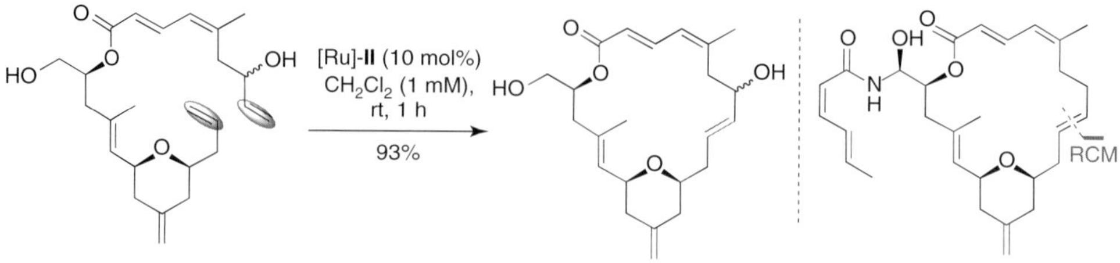

77 78 (20-membered) (−)-Zampanolide (79)

Scheme 5.18

was totally ineffective (Scheme 5.18). In contrast, the conditions allowed the synthesis of the nonnatural enantiomer of the related macrolide dactylolide [42].

Palmerolide A (**82**), a recently disclosed marine natural product that was isolated from *Synoicum adareanum*, possesses potent and selective activity against the melanoma cancer cell line UACC-62. Nicolaou *et al.* reported the total syntheses of five stereoisomers of palmerolide A including the originally proposed (**82**) and the revised [*ent*-(19-*epi*-20-*epi*-**82**)] structures. The RCM cyclization proceeded smoothly under mild conditions (Scheme 5.19) [43].

An effective RCM of **87**, containing no less than 10 double bonds, was successfully used for the synthesis of iejimalides' group – a new group of cytotoxic compounds – recently isolated. RCM reaction gave the desired macrocycles in high yields and *E*-stereoselectivity (Scheme 5.20) [44].

5.3.4
Macrocyclic Musk

Three classes of compounds, namely, nitroarenes, polycyclic benzenoids, and macrocyclic ketones and lactones, possess musky odors. The latter have been used for the longest time in fragrance formulations. Efficient synthetic methods that can be scaled up are necessary to decrease their cost of production. Muscone (**93**) [45] and civetone [46], are two of the most important classical sources of musk odors. Muscone (**93**), the principal musk odorant perfume ingredient from the male musk

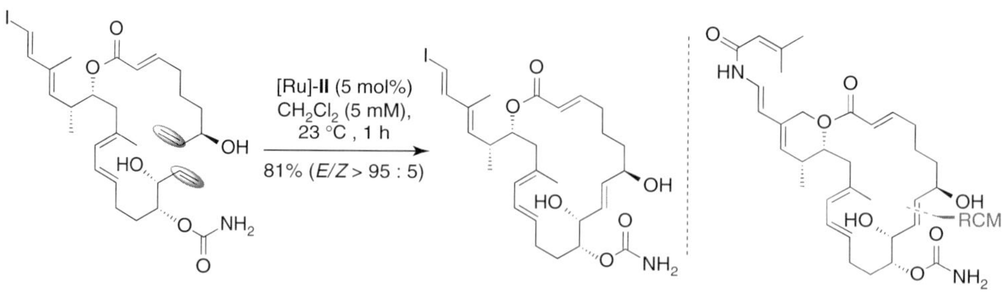

80 81 (20-membered) Palmerolide A (82)

Scheme 5.19

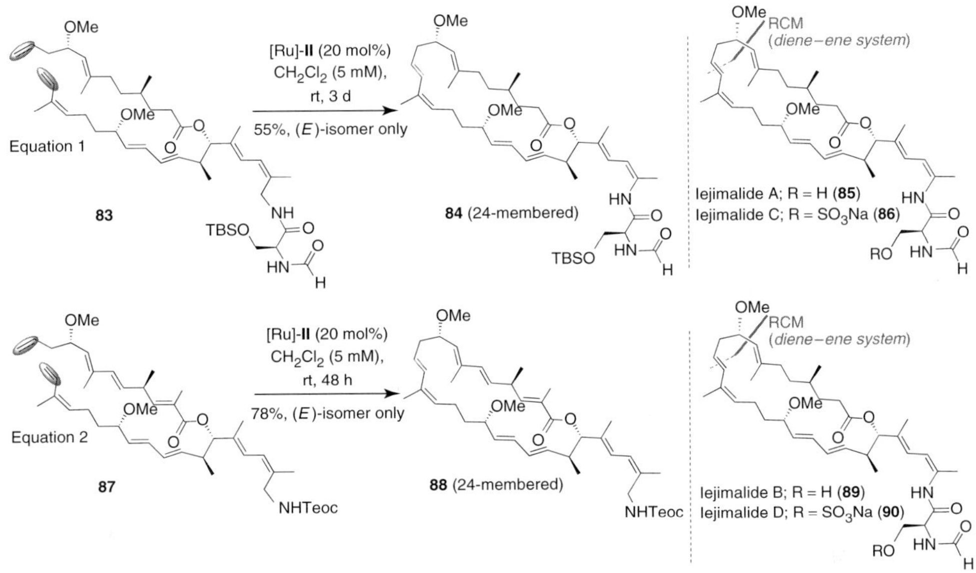

Scheme 5.20

deer (*Moschus moschiferus*) is nowadays produced synthetically by RCM using the conditions shown in Scheme 5.21.

Other relatively simple macrolides synthesized by Fürstner *et al.* are recifeiolide (**96**) – a musk fragrance enhancer isolated from *Cephalosporium recifei* (Equation 1, Scheme 5.22) [47]; exaltolide (**99**) [48] – a musk odorant perfume isolated from *Archangelica officinalis*, which was the first macrolactone to be prepared by RCM (Equation 2, Scheme 5.22); and muscopyridine (**102**) (Equation 3, Scheme 5.22) [49], which is present in the secretion of the ventral glands of male musk deer.

5.3.5
Epothilones

The scope of RCM has been fully illustrated by the synthesis of epothilones (epothilones A–F) and has been reviewed recently [5b, c]. Epothilones are

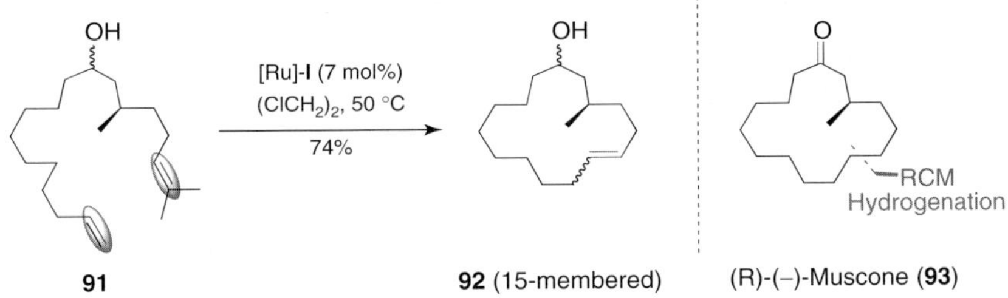

Scheme 5.21

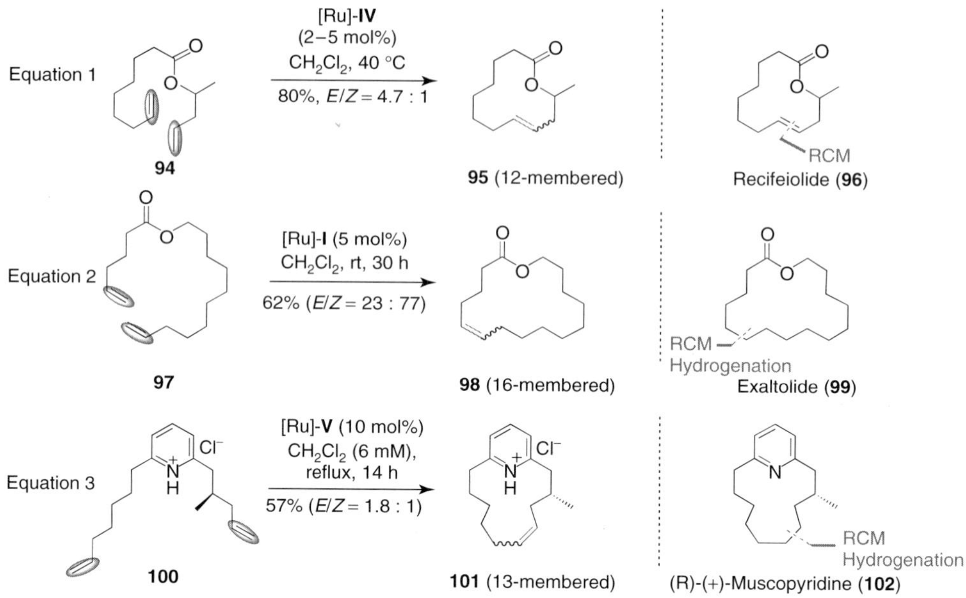

Scheme 5.22

16-membered ring macrolides, first isolated from *Sorangium cellulosum*. Since the discovery of their powerful cytotoxic properties and their taxol-like mode of action, much effort has been undertaken to develop efficient synthesis methodologies that would allow a straight access to these molecules as well as to a variety of structural analogs [50]. In October 2007, the first epothilone derivative, ixabepilone, received FDA approval for the treatment of metastatic breast cancer [51]. The structure of these macrocycles is suitable to RCM-based approaches, although macrolactonization has also been thoroughly used. Two strategies are possible, namely, the C9–C10 and the C12–C13 double bond disconnection. The stereocontrol of the C12–C13 double bond by using RCM is difficult; indeed, none of the catalyst used initially, [Ru]-**I** and [Mo]-**I**, produced a good *Z* selectivity as can be seen from the transformation of **103–104**, which led to a 1 : 1 mixture of isomers. This strategy was used for the synthesis of epothilone C (**105**) (Scheme 5.23) [52].

More recently, a solution was provided by using a silicon-tethered RCM. Hence, a disiloxane was used for the construction of the (*Z*)-C12–C13 double bond via

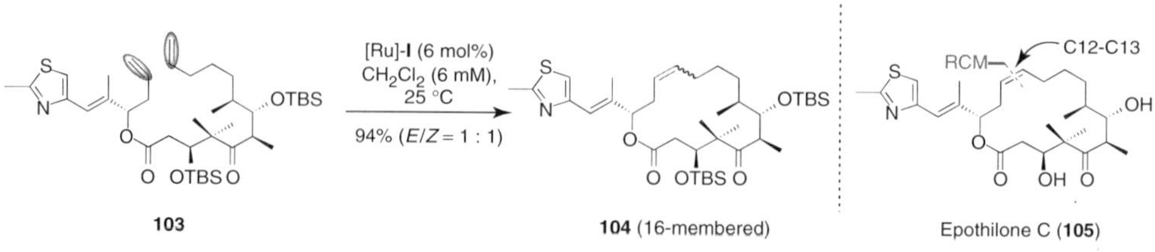

Scheme 5.23

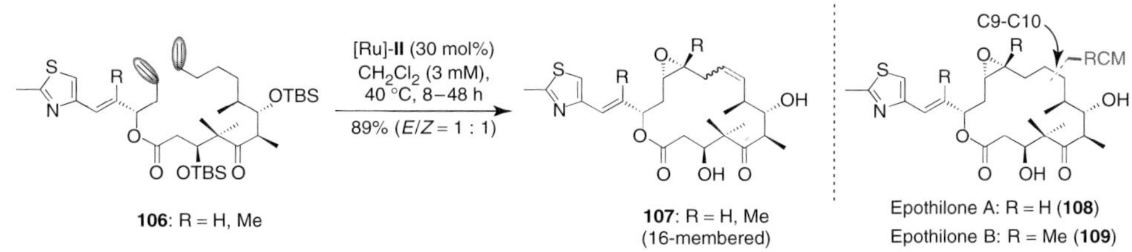

Scheme 5.24

RCM, which led to a nine-membered ring intermediate which was then opened and used for the synthesis of epothilone B and epothilone D [53]. The alternative approach involving the formation of the C9–C10 bond followed by a hydrogenation step was first proposed by Danishefsky *et al.* [52a]. However, it was not until the discovery of more active second-generation ruthenium catalysts, such as [Ru]-**II**, that satisfactory results were obtained. In 2002, the stereoselective total syntheses of epothilone A (**108**) and epothilone B (**109**) by RCM of the epoxy compound **106** in the presence of catalyst [Ru]-**II** was reported [54]. In addition, the epothilone skeleton was also obtained via a diene–ene cyclization using [Ru]-**II** leading to the macrolide in 75% yield (Scheme 5.24).

In a similar approach, Danishefsky *et al.* used **110** as the RCM substrate to obtain epothilone D [55]. Consequently, RCM of **110**, in refluxing toluene in the presence of [Ru]-**II**, gave exclusively the *E*-isomer **111** in high yield. This intermediate was ultimately transformed into epothilone D (**112**), (Scheme 5.25).

An alternative to the previous strategies consisted of an RCM to form the C10–C11 bond of epothilone 490 (**115**), a naturally occurring metabolite. The conditions for RCM were critical in this case, and best results were obtained using [Ru]-**II** as the catalyst in refluxing toluene for a few minutes [56]. These conditions avoided the formation of by-products while increasing the yield of the desired product (Scheme 5.26).

5.3.6 Amphidinolides

Amphidinolides are marine natural products, isolated from symbiotic marine dinoflagellates *Amphidinium* sp. In general, their most outstanding biological activity is their cytotoxic effects. Various subfamilies of these compounds bearing

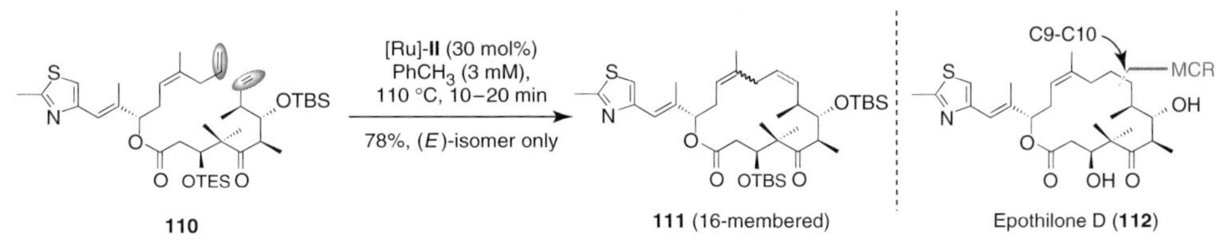

Scheme 5.25

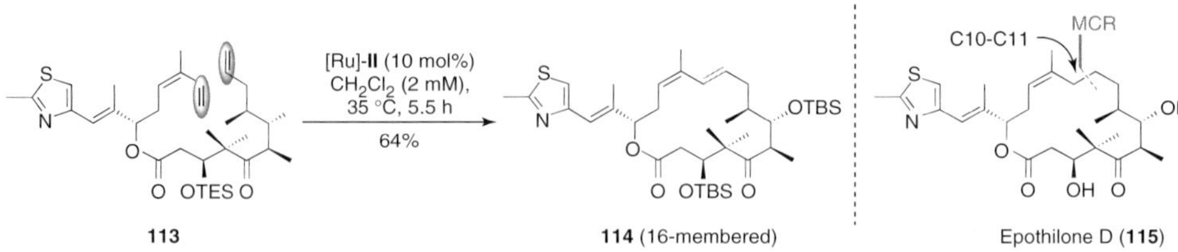

Scheme 5.26

different macrocyclic ring sizes have been described to date. Amphidinolide A (**118**) was synthesized in 57. The highly unsaturated 20-membered ring lactone was formed at a late stage by RCM. To provide regio- and *E*-stereoselectivity, 0.5 equiv. of catalyst [Ru]-**II** was necessary; however, the yield of the cyclized product remained low (35%, Scheme 5.27).

The amphidinolide T subfamily contains a saturated 19-membered ring lactone. The synthesis of a few members of this subfamily has been achieved by Fürstner *et al.* [58]. For the synthesis of amphidinolide T3 (**121**), for instance, good conversion could be attained only in refluxing toluene. The desired ring closure was faster than the decomposition of the catalyst, delivering cycloalkene **120** in 82% yield as an *E*/*Z* mixture of isomers in a 2 : 1 ratio, which was further hydrogenated (Equation 1, Scheme 5.28). Amphidinolides T1, T4, and T5 were prepared in high yield by applying RCM to diene **122** utilizing [Ru]-**VI** in CH_2Cl_2. The corresponding cycloalkene **123** ($E/Z = 6:1$) was then hydrogenated (Equation 2, Scheme 5.28).

More recently, amphidinolide Y was discovered. In the total synthesis of this compound, the formation of the (*E*)-trisubstituted alkene via RCM appeared to be challenging. A critical effect of the remote substituents in RCM was demonstrated. Hence, the cyclization of the C6-ketone **125** in refluxing CH_2Cl_2, using [Ru]-**II** as the catalyst, which was added in portions up to 50 mol%, was successful (Scheme 5.29) [59].

The synthesis of amphidinolide E (**130**) is an example of RCM applied to a polyene. It was necessary to use [Ru]-**I** to avoid the metathesis of an internal bond and decomposition of the starting material, as was observed when using [Ru]-**II** (Scheme 5.30) [60].

In a recent contribution, amphidinolide H (**133**) was constructed from polyene **131** by using RCM cyclization as a key step with the use of [Ru]-**II** catalyst in

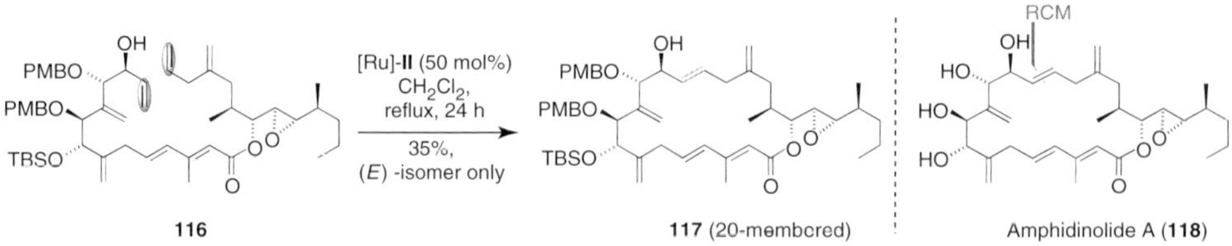

Scheme 5.27

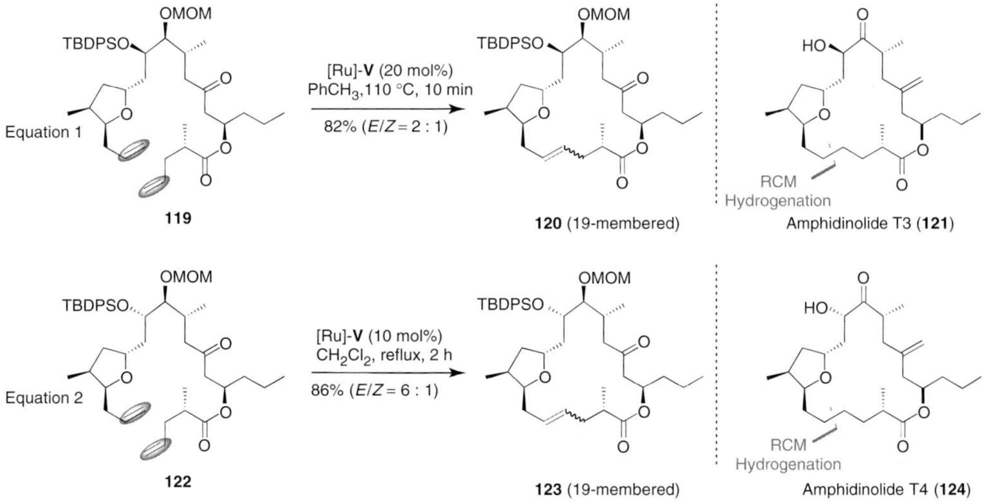

Scheme 5.28

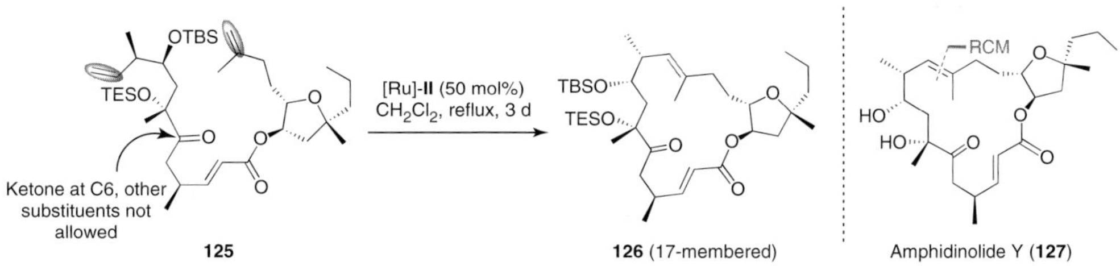

Scheme 5.29

benzene at room temperature. The desired macrocycle **132** was obtained as a single *E*-isomer in 73% yield (Scheme 5.31) [61].

5.3.7 Other Polyketides

Other macrocyclic compounds presumably of polyketide origin such as okilactomycin (**136**) and terpestacin (**139**), both possessing antitumoral activity, were

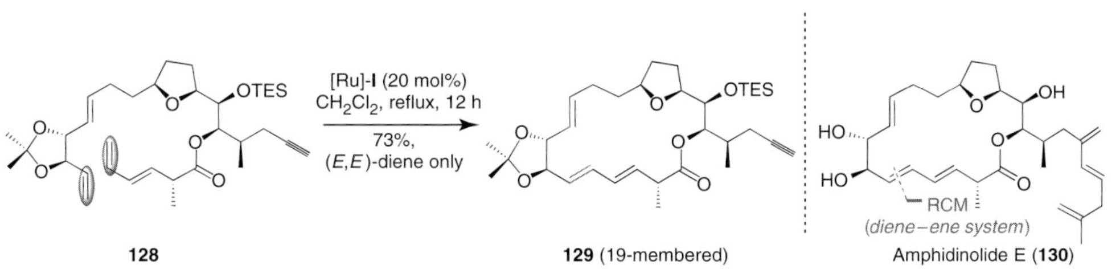

Scheme 5.30

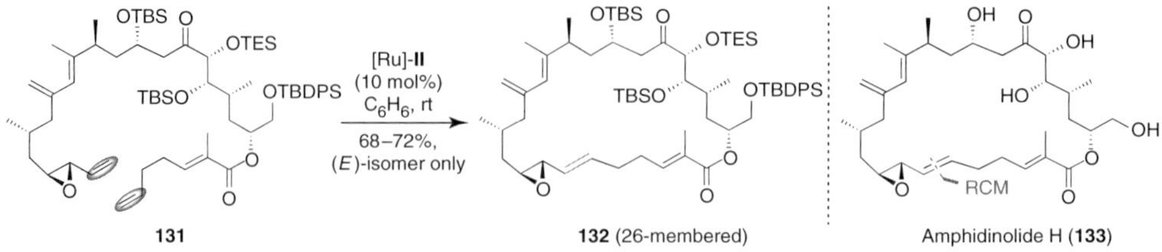

Scheme 5.31

synthesized by RCM utilizing [Ru]-**II** catalyst in benzene (Equations 1 and 2, Scheme 5.32) [62, 63]. In addition, the total synthesis of the IgE antagonist, aspercyclide (**142**), was reported [64]. The key step was a kinetically controlled RCM reaction to form the 11-membered ring cycle **141** (Equation 3, Scheme 5.32).

5.3.8
Natural Cyclophanes

Naturally occurring paracyclophanes are scarce, and cylindrocyclophanes were the first to be isolated. These 22-membered [7,7]-paracyclophanes were found to be the major cytotoxic components in three different strains of the terrestrial blue-green algae *Cylindrospermum lichenforme*. Biosynthetically, the [7,7]-paracyclophane skeleton of the cylindrocyclophanes is presumed to arise from the dimerization of two identical resorcinol fragments of polyketide-originated metabolite. Smith *et al.*

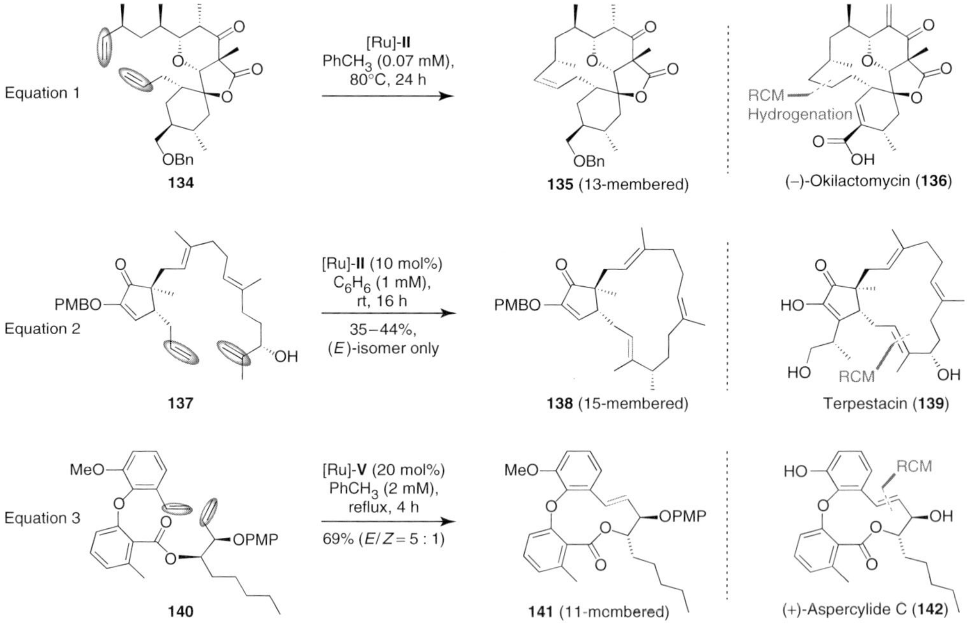

Scheme 5.32

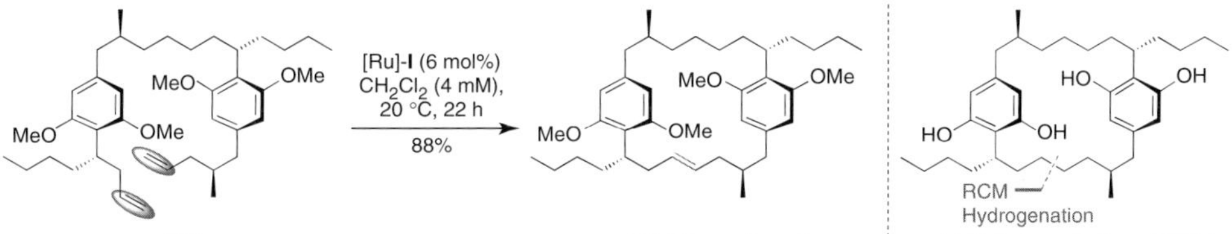

143 **144** (22-membered) (–)-Cylindrocyclophane F (**145**)

Scheme 5.33

reported a huge synthetic work where they used several synthetic strategies to construct the macrocycle, including a double CM (cross-metathesis) dimerization, and RCM of substrate **143** using [Ru]-**I**, which was highly efficient and gave **144** as a single isomer (Scheme 5.33). [Mo]-**I** catalyst was also used and was shown to give almost the same results [65].

5.4 Terpenoids

Terpenoids, especially those constituted by four mevalonate units, are among the most abundant macrocycles in Nature. There are few examples of synthetic approaches toward these compounds using RCM for the cyclization step.

5.4.1 Diterpenoids

One of the most compact and condensed carbon frameworks consisting of multiple interconnected ring systems, coleophomones B–C (**148**, **151**) – isolated from a *Stachybotrys cylindrospora* fungal broth, has an antifungal activity and inhibits the serine protease enzyme, heart chimase. The 11-membered ring of these two compounds was constructed using an impressive olefin metathesis reaction to build the C16–C17 bond (Equations 1 and 2, Scheme 5.34). RCM reactions proceeded with excellent yields and selectivities for both members of the family [66].

The synthesis of clavirolide C (**154**), a member of the dolabellane family of diterpenes isolated from coral *Clavularia viridis*, was recently disclosed. Consequently, treatment of allylic ether **152** (dr = 4 : 1) with [Ru]-**III** (10 mol%) under high dilution conditions and with a slow addition of the substrate to a solution of catalyst gave the desired compound **153** in 70% yield as a single isomer ($E/Z >$ 95 : 5), (Scheme 5.35). Interestingly, other catalysts such as [Ru]-**II** failed to give the cyclized product [67].

Several strategies to construct the macrocyclic diterpene tonantzitlolone (**157**) via RCM have been accomplished [68]. While the C1–C2 double bond could not be formed probably due to steric congestion created by the three allylic methyl groups present in the starting material, macrocycle **156** was eventually obtained from **155** with high *E*-selectivity due to less congested olefins (Scheme 5.36).

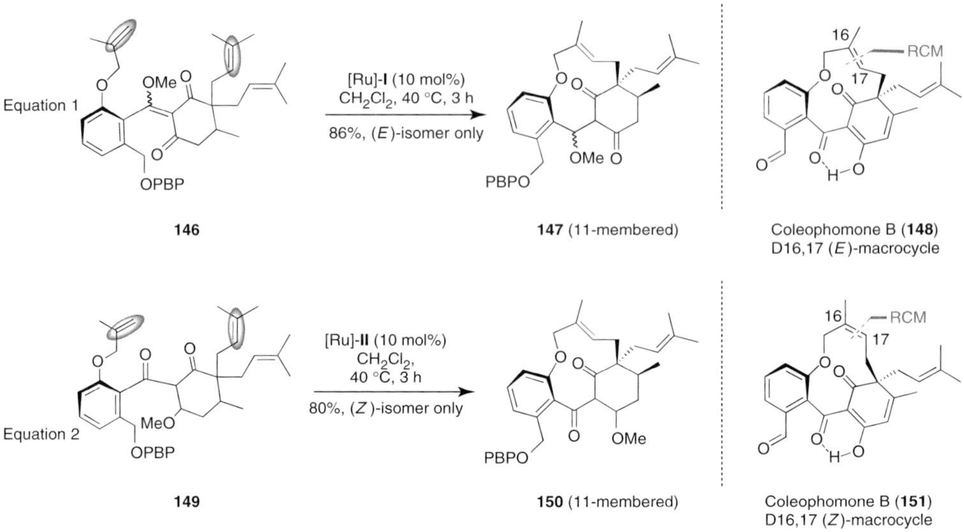

Scheme 5.34

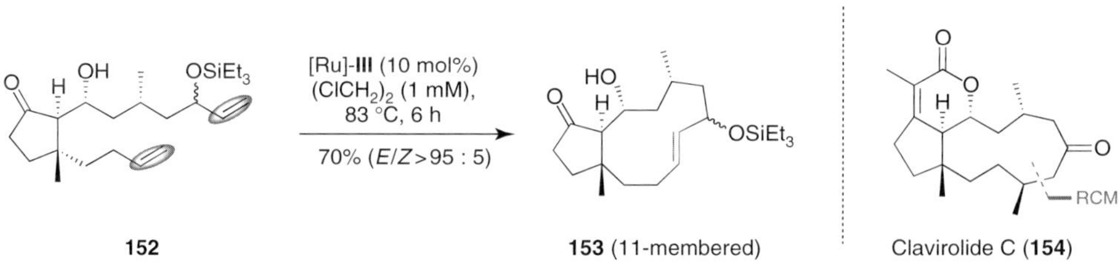

Scheme 5.35

Floresolide B (**160**), a recently discovered natural product isolated from an ascidian of genus *Aplidium* that exhibits cytotoxicity against KB tumor cells, contains an aromatic ring connected to a [10]-metacyclophane and a seven-membered ring lactone. The lactone bridge present in substrate **158** was installed prior to the

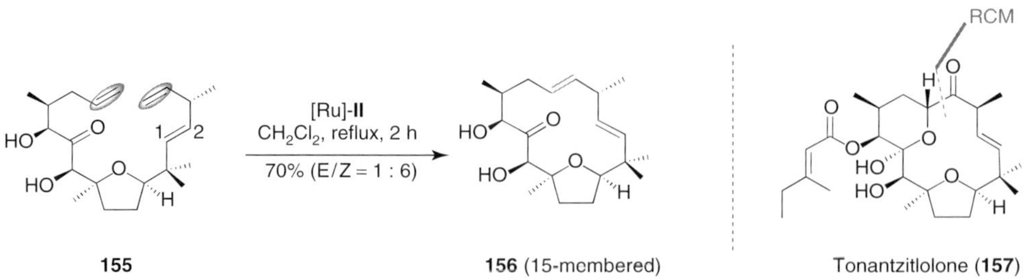

Scheme 5.36

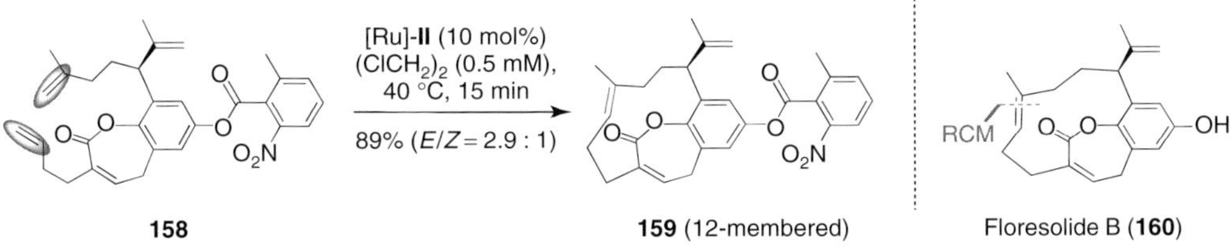

Scheme 5.37

metathesis, as substrates without any constraint gave either dimeric products or low yields (Scheme 5.37) [69].

5.4.2
Macrocyclic Lipids

Large macrocyclic lipids are rarely found in bacterial membranes. 36- and 72-membered ring macrocyclic lipids **163** and **166** were found in *Archaeabacteria*, and built up using an olefin metathesis approach. Consequently, in the presence of [Ru]-**I**, RCM of **161** and **164** proceeded efficiently in 79% yield under high dilution conditions to give the 36-membered ring compound **162** (Equation 1, Scheme 5.38). The 72-membered ring compound **165** was in turn obtained in 45% yield (Equation 2, Scheme 5.38) [70].

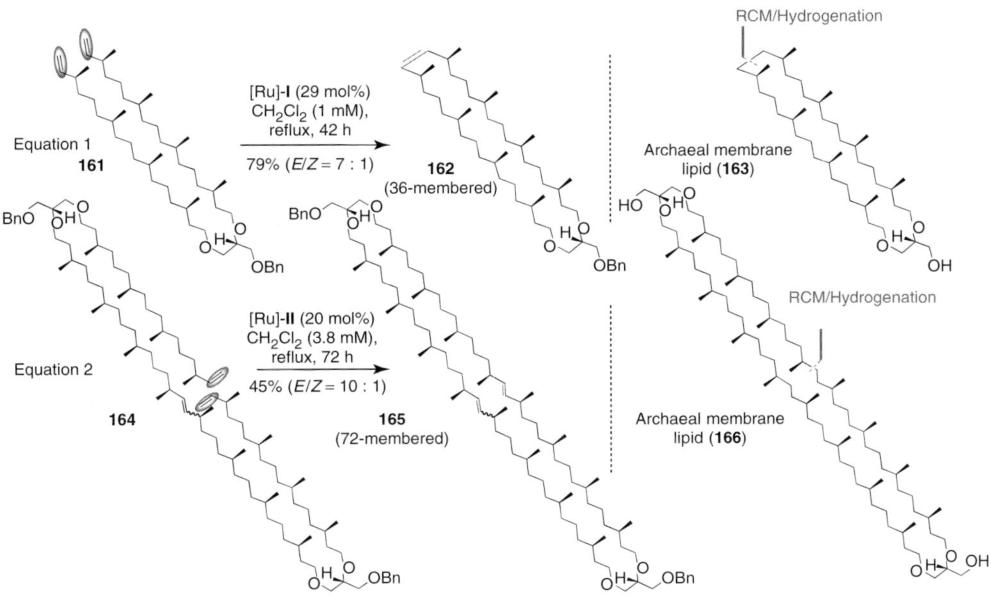

Scheme 5.38

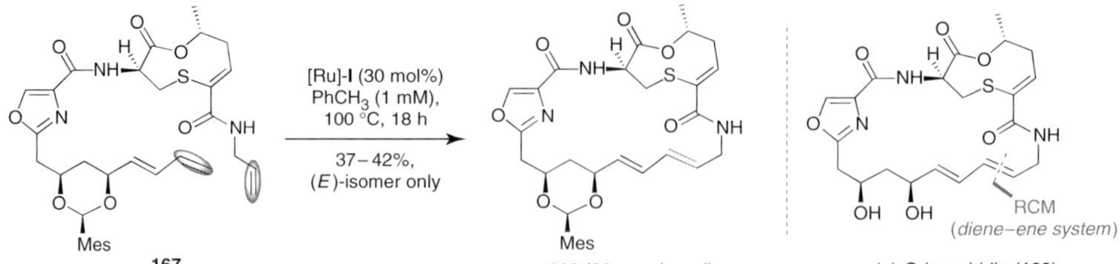

167

168 (20-membered)

(–)-Griseoviridin (169)

Scheme 5.39

5.5 Macrocycles of Amino Acid Origin

5.5.1 Macrolactams

Pharmacologically active macrolactams are highly functionalized. The reaction conditions used for the RCM key macrocyclization step are summarized in Schemes 5.39–5.41. The macrocyclic antibiotic griseoviridin (**169**), a compound with antibacterial properties isolated from *Streptomyces griseus*, has a 23-membered ring cycle. The synthesis of this natural product was described by Meyers *et al.* using a diene–ene RCM. Consequently, the cyclization of precursor **167** using catalyst [Ru]-**I** gave compound **168** with moderate yield but with total *E*-selectivity. A simple deprotection step was necessary to afford griseoviridin (**169**) (Scheme 5.39) [71].

Fluvirucin B1 (**172**) is one of the first natural macrolides being synthesized by RCM, and one of the few examples utilizing [Mo]-**I** efficiently. This antifungal agent, isolated from *Actinomadura* SCC 1777, was obtained through a high yielding and totally stereoselective RCM of compound **170** (Scheme 5.40) [72].

A new class of ansamycin compounds bearing a (*E*,*E*,*E*)-triene within a 21-membered ring lactam were recently described. In particular, cytotrienins A–D (**175**) that were isolated from *Streptomyces* sp. RK95-74 and are important antitumoral compounds. Their synthesis was disclosed by Panek *et al.* and the crucial RCM step was performed on polyenic substrate **173**. Surprisingly, when using [Ru]-**II**, the insertion of the ruthenium took place on one of the disubstituted double bonds thus affording the 19-membered ring. On the other

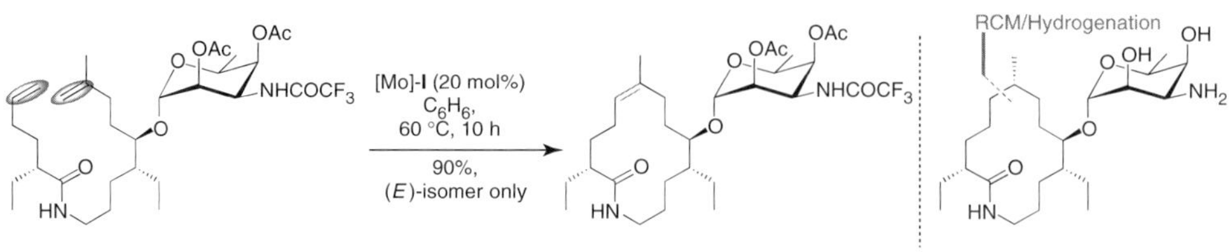

170

171 (14-membered)

Sch 38516 (fluvirucin B1) (**172**)

Scheme 5.40

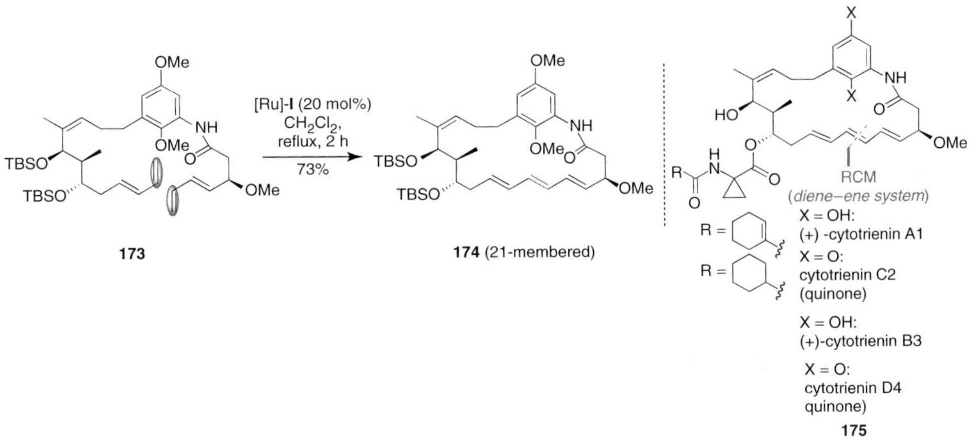

Scheme 5.41

hand, when using [Ru]-**I** in refluxing CH_2Cl_2, the insertion took place on the terminal double bonds affording (*E*,*E*,*E*)-triene **174** with excellent selectivity and yield (Scheme 5.41) [73].

5.5.2 Cyclodepsipeptides

The 13-membered ring lactone and lactam antifungal–antibiotic (−)-PF1163B (**178**), isolated from *Streptomyces*, was synthesized using 12 mol% of Ru-[**II**] (Scheme 5.42) in 60% yield [74]. The natural product was eventually obtained after hydrogenation.

Five total syntheses of the antiproliferative marine metabolite (−)-spongidepsine (**181**) have been reported to date, and most of them use RCM. As a general trend, when RCM is performed on substrates of type **179**, high yields are obtained but with variable *E*/*Z* ratios. This outcome is fortunately not detrimental as the resulting double bond is then hydrogenated (Scheme 5.43) [75].

An RCM approach to the total synthesis of antimitotic marine product arenastatin (**184**) using [Ru]-**I** catalyst is shown in Scheme 5.44 [76].

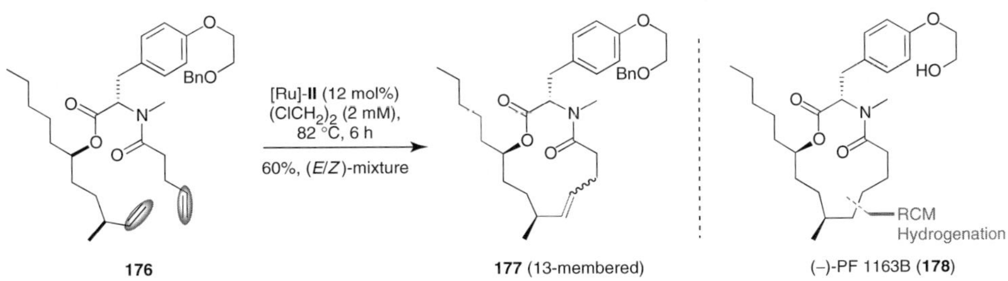

Scheme 5.42

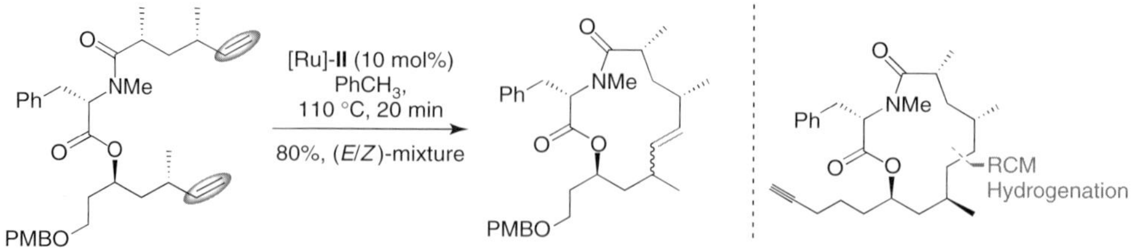

Scheme 5.43

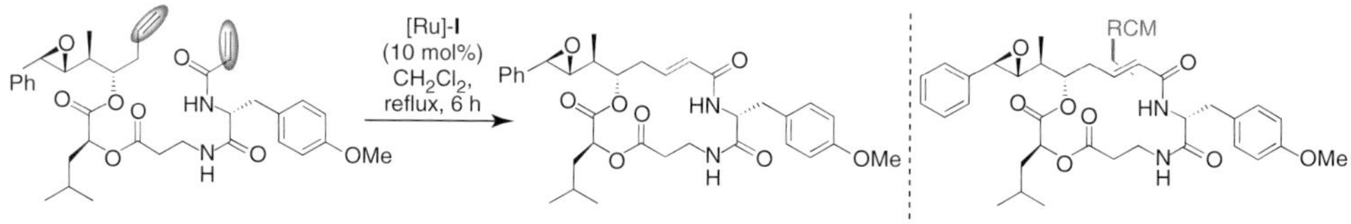

Scheme 5.44

5.5.3
Alkaloids

Roseophilin (**186**) is a 13-membered macrocycle, which was isolated from the culture broth of *Streptomyces griseoviridis*, and possesses high antitumoral activity. This product belongs to the pyrrole alkaloids of the prodigiosin family. The problem in the synthesis of this compound is the incorporation of the macrocycle into an *ansa*-bridged azafulvene. The synthetic efforts were first attempted on precursors bearing fulvalene cycles [77]. However, RCM was successful only when bulky groups were suitably situated to exert conformational constraints that favored the reactive conformation. Unfortunately, these early efforts led to the undesired enantiomer or to racemic products (Scheme 5.45a,b). Better results were obtained when forming the *ansa*-macrocycle prior to the formation of one or both five-membered ring cycles as the strain was partially avoided. Thus, Fürstner *et al.*' acyclic precursor [78], led to the corresponding macrocycle in high yield by using [Ru]-**I** (Scheme 5.45c) while Boger *et al.* obtained *ent*-**186** efficiently by treatment of a monocyclic triene with [Ru]-**I** (Scheme 5.45d) [79]. The first total synthesis of the enantiomerically pure **186** was achieved by Tius *et al.* from a monocyclic diene (**185**) (Scheme 5.45e) [80]. The pyrrole ring was formed subsequently using a Paal–Knorr reaction.

In Scheme 5.46, the syntheses of different macrocyclic natural products classified as alkaloids are summarized. The rich variety of biological activities associated with both plant-derived and animal-derived alkaloids has stimulated many organic synthetic groups. The main problem from a metathesis point of view is the ability of the nitrogen to coordinate to metal-alkylidene complexes and to interfere unproductively with the catalytic activity. In the examples summarized herein, the use of [Ru]-**I**, [Ru]-**II**, and [Ru]-**V** has proven to be effective. Consequently, (−)-sarain A (**189**) (Equation 1, Scheme 5.46), isolated from sponge *Reniera sarai* [81]; ircinal

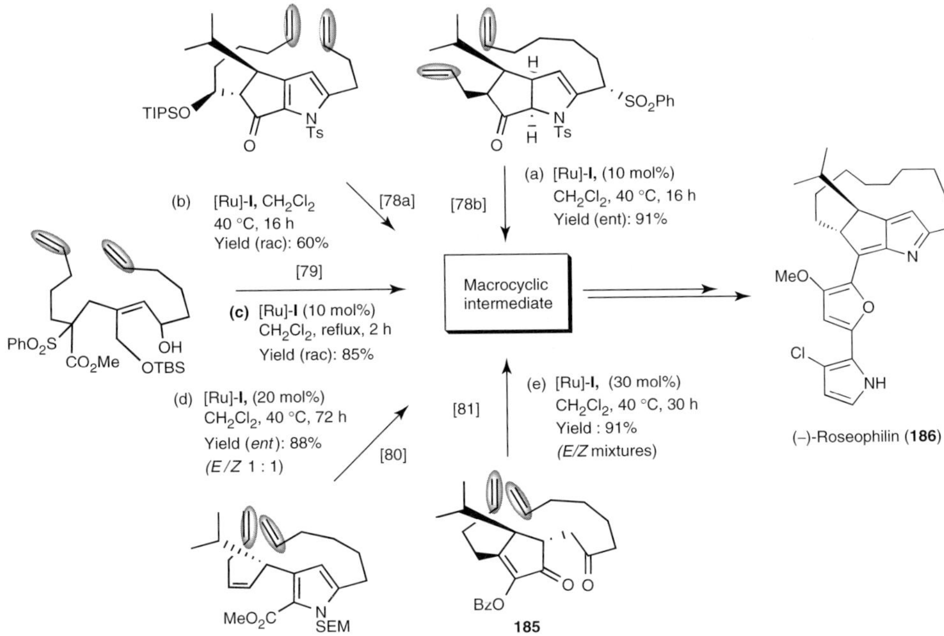

Scheme 5.45

(**192**) and manzamine A (**193**) (Equation 2, Scheme 5.46), isolated from the marine sponges *Haliclona* and *Pellina* [82], respectively; nakadomanine (**196**) (Equation3, Scheme 5.46), isolated from the sponge *Amphimedon* [83]; nonylprodigiosyn (**199**) (Equation 4, Scheme 5.46), isolated from *Actinomadura madurae* [84]; and isooncinotine (**202**) (Equation 5, Scheme 5.46), isolated from Stem bark *Oncinotis nitida* and *Oncinoris tenuiloba* [85], have all been efficiently synthesized using RCM with [Ru]-**I** or [Ru]-**V**. In general, chelation of the catalyst with the nitrogen functionality is avoided by protonation or by RCM of bonds with a remote relative situation from the potentially interfering groups.

5.6
Macrocyclic Glycolipids

The performance and excellent application profile of RCM are illustrated by the total synthesis of various resin glycosides and sugar-based macrodiolides such as tricolorins A and G, woodrosin I, and ipomoeassins B and E (Scheme 5.47). The synthesis of these amphiphilic natural products has resulted in the advancement of carbohydrate chemistry and is due to Fürsner *et al.* who have recently reviewed their effort in this area [86]. In this review, they fully detail the syntheses of tricolorins A (**208**) (Equation 1, Scheme 5.47) and G (**211**) (Equation 2, Scheme 5.47), two natural products isolated from *Ipomea tricolor*, which are plant growth inhibitors

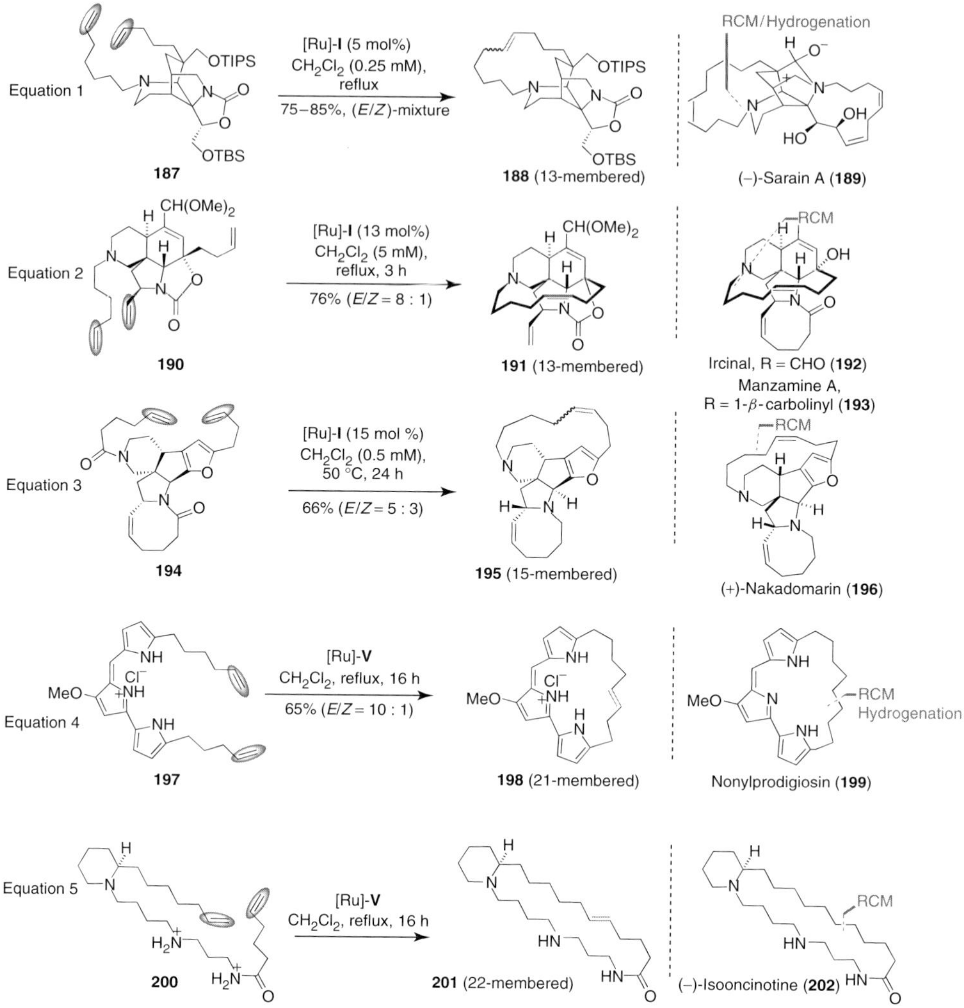

Scheme 5.46

and protect sugarcane against invasive weeds [87] as well as the synthesis of woodrosin I (**214**) (Equation 3, Scheme 5.47) isolated from *Ipomea tuberulosa L* [88]. RCM was carried out with the first-generation catalysts and proceeded in high yields and stereoselectivities.

More recently, the same group has disclosed the total synthesis of cytotoxic agents ipomoeassin B and E, isolated from *Ipomoea squamosa*. Consequently, exposure of diene **215** to [Ru]-**II** (10 mol%) in refluxing CH_2Cl_2 afforded macrocycle **216** in excellent yield as a mixture of both geometrical isomers, which could be selectively hydrogenated using $RhCl(PPh_3)_3$ without affecting the lateral unsaturated esters. The resulting products were transformed into ipomoeassin E (**217**) (Equation 4,

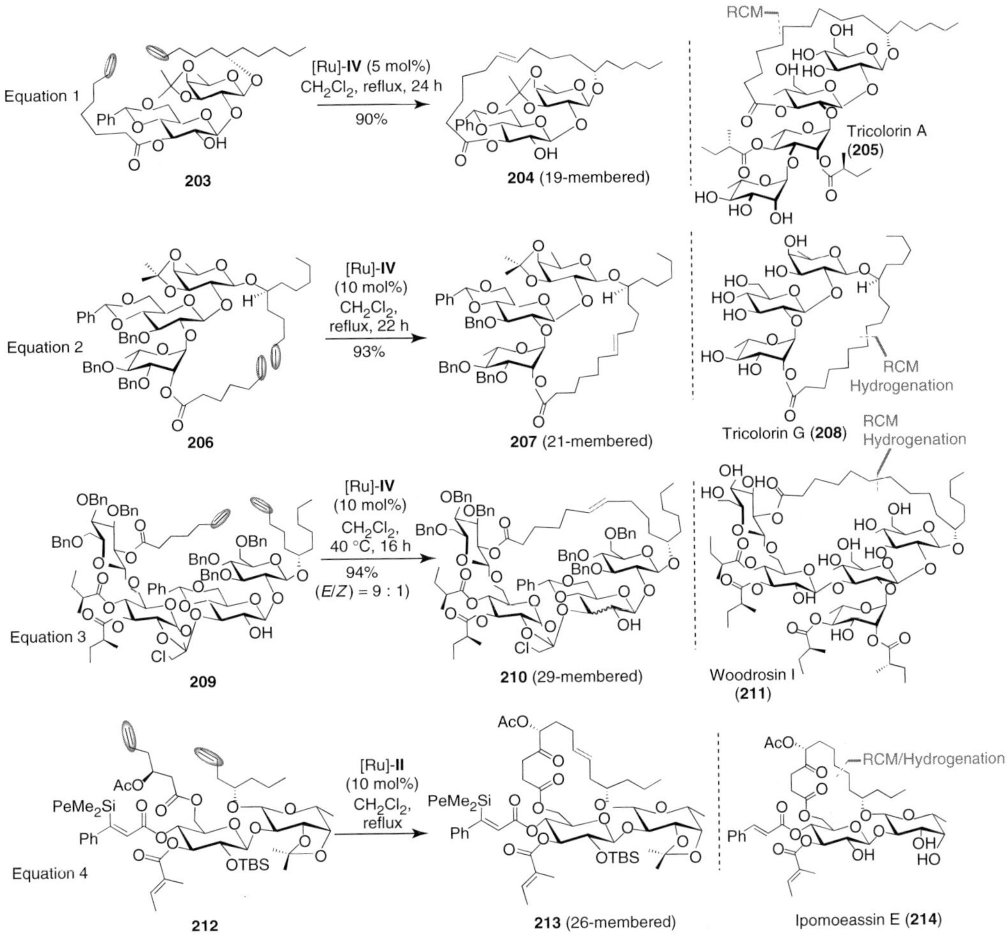

Scheme 5.47

Scheme 5.47). Extension of this methodology to the preparation of ipomoeassin B was straightforward [89].

5.7 Conclusions and Outlook

Since the discovery of ruthenium- and molybdenum catalysts, the metathesis reaction has triggered the attention of synthetic chemists. Even in the early times, RCM process has been used in the construction of macrocycles with great success. Actually, a wide range of examples has been reported using olefin metathesis in the context of natural macrocycles. The works published before 2000 used the only available first-generation catalysts and had some problems with efficiency. These complexes do not stand harsh reaction conditions and are generally unreactive with di- or trisubstituted double bonds. However, after the second-generation complexes were introduced, several examples have shown that some synthetic problems related to regioselectivity could be solved using the first-generation catalysts, especially with polyenic substrates. There is not a clear rule that indicates the best catalyst to use in each case. Subtle variations in the substrate structure and in the type of final product may lead to different outcomes with each type of carbene complexes. In particular, remote substitution and the presence of coordinating heteroatoms have an important influence on the results. In addition, the *E*/*Z* ratio of the product may be influenced by all these factors and by secondary isomerizations mediated by the metathesis catalysts. Unfortunately, there still exists a lack of prediction over the stereochemistry for unsaturated macrocycles obtained by RCM. The catalyst loading varies from one case to another (from 1 to 25 mol%) and is highly dependent on the species used. Other important aspects are the concentration (0.25–8 mM) and the addition time which sometimes have to be slow. The catalyst may be more effective if added in portions. Long reaction times are sometimes required when working at low temperatures.

In conclusion, when planning a new synthesis, it is worth testing different catalysts, temperatures, and concentrations with additional fine tuning of the loading and addition rate. On the other hand, some aspects in the selection of reaction conditions for a macrocycle construction using RCM are quite straightforward. These are solvent selection [toluene, CH_2Cl_2, $(ClCH_2)_2$] and the reaction temperature (from rt to refluxing toluene).

In the near future, the discovery of more powerful and selective catalysts should solve the remaining problems.

References

1 Among the numerous references, we will mention those that cover specifically or in a particular section the diene or ene-diene metathesis in the synthesis of macrocycles (a) Mori, M. (2007) *Adv. Synth. Catal.*, **349**, 121–135; (b) Schmidt, B. (2004) *Eur. J. Org. Chem.*, 1865–1880; (c) Hoveyda, A.H., Gillingham, D.G., Van Veldhuizen, J.J., Kataoka, O., Garber, S.B., Kingsbury, J.S., and Harrity, J.P.A. (2004) *Org. Biomol. Chem.*, **2**, 8–23; (d) Grubbs, R.H. (2004) *Tetrahedron*, **60**, 7117–7140; (e) Prunet, J. (2003) *Angew. Chem. Int. Ed. Engl.*, **42**, 2826–2830; (f) Trnka, T.M., Grubbs, R.H. (2001) *Acc. Chem. Res.*, **34**, 18–29; (g) Fürstner, A. (2000) *Angew. Chem. Int. Ed. Engl.*, **39**, 3012–3043; (h) Maier, M.E. (2000) *Angew. Chem. Int. Ed.*, **39**, 2073–2077; (i) Yet, L. (2000) *Chem. Rev.*, **100**, 2963–3007; (j) Randall, M.L. and Snapper, M.L. (1998) *Strem Chem.*, **17**, 1–9; (k) Armstrong, S.K. (1998) *J. Chem. Soc., Perkin Trans. 1*, 371–388; (l) Grubbs, R.H. and Chang, S. (1998) *Tetrahedron*, **54**, 4413–4450; (m) Mulzer, J. and Öhler, E. (2004) *Topics Organomet. Chem.*, **13**, 269–366; (n) Nicolaou,K.C., Bulger, P.G., and Sarlah, D. (2005) *Angew. Chem. Int. Ed.*, **44**, 4490–4527; (o) Gaich, T. and Mulzer, J. (2005) *Curr. Topics Med.*

Chem., **5**, 1473–1494; (p) Van der Weghe, J. and Eustache, J. (2005) *Curr. Topics Med. Chem.*, **5**, 1495–1519; (q) Deiters, A. and Martin, S.F. (2004) *Chem. Rev.*, **104**, 2199–2238.

2 (a) Newman, D.J., Cragg, G.M., and Snader, K.M. (2003) *J. Nat. Prod.*, **66**, 1022–1037; (b) Henkel, T., Brunne, R.M., Mueller, H., and Reichel, F. (1999) *Angew. Chem. Int. Ed.*, **38**, 643–647.

3 (a) Wessjohann, L.A., Ruijter, E., García-Rivera, D., and Bradt, W. (2005) *Mol. Divers.*, **9**, 171–186; (b) Wessjohann, L.A. and Ruijter, E. (2004) *Topics in Current Chemistry*, vol. 243, Springer, Heidelberg, pp. 137–184.

4 Gournelis, D.C., Laskaris, G.G., and Verpoorte, R. (1997) *Nat. Prod. Rep.*, **14**, 75–82.

5 (a) Altmann K.-H. (2004) *Org. Biomol. Chem.*, **2**, 2137–2152; (b) Zhang, R. and Liu, Z. (2004) *Curr. Org. Chem.*, **8**, 267–290; (c) Rivkin, A., Cho, Y.S., Gabarda, A.E., Yoshimura, F., and Danishefsky, S.J. (2004) *J. Nat. Prod.*, **67**, 139–143; (d) Nicolaou, K.C., Wissinger, N., Pastor, J., Ninkovic, S., Sarabia, F., He, Y., Vourloumis, D., Li, T., Giannakakou, P., and Hamel, E. (1997) *Nature*, **390**, 268–272; (e) Nicolaou, K.C., King, N.P., and He, Y. (1999) *Top. Organomet. Chem.*, **1**, 73–104.

6 Nishioka, T., Iwabuchi, Y., Irie, H., and Hatakeyama, S. (1998) *Tetrahedron Lett.*, **39**, 5597–5600.

7 Borer, B.C., Deeremberg, S., Bierfugel, H., and Pandit, U.K. (1994) *Tetrahedron Lett.*, **35**, 3191–3194.

8 Cabrejas, L.M.M., Rohrbach, S., Wagner, D., Kallen, J., Zenke, G., and Wagner, J. (1999) *Angew. Chem. Int. Ed.*, **38**, 2443–2446.

9 For a review on ruthenium based catalyst, see: Bieniek, M., Michrowska, A., Usanov, D.L., and Grela, K. (2008) *Chem. Eur. J.*, **14**, 806–818.

10 Schrock, R.R. and Czekelius, C. (2007) *Adv. Synth. Catal.*, **349**, 55–77.

11 (a) Katz, L. and Ashley, G.W. (2005) *Chem. Rev.*, **105**, 499–528; (b) Yeung, K.S. and Paterson, I. (2005) *Chem. Rev.*, **105**, 4237–4313; (c) McDaniel, R., Welch, M., and Hutchinson, C.R. (2005) *Chem. Rev.*, **105**, 543–558; (d) Rychnovsky, S.D. (1995) *Chem. Rev.*, **95**, 2021–2040; (e) Norcross, R.D. and Paterson, I. (1995) *Chem. Rev.*, **95**, 2041–2114.

12 For a recent review on the structure, biology and chemistry of RALs see: Winssinger, N. and Barluenga, S. (2007) *Chem. Commun.*, 22–36.

13 (a) Ayer, W.A. and Pena-Rodriguez, L. (1987) *Phytochemistry*, **26**, 1353–1355; (b) Sugawara, F., Kim, K.W., Kobayashi, K., Uzawa, J., Yoshida, S., Murofushi, N., Takahashi, N., and Strobel, G.A. (1992) *Phytochemistry*, **31**, 1987–1990; (c) Sharma, S.V., Agatsuma, T., and Nakano, H. (1998) *Oncogene*, **16**, 2639–2645; (d) Yamamoto, K., Garbaccio, R.M., Stachel, S.J., Solit, D.B., Chiosis, G., Rosen, N., and Danshefsky, S.J. (2003) *Angew. Chem. Int. Ed.*, **42**, 1280–1284.

14 Geng, X., Yang, Z.-Q., and Danishefsky, S.J. (2004) *Synlett*, 1325–1333.

15 Yamamoto, K., Biswas, K., Gaul, C., and Danishefsky, S.J. (2003) *Tetrahedron Lett.*, **44**, 3297–3299.

16 (a) Yang, Z.-Q., Geng, X., Solit, D., Pratilas, C.A., Rosen, N., and Danishefsky, S.J. (2004) *J. Am. Chem. Soc.*, **126**, 7881–7889; (b) Geng, X. and Danishefsky, S.J. (2004) *Org. Lett.*, **6**, 413–416.

17 Barluenga, S., Dakas, P.Y., Fernandin, Y., Meijer, L., and Winssinger, N. (2006) *Angew. Chem. Int. Ed.*, **45**, 3951–3954.

18 (a) Garbaccio, R.M., Stachel, S.J., Baeschlin, D.K., and Danishefsky, S.J. (2001) *J. Am. Chem. Soc.*, **123**, 10903–10908; (b) Couladouros, E.A., Mihou, A.P., and Bouzas, E.A. (2004) *Org. Lett.*, **6**, 977–980.

19 (a) Barluenga, S., López, P., Moulin, E., and Winssinger, N. (2004) *Angew. Chem. Int. Ed.*, **43**, 3467–3470; (b) Barluenga, S., Moulin, E., Lopez, P., and Winssinger, N. (2005) *Chem. Eur. J.*, **11**, 4935–4952.

20 Moulin, E., Barluenga, S., and Winssinger, N. (2005) *Org. Lett.*, **7**, 5637–5639.

21 Moulin, E., Zoete, V., Barluenga, S., Karplus, M., and Wissinger, N. (2005) *J. Am. Chem. Soc.*, **127**, 6999–7004.

22 (a) Fürstner, A., Seidel, G., and Kindler, N. (1999) *Tetrahedron*, **55**, 8215–8230; (b) Fürstner, A., Thiel, O.R., Kindler, N., and Bartkowska, B. (2000) *J. Org. Chem.*, **65**, 7990–7995.

23 For a review on the chemistry and biology of salicylihalamide A and related compounds, see: Yet, L. (2003) *Chem. Rev.*, **103**, 4283–4306.

24 (a) Kunze, B., Jansen, R., Sasse, F., Hofle, G., and Reichenbach, H. (1998) *J. Antibiot.*, **51**, 1075–1080; (b) Galinis, D.L., McKee, T.C., Pannell, L.K., Cardellina, J.H., and Boyd, M.R. (1997) *J. Org. Chem.*, **62**, 8968–8969; (c) Kim, J.W., Shin-ya, K., Furihata, K., Hayakawa, Y., and Seto, H. (1999) *J. Org. Chem.*, **64**, 153–155.

25 Herb, C., Bayer, A., and Maier, M.E. (2004) *Chem. Eur. J.*, **10**, 5649–5660.

26 Smith, A.B.III and Zheng, J. (2002) *Tetrahedron*, **58**, 6455–6471.

27 Snider, B.B. and Song, F. (2001) *Org. Lett.*, **3**, 1817–1820.

28 Labrecque, D., Charron, S., Rej, R., Blais, C., and Lamothe, S. (2001) *Tetrahedron Lett.*, **42**, 2645–2648.

29 (a) Wu, Y., Liao, X., Wang, R., Xie, X.S., and de Brabander, J.K. (2002) *J. Am. Chem. Soc.*, **124**, 3245–3253; (b) Wu, Y., Esser, L., and de Brabander, J.K. (2000) *Angew. Chem. Int. Ed.*, **39**, 4308–4310; (c) Wu, Y., Seguil, O.R., and de Brabander, J.K. (2000) *Org.Lett.*, **2**, 4241–4244.

30 Fürstner, A., Dierkes, T., Thiel, O.R., and Blanda, G. (2001) *Chem. Eur. J.*, **7**, 5286–5298.

31 (a) Wang, X. and Porco, J.A. (2003) *J. Am. Chem. Soc.*, **125**, 6040–6041; (b) Wang, X., Bowman, E.J., Bowman, B.J., and Porco, J.A. (2004) *Angew. Chem. Int. Ed.*, **43**, 3601–3605.

32 Xuan, R., Oh, H.-S., Lee, Y., and Kang, H.-Y. (2008) *J. Org. Chem.*, **73**, 1459–1461.

33 Kanada, R.M., Itoh, D., Nagai, M., Niijima, J., Asai, N., Mizui, Y., Abe, S., and Kotake, Y. (2007) *Angew. Chem. Int. Ed.*, **46**, 4350–4355.

34 Chen, Q. and Du, Y. (2006) *Tetrahedron Lett.*, **47**, 8489–8492.

35 Krishna, P.R. and Srinivas, R. (2008) *Tetrahedron Asymmetry*, **19**, 1153–1160.

36 Fürstner, A. and Langemann, K. (1997) *J. Am. Chem. Soc.*, **119**, 9130–9136.

37 (a) Gaul, C., Njardarson, J.T., Shan, D., Dorn, D.C., Wu, K.-D., Tong, W.P., Huang, X.-Y., Moore, M.A.S., and Danishefsky, S.J. (2004) *J. Am. Chem. Soc.*, **126**, 11326–11337; (b) Gaul, C., Njardarson, J.T., and Danishefsky, S.J. (2003) *J. Am. Chem. Soc.*, **125**, 6042–6043.

38 (a) Dixon, D.J., Foster, A.C., and Ley, S.V. (2000) *Org. Lett.*, **2**, 123–125; See also: ref 6 and (b) Banwell, M.G., and McRae, K.J. (2000) *Org. Lett.*, **2**, 3583–3586.

39 (a) Kang, E.J., Cho, E.J., Lee, Y.E., Ji, M.K., Shin, D.M., Chung, Y.K., and Lee, E. (2004) *J. Am. Chem. Soc.*, **126**, 2680–2681; (b) Kang, E.J., Cho, E.J., Ji, M.K., Lee, Y.E., Shin, D.M., Choi, S.Y., Chung, Y.K., Kim, J.S., Kim, H.J., Lee, S.G., Lah, M.S., and Lee, E. (2005) *J. Org. Chem.*, **70**, 6321–6329; (c) Soltani, O. and de Brabander, J.K. (2005) *Org. Lett.*, **7**, 2791–2793; (d) Bolshakov, S. and Leighton, J.L. (2005) *Org. Lett.*, **7**, 3809–3812; (e) Crimmins, M.T. and Vanier, G.S. (2006) *Org. Lett.*, **8**, 2887–2890.

40 Cheung, L.L., Marumoto, S., Anderson, C.D., and Rychnovsky, S.D. (2008) *Org. Lett.*, **10**, 3101–3104.

41 Kawaguchi, T., Funamori, N., Matsuya, Y., and Nemoto, H. (2004) *J. Org. Chem.*, **69**, 505–509.

42 (a) Ding, F. and Jennings, M.P. (2008) *J. Org. Chem.*, **73**, 5965–5976; (b) Hoye, T.R. and Hu, M. (2003) *J. Am. Chem. Soc.*, **125**, 9576–9577.

43 Nicolaou, K.C., Sun, Y.-P., Guduru, R., Banerji, B., and Chen, D.Y.-K. (2008) *J. Am. Chem. Soc.*, **130**, 3633–3644.

44 Fürstner, A., Nevado, C., Waser, M., Tremblay, M., Chevrier, C., Teplý, F., Aïssa, C., Moulin, E., and Müller, O. (2007) *J. Am. Chem. Soc.*, **129**, 9150–9161.

45 (a) Louie, J., Bielawski, C.W., and Grubbs, R.H. (2001) *J. Am. Chem. Soc.*, **123**, 11312–11313; For the RCM of related substrates and conditions see: (b) Fujimoto, S., Yoshikawa, K., Itoh,

M., and Kitahara, T. (2002) *Biosci. Biotechnol. Biochemi.*, **66**, 1389–1392; (c) Kamat, V.P., Hagiwara, H., Suzuki, T., and Ando, M. (1998) *J. Chem. Soc., Perkin Trans 1*, 2253–2254.

46 Obtained via RCAM methodology: Fürstner, A. and Seidel, G. (2000) *J. Organomet. Chem.*, **606**, 75–78.

47 Fürstner, A. and Lagemann, K. (1997) *Synthesis*, 792–803.

48 Fürstner, A. and Langemann, K. (1997) *J. Am. Chem. Soc.*, **119**, 9130–9136.

49 Fürstner, A. and Leitner, A. (2003) *Angew. Chem. Int. Ed.*, **42**, 308–311.

50 Reviews on epothilones, see: (a) Feyen, F., Cachoux, F., Gertsch, J., Wartmann, M., and Altmann, K.H. (2008) *Acc. Chem. Res.*, **41**, 21–31; (b) Altmann, K.H., Pfeiffer, B., Arseniyadis, S., Pratt, B.A., and Nicolaou, K.C. (2007) *ChemMedChem*, **2**, 396–483; (c) Nicolaou, K.C. (2005) *J. Med. Chem.*, **48**, 5613–5638; (d) Mulzer, J. (2000) *Monatsh. Chem.*, **131**, 205–238; (e) Nicolaou, K.C., Ritzén, A., and Namoto, K. (2001) *Chem. Commun.*, 1523–1524.

51 Mulzer, J., Altmann, K.-H., Höfle, G., Müller, R., and Prantz, K. (2008) *C. R. Chim.*, **11**, 1336–1368.

52 (a) Meng, D., Bertinato, P., Balog, A., Su, D.S., Kamenecka, T., Sorensen, E.J., and Danishefsky, S.J. (1997) *J. Am. Chem. Soc.*, **119**, 10073–10092; (b) Nicolaou, K.C., He, Y., Vourloumis, D., Vallberg, H., Roschangar, F., Sarabia, F., Ninkovic, S., Yang, Z., and Trujillo, J.I. (1997) *J. Am. Chem. Soc.*, **119**, 7960–7973.

53 Gaich, T. and Mulzer, J. (2005) *Org. Lett.*, **7**, 1311–1313.

54 (a) Sun, J. and Sinha, S.C. (2002) *Angew. Chem. Int. Ed.*, **41**, 1381–1383; For previous syntheses, see: (b) Yang, Z., He, Y., Vourlumis, D., Vallberg, H., and Nicolaou, K.C. (1997) *Angew. Chem. Int. Ed.*, **36**, 166–169; For the synthesis of epothilone A derivatives see: (c) Meng, D., Su, D.S., Balog, A., Bertinato, P., Sorensen, E.J., Danishefsky, S.J., Zheng, Y.H., Chou, T.C., He, L., and Horwitz, S.B. (1997) *J. Am. Chem. Soc.*, **119**, 2733–2734.

55 Rivkin, A., Yoshimura, F., Gabarda, A.E., Chou, T.C., Dong, H., Tong, W.P., and Danishefsky, S.J. (2003) *J. Am. Chem. Soc.*, **125**, 2899–2901.

56 Biswas, K., Lin, H., Njardarson, J.T., Chappell, M.D., Chou, T.C., Guan, Y., Tong, W.P., He, L., Horwitz, S.B., and Danishefsky, S.J. (2002) *J. Am. Chem. Soc.*, **124**, 9825–9832.

57 (a) Malecka, E.R. Jr., Terrell, L.R., Geng, F., and Ward, J.S.III (2002) *Org. Lett.*, **4**, 2841–2844; (b) Trost, B., Chisholm, J., Wrobleski, S., and Jung, M. (2002) *J. Am. Chem. Soc.*, **124**, 12420–12421; (c) Lam, H.W. and Pattenden, G. (2002) *Angew. Chem. Int. Ed.*, **41**, 508–511; (d) Hoye, T.R. and Zhao, H. (1999) *Org. Lett.*, **1**, 1123–1126.

58 (a) Fürstner, A., Aissa, C., Riveiros, R., and Ragot, J. (2002) *Angew. Chem. Int. Ed.*, **41**, 4763–4766; (b) Aïssa, C., Riveiros, R., Ragot, J., and Fürstner, A. (2003) *J. Am. Chem. Soc.*, **125**, 15512–15520.

59 Jin, J., Chen, Y., Li, Y., Wu, J., and Dai, W.-M. (2007) *Org. Lett.*, **9**, 2585–2588.

60 Va, P. and Roush, W.R. (2006) *J. Am. Chem. Soc.*, **128**, 15960–15961.

61 Fürstner, A., Bouchez, L.C., Funel, J.-A., Liepins, V., Porée, F.-H., Gilmour, R., Beaufils, F., Laurich, D., and Tamiya, M. (2007) *Angew. Chem. Int. Ed.*, **46**, 9265–9270.

62 Smith, A.B. III, Basu, K., and Bosanac, T. (2007) *J. Am. Chem. Soc.*, **129**, 14872–14874.

63 Trost, B.M., Dong, G., and Vance, J.A. (2007) *J. Am. Chem. Soc.*, **129**, 4540–4541.

64 Fürstner, A. and Müller, C. (2005) *Chem. Commun.*, 5583–5585.

65 Smith, A.B.III, Adams, C.M., Kozmin, S.A., and Paone, D.V. (2001) *J. Am. Chem. Soc.*, **123**, 5925–5937.

66 Nicolaou, K.C., Montagnon, T., Vassilikogiannakis, G., and Mathison, C.J. (2005) *J. Am. Chem. Soc.*, **127**, 8872–8888.

67 Brown, M.K. and Hoveyda, A.H. (2008) *J. Am. Chem. Soc.*, **130**, 12904–12906.

68 Jasper, C., Wittenberg, R., Quitschalle, M., Jakupovic, J., and Kirschning, A. (2005) *Org. Lett.*, **7**, 479–482.

69 Nicolaou, K.C. and Xu, H. (2006) *Chem. Comm.*, 600–602.

70 Arakawa, K., Eguchi, T., and Kakinuma, K. (1998) *J. Org. Chem.*, **63**, 4741–4745.

71 Dvorak, C.A., Schmitz, W.D., Poon, D.J., Pryde, D.C., Lawson, J.P., Amos, R.A., and Meyers, A.I. (2000) *Angew. Chem. Int. Ed.*, **39**, 1664–1666.

72 Xu, Z., Johannes, C.W., Salman, S.S., and Hoveyda, A.H. (1996) *J. Am. Chem. Soc.*, **118**, 10926–10927.

73 Evano, G., Schaus, J.V., and Panek, J.S. (2004) *Org. Lett.*, **6**, 525–528.

74 Bouazza, F., Renoux, B., Bachmann, C., and Gesson, J.-P. (2003) *Org. Lett.*, **5**, 4049–4052.

75 (a) Chen, H.-J. and Forsyth, C.J. (2004) *Angew. Chem. Int. Ed.*, **43**, 2148–2152; See also: (b) Ferrié, L., Reymond, S., Capdevielle, P., and Cossy, J. (2006) *Org. Lett.*, **8**, 3441–3443; (c) Chandrasekhar, S., Yaragorla, S.R., Sreelakshmi, L., and Reddy, C.R. (2008) *Tetrahedron*, **64**, 5174–5183.

76 Tripathy, N.K. and Georg, G.I. (2004) *Tetrahedron Lett.*, **45**, 5309–5311.

77 (a) Kim, S.H., Figueroa, I., and Fuchs, P.L. (1997) *Tetrahedron Lett.*, **38**, 2601–2604; (b) Bamford, S.J., Luker, T., Speckamp, W.N., and Hiemstra, H. (2000) *Org Lett.*, **2**, 1157–1160.

78 Fürstner, A., Gartner, T., and Weintritt, H. (1999) *J. Org. Chem.*, **64**, 2361–2366.

79 Boger, D.L. and Hong, J. (2001) *J. Am. Chem. Soc.*, **123**, 8515–8519.

80 Harrington, P.E. and Tius, M.A. (2001) *J. Am. Chem. Soc.*, **123**, 8509–8514.

81 Becker, M.H., Chua, P., Downham, R., Douglas, C.J., Garg, N.K., Hiebert, S., Jaroch, S., Matsuoka, R.T., Middleton, J.A., Ng, F.W., and Overman, L.E. (2007) *J. Am. Chem. Soc.*, **129**, 11987–12002.

82 (a) Martin, S.F., Humphrey, J.M., Ali, A., and Hillier, M.C. (1999) *J. Am. Chem. Soc.*, **121**, 866–867; (b) Humphrey, J.M., Liao, Y., Ali, A., Rein, T., Wong, Y.-L., Chen, H.-J., Courtney, A.K., and Martin, S.F. (2002) *J. Am. Chem. Soc.*, **124**, 8584–8592.

83 (a) Nagata, T., Nakagawa, M., and Nishida, A. (2003) *J. Am. Chem. Soc.*, **125**, 7484–7485; For other RCM conditions, see also: (b) Young, I.S. and Kerr, M.A. (2007) *J. Am. Chem. Soc.*, **129**, 1465–1469.

84 Fürstner, A., Grabowski, J., and Lehmann, C.W. (1999) *J. Org. Chem.*, **64**, 8275–8280.

85 Scheiper, B., Glorius, F., Leitner, A., and Fürstner, A. (2004) *Proc. Natl. Acad. Sci. U. S. A.*, **101**, 11960–11965.

86 Fürstner, A. (2004) *Eur. J. Org. Chem.*, 943–958.

87 For the original total synthesis paper see: (a) Tricolorin A., Fürstner, A., and Müller, T. (1998) *J. Org. Chem.*, **63**, 424–425; (b) Tricolorin, G., Fürstner, A., and Muller, T. (1999) *J. Am. Chem. Soc.*, **121**, 7814–7821.

88 For the original total synthesis of woodrosin I, see: (a) Fürstner, A., Jeanjean, F., and Razon, P. (2002) *Angew. Chem. Int. Ed.*, **41**, 2097–2101; (b) Fürstner, A., Jeanjean, F., Razon, P., Wirtz, C., and Mynott, R. (2003) *Chem. Eur. J.*, **9**, 320–326; (c) Fürstner, A., Jeanjean, F., Razon, P., Wirtz, C., and Mynott, R. (2003) *Chem. Eur. J.*, **9**, 307–319.

89 Fürstner, A. and Nagano, T. (2007) *J. Am. Chem. Soc.*, **129**, 1906–1907.

6
Synthesis of Natural Products and Related Compounds Using Ene–Yne Metathesis

Miwako Mori

6.1
Introduction

Metathesis is the most important and useful reaction in recent synthetic organic chemistry [1]. In ring-closing olefin metathesis (RCM), formally, bond fission of two double bonds occurs and the new double bond is formed at the same time to produce a cyclic compound (Equation 6.1).

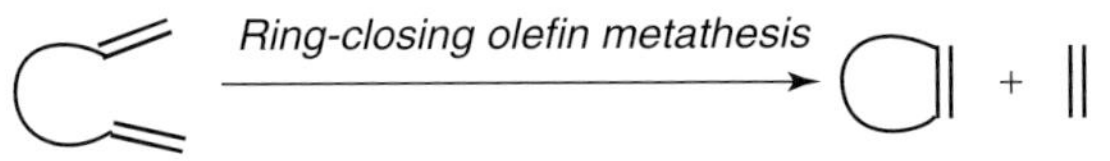

(6.1)

Metathesis of ene–ynes having alkene and alkyne moieties in a molecule is an extremely interesting reaction [2]. In this reaction, the double bond of ene–yne **1** is cleaved and a carbon–carbon bond is formed between the double and triple bonds, and the cleaved alkylidene part of the double bond migrates onto the alkyne carbon to produce a cyclic compound **2** having a 1,3-diene moiety (Equation 6.2).

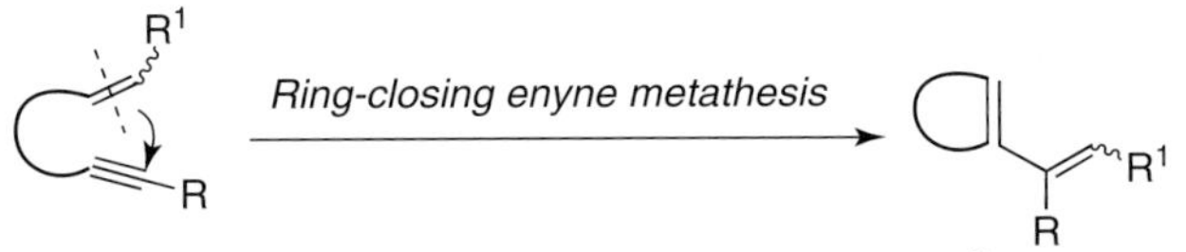

1 **2** (6.2)

The first ene–yne metathesis was reported by Katz *et al.* [3], who used a Fischer tungsten–carbene complex. Then, Mori *et al.* reported a chromium-catalyzed ene–yne metathesis [4]. It was later found that the ruthenium–carbene complexes [Ru]-**IV** and [Ru]-**I** were very effective for ene–yne metathesis [5]. The reaction would proceed via a [2 + 2]-cycloaddition of a ruthenium–carbene complex **3** with an alkyne part to produce ruthenacyclobutene **4**, and ring opening of this latter species affords a ruthenium carbene complex **5**, which reacts with an alkene part to produce ruthenacyclobutane **6**. Subsequent ring opening of **6** gives a cyclized compound **2**, and a ruthenium–carbene complex is regenerated (Scheme 6.1, Route 1). The other mechanism considered also involves at first reaction of the

Metathesis in Natural Product Synthesis: Strategies, Substrates and Catalysts. Edited by Janine Cossy, Stellios Arseniyadis, and Christophe Meyer

ISBN: 978-3-527-32440-8

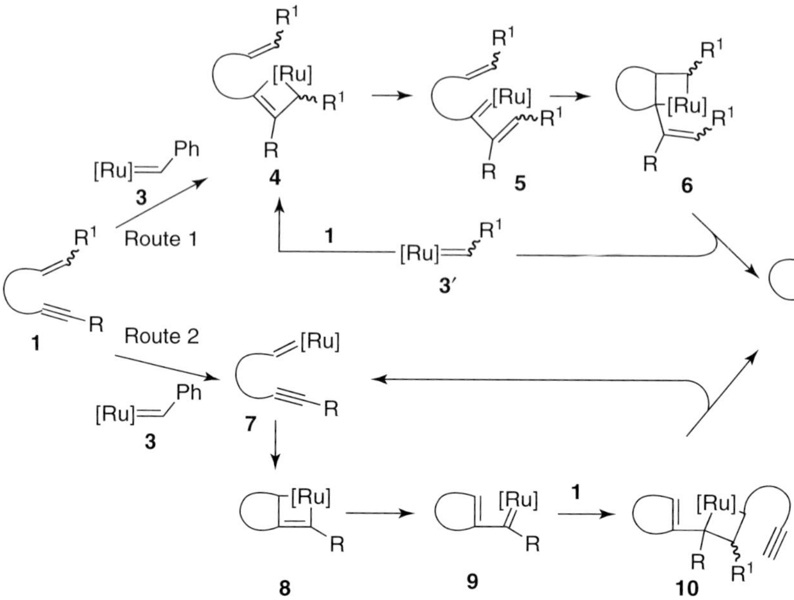

Scheme 6.1 Ring-closing ene–yne metathesis.

alkene part of ene–yne **1** with a ruthenium carbene complex to afford a new ruthenium carbene complex **7**. The latter species reacts with the alkyne part to produce ruthenacyclobutene **8**, and its subsequent ring opening gives ruthenium carbene **9**, which undergoes [2 + 2]-cycloaddition with the alkene part of ene–yne **1** to produce ruthenacyclobutane **10**. Ring opening of **10** gives cyclic compound **2**, and the ruthenium carbene complex **7** is regenerated (Scheme 6.1, Route 2).

Later, the detailed study on the reaction mechanism was shown by Lippstreu and Staub, who described that the reaction would proceed via route 2, and ruthenacyclobutene **4** generated from the alkyne part of ene–yne **2** and ruthenium carbene complex **3** do not exist as local minimum in the catalytic cycle [6]. Using ruthenium carbene complexes [Ru]-**IV** and [Ru]-**I**, various carbo- and heterocycles could be synthesized from the corresponding ene–ynes [5]. Furthermore, diene–yne metathesis, cross ene–yne metathesis, and ring-opening ene–yne metathesis have been developed (Scheme 6.2).

Novel routes for the synthesis of various structurally complex compounds, such as natural products and related biologically active substances, were thus available.

Mori *et al.* reported the synthesis of heterocycles having a diene moiety using ene–yne RCM [5]. Ene–ynes **11a–c** were treated by Grubbs's catalyst [Ru]-**IV** (1 mol%, C_6H_6, rt) at room temperature to afford heterocycles **12a–c** in high yields. Using this procedure, five- to nine-membered heterocycles could be synthesized (Scheme 6.3).

In this reaction, ene–ynes having a terminal alkyne did not give satisfactory results [7]. For example, RCM of enyne **11d** afforded cyclic compound **12d** in only 21% yield. It is reasoned that an alkene part in product **12d** further reacts with

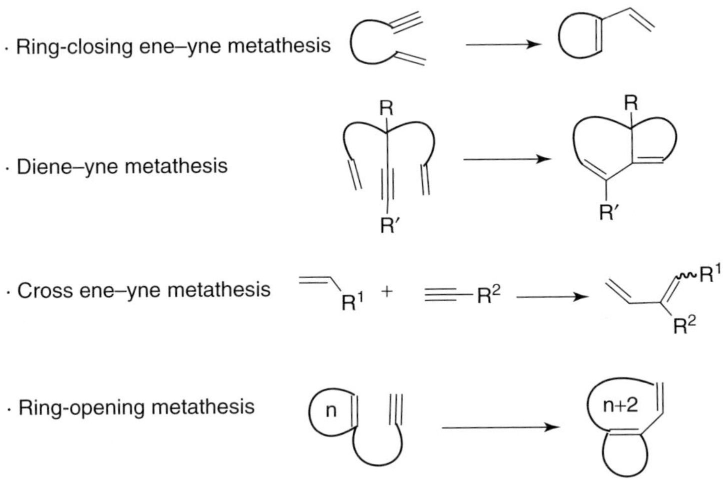

Scheme 6.2 Various ene–yne metathesis.

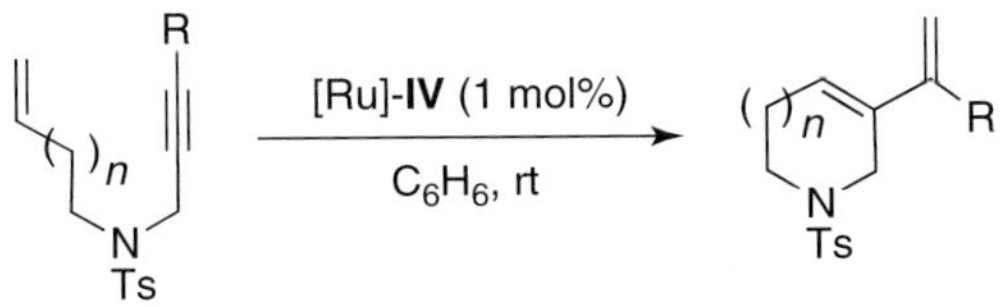

11a	$n = 0$	R = Me	**12a**	91%
11b	$n = 1$	R = CH_2OTBS	**12b**	88%
11c	$n = 2$	R = CH_2OTBS	**12c**	77%

Scheme 6.3 Synthesis of heterocycles using ruthenium catalyst.

methylidene ruthenium carbene to afford ruthenium carbene **14d**, which would be coordinated by the alkene part to produce **15d**. Thus, the catalytic activity would decrease (Scheme 6.4). In fact, when the reaction of **11d** using [Ru]-**I** was carried out under ethylene gas, the catalytic activity was improved to afford **12d** in 90% yield even with the use of 1 mol% of the ruthenium catalyst [Ru]-**I** [7]. The higher reactivity observed in ene–yne RCM in the presence of ethylene gas has often been advantageously explored in applications to natural product synthesis.

6.2
Synthesis of Natural Products and Related Compounds Using Ene–yne Metathesis

The first application of ene–yne RCM was reported in the total synthesis of the natural product (−)-stemoamide [8]. (−)-Pyroglutamic acid was converted into ene–yne **16** having an ester group on the alkyne, and RCM of ene–yne **16** was carried out in the presence of ruthenium–carbene complex [Ru]-**I** (4 mol%) to afford bicyclic compound **17** in 87% yield. Conversion of **17** into **18** followed by

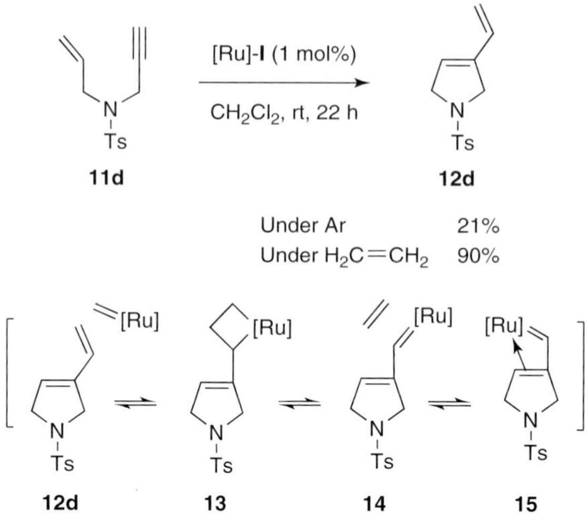

Scheme 6.4 Metathesis of enyne having terminal alkyne under ethylene gas.

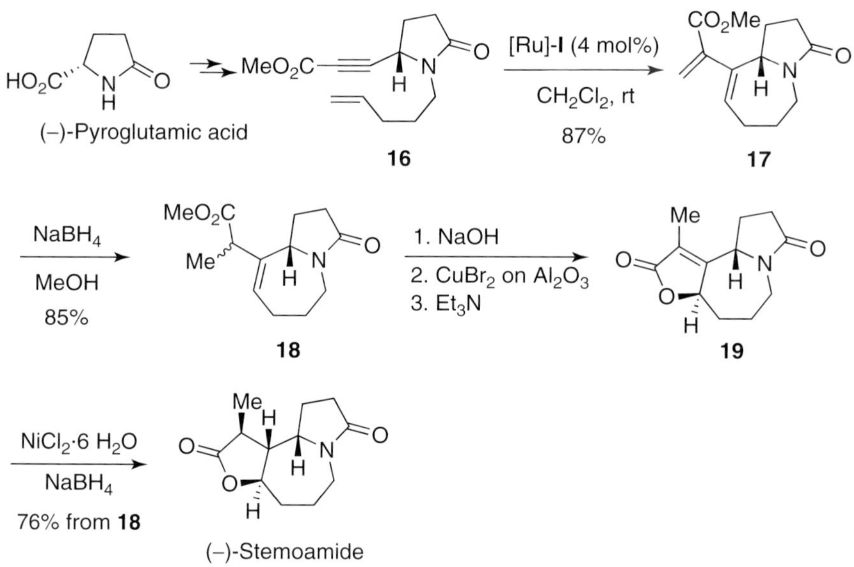

Scheme 6.5 Total synthesis of (−)-stemoamide.

halo-lactonization gave **19**. From this compound **19**, (−)-stemoamide could be synthesized by conjugate reduction (Scheme 6.5).

Carbacephem **22a** and carbapenem **22b** were synthesized, respectively, from ene–ynes **21a** and **21b**, which were prepared from 4-acetoxy-azetidinone **20**. The yield in compound **22b** (29%) is lower compared to that in **22a** (80%) because of the highly strained, fused 4,5-membered ring system (Scheme 6.6) [9].

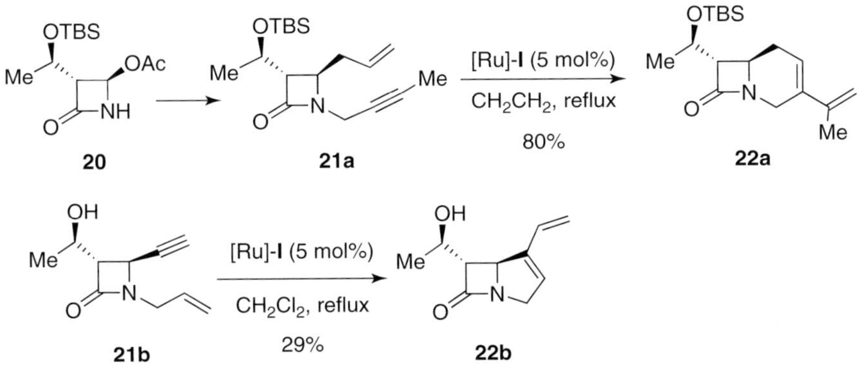

Scheme 6.6 Construction of carbacephem and carbapenem skeleton.

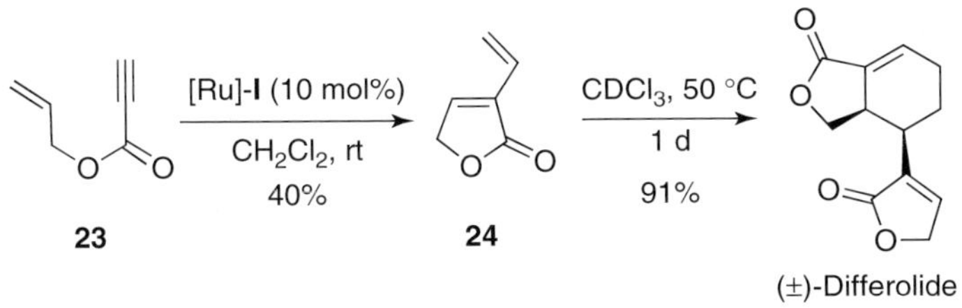

Scheme 6.7 Synthesis of (±)-differolide.

(±)-Differolide was easily synthesized by ene–yne metathesis [10]. Ene–yne **23** was reacted with catalyst [Ru]-**I** to give lactone **24**, which was spontaneously dimerized to afford (±)-differolide (Scheme 6.7).

An enantioselective biomimetic synthesis of (−)-longithorone A was accomplished on the basis of the proposed biosynthesis [11]. Two [12]-paracyclophanes **25** and **26** were synthesized from common intermediate **27** by applying ene–yne metathesis macrocyclization in 42 and 31% yields, respectively. Intermolecular Diels–Alder reaction of **25** and **26** provided **32**. Deprotection followed by oxidation gave **33**, which spontaneously gave longithorone A via transannular Diels–Alder reaction (Scheme 6.8).

The total synthesis of (+)-anatoxin-a was achieved by Martin *et al.* 12a, b and Mori *et al.* [12c, d] by using the same strategy. The key step is the construction of an azabicyclo[4.2.1]nonene ring system. For that purpose, the 2,5-*cis*-disubstituted pyrrolidine derivative **34** was synthesized from (+)-pyroglutamic acid. Ene–yne metathesis applied to **34** was carried out using second-generation catalyst [Ru]-**II** to form the azabicyclic compound **35** [12a, b], which was chemoselectively dihydroxylated to provide **36**. From the latter compound, anatoxin-a was synthesized (Scheme 6.9).

Using a similar procedure, (+)-ferruginine was synthesized from (−)-pyroglutamic acid [13]. Construction of the azabicyclo[3.2.1]octene ring system was carried out by applying RCM to ene–yne **37**. Wacker oxidation of the resultant

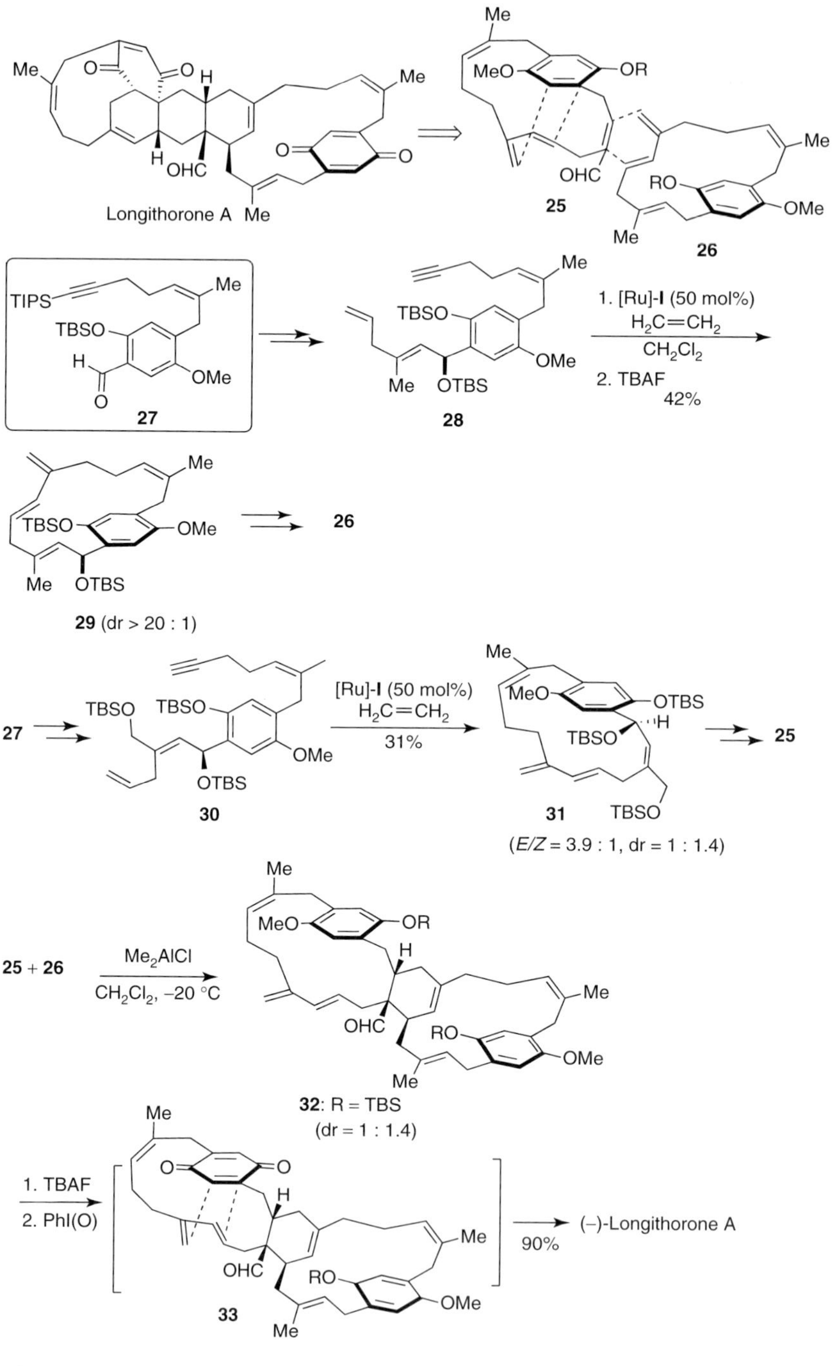

Scheme 6.8 Total synthesis of (−)-longithorone A.

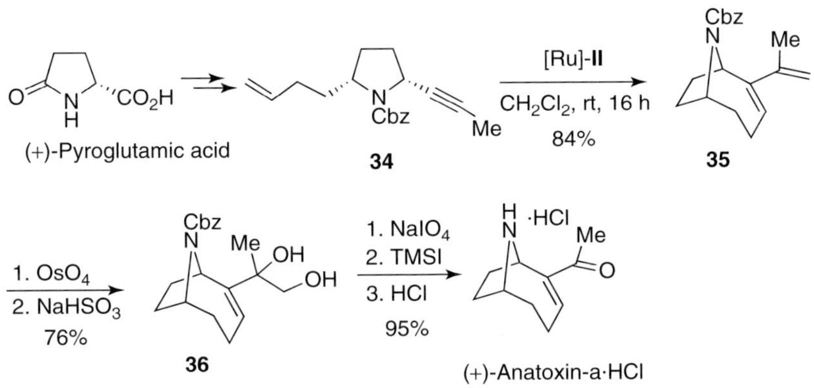

Scheme 6.9 Total synthesis of anatoxin-a.

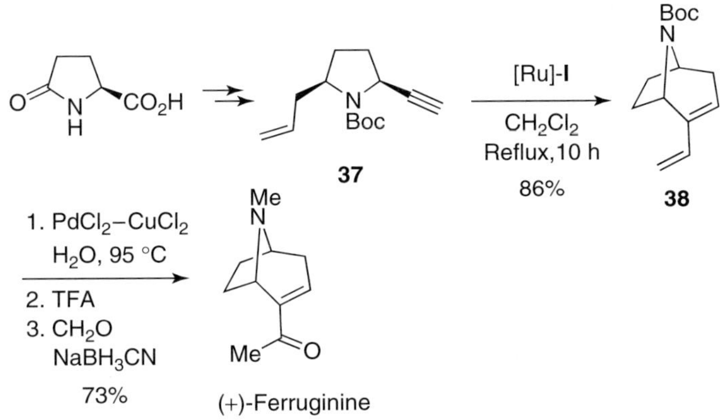

Scheme 6.10 Synthesis of (+)-ferruginine.

diene **38** afforded methyl ketone, and then deprotection followed by methylation gave (+)-ferruginine (Scheme 6.10).

Kozmin *et al.* developed a highly efficient synthesis of cyclic compound **41a** bearing a methyl ketone functionality from ene–yne **39a** having a silyloxy group on the alkyne using RCM [14a]. As an application of this method, they succeeded in the synthesis of eremophilanes [14b]. RCM of ene–yne **39b** having a triisopropylsilyloxy group on the alkyne followed by treatment with HF gave cycloalkene **41b** having the methyl ketone functionality. From this compound **41b**, α- and β-eremophilanes were synthesized (Scheme 6.11).

New *allo*-colchinoids functionalized at C10 or C11 in the C-ring were synthesized using the RCM of ene–yne **43** for the construction of the seven-membered ring [15]. The reaction proceeded smoothly to give **44** using catalyst [Ru]-**II** in 92% yield. Deprotection followed by PCC oxidation gave **45**, which was subjected to a Diels–Alder aromatization sequence to form **46**. Amination followed by acetylation gave *allo*-colchicines (Scheme 6.12).

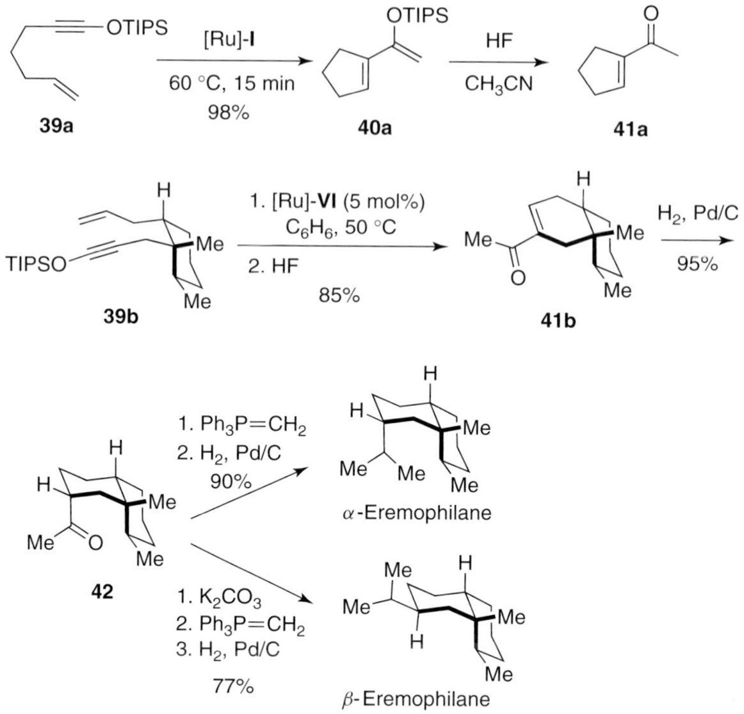

Scheme 6.11 Total synthesis of eremophilane.

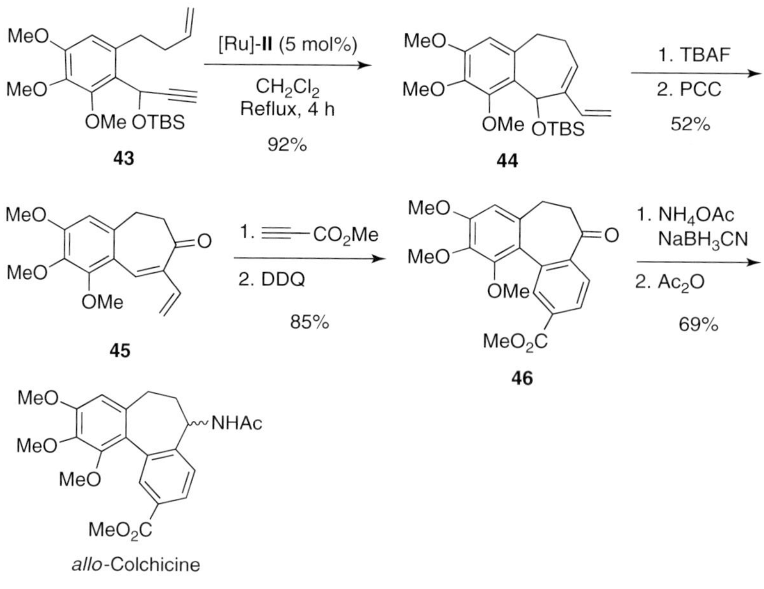

Scheme 6.12 Synthesis of *allo*-colchicine.

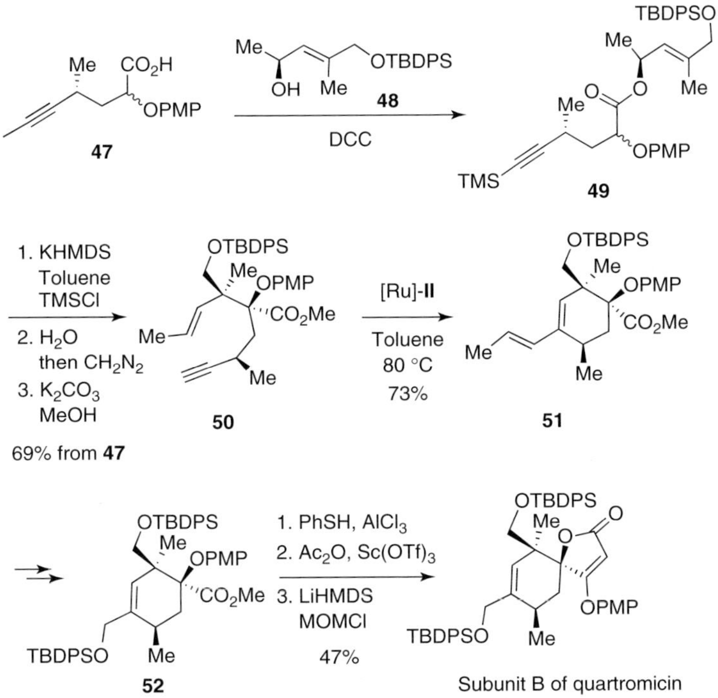

Scheme 6.13 Synthesis of subunit of quartromicin.

The *agalacto*-spirolactone B subunit of quartromicins has been synthesized using the Claisen–Ireland/RCM of ene–yne approach by Haudrechy *et al.* [16]. Compound **49** was converted into ene–yne **50**, which was treated with catalyst [Ru]-**II** (toluene, 80 °C) to afford **51** in 73% yield. From this compound **51**, subunit B of quartromicin was synthesized (Scheme 6.13).

An enantioselective synthesis of (−)-galanthamine was realized in 11 steps starting from isovanillin (Scheme 6.14). The ene–yne **53** (ee = 92%) underwent an efficient RCM reaction in the presence of catalyst [Ru]-**I** (3 mol%) to give diene **54** in 85% yield. Hydroboration and oxidation of the terminal alkene present in **54** gave an intermediate homoallylic alcohol in excellent yield, which underwent sequential Heck reaction with a palladium catalyst and allylic oxidation with SeO_2 to give **56**. From compound **56**, (−)-galanthamine could be synthesized via mesylation of the hydroxyl group, cleavage of *N*-Boc, and intramolecular nucleophilic substitution [17].

For the synthesis of angucyclinone-type natural products, a concise and highly enantioselective route was developed [18]. Chiral vinylcyclohexene derivative **58** was synthesized using RCM applied to ene–yne **57** under an atmosphere of ethylene in high yield. Intermolecular Diels–Alder reaction of **58** and **59** followed by aromatization gave compound **60** having the benz[*a*]anthraquinone skeleton. By utilizing this strategy, the total synthesis of five natural products: YM-181741, (+)-ochromycinone, (+)-rubiginone B, (−)-tetragomycin, and MM-47755, was accomplished (Scheme 6.15).

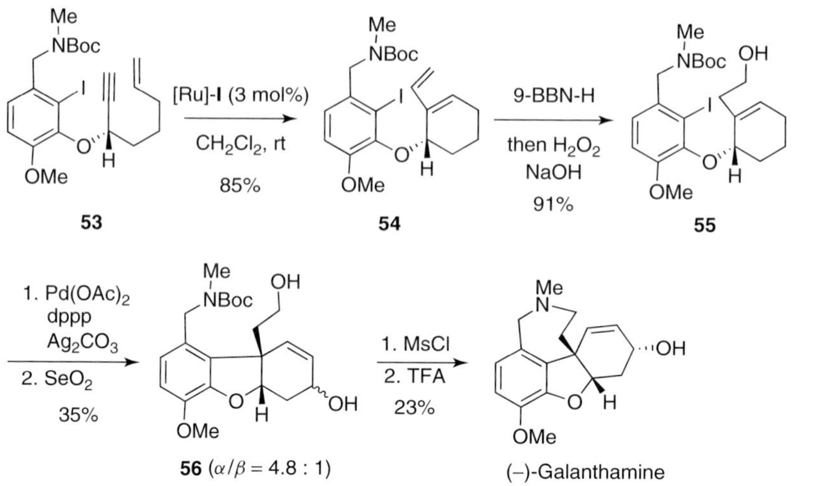

Scheme 6.14 Total synthesis of galanthamine.

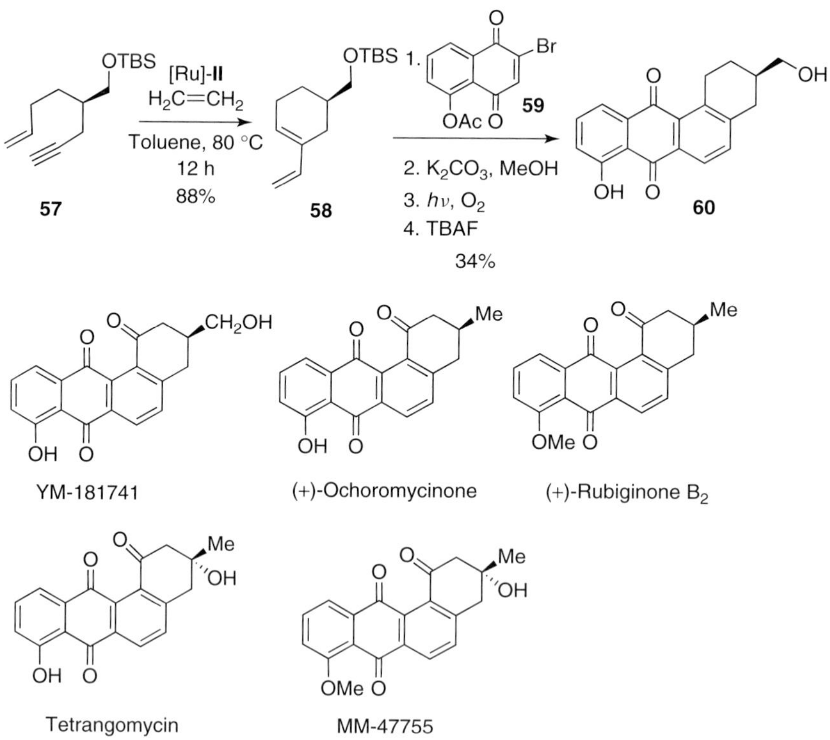

Scheme 6.15 Synthesis of benz[*a*]anthraquinone skeleton.

Synthesis of anthramycin derivative **69a** was achieved using RCM and alkene cross-metathesis (CM) (Scheme 6.16) [19]. L-Methionine was converted to ene–yne **61**, and subsequent RCM with catalyst [Ru]-**I** under an atmosphere of ethylene gave pyrrolidine derivative **62**. After deprotection and condensation with the commercially available acid chloride **63**, the resulting amide **64** underwent reductive cyclization using Zn/AcOH followed by treatment with diluted HCl to

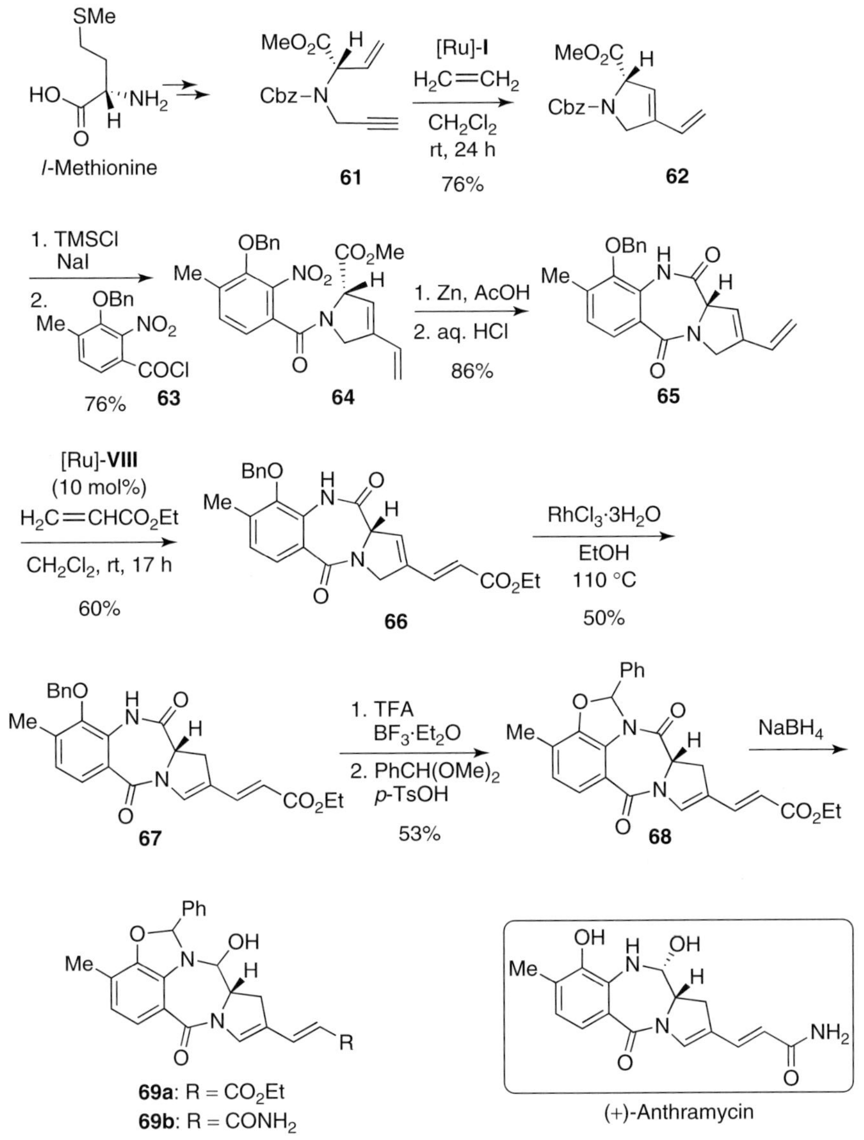

Scheme 6.16 Synthesis of anthramycin derivative.

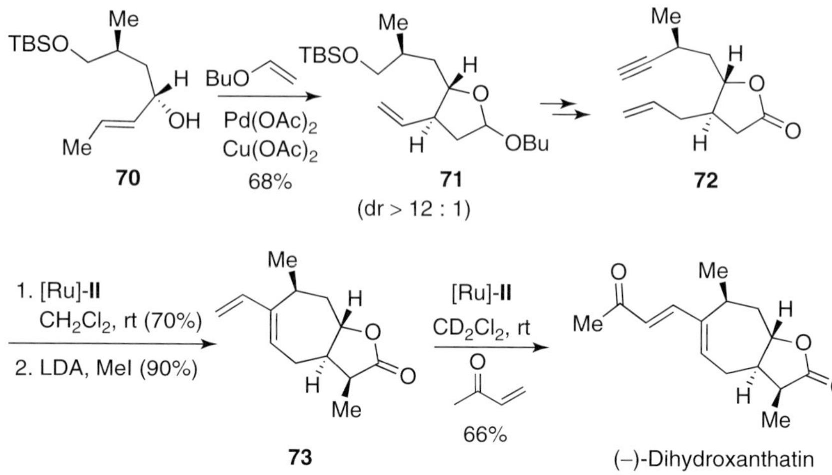

Scheme 6.17 Synthesis of (−)-dihydroxanthatin.

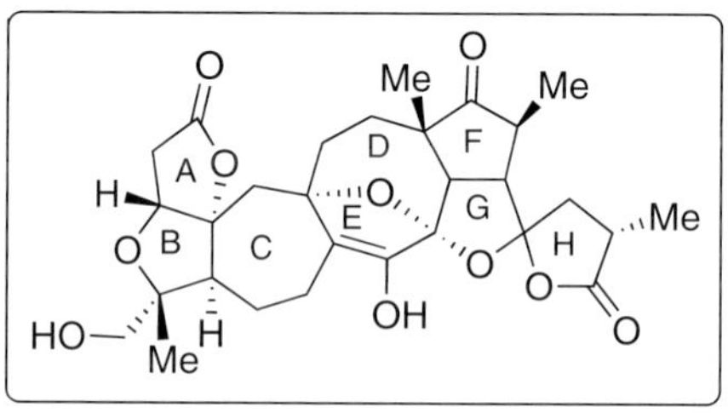

Lancifodilactone G

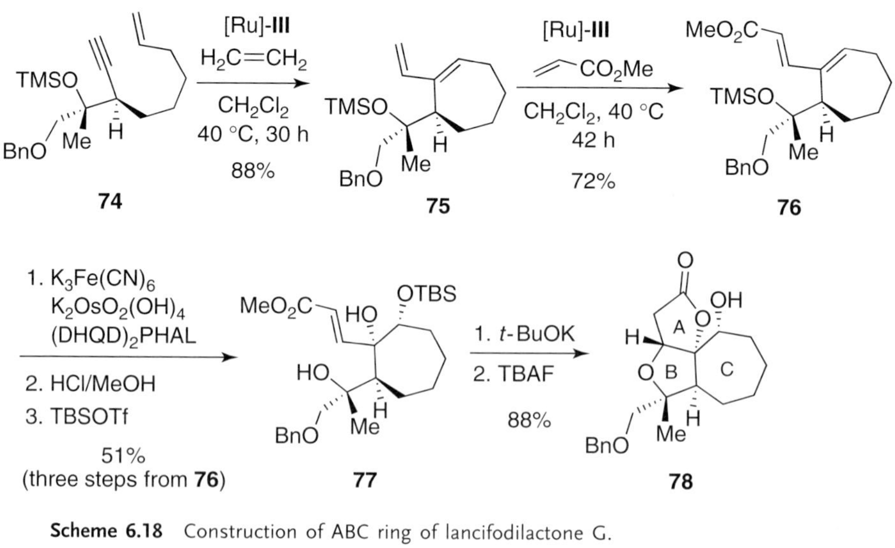

Scheme 6.18 Construction of ABC ring of lancifodilactone G.

provide the pyrrolo-1,4-benzodiazepinone **65**. To convert the vinyl group into an α,β-unsaturated ester group, CM with ethyl acrylate was carried out using catalyst [Ru]-**VIII**. The reaction proceeded smoothly to give compound **66** in 60% yield. Isomerization of the double bond in the pyrrolidine ring using $RhCl_3 \cdot H_2O$ afforded compound **67**. Debenzylation followed by protection gave **68**, the amido group of which was converted into aminal **69a**. Conversion of **69b** into (+)-anthramycin was previously described by Stille *et al.* [20].

Morken and Evans succeeded in the synthesis of (−)-dihydroxanthatin using RCM and CM [21]. Allylic alcohol **70** was converted into tetrahydrofuran **71** by the catalytic Oshima–Utimoto reaction [22]. From compound **71**, lactone **72** was synthesized. Ene–yne metathesis of **72** using catalyst [Ru]-II followed by methylation gave bicyclic compound **73**. CM of **73** and methyl vinyl ketone in the presence of [Ru]-**II** afforded (−)-dihydroxanthatin (Scheme 6.17).

Lancifodilactone G has eight rings with complex cyclic connectivity. Paquette *et al.* synthesized the ABC rings of lancifodilactones G, compound **78**, using ene–yne RCM and then CM as the key steps (Scheme 6.18) [23]. RCM of ene–yne **74** using catalyst [Ru]-**III** in CH_2Cl_2 under an ethylene atmosphere gave cycloheptene derivative **75**, which was subjected to CM with methyl acrylate and catalyst [Ru]-**III** to give α,β-unsaturated ester **76**. From this functionalized seven-membered ring intermediate, the target compound **78** was obtained.

6.3
Synthesis of Natural Products and Related Compounds Using Ene–yne Cross-metathesis (CM)

A novel synthetic procedure of 1,3-diene from alkyne and ethylene using CM was developed in 1997 by Mori *et al.* [24]. When a solution of alkyne **79a** in CH_2Cl_2 was stirred under an atmosphere of ethylene at rt in the presence of catalyst [Ru]-**I**, 1,3-diene **80a** was obtained in 62% yield. It is interesting that, formally, the double bond of ethylene is cleaved and each methylene part is introduced onto the alkyne carbon to produce 1,3-diene **80a** (Scheme 6.19). The possible reaction course is shown in Scheme 6.20. Reaction of methylidene carbene complex **81**, generated from **3** and ethylene, with alkyne **79** gives ruthenacyclobutene **82**, ring opening of which gives ruthenium carbene **83**. It reacts with ethylene to afford ruthenacyclobutane **84**, which undergoes ring opening to give 1,3-diene **80**, and **81** is regenerated. If **81** reacts with ethylene, ruthenacyclobutane would be produced, but this is a nonproductive process as **81** would be regenerated (Scheme 6.20).

However, propargyl ester **79a** or amide **79b** gave good results, whereas homopropargyl amide **80c** led to 1,3-diene **80c** in only 11% yield (Scheme 6.19). Presumably, a heteroatom at the propargylic position is important, and the ruthenium catalyst would be coordinated by the heteroatom at first and then the reaction would proceed. When the second-generation ruthenium–carbene complex [Ru]-**II** was used for this reaction (toluene, 80 °C), alkynes **79d** and **79e** lacking heteroatoms at the propargylic positions, gave 1,3-dienes **80d** and **80e** in good yield [25a, b].

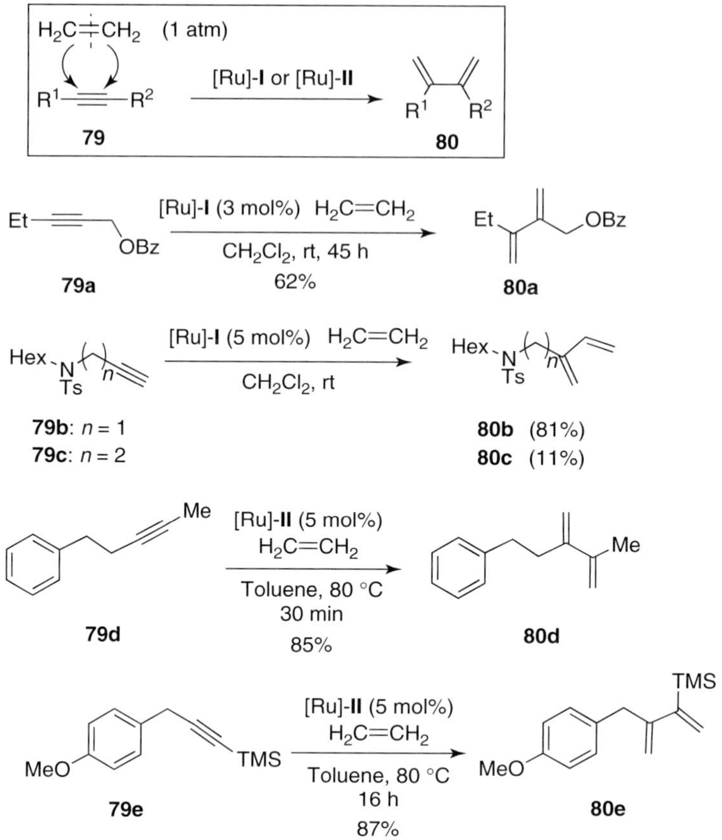

Scheme 6.19 Synthesis of 1,3-diene using cross-metathesis.

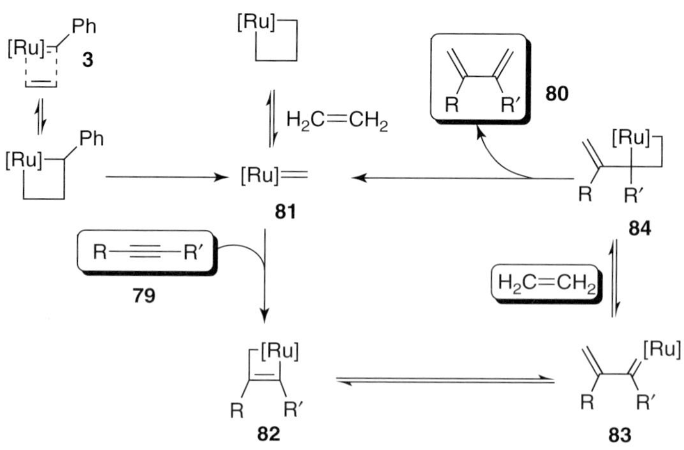

Scheme 6.20 Possible reaction course for the formation of 1,3-diene.

Furthermore, the reaction was more rapid and several substituents on the alkyne were tolerated (TMS, CO_2Me).

Synthesis of natural products using ene–yne CM is interesting because a 1,3-diene moiety can be constructed onto the alkyne carbons at a later stage. Anolignans were synthesized using CM of ene–yne as a key step. 1,3-Diene **86** could be synthesized from alkyne **85** by treatment with catalyst [Ru]-**II** under ethylene gas. Palladium-catalyzed deacetoxylation followed by deprotection gave anolignan A. Anolignan B could be synthesized in a similar manner. It is interesting to note that the two methylene parts of the anolignan skeleton were introduced at a later stage of the total synthesis using CM (Scheme 6.21) [26].

A short and efficient synthesis of highly substituted tetrahydropyridines **88** was achieved from a monosubstituted alkyne, a terminal alkene, and an imine by a combination of ene–yne CM and aza-Diels–Alder reaction under high pressure. CM of terminal alkyne and alkene afforded diene **87**, which reacted with imine to give pipecolic acid derivative **88** in high yield (Scheme 6.22) [27].

The reaction was further extended to intramolecular Diels–Alder reaction, and *cis*-hexahydro-1*H*-indene **90** was synthesized from diene **89** and terminal alkyne in one operation. The possible intermediate **92** was spontaneously converted into **90**. Deprotection of the silyl group followed by pyridinium chloro chromate (PCC) oxidation gave indanone **91** (Scheme 6.23) [28].

New vitamin D receptor antagonists 24,24-ethanovitamine D3-26,23-lactones **93a** and **93b** and their analogs were synthesized (Scheme 6.24) [29]. The CD-ring precursors **99a** and **99b** were efficiently prepared by using a ruthenium-catalyzed intermolecular ene–yne CM applied to **95** in the presence of ethylene as a key step. Cyclopropanation of the enone moiety in the resulting product **96**, followed by treatment with DIBAL-H and then deprotection, gave compounds **98a** and **98b**, epimeric at C23. Chemoselective oxidation of **98a** and **98b** followed by palladium-catalyzed coupling with **100** and then deprotection afforded **93a** and **93b**, respectively.

Lee *et al.* succeeded in the total synthesis of (−)-amphidinolide E, whose side chain was constructed using ene–yne CM [30]. Alkyne **101** was first reacted with ethylene in the presence of catalyst [Ru]-**II** to give **102**, which was further engaged *in situ* in a chemoselective CM with 2-methyl-1,4-pentadiene (**103**) to produce triene **104** in 65% yield along with diene **102** in 19% yield. The isolated diene **102** was recycled and further reacted with **103** in a similar manner to afford triene **104**, which was finally converted to sulfone **105**. Condensation of sulfone **105** and aldehyde **106** afforded compound **107**, which was further elaborated to seco-acid **108**. Lactonization of **108** followed by deprotection completed the total synthesis of amphidinolide E (Scheme 6.25).

6.4
Synthesis of Natural Products Using Skeletal Reorganization

Transition-metal-catalyzed cycloisomerization of ene–ynes **109** occurring with skeletal rearrangement can produce Type I product **111** (similar to ene–yne RCM

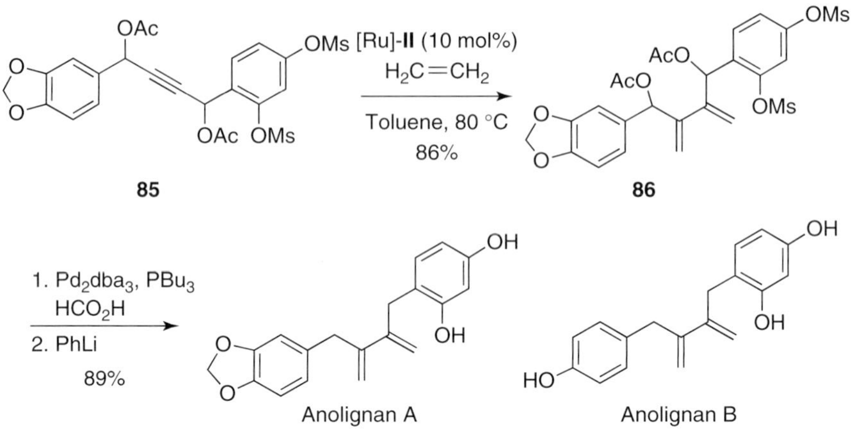

Scheme 6.21 Synthesis of anolignan s using ene–yne CM.

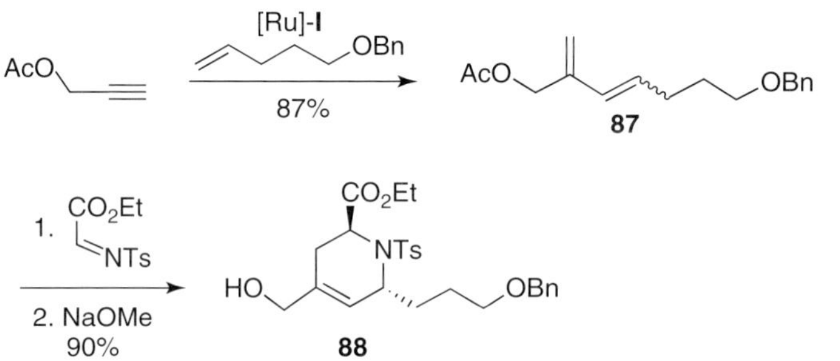

Scheme 6.22 Synthesis of pipecolic acid derivatives.

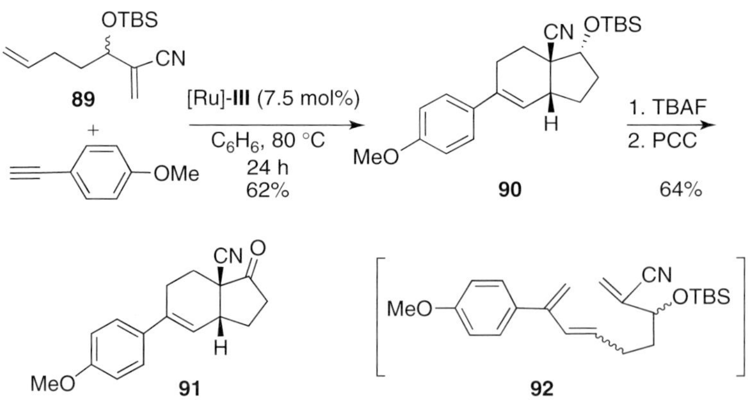

Scheme 6.23 Synthesis of *cis*-fused carbo-bicycles.

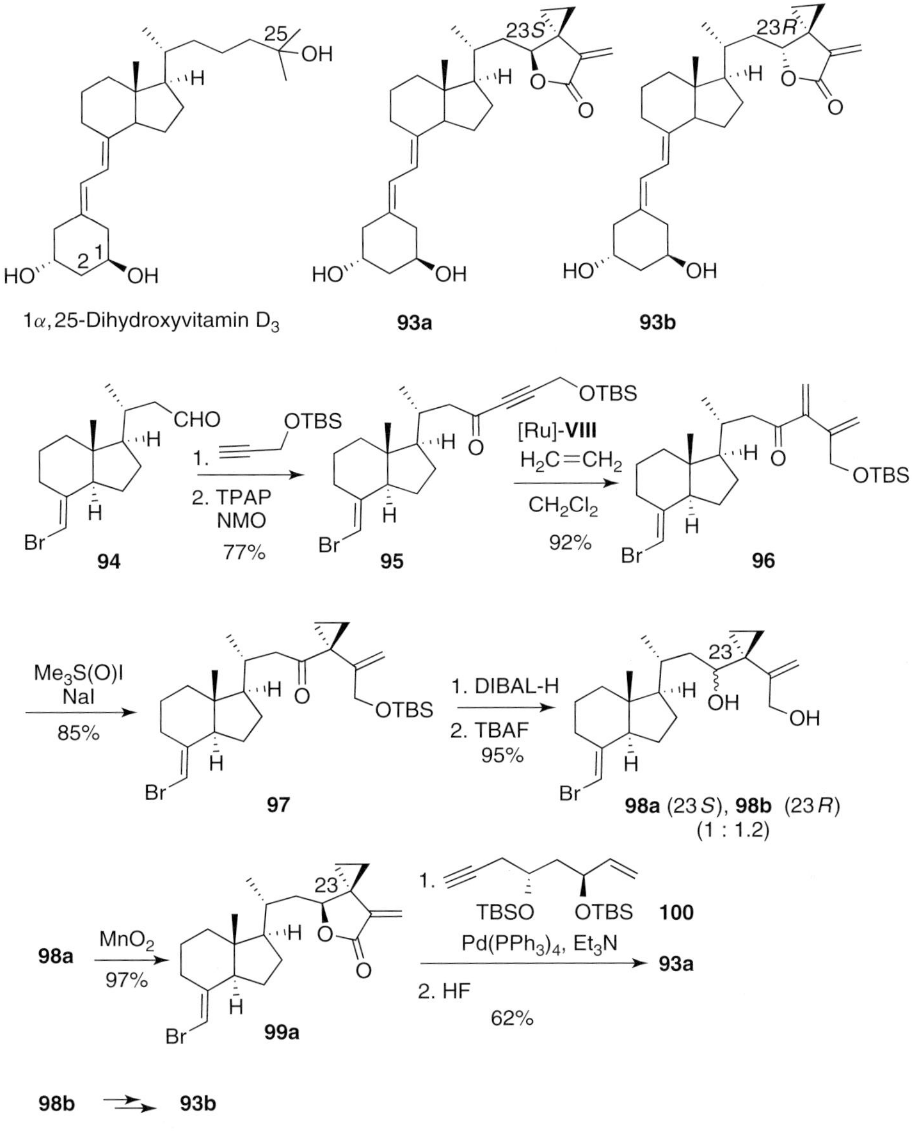

Scheme 6.24 Synthesis of 24,24-ethanovitamin D3 lactones.

product) or Type II product **112** resulting from single and double cleavage from a cyclopropyl metal carbene intermediate (Scheme 6.26) **110** (31, 32).

Fürstner *et al.* achieved the formal total syntheses of the antibiotics metacycloprodigiosin and streptorubin B by a platinum-catalyzed skeletal reorganization reaction (Scheme 6.27) [33]. The key step leading to the *meta*-bridged pyrrole core structures consisted of a "formal" metathesis reaction of electron-deficient

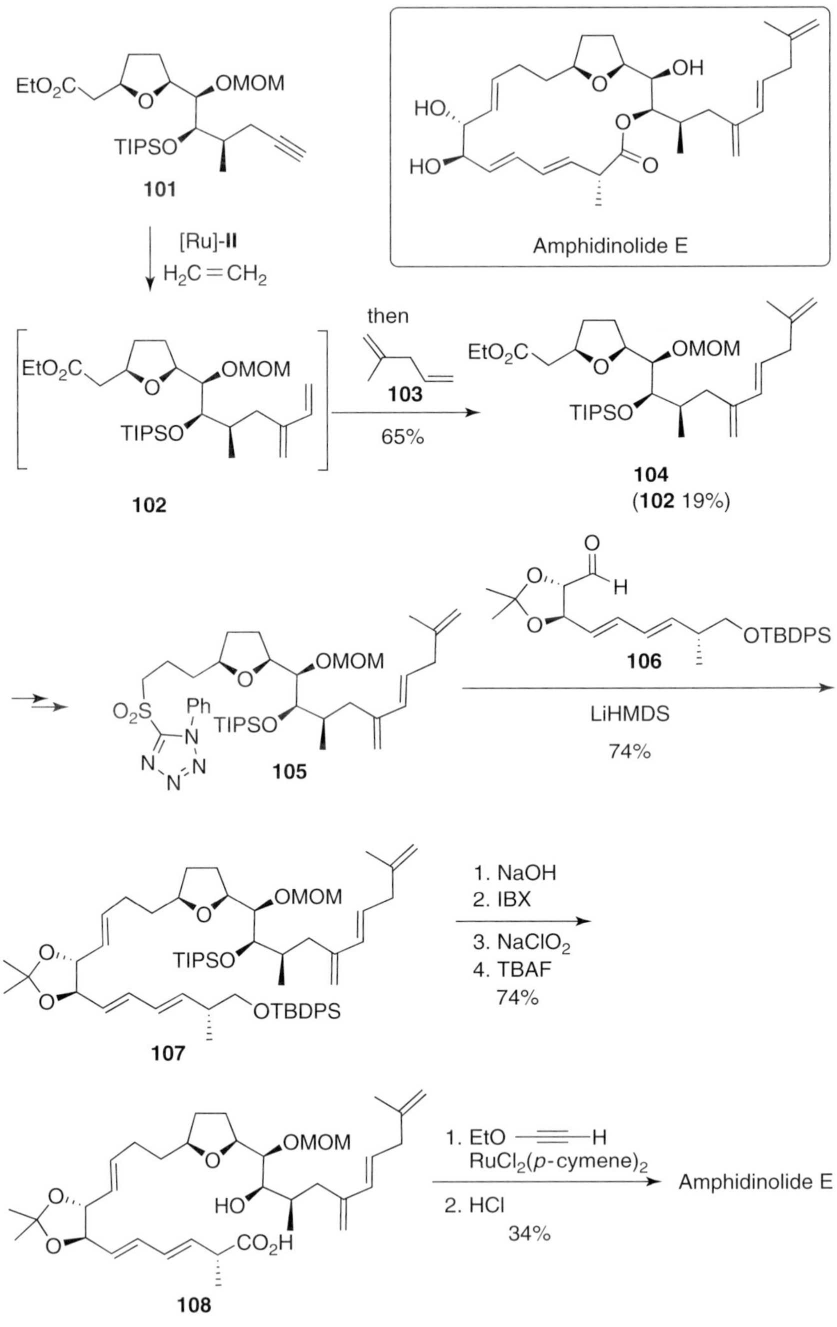

Scheme 6.25 Synthesis of amphidinolide E.

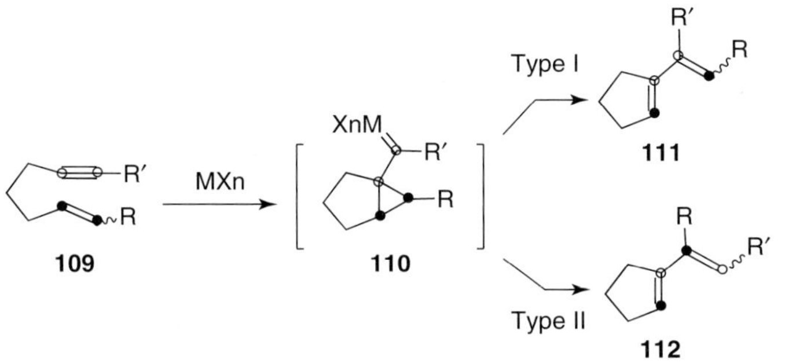

Scheme 6.26 Skeletal reorganization.

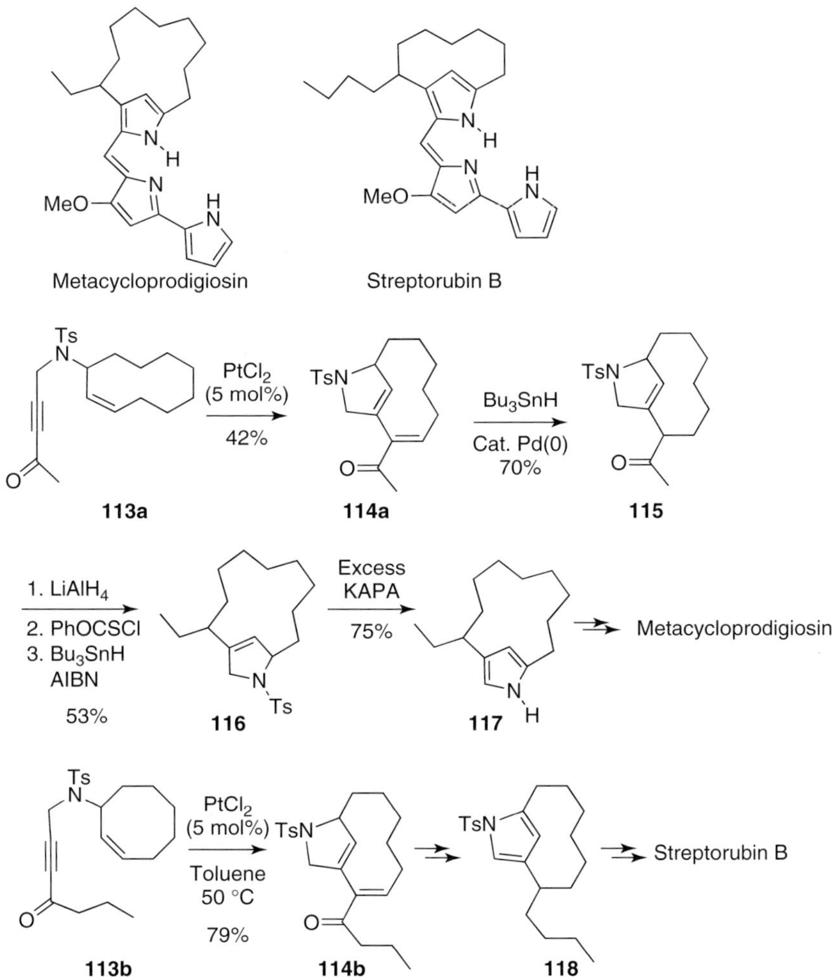

Scheme 6.27 Formal total syntheses of streptorubin B and metacycloprodigiosin.

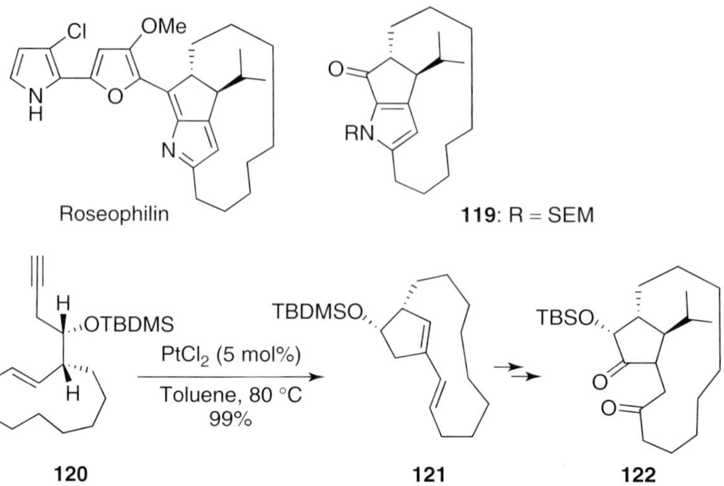

Scheme 6.28 Formal total synthesis of roseophilin.

ene–ynes **113a** and **113b** catalyzed by $PtCl_2$. The skeletal reorganization products **114a** and **114b** were converted into the respective target molecules.

Trost *et al.* succeeded in the formal total synthesis of roseophilin [34]. Macrocyclic compound **121** was synthesized from ene–yne **120** by platinum-catalyzed skeletal reorganization. From **121**, the 1,4-diketone **122** was synthesized and converted to pyrrole derivative whose transformation into roseophilin had previously been reported (Scheme 6.28).

References

1 (a) Grubbs, R.H. (ed.) (2003) *Handbook of Metathesis*, Wiley-VCH Verlag GmbH; (b) Fürstner, A. (ed.) (1998) *Topics in Organometallic Chemistry*, vol. 1, Springer-Verlag, Berlin, Heidelberg; (c) Trunk, T.M. and Grubbs, R.H. (2001) *Acc. Chem. Res.*, **34**, 18–29; (d) Fürstner, A. (2000) *Angew. Chem. Int. Ed.*, **39**, 3012–3043; (e) Herisson, J.-L. and Chauvin, Y. (1971) *Macromol. Chem.*, **141**, 161.

2 Reviews for Enyne Metathesis; (a) Mori, M. (1998) *Top. Organomet. Chem.*, **1**, 133–154; (b) Poulsen, C.S. and Madsen, R. (2003) *Synthesis*, 1–18; (c) Mori, M. (2003) in *Handbook of Metathesis*, vol. 2 (ed. R.H. Grubbs), Wiley-VCH Verlag GmbH, pp. 176–204; (d) Giessert, A.J. and Diver, S.T. (2004) *Chem. Rev.*, **104**, 1317–1382; (e) Mori, M. and Kitamura, T. (2005) Ene–Yne and alkyne metathesis, in *Comprehensive Organometallic Chemistry III, Transition Metal Organometallics in Organic Synthesis* (ed. T. Hiyama), Elsevier, pp. 271–310.

3 (a) Katz, T.J. and Sivavec, T.M. (1985) *J. Am. Chem. Soc.*, **107**, 737–738; (b) Sivavec, T.M., Katz, T.J., Chiang, M.Y., and Yang, G.X.-Q. (1989) *Organometallics*, **8**, 1620–1625; (c) Katz, T.J. and Yang, G.X.-Q. (1991) *Tetrahedron Lett.*, **32**, 5895–5898.

4 (a) Watanuki, S., Ochifuji, N., and Mori, M. (1994) *Organometallics*, **13**, 4129–4130; (b) Watanuki, S., Ochifuji, N., and Mori, M. (1995) *Organometallics*, **14**, 5062–5067; (c) Mori, M. and Watanuki, S. (1992) *J. Chem. Soc., Chem. Commun.*, 1082–1084; (d) Watanuki,

S. and Mori, M. (1993) *Heterocycles*, **35**, 679–682.

5 (a) Kinoshita, A. and Mori, M. (1994) *Synlett*, 1020–1022; (b) Kinoshita, A., Sakakibara, N., and Mori, M. (1999) *Tetrahedron*, **55**, 8155–8167.

6 Lippstreu, J.J. and Straub, B.F. (2005) *J. Am. Chem. Soc.*, **127**, 7444–7457.

7 Mori, M., Sakakibara, N., and Kinoshita, A. (1998) *J. Org. Chem.*, **63**, 6082–6083.

8 (a) Kinoshita, A. and Mori, M. (1996) *J. Org. Chem.*, **61**, 8356–8357; (b) Kinoshita, A. and Mori, M. (1997) *Heterocycles*, **46**, 287–299.

9 (a) Barrett, A.G.M., Baugh, S.P.D., Braddock, D.C., Flack, K., Gibson, V.C., Procopiou, P.A., White, A.J.P., and Williams, D.J. (1998) *J. Org. Chem.*, **63**, 7893–7907; (b) Duboc, R., Henaut, C., Savignac, M., Genet, J.-P., and Bhatnagar, N. (2001) *Tetrahedron Lett.*, **42**, 2461–2464.

10 Hoye, T.R., Donaldson, S.M., and Vos, T.J. (1999) *Org. Lett.*, **1**, 277–280.

11 Layton, M.E., Morales, C.A., and Shair, M.D. (2002) *J. Am. Chem. Soc.*, **124**, 773–775.

12 (a) Brenneman, J.B. and Martin, S.F. (2004) *Org. Lett.*, **6**, 1329–1332; (b) Brenneman, J.B., Machauer, R.M., and Martin, S.F. (2004) *Tetrahedron*, **60**, 7301–7314; (c) Mori, M., Tomita, T., Kita, Y., and Kitamura, T. (2004) *Tetrahedron Lett.*, **45**, 4397–4399; (d) Tomita, T., Kita, Y., Kitamura, T., Sato, Y., and Mori, M. (2006) *Tetrahedron*, **62**, 10518–10527.

13 Aggarwal, V.K., Astle, J., and Rogers-Evans, M. (2004) *Org. Lett.*, **6**, 1469–1472.

14 (a) Reddy, S. and Kozmin, S.A. (2004) *J. Org. Chem.*, **69**, 4860–4862; (b) Schramm, M.P., Reddy, D.S., and Kozmin, S.A. (2001) *Angew. Chem. Int. Ed.*, **40**, 4274–4277.

15 Boyer, F.-D. and Hanna, I. (2007) *Org. Lett.*, **9**, 715–718.

16 Bedel, O., Francais, A., and Haudrechy, A. (2005) *Synlett*, 2313–2316.

17 Satcharoen, V., McLean, N.J., Kemp, S.C., Camp, N.P., and Brown, R.C.D. (2007) *Org. Lett.*, **9**, 1867–1869.

18 Kaliappan, K.P. and Ravikumar, V. (2007) *J. Org. Chem.*, **72**, 6116–6126.

19 Kitamura, T., Sato, Y., and Mori, M. (2004) *Tetrahedron*, **60**, 9649–9657.

20 Pena, M.R. and Stille, J.K. (1989) *J. Am. Chem. Soc.*, **111**, 5417.

21 Evans, M.A. and Morken, J.P. (2005) *Org. Lett.*, **7**, 3371–3374.

22 (a) Fugami, K., Oshima, K., and Utimoto, K. (1987) *Tetrahedron Lett.*, **28**, 809; (b) Fugami, K., Oshima, K., and Utimoto, K. (1989) *Bull. Chem. Soc. Jpn.*, **62**, 2050.

23 Paquette, L.A. and Lai, K.W. (2008) *Org. Lett.*, **10**, 2111–2113.

24 Kinoshita, A., Sakakibara, N., and Mori, M. (1997) *J. Am. Chem. Soc.*, **119**, 12388–12389.

25 (a) Mori, M., Tonogaki, K., and Kinoshita, A. (2004) *Org. Synth.*, **81**, 1–13; (b) Tonogaki, K. and Mori, M. (2002) *Tetraheron Lett.*, **43**, 2235–2238; (c) Smulik, J.A. and Diver, S.T. (2000) *J. Org. Chem.*, **65**, 1788–1792; (d) Smulik, J.A. and Diver, S.T. (2000) *Org. Lett.*, **2**, 2271–2274; (e) Smulik, J.A., Giessert, A.J., and Diver, S.T. (2002) *Tetrahedron Lett.*, **43**, 209–211.

26 Mori, M., Tonogaki, K., and Nishiguchi, N. (2002) *J. Org. Chem.*, **67**, 224–226.

27 Schürer, S.S. and Blechert, S. (1999) *Tetrahedron Lett.*, **40**, 1877–1880.

28 Mix, S. and Blechert, S. (2005) *Org. Lett.*, **7**, 2015–2018.

29 Saito, N., Masuda, M., Saito, H., Takenouchi, K., Ishizuka, S., Namekawa, J., T-Kamimura, M., and Kittaka, A. (2005) *Synthesis*, 2533–2543.

30 (a) Kim, C.H., An, H.J., Shin, W.K., Yu, W., Woo, S.K., Jung, S.K., and Lee, E. (2008) *Chem. Asian J.*, **3**, 1523–1534; (b) Kim, C.H., An, H.J., Shin, W.K., Yu, W., Woo, S.K., Jung, S.K., and Lee, E. (2006) *Angew. Chem. Int. Ed.*, **45**, 8019–8021.

31 (a) Trost, B.M. and Tanoury, G.J. (1988) *J. Am. Chem. Soc.*, **110**, 1636–1638; (b) Trost, B.M. and Trost, M.K. (1991) *J. Am. Chem. Soc.*, **113**, 1850–1852; (c) Trost, B.M. and Trost, M.K. (1991) *Tetrahedron Lett.*, **32**, 3647–3650; (d) Trost, B.M., Ynai, M., and Hoogsteen, K. (1993) *J. Am. Chem. Soc.*, **115**, 5294–5295; (e) Trost, B.M., Hashmi, A.S.K. (1994)

J. Am. Chem. Soc., **116**, 2183–2184; (f) Trost, B.M., Hashmi, A.S.K. (1993) *Angew. Chem. Int. Ed.*, **32**, 1085–1087; (g) Trost, B.M. and Chang, V.K. (1993) *Synthesis*, 824–832.

32 (a) Chatani, N., Furukawa, N., Sakurai, H., and Murai, S. (1996) *Organometallics*, **15**, 901–903; (b) Chatani, N., Morimoto, T., Muto, T., and Murai, S. (1994) *J. Am. Chem. Soc.*, **116**, 6049–6050; (c) Chatani, N., Kataoka, K., and Murai, S. (1998) *J. Am. Chem. Soc.* **120**, 9104–9105; (d) Lee, S.I. and Chatani, N. (2009) *Chem. Commun.*, 371–384.

33 Fürstner, A., Szillat, H., Gabor, B., and Mynott, R. (1998) *J. Am. Chem. Soc.*, **120**, 8305–8314.

34 Trost, B.M. and Doherty, G.A. (2000) *J. Am. Chem. Soc.*, **122**, 3801–3810.

7
Ring-closing Alkyne Metathesis in Natural Product Synthesis

Paul W. Davies

7.1
Introduction

There is ever increasing societal and legislative pressure to minimize waste production and energy consumption while retaining or improving our current standard of living. The chemical community can contribute significantly toward achieving these goals, not least as the synthesis and manipulation of molecules is of fundamental importance across our developing societies. A significant contribution can therefore be achieved by performing molecular synthesis as efficiently as possible to diminish the associated energy consumption and waste production footprint. This can entail developing new and more rapid strategies to construct molecules, and to employ reactions which are inherently efficient; produce minimal waste; and use simple, readily introduced functionality so as to accelerate precursor preparations. The metal-catalyzed alkene metathesis processes discussed throughout this book are shining examples of such advances, and have rapidly been integrated and proven within the crucible of complex natural product synthesis. In this chapter we will look at the application of a sister reaction: alkyne metathesis.

7.2
Alkyne Metathesis

Since its discovery, alkyne metathesis has become a potent catalytic tool applied in the preparation of materials and organometallic complexes as well as bioactive organic molecules including natural products. In an analogous fashion to alkene metathesis, the process involves exchange of methine units between two carbon–carbon triple bond systems (Scheme 7.1).

This chapter focuses on one aspect of alkyne metathesis: ring-closing alkyne metathesis (RCAM) in natural product synthesis. A brief background precedes an overview of RCAM, the catalyst systems used in RCAM-containing approaches to natural products, and a discussion of its potential as a synthetic strategy. The value

Metathesis in Natural Product Synthesis: Strategies, Substrates and Catalysts. Edited by Janine Cossy, Stellios Arseniyadis, and Christophe Meyer

ISBN: 978-3-527-32440-8

Scheme 7.1 Alkyne metathesis.

of alkyne metathesis in macrocyclizations is then explored through examples of its successful application in a range of natural product syntheses.

7.2.1 Background to Alkyne Metathesis

After the initial discovery of the first effective catalytic alkyne metathesis using a heterogeneous mixture of tungsten oxides and silica at very high temperature (about 200–450 °C) [1], Mortreux *et al.* reported alkyne metathesis by a homogeneous catalyst system in 1974 [2a]. This catalytically active system derived from heating a mixture of $Mo(CO)_6$ (or a related molybdenum source) and simple phenol additives such as resorcinol, in high boiling solvents [2]. This system is still employed, even in complex molecule synthesis, and there have been a number of significant refinements made to it. Additionally, well-defined and isolable metal alkylidynes and other precatalysts have been introduced. Progress across this area has been reviewed recently and will not be discussed here in further depth [3]. A major contribution to the field was provided by Katz and McGinnis when they proposed the now-accepted mechanism for alkyne metathesis at an early stage of the reactions development [4]. A metal carbyne-promoted metathesis of alkynes proceeds through a sequence of formal [2+2]-cycloaddition and cycloreversion steps *via* metallacyclobutadiene intermediates (Scheme 7.2). Experimental evidence for this proposal was provided by Schrock's use of high-valent metal alkylidynes: with isolated metallacyclobutadiene intermediates acting as catalytically competent species for alkyne metathesis [5]. More recently, theoretical investigation of tungsten and molybdenum alkylidynes as catalysts for alkyne metathesis has been performed to shed light onto orders of relative reactivity [6].

These intertwined advances in mechanistic understanding and catalyst development have progressed the field to the stage that it is now applicable to molecules of significant complexity and fragility.

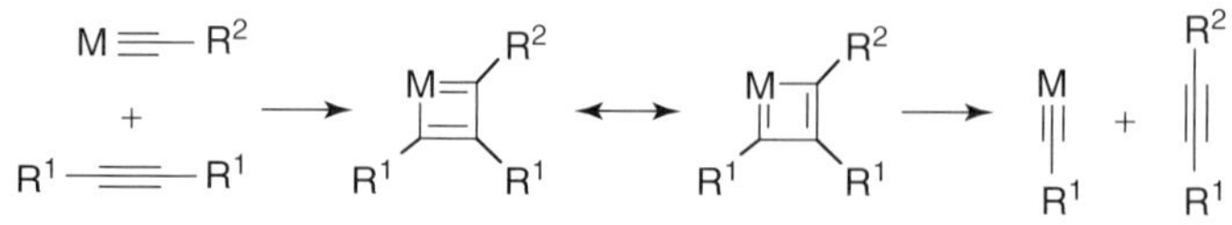

Scheme 7.2 The mechanism of alkyne metathesis.

7.3
Ring-closing Alkyne Metathesis

RCAM is the reaction of an acyclic diyne to afford a cyclic alkyne and an acyclic alkyne side product (Scheme 7.3). As with ring-closing (alkene) metathesis (RCM), the reaction is attractive in synthetic terms, as it requires relatively simple and mild reaction conditions and precludes a need to prepare and use highly reactive functional groups.

The first examples of catalytic RCAM for the synthesis of functionalized macrocycles were reported by Fürstner and Seidel [7]. Two catalyst systems were employed in this study: the "instant" system prepared by reaction of catalytic amounts of $Mo(CO)_6$ **1** with stoichiometric quantities of phenols, and the well-defined tungsten–alkylidyne **2** first prepared by Schrock in 1981 [5, 8]. Subsequent applications of RCAM, particularly in natural product synthesis, have either utilized these two systems or a molybdenum system developed by Fürstner that employs *in situ* reaction of trisamido molybdenum complex **3** [9] and CH_2Cl_2 (Figure 7.1) [10].

The choice of a catalyst system for any particular reaction is generally dictated by substrate compatibility. Although not yet at the same level of applicability as the catalysts available for alkene metathesis, these catalyst systems show some excellent functional group tolerance [3d]. The "instant" system represents a very practical and easily applicable method as it employs cheap, commercially available, and stable reagents without the need to use rigorously purified solvents or an inert atmosphere. Since its original discovery, modifications have led to improved functional group tolerance and a reduced reaction temperature (~140 °C) [3b, d].

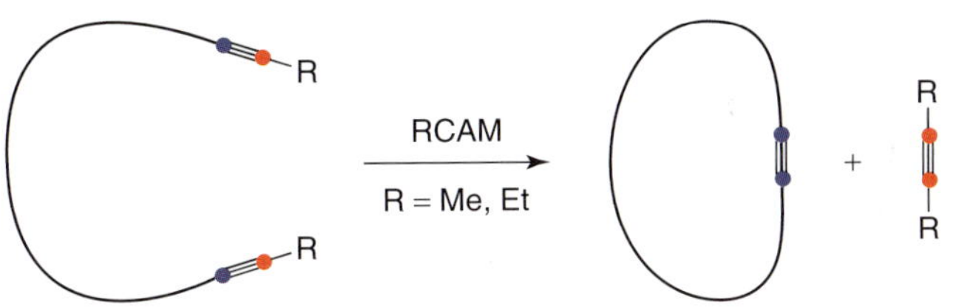

Scheme 7.3 Ring-closing alkyne metathesis (RCAM).

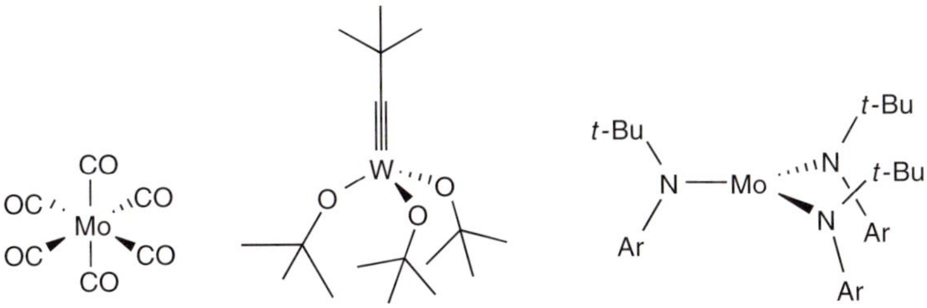

1 2 **3**: Ar = 3,5-dimethylphenyl

Figure 7.1 Key metal complexes employed in RCAM.

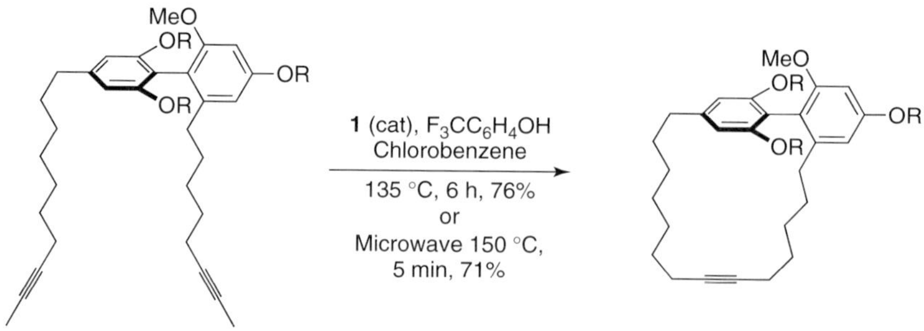

4: R = PMB

5

Scheme 7.4 Microwave heating in RCAM.

However, this process is generally less applicable in comparison to the other available systems, as the reaction conditions remain relatively harsh and the catalyst displays lower activity.

In cases where a substrate is suitably robust, the "instant" Mortreux-type system of **1** and phenols can be ideal. A modified Mortreux-type reaction system was successfully employed in the synthesis of turrianes, which are naturally occurring cyclophane derivatives. The RCAM approach can also be combined with modern synthetic techniques; in the cyclization of **4** the use of microwave heating resulted in a reduction of the reaction time from 6 hours under conventional heating to 5 minutes (Scheme 7.4) [11].

The commercially available complex **2** shows higher activity and is metathesis active under mild conditions, typically ambient temperature up to 90 °C [3c, h]. RCAM reactions with **2** are often run at 80 °C in toluene or a related aromatic solvent. However, rigorously inert (anhydrous and oxygen-free) conditions are required. The core structure is amenable to tuning by ligand modifications, reflected in a recent report which demonstrated that the effective RCAM of diyne **6** can be performed at room temperature to afford cyclic alkyne **8** in excellent yield when imidazolin-2-iminato tungsten alkylidyne catalyst **7** was employed (Scheme 7.5) [12]. The same catalyst has also been employed in the preparation of cyclophanes [13].

The other commonly used system employs molybdenum amide **3** as a precatalyst [5c, 14]. In the presence of CH_2Cl_2, **3** reacts to form a mixture of molybdenum chloride **9** and terminal alkylidyne **10** (Scheme 7.5) [10, 14]. Interestingly, **9** is the catalytically competent component at 80 °C, whereas the metal alkylidyne **10** effected only one turnover in alkyne metathesis. Alternatively, the use of a higher order alkyl chloride gives rise to catalytically competent nonterminal alkylidyne **11** (Scheme 7.6). This approach can be coupled with a "reductive-recycle" strategy to convert chloride **9** back to **3** [15, 16]. As with the tungsten analogues, molybdenum alkylidynes of type **11** can be further modified by ligand exchange to effect their properties and reactivity; the use of alcohols affords trisalkoxy molybdenum alkylidynes [15–17].

To date, applications of this system in natural product synthesis have employed the *in situ* generation of the catalytically active species. Molybdenum complex **3** is

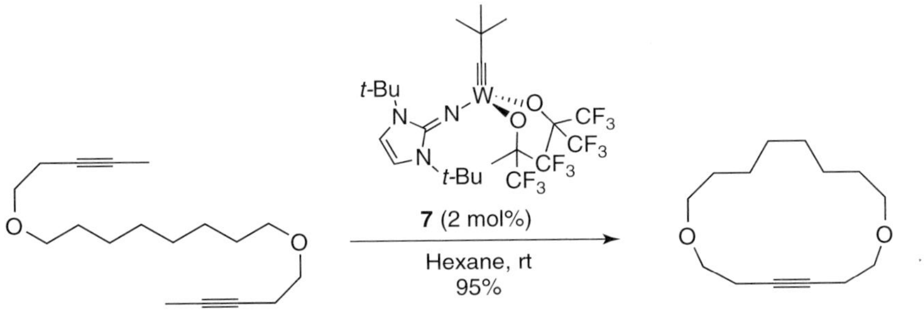

6

8

Scheme 7.5 Use of imidazolin-2-iminato-substituted tungsten alkylidyne **7** in RCAM.

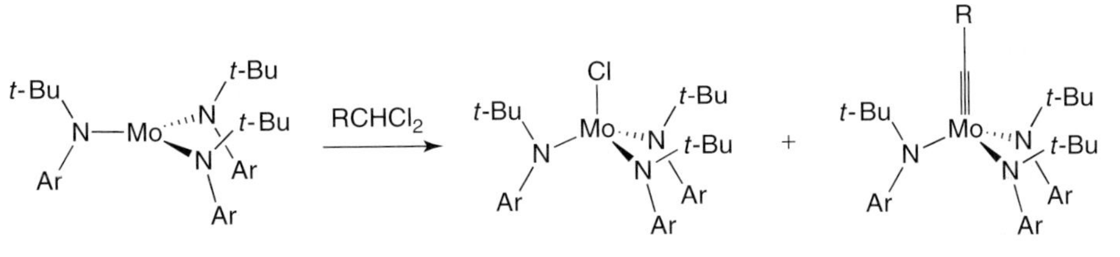

3: Ar = 3,5-dimethylphenyl

9

10: R = H
11: R = Et

Scheme 7.6 Activation of the molybdenum trisamido complex **3** to form metathesis-active complexes.

typically reacted with the diyne in a toluene/CH_2Cl_2 mixture at ~80 °C under inert atmosphere.

Taken alongside comparative studies [3d, 7], the applications discussed below show that the catalyst formed from **3**/CH_2Cl_2 displays the widest general applicability and functional group tolerance of the three most used systems. Functionalities such as esters, isolated double bonds, silyl ethers, sulfones, aldehydes, nitro groups, ketones, alkyl chlorides, acetals, and nitriles can be present in reactions using **3**/CH_2Cl_2 [3d]. "Acidic" protons, such as those of secondary amides, are incompatible with **3**, although tertiary amides are fully compatible. While the "instant" system is effective for robust substrates, the reaction conditions are too harsh for more labile units. Tungsten alkylidyne complex **2** tolerates units that include acetal, alkene, amide, carbonyl, enoate, ester, ether, furan, ketone, silyl ether, sulfonamide, and sulfone and is compatible with secondary amides. The high Lewis acidity of the tungsten center in **2** does, however, render substrates containing thioether, basic nitrogen groups, and polyethers incompatible [3d]. The decreased effective Lewis acidity and hence functional group tolerance of **3** is proposed to result from the close-packing amido ligands efficiently shielding the central molybdenum in the catalytically active template and preventing coordination of potential donor substrates [14].

7.3.1 RCAM as a Synthetic Strategy

RCAM is intrinsically less applicable than the equivalent RCM strategy, as the cyclic products of RCAM must be of sufficient size to be able to accommodate a linear alkyne unit. In such cases, however, the RCAM approach shares many of the positive characteristics of RCM, and may indeed be more suitable in certain situations. RCAM requires a reaction system of substrate, catalyst, and solvent only; it converts relatively straightforward and handleable functionality to make new carbon–carbon bonds; and it shows good functional group tolerance. Importantly, and in contrast to alkene metathesis, RCAM shows excellent chemoselectivity toward other carbon–carbon π-systems such as alkenes. A limitation of current RCAM methods is that end-capped alkynes must be used to achieve efficient catalysis. Terminal alkynes are incompatible with the catalyst systems in use [7, 8, 18]. This proviso can impact negatively on a synthetic strategy, as an additional step may be required to cap terminal alkynes. Furthermore, the RCAM of capped alkynes leads to less volatile side products (but-2-yne or hex-3-yne) which may be more difficult to purge from the system, particularly under the milder conditions associated with the more reactive catalysts. The reactions may therefore be performed under reduced pressure to aid the removal of side products. This helps drive the desired metathesis toward completion and reduces the possibility of a catalyst deactivation route known to occur with the polymerization of but-2-yne by ring expansion in the presence of Mo(VI) and W(VI) complexes [3a, 15].

Despite this current complication, the RCAM strategy provides a useful macrocyclization technique, which can also be considered as a solution to the vexing issue of stereoselectivity in ring-closing macrocyclization metathesis. When the stereochemistry of the macrocyclic alkene is important, the otherwise superb RCM strategy can become impractical because of the absence of any predictable control over alkene geometry. This problem has led to the formation of the undesired alkene isomer in significant quantities and even as the major product in several advanced syntheses [19]. The use of RCAM provides a route around this issue, yet employs a similar retrosynthetic logic to RCM, as stereoselective partial reduction of alkynes may be performed in a predictable fashion (Scheme 7.7). The potential of this strategy was recognized at the outset of RCAM development. Indeed, the combination of RCAM and Lindlar reduction to access (*Z*)-alkenes is the most widely employed alkyne metathesis strategy within natural product synthesis [3d, 7]. The combination of RCAM and semi-reduction to an (*E*)-alkene has been less widely applied to date, possibly because the semi-reduction strategy itself is less well established. However, a useful stereoselective semi-reduction method has been established, which employs a ruthenium-catalyzed *trans*-hydrosilylation followed by a protodesilylation [20–22].

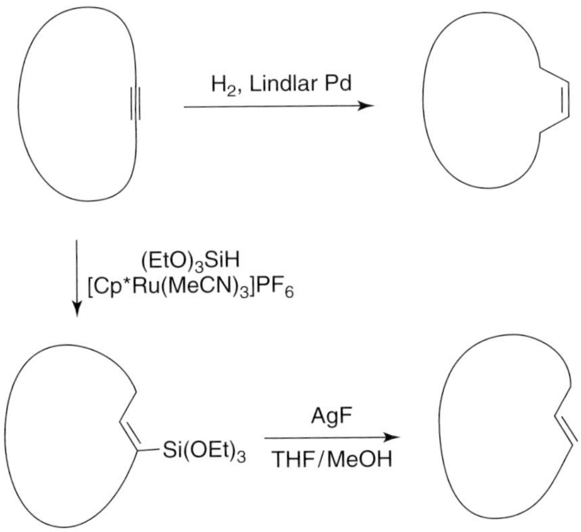

Scheme 7.7 Post RCAM entry to stereodefined macrocyclic alkenes.

7.4 Applications of RCAM in Natural Product Synthesis

The RCAM strategy has been employed toward a variety of biologically relevant synthetic targets including materials such as diaminosuberic acid derivatives [23], oligopeptides [16, 17], glycophanes [24], cyclic β-turn [25], and lantibiotic nisin Z mimics [26], and alkyne-linked glycoamino acids [27]. For the remainder of this chapter, we will only explore natural product applications with a focus on the RCAM macrocyclization step and the manipulation of the resulting alkyne.

7.4.1 RCAM/Hydrogenation Strategies

7.4.1.1 Macrocyclic Musks

The macrocyclic musks and fragrances, yuzu lactone (**12**) [28], ambrettolide (**13**) [28], and civetone (**14**) [29], which all contain a (*Z*)-configured alkene, were chosen as early targets for application of an RCAM/Lindlar reduction strategy (Figure 7.2). In all cases, application of the Mortreux "instant" system was successful, and the desired (*Z*)-alkene isomers were obtained exclusively from hydrogenation [28, 29].

Good overall yields for the formation of cyclic (*Z*)-alkene **14** from **15** were achieved in the presence of an unprotected carbonyl functionality using either tungsten complex **2** or the user-friendly and low-tech "instant" RCAM catalyst system (Scheme 7.8).

The insect repellent alkaloids epilachnene **19** and homologues as well as the cytotoxic sponge extract motuporamine C were prepared in an analogous fashion by an RCAM/Lindlar hydrogenation approach [28, 30]. All of these early examples demonstrate the complementary nature of the RCAM/semi-reduction manifold

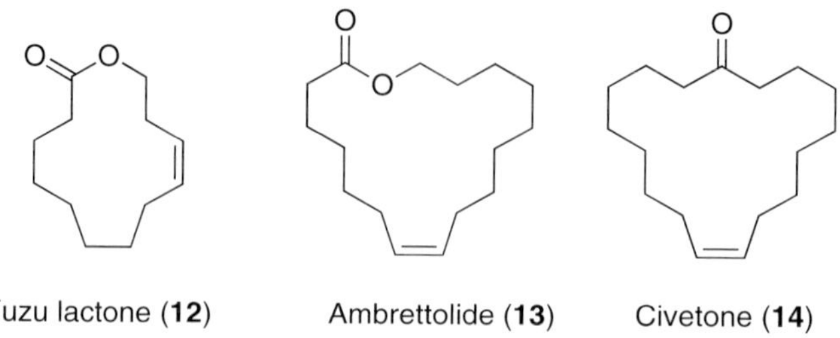

Figure 7.2 Macrocyclic musks, which were early natural product targets for the RCAM/Lindlar hydrogenation synthetic strategy.

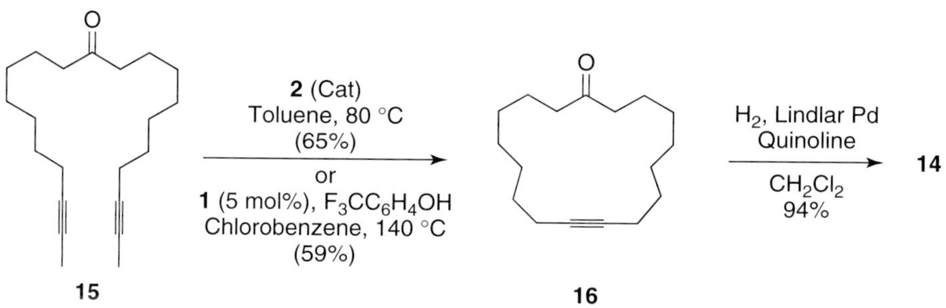

Scheme 7.8 The RCAM/Lindlar hydrogenation manifold in the synthesis of civetone.

to an RCM strategy that often leads to a mixture of alkene geometries [31]. For instance, in the synthesis of epilachnene, high yielding formation of the macrocyclic (*Z*)-alkene **19** in a fully stereoselective manner is achieved by an RCAM/semi-reduction [28]. This contrasts with the RCM approach which gives the undesired (*E*)-isomer **21** as the major constituent when starting from diene **20** (Scheme 7.9) [31a, 32].

The same key processes to install macrocycles with a (*Z*)-alkene were recently employed in the preparation of some novel designed musk compounds [33].

7.4.1.2 **Prostaglandin Lactones**

While the "instant" system was suitable for the previous examples, it was found to be incompatible with more elaborate applications. For example, no product was obtained from the cyclization of diyne **22**. In contrast, the **3**/CH_2Cl_2 catalyst system was found to be highly effective for the cyclization of **22** to **23** (Scheme 7.10). In this case, a slightly higher yield of 77% in the RCAM step could be achieved by modifying the aromatic unit on the amido ligand of **3** to 4-fluorophenyl [34]. As a comparison, the use of **2** required longer reaction times and resulted in lower conversion of material. Relative efficacies of the catalyst systems were compared in the cyclization of the C15 epimer of **22** and showed that for substrates of this complexity, the molybdenum amide systems **3** (81–87% in RCAM) were more effective than the tungsten alkylidyne **2** (65%) [34].

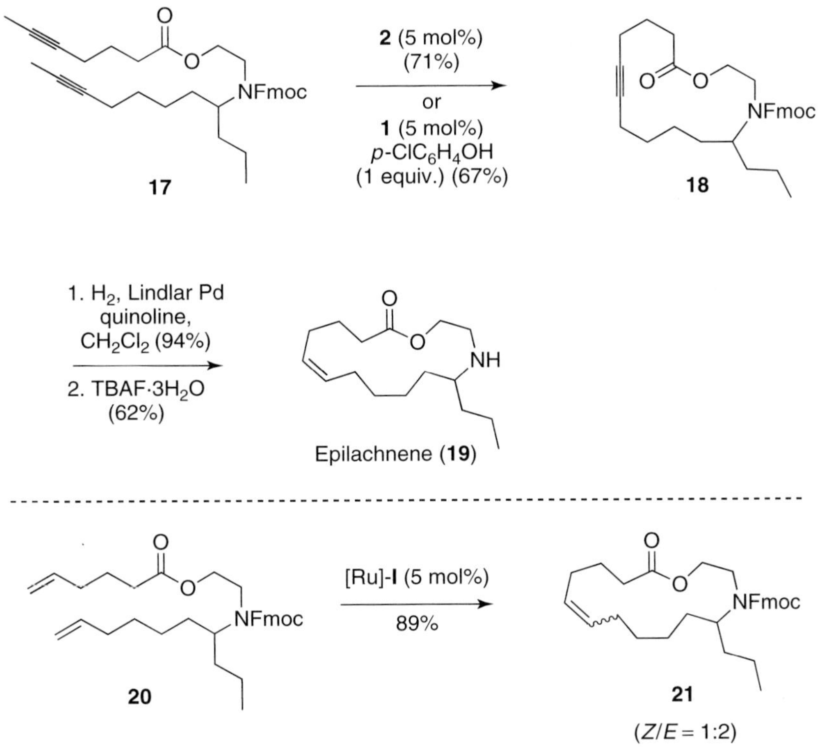

Scheme 7.9 A comparison of the RCAM/Lindlar hydrogenation manifold and the RCM approach in the synthesis of epilachnene.

7.4.1.3 Sophorolipid Lactone

The synthesis of sophorolipid lactone **28** demonstrated the low Lewis acidity of the metal center in **3** and the resulting catalytically active species. A high yield of **26** was achieved in RCAM of diyne **25** despite the presence of acid-labile PMB ethers and glycidic linkages (Scheme 7.11) [35]. Once again, stereoselective semi-reduction resulted in quantitative yield of the required (*Z*)-olefin **27**.

7.4.1.4 Epothilone A

The potent biological activity of the epothilones rendered them hot targets for synthesis teams and it is significant that the first three syntheses of epothilone A (**33**) used RCM to form the macrocycle. This served to highlight the confidence of the synthetic community, with what was then a relatively new tool. However, despite the effective cyclizations, the absence of any significant selectivity for the desired (*Z*)-olefin geometry represented a major issue at a late stage of the synthesis [19a–c].

The epothilones serve as a similarly enlightening example of the potential applicability of a RCAM/Lindlar reduction; selective access into the (*Z*)-alkene **31**

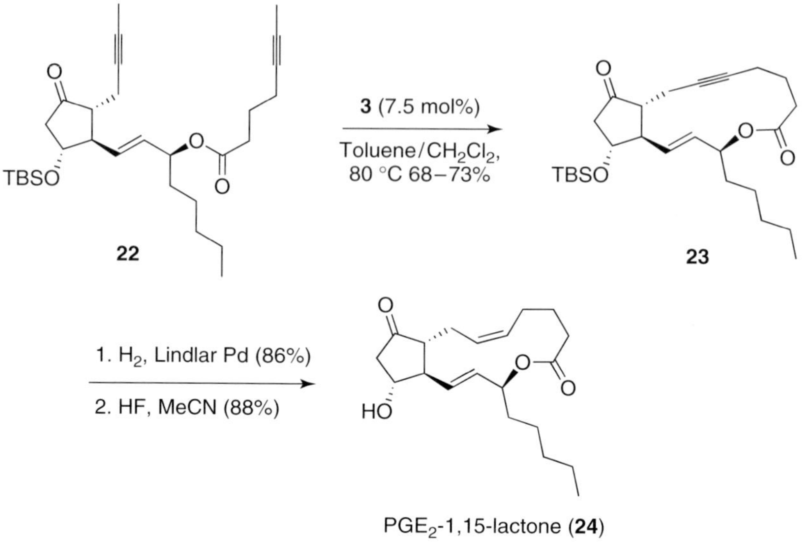

Scheme 7.10 Prostaglandin lactone synthesis.

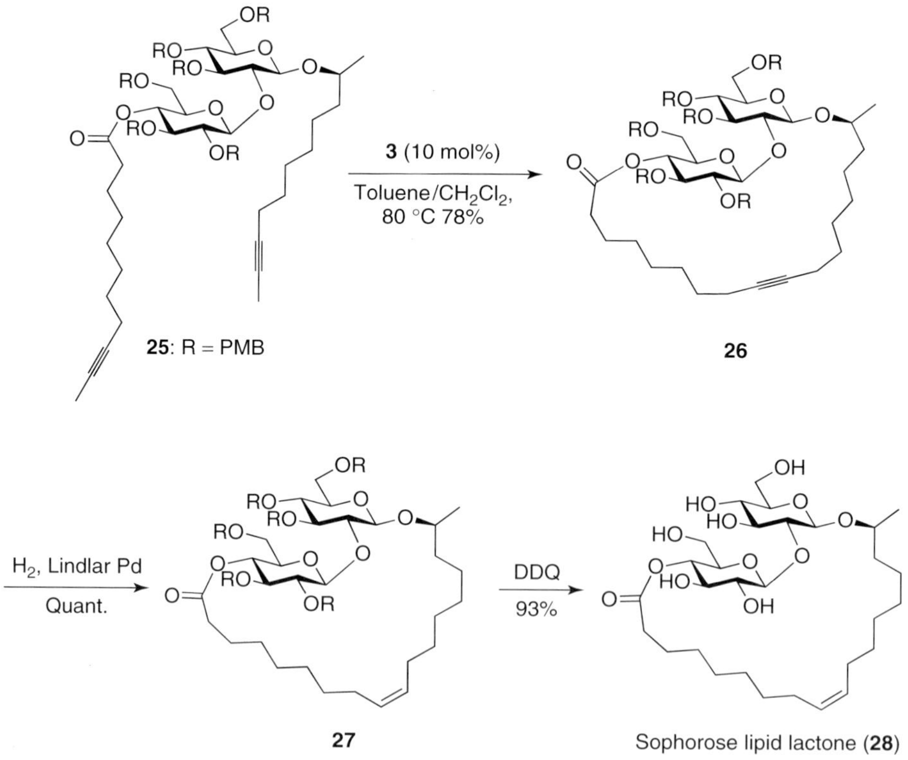

Scheme 7.11 Synthesis of sophorose lipid lactone.

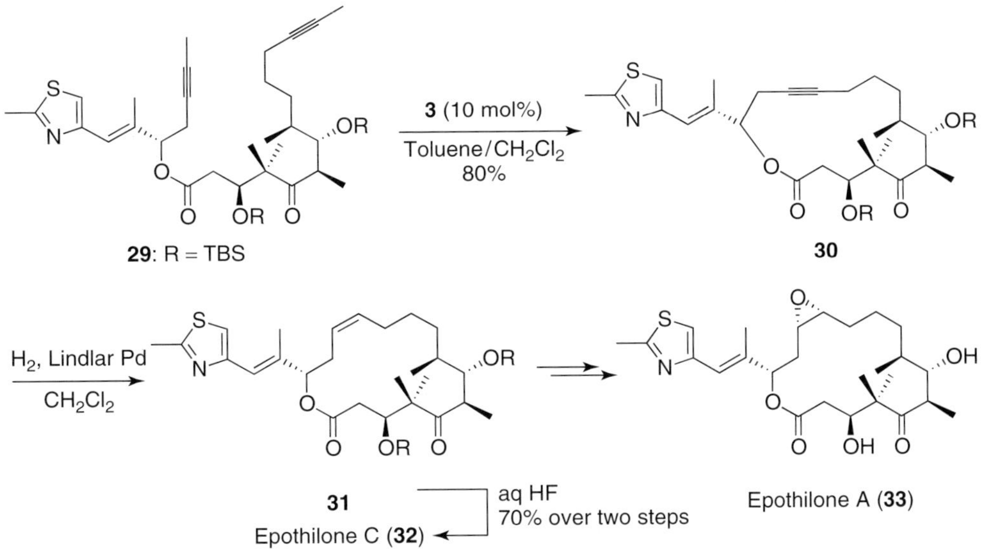

Scheme 7.12 RCAM-based synthesis of epothilones.

was achieved in high yield (Scheme 7.12). Aside from the issue of alkene geometry, this synthesis highlighted the capability of RCAM to be applied to molecules of similar complexity to those RCM is used for. Labile functionalities such as the aldol substructure were left intact, no epimerization of the stereogenic center α to the carbonyl was seen, and the reaction proceeded even in the presence of the basic nitrogen and sulfur atoms of the thiazole [36].

7.4.1.5 **Cruentaren A**

Complex RCAM processes are mediated by both **2** and **3**. The macrocyclic salicylate lactone cruentaren A (**36**) has been prepared independently by two groups, Maier [37, 38] and Fürstner [39], using a RCAM approach. Their chosen cyclization precursors, **34a** and **34b**, differ only by the choice of a particular silyl protecting group and in the extent of functionalization on the side chain (Scheme 7.13). Both RCAM processes could be performed in high yield. Complex **3**/CH_2Cl_2 had to be used for the cyclization of **34b**, as its acid-labile OTHP functionality was incompatible with complex **2**.

It is interesting to note that in the subsequent transformations to the target molecule **36**, the Lindlar reduction was carried out immediately after the RCAM by Fürstner and coworkers, whereas it was applied at the final step of the synthesis by Vintonyak and Maier, thus demonstrating some flexibility in the RCAM/semi-reduction approach. The success of the key RCAM strategy and endgame led to its retention in a second-generation synthesis by Maier *et al.* [40].

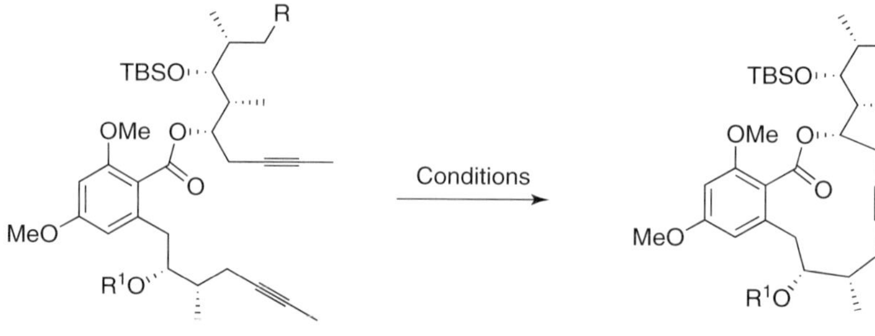

34 **35**

34a: R = CH_2ODMB, R^1 = TIPS — **2** (10 mol%), Toluene, 80 °C (91%)

34b: R = ⁄=╲OTHP, R^1 = TIPS — **3** (10 mol%), Toluene/CH_2Cl_2, 80 °C (87%)

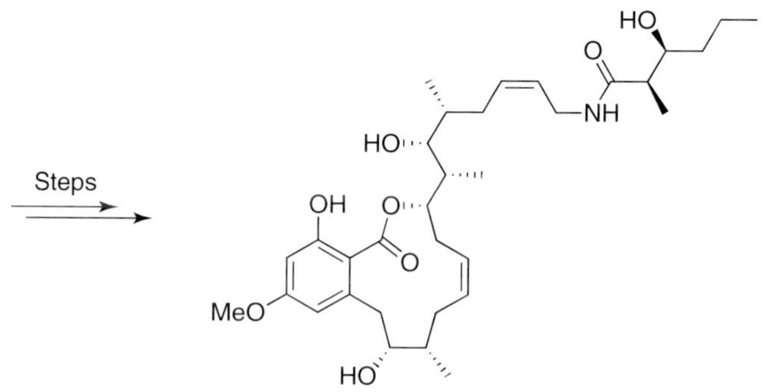

Cruentaren A (**36**)

Scheme 7.13 The synthesis of cruentaren A by RCAM/Lindlar reduction strategies.

7.4.1.6 **Latrunculins A, B, C, M, and S**

A major application of RCAM is seen in the synthesis of a range of actin binding macrolides by Fürstner *et al.* First employed in the synthesis of latrunculin B (**39**) [41], the RCAM/Lindlar reduction manifold was subsequently utilized to prepare latrunculin C, M, and other analogs [42]. A mixed ketal, a thioazolidinone unit, and a branching α-substituent to one of the alkyne units of diyne **37** proved no obstacle to successful RCAM with either precatalyst **3** or the tungsten complex **2**, with the former giving slightly higher yield of **38** (Scheme 7.14).

The latrunculins A (**42**) and S, however, represented a new challenge for application of RCAM. The presence of a conjugated diene in the target compounds required the first use of a chemoselective enyne–yne metathesis in natural product synthesis [42, 43].

A model study employing tungsten complex **2** had shown that the metathesis of simple enyne–yne systems was viable [21b]. However, latrunculin A (**42**) contains a cyclic structure which is smaller that those previously prepared by RCAM (a

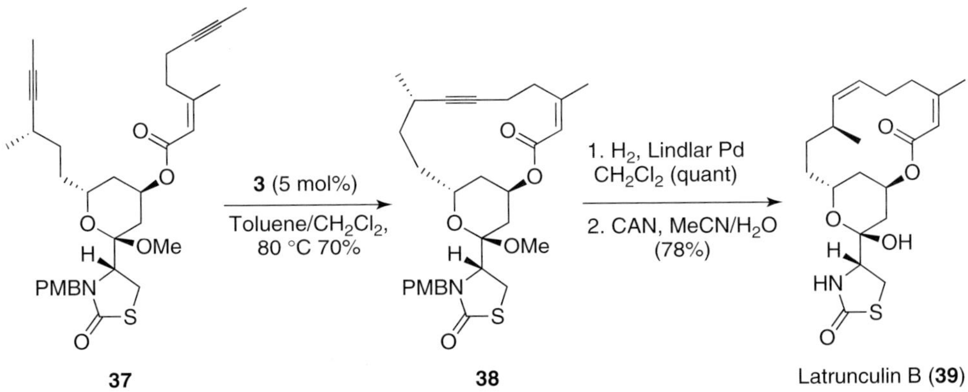

Scheme 7.14 Synthesis of latrunculin B via RCAM.

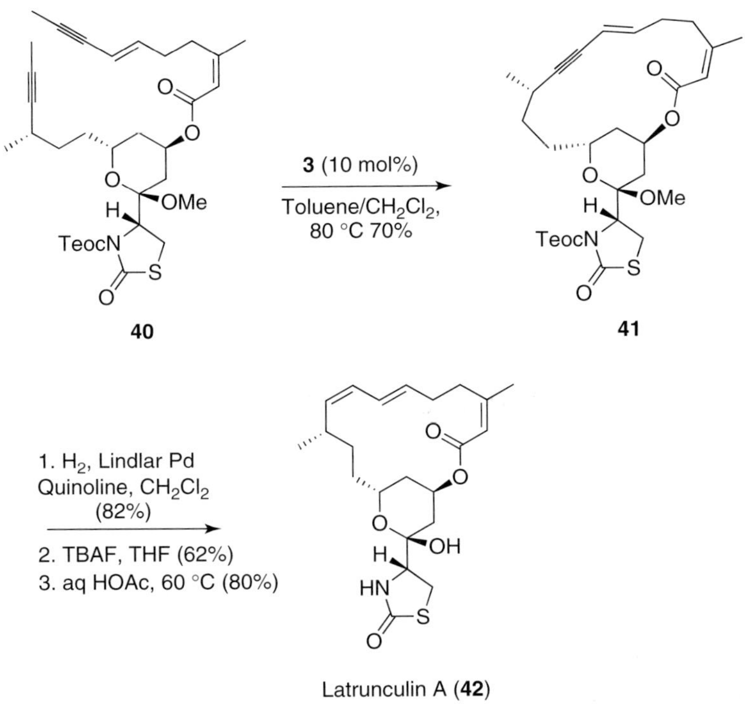

Scheme 7.15 An RCAM/Lindlar hydrogenation approach to latrunculin A involving enyne–yne coupling.

16-membered ring). Despite this issue and the increased complexity of enyne–yne **40**, no problems were observed in the effective RCAM of the conjugated alkyne, demonstrating the orthogonality of the reaction system derived from **3**/CH_2Cl_2 toward alkenes (Scheme 7.15). Although the PMB functionality is compatible with RCAM, a late-stage protecting group exchange on the thiazolidinone to Teoc had to be performed before RCAM as a consequence of the problematic deprotection of the PMB group after RCAM. Finally, the required conjugated (*Z*)-olefin of **42** was obtained by Lindlar hydrogenation of **41** [43].

7.4.1.7 Myxovirescin A_1

The 28-membered ring macrocycle of the antibiotic macrolide myxovirescin A_1 was also assembled using an enyne–yne RCAM/semi-reduction approach. However, the presence of a conjugated (*E*)-olefin in the target compound required an alternative manipulation of the cyclic alkyne to the Lindlar reduction approach that had previously been employed to access conjugated (*Z*)-alkenes. Although enyne–yne **45** features a methyl ether close to the reaction site in a potentially ligating position, the RCAM step proceeded smoothly with the **3**/CH_2Cl_2 system (Scheme 7.16). In this case, the tungsten alkylidyne **2** was less effective than the **3**/CH_2Cl_2 catalyst system. However, the presence of a secondary amide linkage meant that a high catalyst loading was required to achieve a high yield of the cyclic alkyne **46** [44].

The following step required the semi-reduction of the cyclic enyne to a conjugated (*E*)-olefin, which is much less well established than the proven technology for stereospecific semi-reduction to (*Z*)-olefins. Although model studies on the hydrosilylation of cyclic enyne system had given the desired outcome with very high stereoselectivity [21b], the reaction of **46** was poorly stereoselective. Indeed, three isomeric vinyl silanes **47**, **48a**, and **48b** were obtained even when using the less sterically imposing $[CpRu(MeCN)_3]PF_6$ catalyst in place of the more commonly applied $[Cp^*Ru(MeCN)_3]PF_6$. The formation of the (*Z,Z*)-configured isomer **50** along with the desired (*E,Z*)-configured **49** major product, after protodesilylation of the vinyl silanes **47**, **48a**, and **48b**, demonstrates that the hydrosilylation methodology requires further improvement for such highly complex substrates.

7.4.2 RCAM and Alternative Alkyne Manipulations

While macrocylic structures containing (*Z*)-olefins would now appear to suggest an obvious site for an RCAM-based disconnection strategy, the alkyne can also be employed for other means. Indeed, the use of alkynes as reactive units is growing rapidly [45] and application of such processes alongside an RCAM strategy may well be a fruitful approach in the future. The syntheses of two natural products in which the alkyne introduced by RCAM is employed in reactions other than semi-reduction of the alkyne are shown below.

7.4.2.1 Citreofuran

In the synthesis of the polyketide metabolite citreofuran (**57**), RCAM was incorporated into the retrosynthetic analysis by setting an alkyne–one as the precursor for the furan ring. The diyne precursor **55** was rapidly assembled from acid **51** by its reaction with chloroenamine **52** to yield the alkoxyisocoumarin **53** upon treatment with Et_3N. The addition of homopropargylic Grignard **54** to unstable **53** resulted in

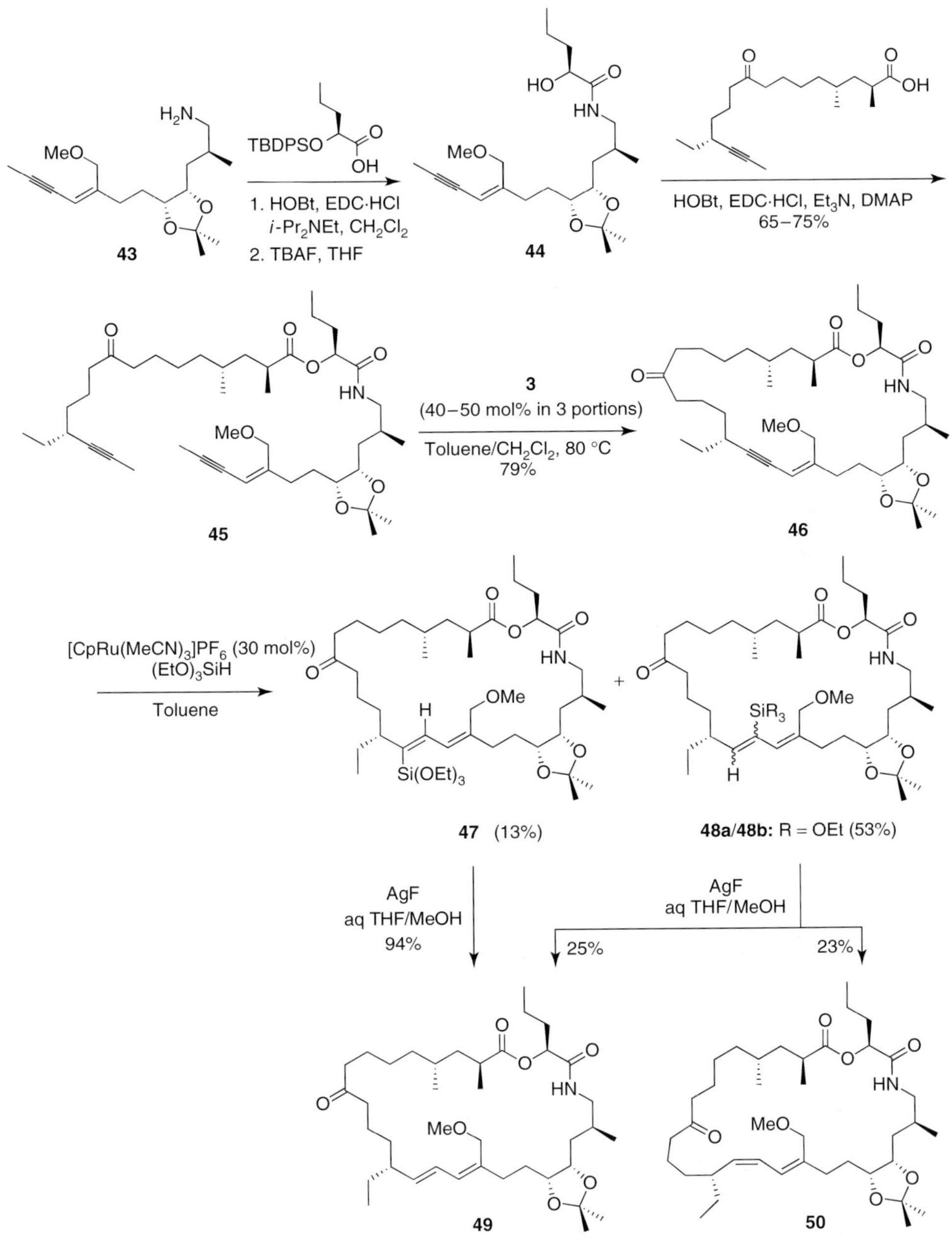

Scheme 7.16 The enyne–yne RCAM/hydrosilylation/protodesilylation approach to myxovirescin A_1.

the desired RCAM precursor diyne **55** when TMSCl was used as a mild activating agent [46].

RCAM of **55** using complex **2** then worked well despite the presence of an unprotected ketone. It was specifically noted that the precursors must be of high purity to ensure good conversion. In particular, traces of amine can act as catalyst poison. Once the macrocycle **56** was formed, an acid-catalyzed cycloisomerization between the ketone and the alkyne afforded the furan in good yield with the natural product **57** reached after demethylation of the bisphenol (Scheme 7.17) [46].

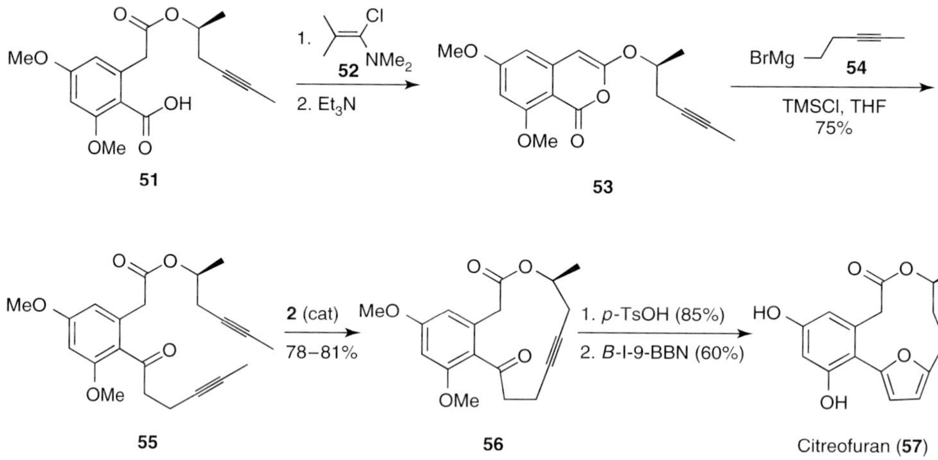

Scheme 7.17 An RCAM/cycloisomerization approach to citreofuran.

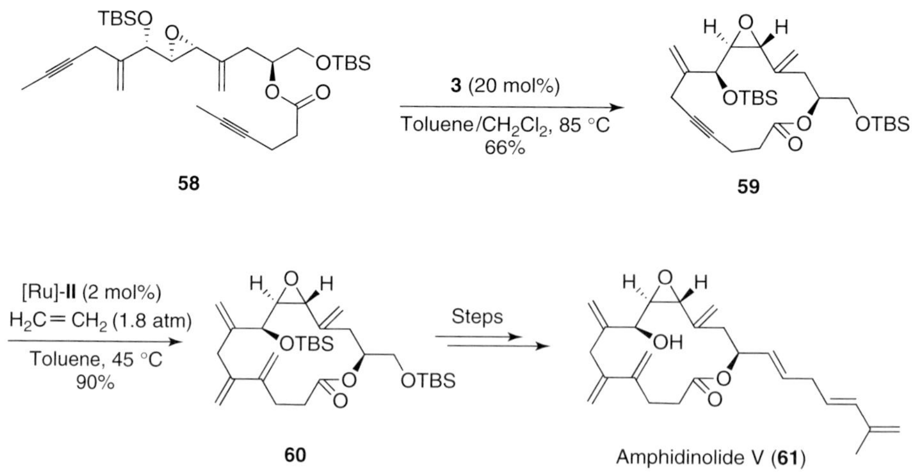

Scheme 7.18 The synthesis of amphidinolide V using complementary metathesis methods.

7.4.2.2 Amphidinolide V

An alternative use of an alkyne installed by RCAM was demonstrated in the synthesis of the proposed structure of amphidinolide V when a complementary ene–yne metathesis reaction was employed (Scheme 7.18) [47]. RCAM of diyne **58** using **3**/CH_2Cl_2 proceeded smoothly despite the labile *trans*-configured vinyl epoxide, the allylic alcohol, and the apparent ring strain of the product. The resulting cyclic alkyne **59** was then subjected to a ruthenium-catalyzed ene–yne metathesis under ethylene atmosphere with [Ru]-**II**. This step installed the two *exo*-methylene units required in the natural product (Scheme 7.18). On completion of the synthesis, the NMR data of **61** corresponded to those of the isolated natural product, apart from a discrepancy with a singular resonance. Importantly, the RCAM/ene–yne metathesis synthetic approach was flexible enough to allow this issue to be explored through the preparation of several isomers of **61**. All of these isomers matched the reported data less closely than the originally prepared isomer [47].

7.5 Conclusions

The examples shown throughout this chapter demonstrate that RCAM is a powerful methodology for the preparation of complex natural compounds. The technique is both robust and able to mediate transformations of complex structures bearing labile functionalities. A significant use of the RCAM involves its combination with a subsequent stereoselective semi-reduction of the alkyne. This approach provides a predictable alternative to the use of RCM in macrocyclizations, and has been verified by successful application in a range of natural products. The strategy accommodates a range of differing functionalities with a level of complexity closely comparable to that associated with RCM.

Further advances in catalyst development are likely to improve the uptake and applicability of the RCAM strategy. In particular, the design of a catalyst system that mediates effective RCAM of terminal alkynes under mild conditions with good functional group tolerance is extremely desirable and would have great effect on the synthetic impact of this already highly employable reaction.

References

1 Pennella, F., Banks, R.L., and Bailey, G.C. (1968) *Chem. Commun.*, 1548–1549.

2 (a) Mortreux, A. and Blanchard, M. (1974) *J. Chem. Soc., Chem. Commun.*, 786–787; (b) Mortreux, A., Dy, N., and Blanchard, M. (1975/76) *J. Mol. Catal.*, **1**, 101–109; (c) Mortreux, A., Petit, F., and Blanchard, M. (1978) *Tetrahedron Lett.*, **19**, 4967–4968; (d) Bencheick, A., Petit, M., Mortreux, A., and Petit, F. (1982) *J. Mol. Catal.*, **15**, 93–101; (e) Mortreux, A., Delgrange, J.C., Blanchard, M., and Lubochinsky, B. (1977) *J. Mol. Catal.*, **2**, 73–82; (f) Mortreux, A., Petit, F., and

Blanchard, M. (1980) *J. Mol. Catal.*, **8**, 97–106.

3 (a) Zhang, W. and Moore, J.S. (2007) *Adv. Synth. Catal.*, **349**, 93–120; (b) Mortreux, A. and Coutelier, O. (2006) *J. Mol. Catal. A: Chem.*, **254**, 96; (c) Schrock, R.R. and Czekelius, C. (2007) *Adv. Synth. Catal.*, **349**, 55–77; (d) Fürstner, A. and Davies, P.W. (2005) *Chem. Commun.*, 2307–2320; (e) Fürstner, A. (2003) in *Handbook of Metathesis*, vol. 2 (ed. R.H. Grubbs), Wiley-VCH Verlag GmbH, Weinheim, pp. 432–462; (f) Schrock, R.R. (2003) in *Handbook of Metathesis*, vol. 1 (ed. R.H. Grubbs), Wiley-VCH Verlag GmbH, Weinheim, pp. 173–189; (g) Gradillas, A., Pérez-Castells, J. (2006) *Angew. Chem. Int. Ed.*, **37**, 6086–6101; (h) Schrock, R.R. (2002) *Chem. Rev.*, **102**, 145–180.

4 Katz, T.J. and McGinnis, J. (1975) *J. Am. Chem. Soc.*, **97**, 1592–1594.

5 (a) Wengrovius, J.H., Sancho, J., and Schrock, R.R. (1981) *J. Am. Chem. Soc.*, **103**, 3932–3934; (b) Pedersen, S.F., Schrock, R.R., Churchill, M.R., and Wasserman, H.J. (1982) *J. Am. Chem. Soc.*, **104**, 6808–6809.

6 Zhu, J., Jiu, G., and Lin, Z. (2006) *Organometallics*, **25**, 1812–1819.

7 Fürstner, A. and Seidel, G. (1998) *Angew. Chem. Int. Ed.*, **37**, 1734–1736.

8 (a) Schrock, R.R., Clark, D.N., Sancho, J., Wengrovius, J.H., Rocklage, S.M., and Pedersen, S.F. (1982) *Organometallics*, **1**, 1645–1651; (b) Freudenberger, J.H., Schrock, R.R., Churchill, M.R., Rheingold, A.L., and Ziller, J.W. (1984) *Organometallics*, **3**, 1563–1573; (c) Listemann, M.L. and Schrock, R.R. (1985) *Organometallics*, **4**, 74–83; (d) Fischer, H., Hofmann, P., Kreissl, F.R., Schrock, R.R., Schubert, U., and Weiss, K. (eds) (1988) *Carbyne Complexes*, Wiley-VCH Verlag GmbH, Weinheim; (e) Schrock, R.R. (1986) *Acc. Chem. Res.*, **19**, 342–348; (f) Schrock, R.R. (1995) *Polyhedron*, **14**, 3177–3195.

9 (a) Laplaza, C.E., Odom, A.L., Davis, W.M., Cummins, C.C., and Protasiewicz, J.D. (1995) *J. Am. Chem. Soc.*, **117**, 4999–5000; (b) Laplaza, C.E. and Cummins, C.C. (1995) *Science*, **268**, 861–863; (c) Cummins, C.C. (1998) *Chem. Commun.*, 1777–1786.

10 Fürstner, A., Mathes, C., and Lehmann, C.W. (1999) *J. Am. Chem. Soc.*, **121**, 9453–9454.

11 Fürstner, A., Stelzer, F., Rumbo, A., and Krause, H. (2002) *Chem. Eur. J.*, **8**, 1856–1871.

12 Beer, S., Hrib, C.G., Jones, P.G., Brandhorst, K., Grunenberg, J., and Tamm, M. (2007) *Angew. Chem. Int. Ed.*, **46**, 8890–8894.

13 Beer, S., Brandhorst, K., Grunenberg, J., Hrib, C.G., Jones, P.G., and Tamm, M. (2008) *Org. Lett.*, **10**, 981–984.

14 Fürstner, A., Mathes, C., and Lehmann, C.W. (2001) *Chem. Eur. J.*, **7**, 5299–5317.

15 Zhang, W., Kraft, S., and Moore, J.S. (2004) *J. Am. Chem. Soc.*, **126**, 329–335.

16 Zhang, W., Kraft, S., and Moore, J.S. (2003) *Chem. Commun.*, 832–833.

17 (a) McCullough, L.G. and Schrock, R.R. (1984) *J. Am. Chem. Soc.*, **106**, 4067–4068; (b) McCullough, L.G., Schrock, R.R., Dewan, J.C., and Murdzek, J.C. (1985) *J. Am. Chem. Soc.*, **107**, 5987–5998; (c) Tsai, Y.-C., Diaconescu, P.L., and Cummins, C.C. (2000) *Organometallics*, **19**, 5260–5262.

18 For a recent report on the use of terminal alkynes in alkyne metathesis see Coutelier, O., Nowogrocki, G., Paul, J.-F., and Mortreux, A. (2007) *Adv. Synth. Catal.*, **349**, 2259–2263.

19 For some examples (a) Bertinato, P., Sorensen, E.J., Meng, D., and Danishefsky, S. (1996) *J. Org. Chem.*, **61**, 8000–8001; (b) Nicolaou, K.C., He, Y., Vourloumis, D., Vallberg, H., Roschangar, F., Sarabia, F., Ninkovic, S., Yang, Z., and Trujillo, J.I. (1997) *J. Am. Chem. Soc.*, **119**, 7960–7973; (c) Schinzer, D., Limberg, A., Bauer, A., Böhm, O.M., and Cordes, M. (1997) *Angew. Chem. Int. Ed. Engl.*, **36**, 523–524; (d) Ono, K., Nakagawa, M., and Nishida, A. (2004) *Angew. Chem. Int. Ed.*, **43**, 2020–2023.

20 Trost, B.M., Ball, Z.T., and Jöge, T. (2002) *J. Am. Chem. Soc.*, **124**, 7922–7923.

21 Fürstner, A. and Radkowski, K. (2002) *Chem. Commun.*, 2182–2183; (b) Lacombe, F., Radkowski, K., Seidel, G., and Fürstner, A. (2004) *Tetrahedron*, **60**, 7315–7324.

22 Trost, B.M. and Ball, Z.T. (2001) *J. Am. Chem. Soc.*, **123**, 12726–12727.

23 Aguilera, B., Wolf, L.B., Nieczypor, P., Rutjes, F., Overkleeft, H.S., van Hest, J.C.M., Schoemaker, H.E., Wang, B., Mol, J.C., Fürstner, A., Overhand, M., van der Marel, G.A., and van Boom, J.H. (2001) *J. Org. Chem.*, **66**, 3584–3589.

24 Doyle, D. and Murphy, P.V. (2008) *Carbohydr. Res.*, **343**, 2535–2544.

25 IJsselstijn, M., Aguilera, B., van der Marel, G.A., van Boom, J.H., van Delft, F.L., Schoemaker, H.E., Overkleeft, H.S., Rutjes, F.P.J.T., and Overhand, M. (2004) *Tetrahedron Lett.*, **45**, 4379–4382.

26 (a) Ghalit, N., Poot, A.J., Fürstner, A., Rijkers, D.T.S., and Liskamp, R.M.J. (2005) *Org. Lett.*, **7**, 2961–2964; (b) Ghalit, N., Rijkers, D.T.S., and Liskamp, R.M.J. (2006) *J. Mol. Catal. A*, **254**, 68–77.

27 Groothuys, S., van den Broek, S.A.M.W., Juijpers, B.H.M., IJsselstijn, M., van Delft, F.L., and Rutjes, F.P.J.T. (2008) *Synlett*, 111–115.

28 Fürstner, A., Guth, O., Rumbo, A., and Seidel, G. (1999) *J. Am. Chem. Soc.*, **121**, 11108–11113.

29 Fürstner, A. and Seidel, G. (2000) *J. Organomet. Chem.*, **606**, 75–78.

30 Fürstner, A. and Rumbo, A. (2000) *J. Org. Chem.*, **65**, 2608–2611.

31 (a) Fürstner, A. and Langemann, K. (1997) *Synthesis*, 792–803; (b) Fürstner, A. and Langemann, K. (1996) *J. Org. Chem.*, **61**, 3942–3943; (c) Fürstner, A., Koch, D., Langemann, K., Leitner, W., and Six, C. (1997) *Angew. Chem. Int. Ed. Engl.*, **36**, 2466–2469.

32 Goldring, W.P.D. and Weiler, L. (1999) *Org. Lett.*, **1**, 1471–1473.

33 Kraft, P. and Berthold, C. (2008) *Synthesis*, 543–550.

34 (a) Fürstner, A. and Grela, K. (2000) *Angew. Chem. Int. Ed.*, **39**, 1234–1236; (b) Fürstner, A., Grela, K., Mathes, C., and Lehmann, C.W. (2000) *J. Am. Chem. Soc.*, **122**, 11799–11805.

35 (a) Fürstner, A., Radkowski, K., Grabowski, J., Wirtz, C., and Mynott, R. (2000) *J. Org. Chem.*, **65**, 8758–8762; (b) Fürstner, A. (2004) *Eur. J. Org. Chem.*, 943–958.

36 Fürstner, A., Mathes, C., and Grela, K. (2001) *Chem. Commun.*, 1057–1059.

37 (a) Vintonyak, V.V. and Maier, M.E. (2007) *Angew. Chem. Int. Ed.*, **46**, 5209–5211; (b) Vintonyak, V.V. and Maier, M.E. (2007) *Org. Lett.*, **9**, 655–658.

38 Vintonyak, V.V., Calà, M., Lay, F., Kunze, B., Sasse, F., and Maier, M.E. (2008) *Chem. Eur. J.*, **14**, 3709–3720.

39 Fürstner, A., Bindl, M., and Jean, L. (2007) *Angew. Chem. Int. Ed.*, **46**, 9275–9278.

40 Maier, M.E. and Ritschel, J. (2008) *Arkivoc*, 314–329.

41 Fürstner, A., De Souza, D., Parra-Rapado, L., and Jensen, J.T. (2003) *Angew. Chem. Int. Ed.*, **42**, 5358–5360.

42 (a) Fürstner, A., Kirk, D., Fenster, M.D.B., Aissa, C., and De Souza, D. (2005) *Proc. Natl. Acad. Sci. U.S.A.*, **102**, 8103–8108; (b) Fürstner, A., De Souza, D., Turet, L., Fenster, M.D.B., Parra-Rapado, L., Wirtz, C., Mynott, R., and Lehmann, C.W. (2007) *Chem. Eur. J.*, **13**, 115–134; (c) Fürstner, A., Kirk, D., Fenster, M.D.B., Aissa, C., De Souza, D., Nevado, C., Tuttle, T., Thiel, W., and Müller, O. (2007) *Chem. Eur. J.*, **13**, 135–149.

43 Fürstner, A. and Turet, L. (2005) *Angew. Chem. Int. Ed.*, **44**, 3462–3466.

44 Fürstner, A., Bonnekessel, M., Blank, J.T., Radkowski, K., Seidel, G., Lacombe, F., Gabor, B., and Mynott, R. (2007) *Chem. Eur. J.*, **13**, 8762–8783.

45 For an example, see the recent advances in platinum and gold activation of alkynes: Fürstner, A. and Davies, P.W. (2007) *Angew. Chem. Int. Ed.*, **46**, 3410–3449.

46 Fürstner, A., Castanet, A.S., Radkowski, K., and Lehmann, C.W. (2003) *J. Org. Chem.*, **68**, 1521–1528.

47 Fürstner, A., Larionov, O., and Flügge, S. (2007) *Angew. Chem. Int. Ed.*, **46**, 5545–5548.

8
Temporary Silicon–Tethered Ring–Closing Metathesis Reactions in Natural Product Synthesis

P. Andrew Evans

8.1
Introduction

Temporary tethered reactions provide an important strategy for target-directed synthesis, since they generally circumvent the problems encountered with entropically unfavorable reactions (Scheme 8.1) [1]. For example, although Lewis-acid- and transition metal-catalyzed reactions have advanced considerably in the last few decades, the intermolecular reaction manifold is often considerably more challenging than the intramolecular version [2]. Nishiyama and Stork independently reported the concept of utilizing a temporary or disposal tether in the context of a free radical addition [3], which has resulted in the application of this strategy to an array of challenging synthetic transformations [4]. Interestingly, the temporary tether has been compared to the formation of an enzyme–substrate complex in enzyme catalysis [5], since it provides a high effective concentration of the reactants [6], which reduces the demands of a challenging intermolecular reaction through the provision of higher reaction rates under milder reaction conditions. Furthermore, this approach generally reduces the degrees of freedom in the transition state (TS), which generally results in improved selectivity and enables the functionalization of these derivatives into useful intermediates for target-directed synthesis.

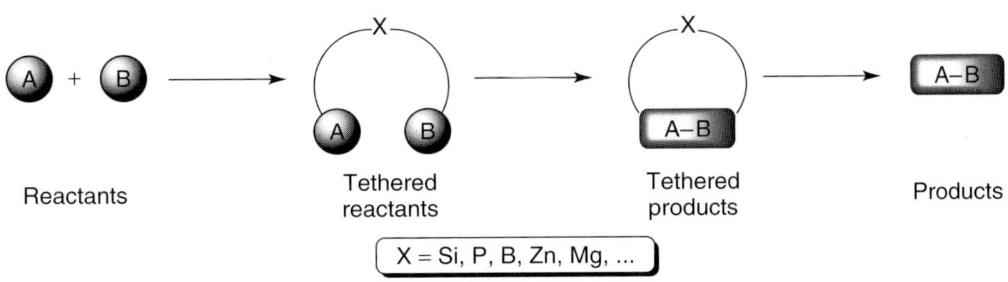

Scheme 8.1 Temporary tethered strategy in organic synthesis.

Metathesis in Natural Product Synthesis: Strategies, Substrates and Catalysts. Edited by Janine Cossy, Stellios Arseniyadis, and Christophe Meyer

ISBN: 978-3-527-32440-8

Despite extensive studies with various temporary tethers, silicon has been proven the most versatile group since 1) the reactive components can be easily tethered; 2) the tether is generally stable to the reaction conditions, and; 3) it can be readily cleaved following the key transformation (Scheme 8.1) [7]. Alternatively, the silicon tether can be functionalized using a variety of methods, including oxidation, group transfer, and transmetallation, which further enhances the synthetic opportunities [8]. While the silicon variant of the temporary tether concept is now well established and has been the subject of several reviews [4], this chapter provides thc first account of this approach in the context of ring-closing metathesis (RCM) reactions that are specifically applied to natural product synthesis.

8.2
Temporary Silicon–Tethered Ring–Closing Metathesis Reactions

Olefin metathesis represents one of the most important carbon–carbon bond-forming reactions in synthetic organic chemistry. Ring-closing metathesis involves the metal-mediated exchange of a metal alkylidene complex with a tethered diene, which has proven highly effective for the construction of an array of carbo- and heterocycles using relatively mild reaction conditions. Hence, it is not surprising that this process has emerged as one of the most popular ring-forming reactions, which has featured so heavily in countless total syntheses [9]. However, a critical limitation with the reaction has been the ability to control alkene geometry, since the reaction generally prefers the formation of the thermodynamically more stable *E*-alkene. This issue is particularly evident in cross-metathesis (CM) and ring-closing reactions that produce large rings, where there is no formal conformational preference for the *Z*-isomer. The temporary silicon-tethered ring-closing metathesis (TST-RCM) overcomes this limitation using ring strain to control the formation of the *Z*-olefin in small and several medium rings, which provides intermediates that can then be functionalized and incorporated into synthetic targets (see below). This process also facilitates the cross-coupling of electronically similar alkenes (Type I), which remain challenging for CM, particularly for the formation of the *Z*-olefin, making this an important method in the context of total synthesis [10]. This chapter is organized chronologically by the type of tether, metathesis reaction, and relevant application to provide a unique prospective in the field over the last decade in the context of total synthesis of natural products.

8.2.1
O–SiR_2–O Tethered Substrates: Symmetrical Silaketals

The first TST-RCM sequence was reported by Grubbs and Fu for the construction of an achiral 1,4-diol, which circumvented the isolation of the silaketal intermediate in order to simplify purification (Equation 8.1) [11]. Treatment of the *bis*-alkoxysilane **1** with Shrock's catalyst [Mo]-**I** furnished after treatment with tetra-*n*-butylammonium fluoride (TBAF) the (*Z*)-1,4-diol **2** in 71% overall yield. A particularly important feature with this approach is that it demonstrated that the sensitive *bis*-allylic siloxane is tolerated, since this functionality is often prone to Lewis acid activation and the decomposition of the catalyst through elimination of the metal alkylidene intermediate [12].

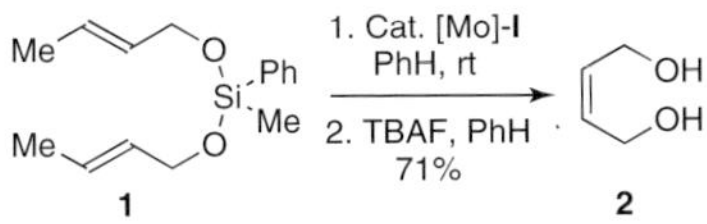

(8.1)

8.2.1.1 C_2-Symmetrical Silaketals and Applications

Evans and Murthy were the first to demonstrate the synthetic potential of silaketal intermediates in the TST-RCM with the construction of C_2-symmetrical derivatives (Table 8.1) [13]. A series of enantiomerically enriched allylic alcohols were treated with dichlorodiphenylsilane and 2,6-lutidine to furnish the *bis*-alkoxysilane intermediates, which were subjected to RCM utilizing Grubbs' first-generation catalyst [Ru]-**I**, to provide the corresponding silaketals in excellent yield (entries 1–5). This study demonstrated that the diphenyl-substituted silaketals were stable and that the reaction tolerates a variety of allylic alcohol substituents to provide a general approach for the construction of C_2-symmetrical silaketals.

The synthetic utility of this methodology was demonstrated with an application to a concise, total synthesis of the reduced carbohydrate D-altritol (Scheme 8.2) [13]. Dihydroxylation of the C_2-symmetrical silaketal **3** with catalytic osmium tetroxide and *N*-methylmorpholine (NMO), followed by deprotection of the silyl ethers and

Table 8.1 Synthesis of C_2-symmetrical silaketals.

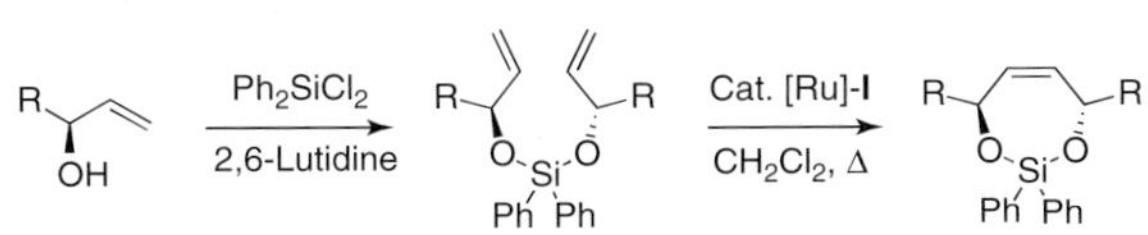

Entry	R	BAS (%)	Reaction Time (h)	DPSK (%)
1	$TBDPSOCH_2$	87	32	88
2	$PhCH_2$	86	26	95
3	*c*-Hex	95	26	90
4	Ph	94	42	93
5	Npth	88	25	84

BAS, *bis*-alkoxysilane; DPSK, diphenyl silaketal; Npth, 2-Naphthyl; TBDPS = *t*-BuPh_2Si.

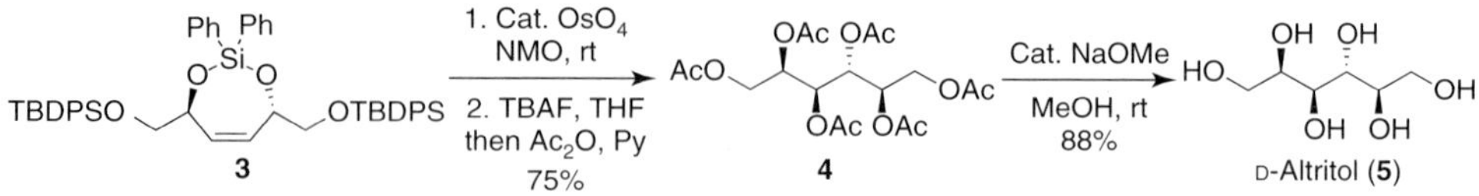

Scheme 8.2 Asymmetric synthesis of D-altritol.

peracetylation of the polyol to expedite isolation, furnished the hexaacetate **4** in 75% overall yield. Saponification of **4** with sodium methoxide in methanol provided D-altritol (**5**) in 88% yield.

Ariza *et al.* utilized the TST-RCM in conjunction with the Ireland–Claisen rearrangement to facilitate the total synthesis of (−)-phaseolinic acid (Scheme 8.3) [14]. The C_2-symmetrical silaketal **6** was prepared in 58% overall yield using an adaptation of the protocol described by Evans and Murthy [13], which employed enantiomerically enriched propargylic alcohols to form the symmetrical *bis*-alkoxysilane rather than allylic alcohols. Selective reduction of the *bis*-alkyne with Lindlar catalyst, followed by RCM with catalyst [Ru]-**I**, afforded the silaketal **6** in good overall yield. Cleavage of the silaketal **6** with TBAF followed by the acetylation of the diol furnished the *bis*-acetate **7** in 96% yield. This set the stage for the Ireland–Claisen rearrangement with the homotopic esters. Treatment of **7** with potassium hexamethyldisilazide followed by *tert*-butyldimethylsilyl chloride at −78 °C provided the silylketene acetal, which underwent rearrangement in refluxing toluene to furnish the *tert*-butyldimethylsilyl ester. Hydrolysis of the secondary alcohol followed by acid-catalyzed cyclization provided the γ-lactones **8**/**9** in 85% overall yield from **7**, which were epimeric at the α-methyl group (dr = 4 : 1). The mixture of stereoisomers originates from the formation of the *E(O)*- versus the preferred *Z(O)*-enolate, which undergo stereospecific rearrangement as the silyl ketene acetals (Scheme 8.3). The synthesis was then completed *via* the ozonolysis of **8** and oxidation of the resulting aldehyde to provide

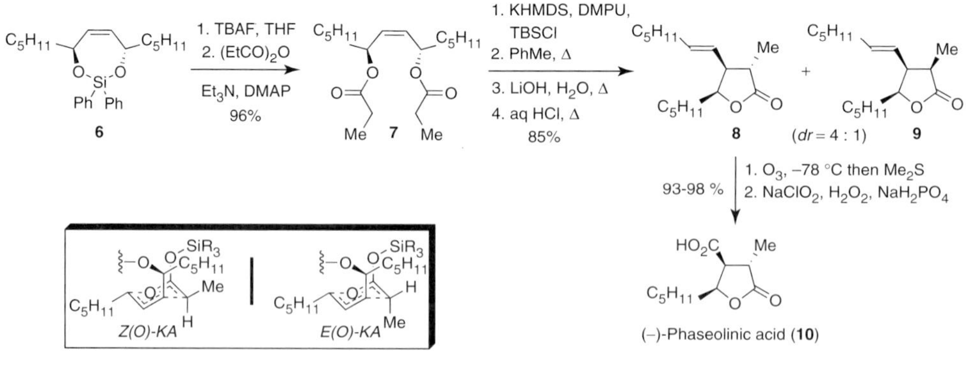

Scheme 8.3 Asymmetric synthesis of (−)-phaseolinic acid.

(−)-phaseolinic acid (**10**) in 40% overall yield from the commercially available propargylic alcohol.

8.2.1.2 **Achiral and Racemic Silaketals**

Hoye and Promo examined the effect of ring size and alkene substitution on *E*/*Z*-selectivity and rate of reaction (Table 8.2) [15]. This study also employed the diphenylsilane as the linker for the construction of symmetrical *bis*-alkoxysilanes. The optimal reaction conditions involved the slow addition of catalyst [Ru]-**I** *via* a syringe pump at room temperature to avoid isomerization of the alkene, which led to truncated ring systems after RCM. Although the reaction favors the formation of the *Z*-isomer, the selectivity does diminish with increasing ring size (Table 8.2, entries 1–3), whereas substituted alkenes result in lower efficiency and selectivity (entry 1 vs 4). Additional studies examined the effect of a racemic secondary and achiral tertiary allylic alcohol in the TST-RCM process. Interestingly, the former affords an equal mixture of *meso*- and *d*,*l*-silaketals, indicating no obvious diastereoselectivity for this particular tether (entry 5), whereas the latter proved completely unreactive (entry 6).

Table 8.2 Synthesis of symmetrical silaketals.

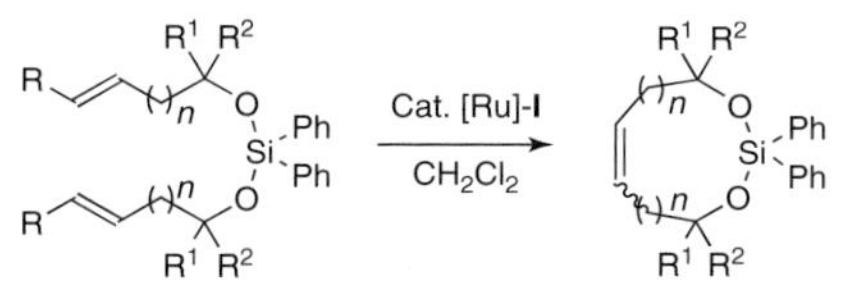

Entry	*n*	R	R[1]	R[2]	Addition Time for [Ru]-I (h)[a]	Temperature (°C)	*Z*/*E*[b]	Yield (%)
1	0	H	H	H	1	25	97 : 3	84
2	1	H	H	H	10	25	95 : 5	85
3	2	H	H	H	10	25	92 : 8	78
4	0	Me	H	H	1	45	96 : 4	61
5	0	H	Et	H	1	55	≥99 : 1	91
6	0	H	Me	Me	1	80	–	NR

[a] Catalyst added in the time indicated *via* syringe pump.
[b] GC–MS ratios.

8.2.1.3 **Related Applications and Developments**

In related studies, Lobbel and Köll utilized this approach in the construction of C_2-symmetrical carbon-linked disaccharides [16]. A critical feature of this approach was the relatively poor efficiency of the RCM, which served to highlight the limitations of the first-generation catalyst [Ru]-**I** for challenging synthetic applications. Mioskowski *et al.* later demonstrated that [Ru]-**VI** provided a more efficient catalyst for simple TST-RCM reactions [17]; however, the Grubbs's second-generation catalyst [Ru]-**II** has proven the optimal catalyst for many transformations [18]. Nevertheless, the ability to utilize different catalysts provides a unique opportunity to control selectivity based on catalyst reactivity (see below).

8.2.2
O-SiR$_2$-O Tethered Substrates: Unsymmetrical Silaketals

Hoye and Promo described the first example of the TST-RCM reaction for the preparation of unsymmetrical silaketals (Scheme 8.4) [15]; however, the concept was independently described by Evans and Murthy in their approach to the annonaceous acetogenin (−)-mucocin [19]. The unsymmetrical *bis*-alkoxysilane **12** was prepared through the sequential reaction of equimolar amounts of dichlorodiphenylsilane, pyridine, and the secondary alcohol *rac*-**11** followed by the addition of allyl alcohol. Treatment of the *bis*-alkoxysilane **12** in an analogous manner to the symmetrical derivatives provided the unsymmetrical silaketal **13** in 62% yield ($Z/E \geq 19:1$). Although the construction of unsymmetrical *bis*-alkoxysilanes remains challenging, it has featured as an important cross-coupling in a number of important total syntheses (see below).

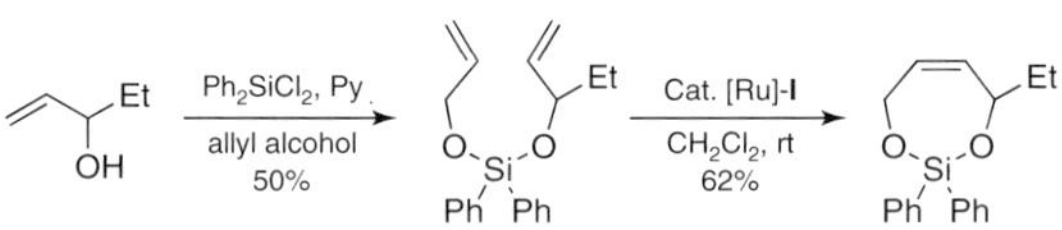

Scheme 8.4 Preparation of unsymmetrical silaketals.

8.2.2.1 Spiroketals

Eustache *et al.* applied this strategy to the synthesis of the eastern spiro[5.5]ketal fragment of okadaic acid (Figure 8.1) [20]. Sequential coupling of the enantiomerically enriched secondary alcohol **14** and racemic allylic alcohol *rac*-**15** with dichlorodimethylsilane at low temperature furnished the unsymmetrical *bis*-alkoxysilane **16** in excellent yield as an equal mixture of diastereoisomers (Scheme 8.5). Treatment of **16** with catalyst [Ru]-**I** furnished the silaketal **17** in 46% yield as a single diastereoisomer, favoring the *cis*-isomer. Interestingly, this tentative stereochemical assignment contradicts the later studies by Verdine and Evans

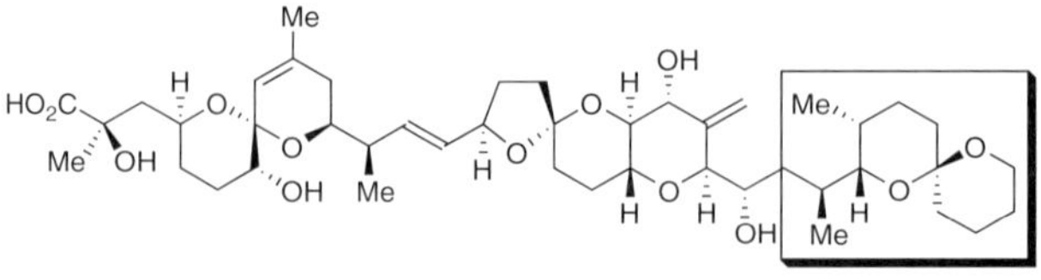

Figure 8.1 Okadaic acid.

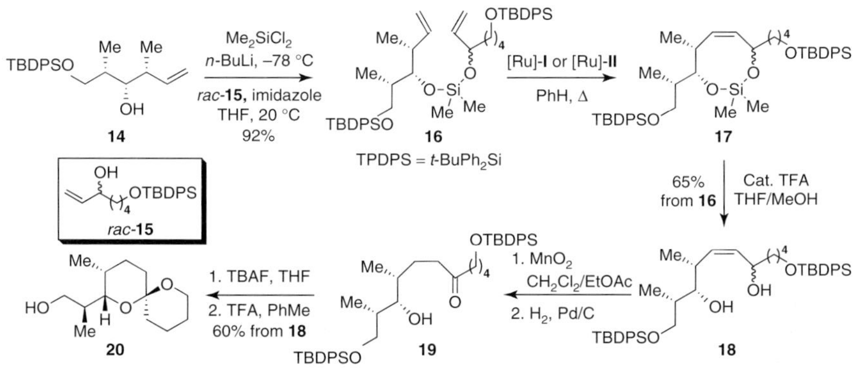

Scheme 8.5 TST-RCM approach to the spiro[5.5]ketal of okadaic acid.

[21, 22], which demonstrated that eight-membered silaketals favor the *trans*-isomer (see below). Nonetheless, the diastereomeric preference could be avoided using the more reactive [Ru]-**II** catalyst, which resulted in the formation of the epimeric diols **18** in 65% overall yield after acid-catalyzed deprotection of the tether. Selective oxidation of the allylic alcohol **18** with MnO_2 followed by the reduction of the alkene of the α,β-unsaturated ketone affords the hydroxy ketone **19**. Deprotection of the primary *tert*-butyldiphenylsilyl ethers in **19** with TBAF followed by acid-catalyzed spiroketalization provided the spiro[5.5]ketal **20** in 60% overall yield, which is directly applicable to okadaic acid.

Additional studies by the same group utilized a similar sequence for the construction of the spiro[5,4]ketal present in attenol A (Scheme 8.6) [23]. The unsymmetrical *bis*-alkoxysilane **23** was prepared from the epimeric allylic alcohol **21** and the secondary alcohol **22** as a mixture of diastereoisomers. The mixture of epimers was again justified on the basis that this stereogenic center would be oxidized to the ketone, and the previous study had demonstrated that both stereoisomers could be converted to the silaketal. Surprisingly, treatment of the *bis*-alkoxysilanes **23** with catalyst [Mo]-**I** furnished the silaketal **24** as a single stereoisomer, favoring the *trans*-diastereoisomer. In contrast to the previous studies, the diastereomeric bias could not be overcome, which provided the diol **25** in 22% overall yield after acid-catalyzed deprotection of the silyl tether. Although the unreactive *bis*-alkoxysilane **23'** enriched in the opposite stereoisomer could be recycled, the deprotection of the tether to facilitate the stereochemical inversion was tedious. The diol **25** was then converted to the spiro[5,4]ketal **26** in 72% overall yield using a sequence similar to that described earlier (see above) [20], which allowed the completion of the total synthesis of attenol A (**27**) using a five-step reaction sequence. Overall, this work demonstrates that there is an inherent preference for the formation of a specific stereoisomer in the TST-RCM in eight-membered rings, which has significant implications for the implementation of this methodology in total synthesis.

Kozmin and Marjanovic described an ingenious application of the TST-RCM strategy for controlling spiroketal formation in the total synthesis of spirofungin

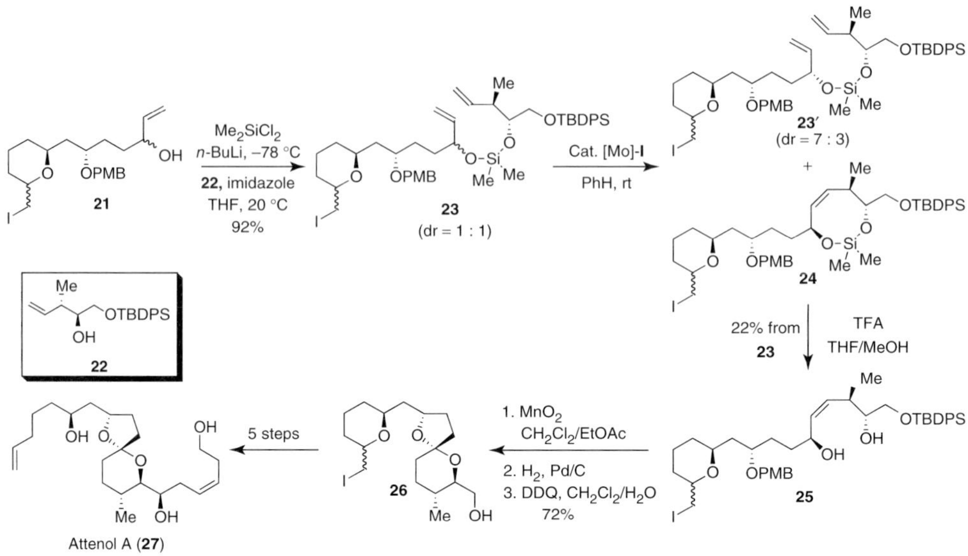

Scheme 8.6 Total synthesis of attenol A.

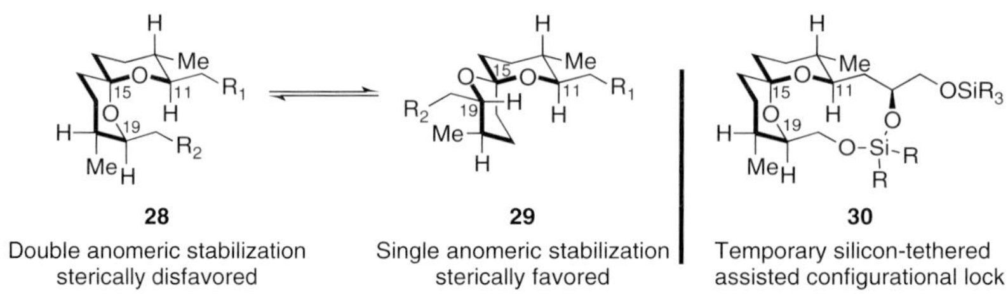

Figure 8.2 Rationale for the TST-RCM spiroketalization.

A over its spiroketal epimer spirofungin B (Figure 8.2) [24]. On the basis of structural analysis with the reveromycins, it was rationalized that spirofungin A was the most likely to display antiproliferative activity in mammalian cells through the inhibition of isoleucyl-*t*RNA synthase. Although the desired spiroketal **28** is stereoelectronically favored by double anomeric stabilization, there is a severe steric interaction between the axial substituent at C19 with the group at C11, whereas the undesired spiroketal **29** possesses single anomeric stabilization but lacks the steric congestion. This is presumably the reason why there is no thermodynamic preference for their formation. In order to circumvent this problem, a TST-RCM strategy was devised that would be able to take advantage of the relative orientation of the groups at C19 and C11 to lock the configuration in a tethered structure **30**.

The tethered trienone **33** was prepared in 65% overall yield *via* the sequential coupling of the primary and secondary alcohols **31** and **32** with dichlorodiisopropylsilane and imidazole, followed by the chemoselective removal of the 1,3-dioxane

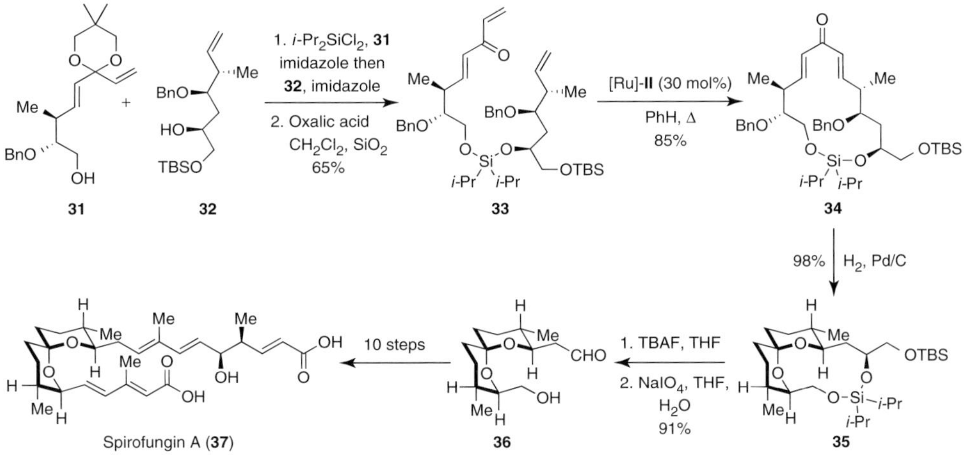

Scheme 8.7 Total synthesis of spirofungin A.

protecting group (Scheme 8.7). Treatment of **33** with catalyst [Ru]-**II** furnished the cyclic dienone **34** in 85% yield, with excellent chemoselectivity for the terminal olefins. Hydrogenation of the dienone **34** and concomitant hydrogenolysis of the secondary benzyl ethers resulted in spontaneous spiroketalization to afford the tethered spiroketal **35** in 98% yield as a single stereoisomer. Another striking feature of this approach is the ability to differentiate the substituents to facilitate the introduction of the side chains. Fluoride-mediated cleavage of the silyl ethers in **35**, followed by oxidative cleavage of the 1,2-diol, furnished the hydroxy aldehyde **36** in 91% yield. An additional 10 steps were required to elaborate this intermediate and thus complete the stereoselective synthesis of spirofungin A (**37**).

8.2.2.2 Long-range Asymmetric Induction

Evans *et al.* devised a novel approach to long-range asymmetric induction using the TST-RCM reaction [22]. It was envisioned that the TST-RCM reaction of the unsymmetrical *bis*-alkoxysilane **40** would preferentially provide the *cis*-diastereoisomer **41** (Scheme 8.8), since the reaction should proceed through the favored TS outlined in Figure 8.3. A critical feature of this approach was the ability to suppress intermolecular alkylidene exchange with the prochiral propenyl groups and to optimize the nonbonding interactions using the substituents on the silicon tether. This work demonstrated that, although the steric nature of the tether is important, the catalyst has the most significant influence. For example, catalyst [Ru]-**I** afforded optimum selectivity, whereas the more reactive [Mo]-**I** and [Ru]-**II** catalysts provided inferior yields and selectivity, which is consistent with related studies by Verdine *et al.* [21]. The diastereoselective TST-RCM of **40** with catalyst [Ru]-**I** proved remarkably tolerant to a range of aryl, alkyl, and hydroxymethyl derivatives, each of which afforded excellent stereocontrol in favor of the *cis*-diastereoisomer **41**.

This concept was also applied to the higher homologs to determine the scope and limitations of this process (Table 8.3) [22]. Interestingly, Grubbs's

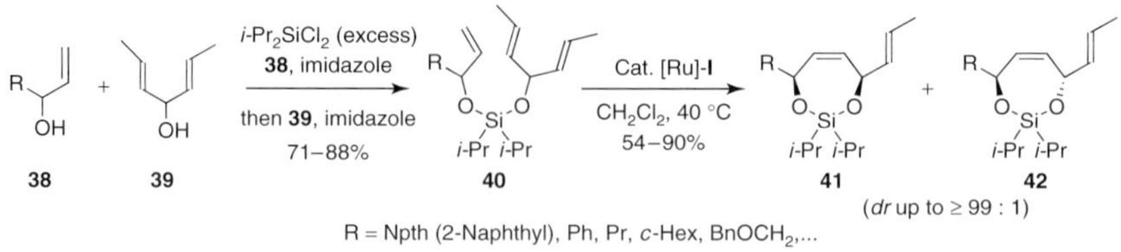

Scheme 8.8 Diastereoselective TST-RCM for 1,4-stereocontrol.

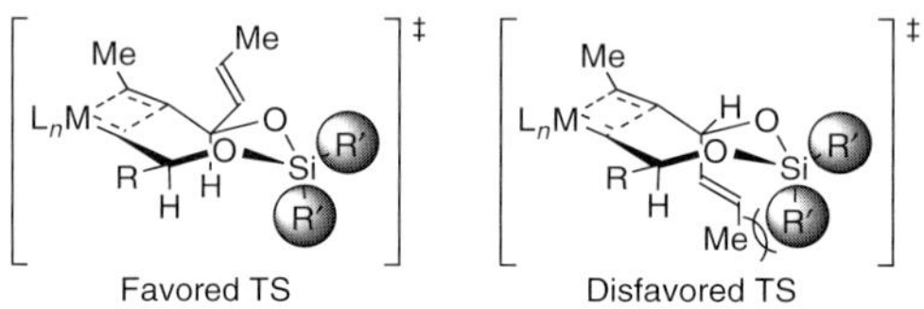

Figure 8.3 Proposed transition states for 1,4-stereocontrol.

Table 8.3 Diastereoselective TST-RCM reactions with homologated alkenyl alcohols.

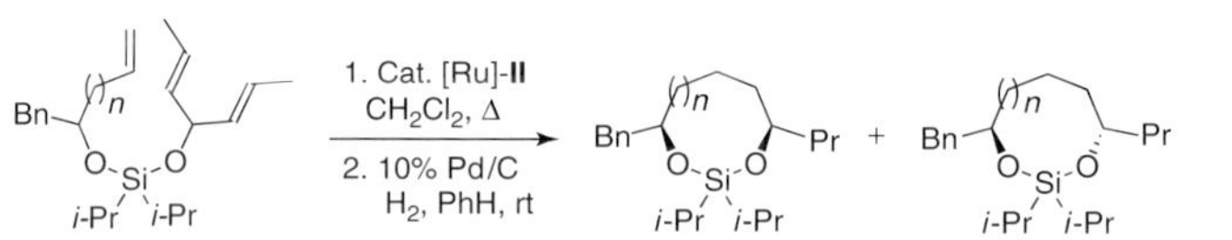

Entry	*n*	Yield of Silaketal (%)	*cis* : *trans*	Yield (%)
1	1	92	1 : 11	90
2	2	68	1 : 27	92
3	3	75	1 : 3	75
4	4	87	1 : 3	73

second-generation catalyst [Ru]-**II** proved optimal for the formation of the medium ring silaketals, which were hydrogenated to remove *E*/*Z*-isomers and simplify analysis. Although the higher homologs favor the formation of the *trans*-diastereoisomer (Table 8.3, entries 1–4), this observation is fully consistent with the model for diastereocontrol in which the *pseudo*-axial/equatorial positions are reversed (Figure 8.4). Hence, the preferred TS for these rings also has both substituents *pseudo*-equatorial to avoid the steric interaction derived from the axial isopropyl group on the silicon tether. Additional studies demonstrated that this process is fully reversible, in which the diastereoselectivity both increases and decreases for several substrates (Table 8.3, entries 1 and 2), thereby permitting the level of stereocontrol to be tuned accordingly.

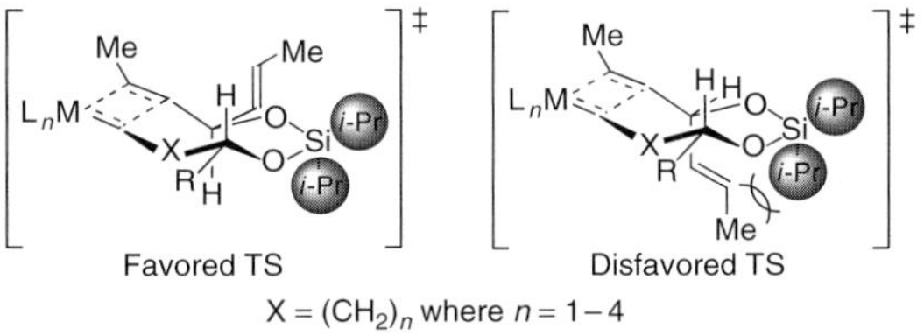

Figure 8.4 Proposed transition states for 1, ($n + 4$)-stereocontrol (where $n = 1$–4).

Pickett *et al.* utilized the long-range asymmetric induction strategy in an improved synthesis of the sex pheromone (2*S*,7*S*)-dibutyroxynonane produced by *Sitodiplosis mosellana* (Scheme 8.9) [25]. Preliminary studies demonstrated that the TST-RCM reactions with the diphenyl- and diisopropyl tethers furnished the *trans*-diastereoisomer with excellent stereocontrol (*dr* = 17 : 1 and 21 : 1, respectively), albeit contaminated with the symmetrical silaketals that were difficult to purify using standard chromatography. It was envisioned that a more sterically hindered silyl tether would reduce the necessity for a large excess of the coupling agent and minimize the formation of the symmetrical silaketal while further improving diastereocontrol. Treatment of the enantiomerically enriched secondary alcohol **43** with di-*tert*-butylsilyl *bis*(trifluoromethanesulfonate), pyridine, and dimethylaminopyridine at −78 °C, followed by warming to room temperature and subsequent addition of divinylcarbinol, furnished the *bis*-alkoxysilane **44**. RCM with catalyst [Ru]-**I** afforded the nine-membered silaketal **45** in 70% overall yield, with improved diastereoselectivity (*dr* = 32 : 1) and less than 4% of the symmetrical silaketal. Although the more sterically hindered tether successfully overcame the previous problems, the deprotection of the silaketal **45** proved significantly more challenging due to the increased steric congestion at silicon. Nonetheless, the synthesis was completed through the deprotection of silyl tether in **45** by heating with excess TBAF, followed by the hydrogenation of the diene and acylation of the diol to provide the sex pheromone (2*S*,7*S*)-dibutyroxynonane (**46**) in 22% overall yield from **43**.

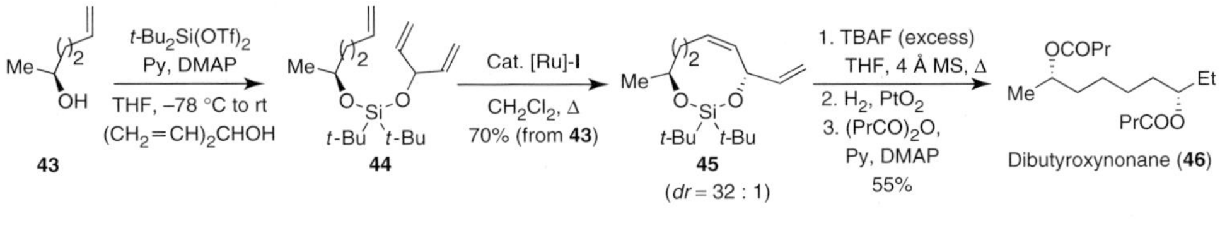

Scheme 8.9 Total synthesis of the sex pheromone (2*S*,7*S*)-dibutyroxynonane.

Harvey *et al.* recently described an adaptation to the long-range asymmetric induction strategy for the construction of the C12–C24 fragment of peloruside A (Figure 8.5), which simultaneously controls the configuration of the trisubstituted *Z*-olefin and the C18 stereogenic center (Scheme 8.10) [26]. Sequential coupling of the racemic homoallylic alcohol *rac*-**47** and allylic alcohol **48** with

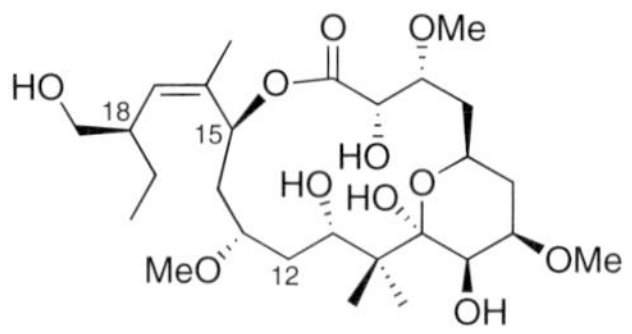

Figure 8.5 Peloruside A.

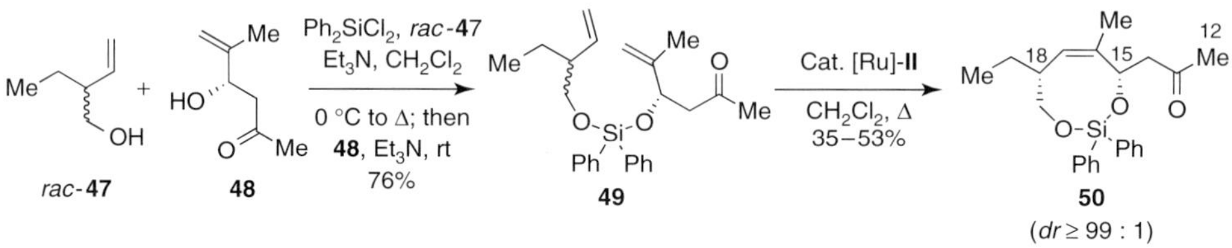

Scheme 8.10 TST-RCM diastereoisomer-discrimination approach to the C12–C24 fragment of peloruside A.

dichlorodiphenylsilane furnished the unsymmetrical *bis*-alkoxysilane **49** in 76% yield as an equal mixture of diastereoisomers. Interestingly, the presence of the C18 ethyl group resulted in a more challenging RCM. Treatment of the unsymmetrical *bis*-alkoxysilane **49** with the slow addition of [Ru]-**II** catalyst in refluxing CH_2Cl_2 afforded the silaketal **50** in 35–53% yield with ≥99 : 1 diastereoselectivity favoring the *cis*-diastereoisomer. The ability to utilize the β-hydroxy ketone **48** in the TST-RCM reaction is fairly significant given the propensity of this type of intermediate to undergo retro-aldol and β-elimination.

The rationalization of the stereochemical outcome is consistent with the Evans' model for 1,5-stereocontrol in an eight-membered silaketal (Figure 8.4), which places the C8 substituent in a *pseudo*-equatorial orientation to avoid nonbonding interactions with the *pseudo*-axial phenyl group on the silicon tether (Figure 8.6) [22]. Nonetheless, an interesting feature with this process is that the alkylidene exchange occurs with the homoallylic derivative, which is contrary to the previous cases that employ a prochiral alkene. Overall, this provides another example of the inherent versatility of this strategy, which enables remote diastereoselectivity to be controlled.

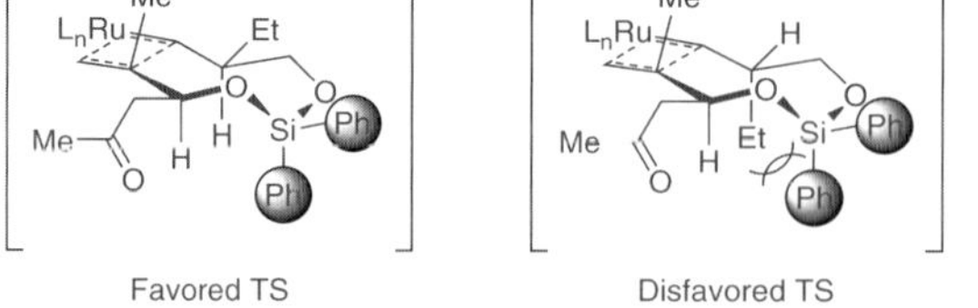

Figure 8.6 Proposed transition states for diastereoisomer discrimination.

8.2.2.3 Annonaceous Acetogenins

Evans *et al.* utilized the TST-RCM strategy in an expeditious total synthesis of the annonaceous acetogenins (−)-mucocin, which has potent antitumor activity (Scheme 8.11) [27]. The unsymmetrical *bis*-alkoxysilane **53** was obtained in 74% yield *via* the sequential coupling of the allylic alcohols **51** and **52** under standard conditions. The RCM was particularly challenging, due to the *pseudo*-axial orientation of one of the substituents. Unfortunately, reducing the steric bias of the tether introduced additional problems with respect to isolation and purification of the silaketal. Hence, treatment of the unsymmetrical *bis*-alkoxysilane **53** with the syringe pump addition of excess [Ru]-**II** in refluxing 1,2-dichloroethane (DCE) furnished the silaketal **54** in 83% yield. Fluoride-mediated global cleavage of the silyl ethers followed by a chemoselective diimide reduction of the alkyne and acyclic alkene furnished (−)-mucocin (**55**) in 12.5% overall yield. In a related study, Brown *et al.* employed a similar strategy to complete a total synthesis of a related non-adjacent tetrahydrofuran (THF) containing acetogenin, (+)-*cis*-sylvaticin (Figure 8.7) [28]. An interesting feature of this synthesis is that a catalytic amount of catalyst [Ru]-**II** may be utilized when the solvent is switched from 1,2-dichloromethane to toluene and the reaction heated at 75 °C.

Hoye *et al.* extended the synthetic utility of the TST-RCM reaction with an elegant and highly convergent total synthesis of (+)-gigantecin and its constitutional isomer (+)-14-deoxy-9-oxygigantecin (Scheme 8.12) [29]. This study examined the merit of combining the TST-RCM reaction in conjunction with CM to facilitate the rapid assembly of this class of nonadjacent THF containing annonaceous acetogenins.

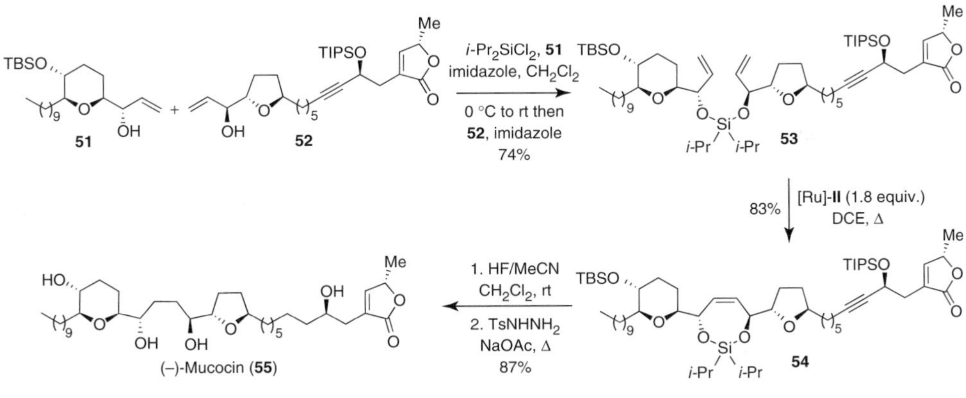

Scheme 8.11 Total synthesis of (−)-mucocin.

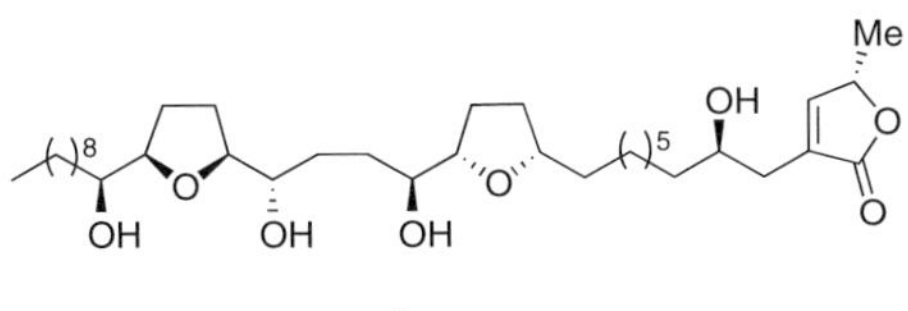

Figure 8.7 (+)-*cis*-Sylvaticin.

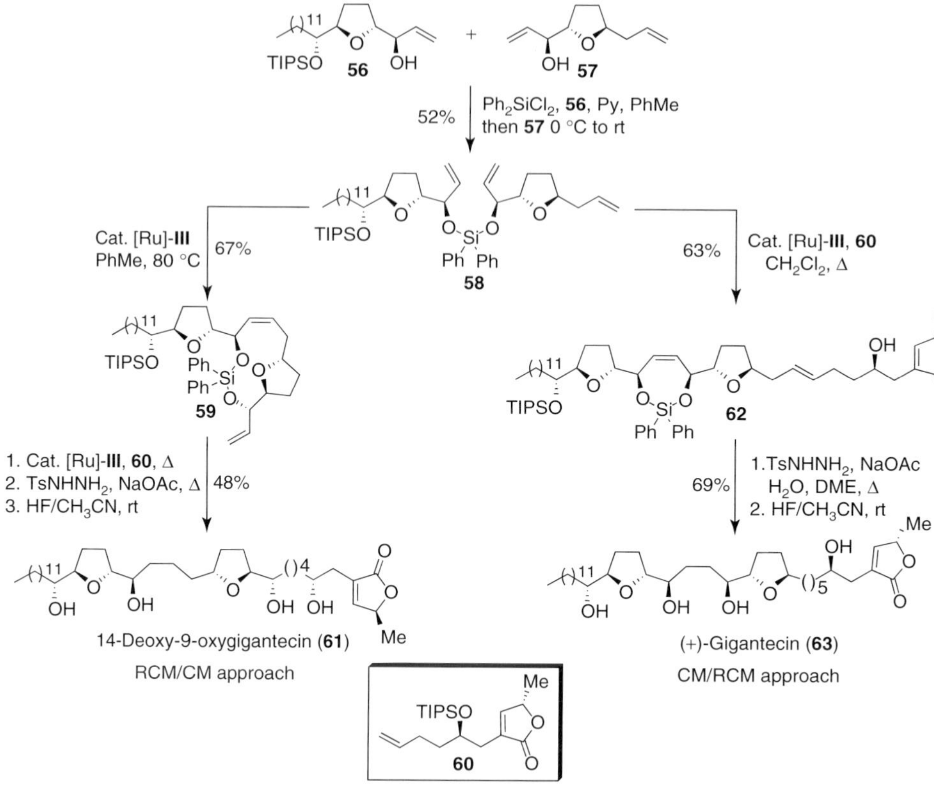

Scheme 8.12 Total synthesis of 14-deoxy-9-oxygigantecin and (–)-gigantecin.

The *bis*-alkoxysilane **58** was prepared through the sequential loading of the allylic alcohols **56** and **57** on to dichlorodiphenylsilane. Interestingly, the RCM of **58** with Hoveyda–Grubbs' second-generation catalyst [Ru]-**III** furnished the 11-membered silaketal **59** rather than the desired 7-membered derivative. Although this anomaly was not immediately obvious, since it is a constitutional isomer of the desired product, it is presumably due to preferential reaction with the less hindered terminal Type I alkene [10]. Butenolide **60** was then installed with a CM with **59**, followed by the chemoselective reduction and global cleavage of the silyl ethers to afford 14-deoxy-9-oxygigantecin (**61**) in 48% overall yield.

Interchanging the order of the CM and TST-RCM reactions allows for the preferential functionalization of the less hindered and more reactive Type I alkene and thus permits the formation of the seven-membered silaketal without competitive RCM. A critical feature for the success of this strategy is that the formation of the *cis*-silaketal is preferred over the *trans*-derivative, which provides optimal rates of reaction [22]. Treatment of a mixture of the unsymmetrical *bis*-alkoxysilane **58** and butenolide **60** (1 : 4 ratio) with catalyst [Ru]-**III** provided the gigantecin skeleton **62** in 63% yield in a single step. Additionally, the formation of the homodimer of butenolide **60** is inconsequential since it can be recycled through a standard CM with ethylene, and the formation of the 11-membered silaketal is significantly reduced (10%) albeit having undergone a CM with butenolide **60** to provide direct access to the constitutional isomer. The synthesis, of (+)-gigantecin (**61**) was completed in a manner analogous to that described for the constitutional isomer in 69% overall yield [29]. Overall, these total syntheses serve to highlight the strategic advantages of the TST-RCM/CM approach for the expeditious syntheses of these agents, and the importance of understanding the rates of reaction for these processes.

8.2.2.4 **Trisubstituted Alkenes**

Mulzer *et al.* utilized the TST-RCM strategy to prepare the *Z*-trisubstituted C12–C13 alkene of the northern fragment of epothilone B and D (Figure 8.8) [30]. Preliminary studies demonstrated that acyclic dienyl esters were unsuitable substrates for the RCM that provide eight-membered lactones. It was envisioned that the strain in the TS could be relieved using a silicon tether, as the silicon atom is larger and has more polarizable soft d-orbitals that allow for more facile bond distortion. Hoye and Promo had previously demonstrated that a nine-membered silaketal is accessible in excellent yield and with 95 : 5 selectivity for the *Z*-isomer, thereby illustrating proof-of-concept (Table 8.2) [15]. The unsymmetrical *bis*-alkoxysilane **66** was prepared *via* the sequential silylation of the homoallylic alcohols **65** and **64** with dichlorodimethylsilane (Scheme 8.13). The RCM of the unsymmetrical *bis*-alkoxysilane **66** required the *slow addition* of either [Ru]-**II** or [Ru]-**III** second-generation catalysts, which afforded the silaketal **67** in 98% yield, as a 5 : 1 mixture favoring the *Z*-isomer. Interestingly, the addition of the catalyst in a single portion leads to a significant reduction in the yield due to its decomposition. Cleavage of the silyl tether in **67**, followed by a standard differential protection sequence, provided the primary alcohol **68** in 88% overall yield. Sequential functionalized of the termini of **68** using a four-step sequence to homologate the alcohol and install the thiazole substituted *E*-trisubstituted alkene furnished **69** in 94% overall yield. The key aldehyde **70**, which had previously undergone fragment coupling *en route* to epothilone B and D, was prepared in an additional four steps. Although the selectivity in the formation of the C12–C13 olefin is suboptimal, it serves to demonstrate the potential of the TST-RCM approach for the preparation of challenging trisubstituted olefins.

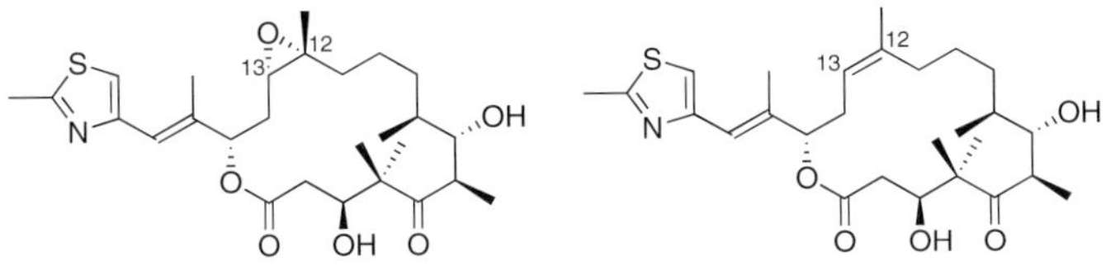

Figure 8.8 Epothilone B and D.

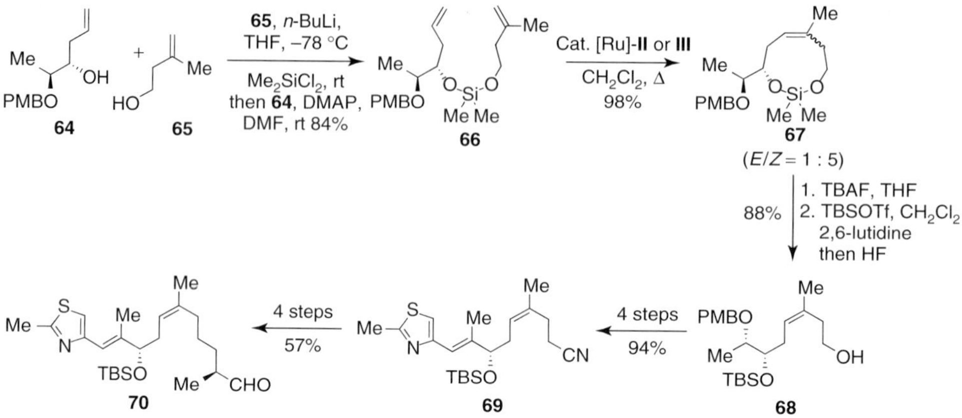

Scheme 8.13 Total synthesis of epothilone B and D.

8.2.2.5 Related Applications and Developments

In related studies, the TST-RCM has been employed to prepare 2-ene-1,5-diols that can be converted to functionalized THF's present in acetogenins using an electrophilic selenocyclization [31]. This strategy is also featured in a novel approach to the pyrrolo[3,2-c]quinoline core of martinelline and martinellic acid natural products [32]. Additionally, Verdine *et al.* utilized this methodology to prepare stereodiversified libraries of 1,4- and 1,5-enediols, which enabled the identification of a low nanomolar ligand for the mu opioid receptor (MOR) [21]. Finally, Postema and Piper revisited the construction of carbon-linked disaccharides using the TST-RCM with the more active Grubbs's second-generation complex, which demonstrated that the RCM of hindered carbohydrates remained problematic for fully functionalized carbohydrates [33].

8.2.3 Dienyne TST-RCM: Symmetrical and Unsymmetrical Silanes

Lee *et al.* examined the merit of the dienyne TST-RCM with symmetrical and unsymmetrical *bis*-alkoxyalkynylsilanes for the stereoselective construction of 1,3-butadienes which are not accessible through classical enyne metathesis reactions [34–36]. Table 8.4 outlines the key findings with the symmetrical *bis*-alkoxyalkynylsilanes, which were prepared through the base-induced alcoholysis of trialkynylsilanes with various racemic and enantiomerically enriched secondary alcohols [36], and subjected to the enyne TST-RCM using catalyst [Ru]-**II** to afford the silaketals in excellent yield (Table 8.4, entries 1–6). Removal of the silyl tether under relatively forcing conditions, furnished the 1,3-butadiene derivative with retention of the alkene configuration. This study demonstrated the importance of tailoring the reactivity and stability of the trialkynylsilane with the volatility of

Table 8.4 Stereoselective synthesis of 1,3-dienes *via* the dienyne TST-RCM of symmetrical bis-alkoxyalkynylsilanes.

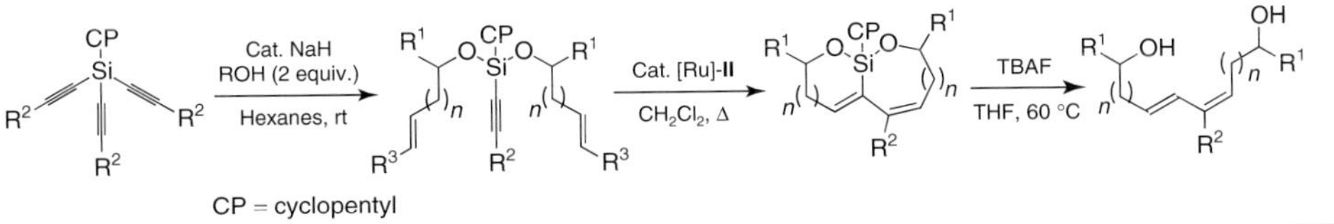

Entry	R^1	R^2	R^3	*n*	BAAS (%)	Method	BCS (%)	Diol (%)
1	Me	CH_2OMe	H	1	70	A	75	68
2	$Ph(CH_2)_2$	CH_2OMe	H	1	74	A	89	78
3	(*R*)-$BnOCH_2$	CH_2OMe	H	1	67	A	83	68
4	$Ph(CH_2)_2$	Me	H	1	82	A	88	65
5	$Ph(CH_2)_2$	CH_2OMe	Me	1	81	B	88	83
6	Me	Me	*n*-Pent	2	66	B	76	50

Method A: CH_2Cl_2, 40 °C; Method B: toluene, 110 °C.
BAAS, *bis*-alkoxyalkynylsilane; BCS, bicyclic siloxane.

the displaced alkyne to reduce the formation of the trialkoxysilane (entry 2 vs. 4). It also demonstrated that there is no diastereoisomer discrimination with the racemates and that the enantiomerically enriched derivatives furnish a mixture of diasteroisomers at silicon (entries 1/2 vs. 3). Additional alkene substitution required more forcing reaction conditions (entry 5) and, lastly, the process was amenable to the formation of larger rings (entry 6).

Additional studies expanded the scope of this process to unsymmetrical *bis*-alkoxyalkynylsilanes since they are more likely to be of synthetic utility in the context of a total synthesis [36]. These systems represent a significant synthetic challenge with regard to the preparation of the unsymmetrical derivatives and the ability to initiate a selective enyne metathesis reaction to avoid the formation of regioisomers. The unsymmetrical *bis*-alkoxyalkynylsilane **72** was prepared through an adaptation of the protocol described for the symmetrical derivatives (Scheme 8.14). Sequential treatment of the trialkynylsilane **71** with a tertiary alcohol followed by a primary alcohol in the presence of a catalytic amount of

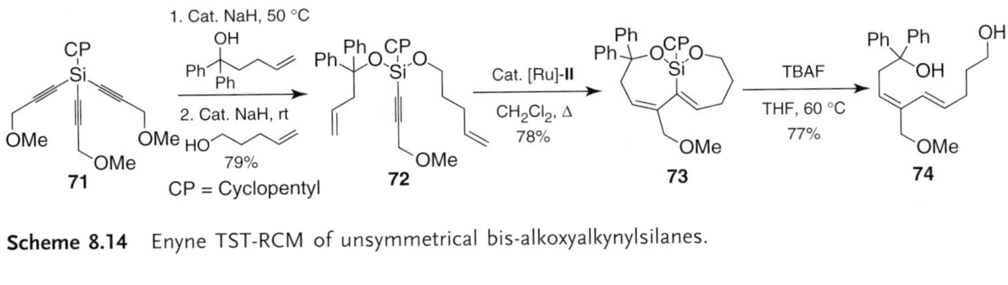

Scheme 8.14 Enyne TST-RCM of unsymmetrical bis-alkoxyalkynylsilanes.

base furnished the unsymmetrical *bis*-alkoxyalkynylsilane **72** in 79% overall yield. Critical to the success of this process is the addition of the tertiary alcohol followed by the primary derivative, which provides the necessary steric differentiation between the two pendant alkenes to facilitate initiation at the less hindered olefin. Hence, the sequential enyne metathesis of **72** furnished the bicyclic siloxane **73**, which upon cleavage of the silyl tether provided the unsymmetrical 1,3-diene **74** in 60% overall yield from **72**. Alternatively, the enyne TST-RCM of **75** provides a mixture of bicyclic siloxanes **76**/**77** in 81% yield, as 1.5 : 1 mixture (Equation 8.2), which illustrates the importance of being able to differentiate the alkene groups in the initiation step.

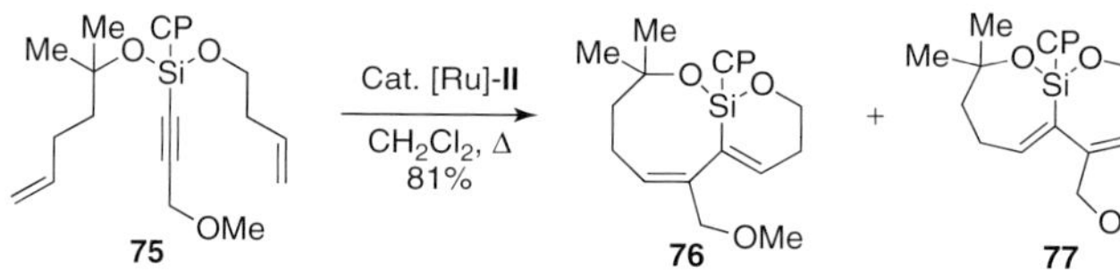

(8.2)

8.2.3.1 Macrolide Antibiotics

Kim and Lee utilized this strategy for the construction of the *E*/*Z*-butadiene motif at C14–C15 in the boron-containing C_2-symmetrical macrodiolide tartrolon B (Figure 8.9), which belongs to the class of ion-carrier antibiotics active against Gram-positive bacteria [37]. Interestingly, tartrolon A1–3 and B are all biologically active, which indicates that the presence of boron is not necessary for its antibiotic activity. Another striking feature of this metabolite is the homology of the C1–C7 segment, which is also present in the structurally related boron-containing antibiotics boromycin, aplasmomycin, and borophycin. In light of the problems encountered with group differentiation, the alkenes were differentially substituted to garner the necessary regiocontrol in the enyne TST-RCM reaction.

The unsymmetrical *bis*-alkoxyalkynylsilane **80**, applicable to the natural product, was prepared as a mixture of diastereoisomers at silicon in 58% yield by the addition of the secondary alcohol **78** to the *bis*-alkynylsilane **79** in the presence of catalytic sodium hydride (Scheme 8.15). Treatment of **80** with catalyst [Ru]-**II** furnished

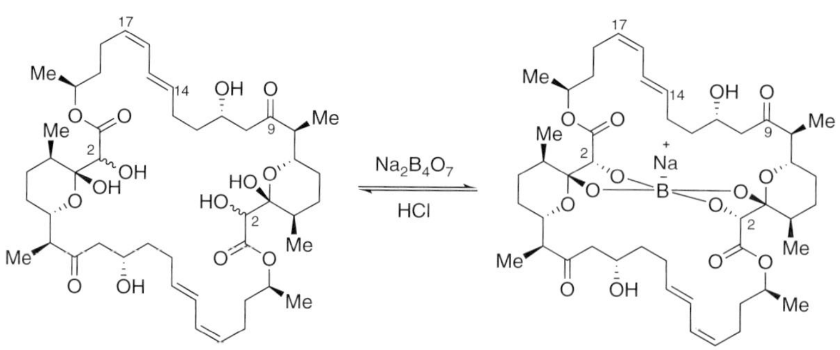

Figure 8.9 Tartrolons A1–3 and B.

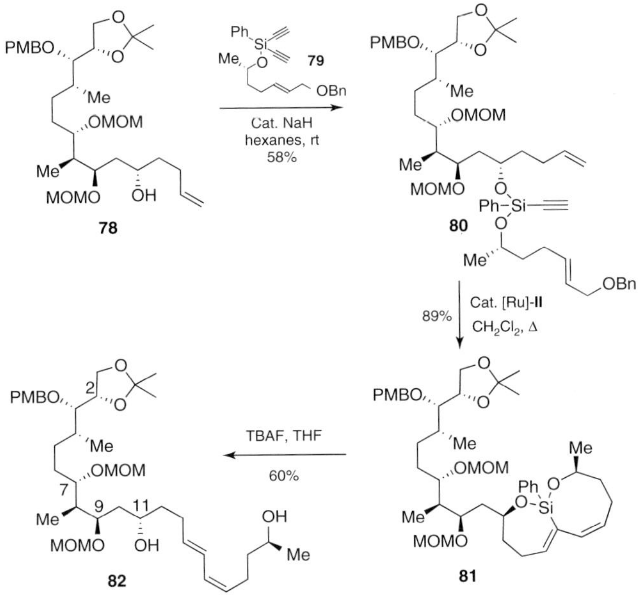

Scheme 8.15 Synthesis of C1–C21 fragment of tartrolon B.

the bicyclic siloxane **81**, which was deprotected with TBAF to afford the C1–C21 fragment applicable to tartrolon B [37a]. Unfortunately, the stereochemical assignments at C2, C7, C8, C9, and C11 had the opposite stereochemistry to the natural product, since the incorrect enantiomer of *B*-methoxydiisopinocampheylborane was employed for the construction of **78** [37b]. Nonetheless, this method demonstrated the synthetic utility of this approach for the construction of stereodefined 1,3-butadienes.

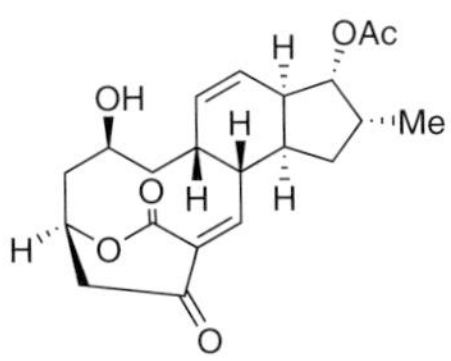

Figure 8.10 (−)-Cochleamycin A.

More recently, Lee *et al.* have also employed this strategy for the construction of the *E*/*Z*-1,3-butadiene motif required for the synthesis of (−)-cochleamycin A using a biomimetic type transannular Diels–Alder reaction, (Figure 8.10) [38]. The enantiomer exhibits antimicrobial activity against Gram-positive bacteria and has modest cytotoxic activity. The unsymmetrical *bis*-alkoxyalkynylsilane **85** was prepared as a mixture of diastereoisomers at silicon by the addition of the secondary alcohol **84** to the *bis*-alkynylsiloxane **83** (Scheme 8.16). Although the

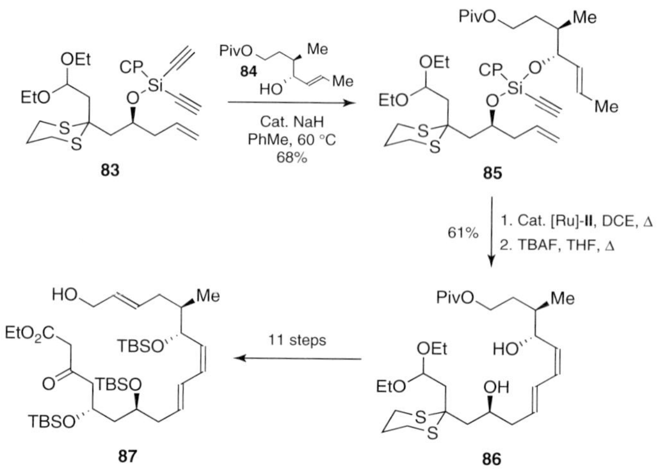

Scheme 8.16 Synthesis of C1–C19 fragment of (−)-cochleamycin A.

bis-alkynylsiloxane **83** was prepared from the trialkynylsilane in 81% yield without the formation of the corresponding symmetrical silaketal, the preparation of the unsymmetrical silaketal **85** required elevated temperature, which also led to problems with the double addition of the secondary alcohol **84**. Treatment of the unsymmetrical *bis*-alkoxyalkynylsilane **85** with catalyst [Ru]-**II** catalyst followed by the deprotection of the bicyclic siloxane using TBAF furnished the *E*/*Z*-butadiene **86** in 61% yield. Although the enyne TST-RCM proceeds with exquisite group selectivity, premature termination at the monocyclic siloxane remained problematic even with longer reaction times and higher catalyst loadings. The elaboration of this fragment using an 11-step sequence provided the same intermediate employed by Roush *et al.*, which completes a formal synthesis of the enantiomer of the natural product.

8.2.4
O–SiR$_2$–C Tethered Substrates: Allyl and Vinylsiloxanes

The first TST metathesis sequence involving a mono-alkoxysilane was reported by Grubbs and Chang for the construction of achiral (*Z*)-1,*n*-diols and hydroxy aldehydes (Table 8.5 and Equation 8.3) [39]. The formation of the mono-alkoxysilanes using an alkyl or vinylsilane is well established and comparatively trivial as compared to the *bis*-alkoxysilanes described earlier in this chapter. However, the oxidative and reductive functionalization of the products from the RCM remains a significant problem. Table 8.5 outlines the preliminary work that inspired the subsequent applications to natural product synthesis. Although Grubbs's first-generation catalyst [Ru]-**I** is effective for the allylsiloxane (Table 8.5, entries 1–2), the more sterically hindered vinylsiloxanes required the more reactive Schrock catalyst [Mo]-**I** (entries 3–4). Nevertheless, these represent the first examples of vinylsiloxanes undergoing RCM, and ironically the catalyst requirement has not changed despite the

Table 8.5 TST-RCM reactions of acyclic mono-alkoxysilanes.

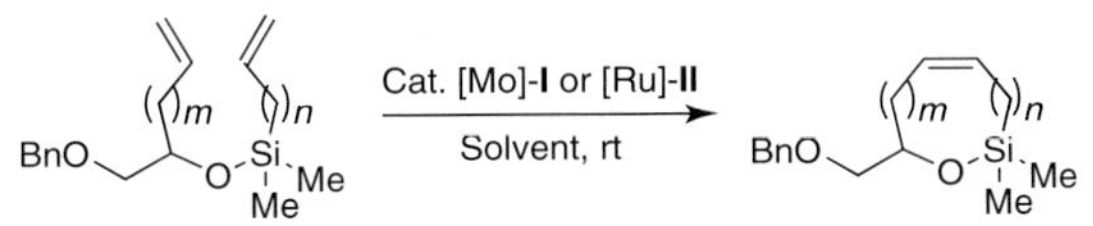

Entry	*m*	*n*	Catalyst (mol%)	Solvent	Yield (%)
1	1	1	[Ru]-**I** (3)	CH_2Cl_2	88
2	2	1	[Ru]-**I** (5)	CH_2Cl_2	96
3	1	0	[Mo]-**I** (3)	PhH	84
4	2	0	[Mo]-**I** (5)	PhH	91

advent of new, more reactive ruthenium complexes. Another interesting feature is the necessity to slightly increase the amount of catalyst used upon increasing ring size; however, the process does not require high dilution for the formation of an eight-membered ring (entry 2).

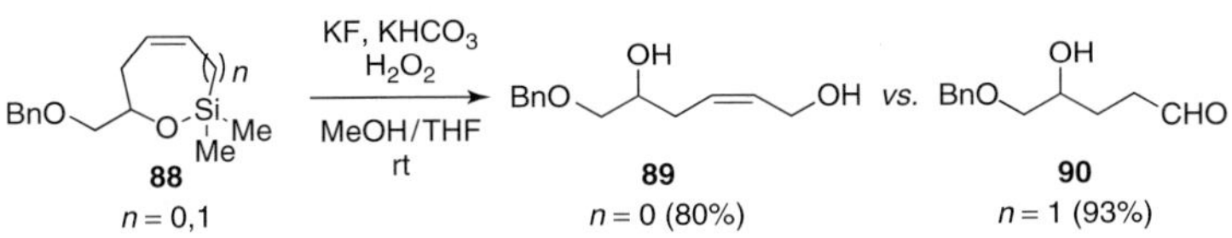

(8.3)

In additional studies, this work also demonstrated the oxidative cleavage of these silacycles using the Tamao and Fleming oxidation (Equation 8.3). Interestingly, the allylsilane **88** ($n = 1$) provided the (*Z*)-1,5-diol **89**, whereas the vinylsilane **88** ($n = 0$) furnished the γ-hydroxy aldehyde **90** [39].

8.2.4.1 **Lignans from Allylsiloxanes**

Cossy and Meyer independently described a similar TST-RCM of mono-alkoxysilanes **91** for the preparation of a variety of allylsiloxanes **92** (Scheme 8.17) [40]. Although the results were analogous to those reported by Grubbs, this study demonstrated that the larger rings required more forcing reaction conditions and that allylic substituents significantly reduce the overall efficiency, illustrating the sensitivity of catalyst [Ru]-**I** to steric effects. Nonetheless, a striking feature of this work is the ability to combine the TST-RCM sequence with a Sakurai reaction using aldehydes and ketones to facilitate the stereoselective construction of tri- and tetrasubstituted THF's and THP's **93/94** (n = 1–2). In concurrent studies, Marsden *et al.* reported an analogous reaction sequence, albeit using a different Lewis acid, in a study that focused entirely on the construction of trisubstituted THF's **93/94** ($n = 1$) [41]. Interestingly, these studies demonstrated that the Sakurai reaction can be effected with either boron trifluoride etherate or trimethylsilyl triflate, wherein straight chain alkyl aldehydes provide optimal stereocontrol over

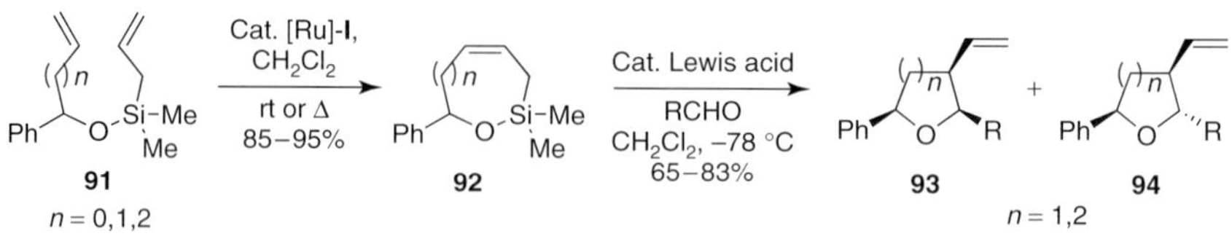

Scheme 8.17 Stereoselective construction of cyclic ethers using the TST-RCM in conjunction with the Sakurai reaction.

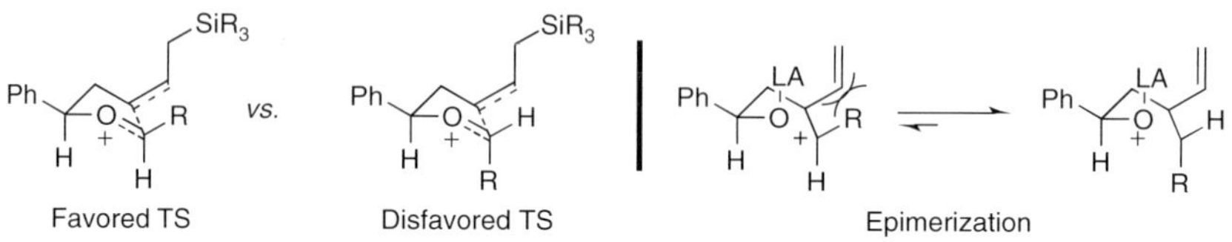

Figure 8.11 Origin of selectivity and equilibration in the Sakurai reactions.

branched alkyl and aryl derivatives (*dr* = 3.3 : 1–49 : 1). Figure 8.11 illustrates the model for stereocontrol, which involves the initial condensation of silyl ether with the aldehyde to provide the (*E*)-oxonium ion that undergoes cyclization through a chair-like TS (favored TS). Marsden *et al.* demonstrated that the formation of the minor diastereoisomer is a result of the ionization of the cyclic ether, which places the C2 substituent *pseudo*-axial to avoid the *gauche* interaction with the vinyl group. This insight explained that the extent of equilibration could be attributed to the reaction temperature and the nature of the Lewis acid, which provided reaction conditions that facilitate the selective formation of either diastereoisomer [41b].

Additional studies by Marsden *et al.* demonstrated the versatility of this strategy for the construction of aryltetralins and lignans [42]. The highlight of this work was a concise total synthesis of the lignan (+)-virgatusin, which is a 2,5-diaryl-3,4-di(alkoxymethyl)THF and the enantiomer of the natural product (Scheme 8.18). Although Cossy and Meyer had demonstrated that this substitution pattern undergoes the Sakurai reaction, they also demonstrated that the presence of allylic substituents were challenging for the RCM [40]. Treatment of the *syn*-aldol adduct **95** with allyldimethylsilyl chloride and imidazole afforded the allylsiloxane, which was subjected to RCM using Grubbs's first-generation catalyst [Ru]-**I** to provide the allylsiloxane **96** in 70% overall yield. The origin of the improved efficiency of the metathesis reaction can be ascribed to the *anti*- versus the *syn*-stereochemical arrangement in the product, which minimizes unfavorable *gauche* interactions. Treatment of the cyclic allylsiloxane **96** with piperonal in the presence trimethylsilyl triflate at −78 °C furnished the tetrasubstituted THF **97** in 63% yield as an inseparable mixture of diastereoisomers (*dr* = 10 : 1). Conversion of this intermediate to (+)-virgatusin (**98**) using a four-step sequence provided the enantiomeric natural product with a significantly reduced diastereomeric ratio (*dr* = 3 : 1), which was attributed to chromatographic resolution rather than equilibration.

Rodríguez-Garcia *et al.* extended the TST-RCM to the pterocarpans, which have a *cis*-fused benzofuran–benzopyran skeleton (Scheme 8.19) [43]. These agents are

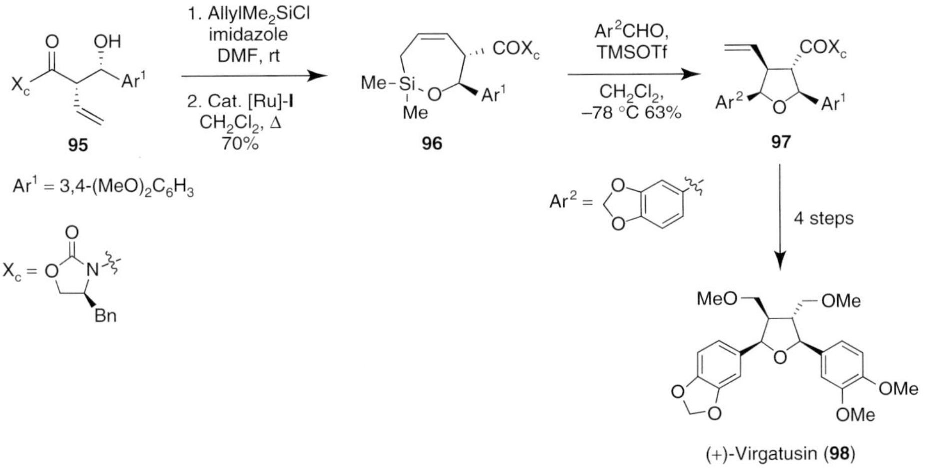

Scheme 8.18 Concise total synthesis of (+)-virgatusin using the TST-RCM and Sakurai reaction.

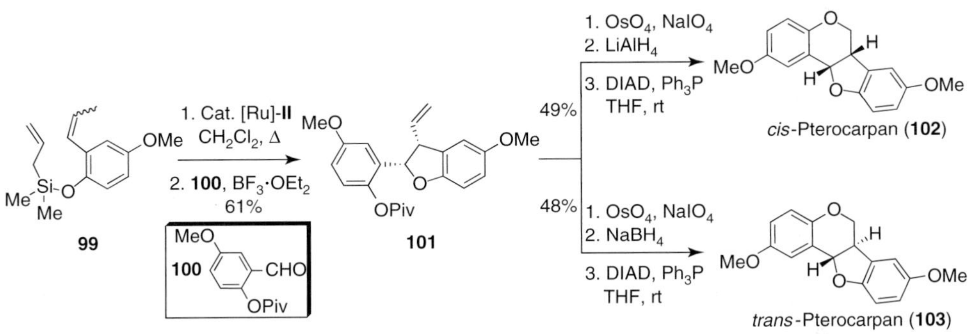

Scheme 8.19 Stereoselective synthesis of *cis*- and *trans*-pterocarpans.

part of a large class of isoflavonoids that have an array of biological activity, which makes them attractive synthetic targets. Although the synthetic strategy is directly analogous to the work of Cossy and Marsden, there are a number of interesting observations that enhance the understanding of this process [40, 41]. For example, the RCM of the allylsiloxane **99** proceeds more efficiently with catalyst [Ru]-**II** and the Sakurai condensation of the resulting cyclic allylsiloxane with the aldehyde **100** in the presence of boron trifluoride etherate furnished the *cis*-2,3-disubstituted benzofuran **101** with excellent stereocontrol ($dr \geq 19:1$) irrespective of the reaction temperature. Although the improved efficiency of the RCM is expected with the newer catalyst; the level of stereocontrol is somewhat intriguing given the work of Marsden *et al.*, which demonstrated that these derivatives are prone to ionization and equilibration [41b]. The synthesis was completed through a three-step sequence that culminates with a Mitsunobu cyclization of the hydroxyphenol derived from

101 *via* the oxidative cleavage of the vinyl group and reduction of the aldehyde with concomitant reductive cleavage of the pivaloyl group to afford the *cis*- and *trans*-pterocarpans (**102**) and (**103**) in modest overall yield. Although the oxidative cleavage of the vinyl group does not equilibrate the aldehyde, the chromatographic purification leads to a mixture of epimers favoring the *trans*-diastereoisomer ($dr = 6:1$). The epimerization can be avoided by reducing the crude aldehyde, which provides a stereodivergent synthesis. Finally, the origin of the stereocontrol was rationalized by the direct allylation of the aldehyde through a closed TS, followed by displacement of the secondary carbinol with the phenoxide, which is similar to the mechanism initially proposed by Marsden *et al.* [41a].

8.2.4.2 Z-Trisubstituted Alkenes from Allylsiloxanes

Barrett *et al.* employed the TST-RCM strategy to prepare a series of C19 analogs from 1α-hydroxy-5,6-*trans*-vitamin D_2 (Figure 8.12) [44]. Although the hormonally active form of vitamin D_3 promotes cell differentiation and inhibits tumor cell proliferation, it also leads to hypercalcemia at therapeutically significant dose levels. Interestingly, several studies have demonstrated that structural changes to this agent provide analogs that discriminate between these activities, albeit there are relatively few reports on C19 modifications.

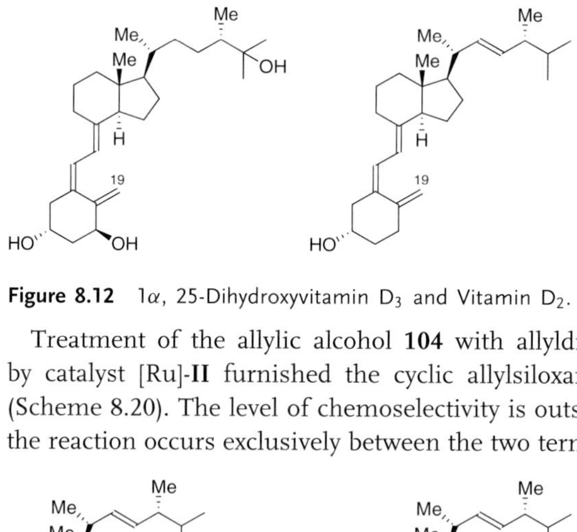

Figure 8.12 1α, 25-Dihydroxyvitamin D_3 and Vitamin D_2.

Treatment of the allylic alcohol **104** with allyldimethylsilyl chloride followed by catalyst [Ru]-**II** furnished the cyclic allylsiloxane **105** in good overall yield (Scheme 8.20). The level of chemoselectivity is outstanding in this process, since the reaction occurs exclusively between the two terminal alkenes of the pentaene,

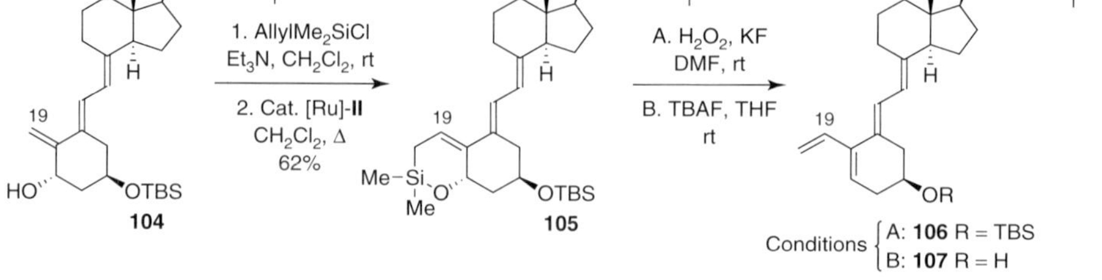

Scheme 8.20 Synthesis of vitamin D_2 analogs *via* the TST-RCM.

which is presumably the result of group differentiation in the initiation step. This study also examined the merit of the phosphonate tethers, which behave analogously in the RCM. Unfortunately, the attempted Tamao–Fleming oxidation of the allylsiloxane **105** to the diol provided the pentaene **106**, whereas the treatment of **105** with TBAF furnished the deprotected pentaene **107** in 50% and 64% yield, respectively, through a preferential Peterson olefination. Finally, the product pentaene **107** provides a versatile intermediate for cycloaddition reactions to demonstrate the utility of this strategy, despite the fact that it also highlights the limitations with the functionalization of trialkysiloxanes.

Miller and Li utilized the TST-RCM reaction to overcome the poor selectivity that had previously been obtained in the installation of the challenging exocyclic ethylidene using a Wittig reaction in the total syntheses of (+)-streptazolin (Scheme 8.21) [45]. This alkaloid, isolated by Drautz and Zahner from cultures of *Streptomyces viridochromogenes*, has both antibiotic and antifungal activity and possesses an unusual skeleton due to an internal carbamate unit, which requires the partial hydrogenation to allow purification and characterization due to its propensity to undergo polymerization. Preliminary studies demonstrated that the internal urethane unit impacts the ability to affect the oxidative cleavage of the cyclic allylsiloxane in a manner similar to that described by Barrett *et al.* for the vitamin D_2 analogs [44]. Furthermore, the formation of the unsymmetrical *bis*-alkoxysilane led to the formation of dimers with either [Ru]-**I** or [Ru]-**II** catalysts, which was attributed to a less ordered TS. Saponification of the secondary allylic acetate **108** followed by the silylation of the resulting allylic alcohol with allyldimethylsilyl chloride furnished the allylsiloxane **109** in 74% overall yield. RCM using catalyst [Ru]-**II**, furnished the six-membered ring-fused silacycle **110** in quantitative yield, which was treated with potassium fluoride and $KHCO_3$ to afford the exocyclic ethylidene **111** in 50% yield. Although the yield is somewhat modest, this represents one of a few examples of the protodesilylation of a conjugated and fused allylsilane.

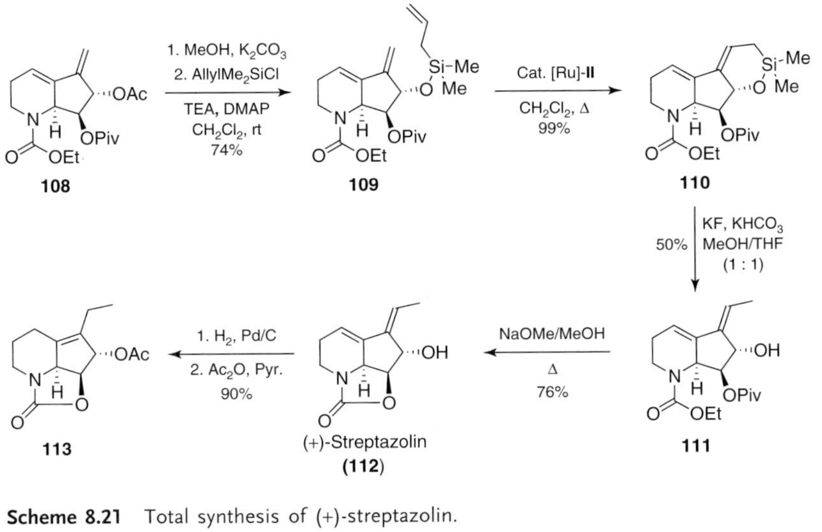

Scheme 8.21 Total synthesis of (+)-streptazolin.

Interestingly, the oxidation of the allylsiloxane **110** provided the allylic alcohol in 55% yield, however the conversion to **111** proved problematic. The synthesis was completed with the removal of the pivaloyl group in **111** followed by *in situ* carbamate formation to afford (+)-streptazolin (**112**) in 76% yield, which was converted to the crystalline allylic acetate **113** for characterization.

The phytotoxic agents cornexistin and 14-hydroxycornexistin were independently isolated from the fungus *Paecilomyces variotii* Bainier strain SANK 21086, and are members of the nonadride family that are biosynthetically derived from the combination of nine maleic anhydride groups (Figure 8.13). These agents display potent herbicidal activity at the same concentration against a number of grass and broadleaf weed species with minimal phytotoxicity toward maize, making them significantly important as potential herbicides. Additionally, cornexistin and 14-hydroxycornexistin display complementary activity, in which the former tends to be more active against grass weeds, while the latter is significantly more active against the broadleaf weeds.

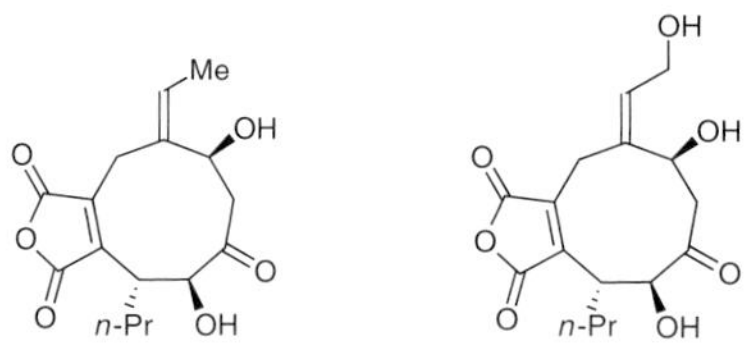

Figure 8.13 Cornexistin and 14-hydroxycornexistin.

These features inspired Taylor *et al.* to develop a unified strategy for the preparation of these agents using the TST-RCM reaction to install the critical *Z*-trisubstituted alkene at C7 (Scheme 8.22) [46]. Interestingly, the relative stereochemistry at C8 is critical for the success of this strategy, since the allylsiloxane derived from the secondary alcohol **114** furnished dimeric products irrespective of the catalyst employed. Inversion of the C8 stereogenic center using a two-step oxidation/reduction sequence provided the epimer **115** (*dr* = 4 : 1), whereas the attempted Mitsunobu reaction led to elimination. Treatment of the secondary alcohol **115** with allyldimethylsilyl chloride provided the allylsiloxane, and subsequent exposure to [Ru]-**II** catalyst gave the cyclic allylsiloxane **116** in modest overall yield due to the relatively modest silylation yield (50%). Although the attempted protodesilylation of **116** required for cornexistin was unsuccessful due to the lack of reactivity, the Tamao–Fleming oxidation afforded the diol **117** in quantitative yield.

8.2.4.3 Di- and Trisubstituted Alkenes from Vinylsiloxanes

The advantage of the TST-RCM strategy is the ability to rationally prepare sterically hindered and inactivated *Z*-olefins, which remain the most challenging substrates for CM. Barrett *et al.* outlined a series of novel ring-opening reactions that permit the stereoselective construction of *E*- and *Z*-olefins (Scheme 8.23) [47]. The vinylsiloxane **118** was prepared in a manner analogous to that described by Grubbs using the Schrock catalyst [Mo]-**I** in excellent yield [39].

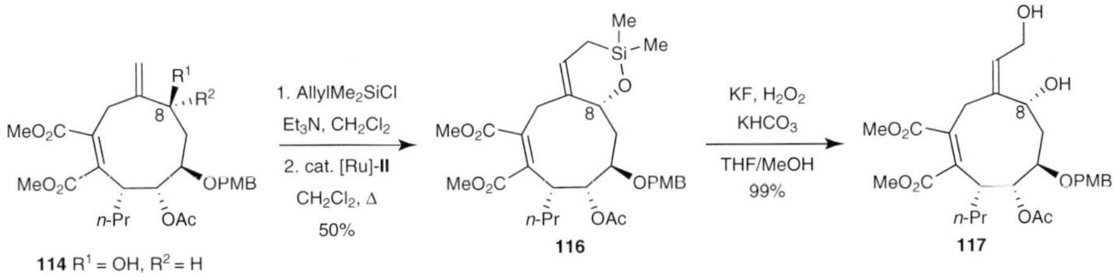

114 R^1 = OH, R^2 = H
115 R^1 = H, R^2 = OH

Scheme 8.22 Stereoselective synthesis of Z-trisubstituted alkene in cornexistin and 14-hydroxycornexistin using the TST-RCM strategy.

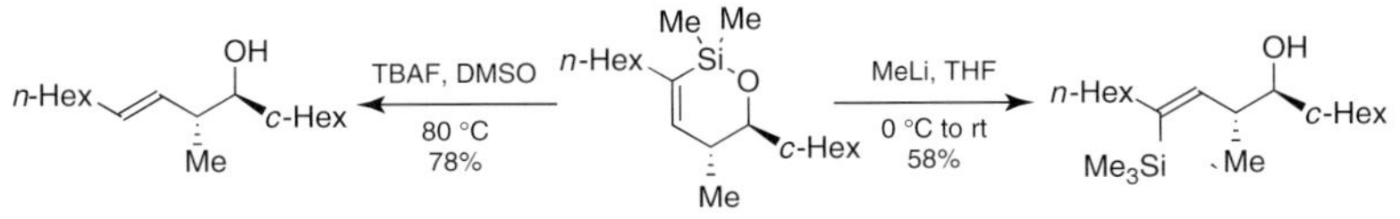

119 118 120

Scheme 8.23 Stereoselective construction of E- and Z-homoallylic alcohols.

Treatment of the vinylsiloxane **118** with TBAF in DMSO at 80 °C furnished the *E*-homoallylic alcohol **119**, whereas the analogous substrate with methyllithium (or other organolithium reagents) afforded the *Z*-homoallylic alcohol **120** as the exclusive products.

Barrett *et al.* employed the organolithium ring-opening strategy for the synthesis of *E*-alkene in a total synthesis of D,L-glucosylceramide, which is a member of the glycosphingolipid class of natural products (Scheme 8.24) [48]. These agents are associated with formation of many structures in biological systems, as they provide for the shape and support for cell membranes. Hence, they are intimately involved in cellular recognition, which has implications for many biological processes, including cancer, allergy, and inflammation. The preparation of these agents present several challenges, the most notable being the regio- and stereoselective construction of the polysaccharide unit through the *O*-glycosylation and the stereoselective installation of the *E*-alkene. The examination of the TST-RCM was prompted after preliminary studies demonstrated that CM, was unsuccessful due to the presence of an allylic substituent [10]. The acyclic vinylsiloxane **122** was prepared from the secondary alcohol **121** using the silyl triflate formed *in situ* from the silyl chloride with silver triflate. Treatment of the vinylsiloxane **122** with catalyst [Mo]-**I** followed by phenyllithium afforded the *Z*-vinylsilane **123** in 83% overall yield, which was converted to D,L-glucosylceramide (**124**) using a four-step sequence. The ability to chemoselectively manipulate the cyclic vinylsiloxane intermediate in the presence of additional silyl ethers illustrates the inherent versatility of this strategy.

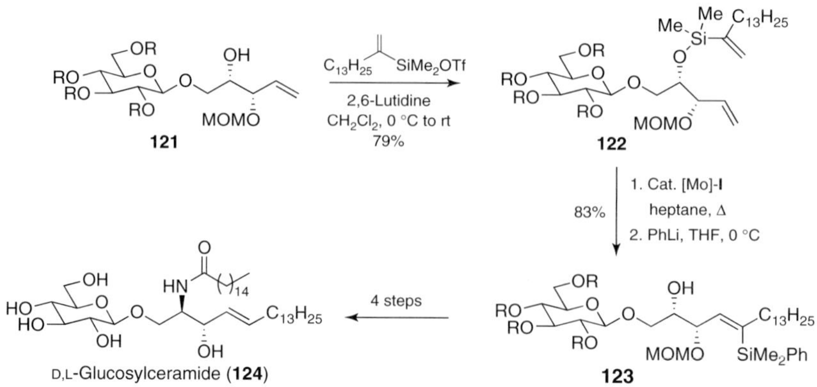

Scheme 8.24 Stereoselective synthesis of D,L-glucosylceramide (RO = TBSO).

Denmark and Yang demonstrated that vinylsiloxanes participate in a series of intermolecular silicon-assisted metal-catalyzed cross-couplings [49a,c]. Interestingly, the RCM of the alkoxysilane **125** (n = 0–2) with [Mo]-**I** furnished the vinylsiloxane **126**, whereas the analogous process with Grubbs's second-generation catalyst [Ru]-**II** was ineffective (Scheme 8.25). Sequential treatment of the vinylsiloxane **126** with TBAF followed by the requisite aryl halide and the palladium catalyst furnished the *Z*-alkenyl alcohols **127** in excellent yield. Interestingly, electron-poor and electron-rich aryl halides exhibit similar reactivity in the cross-coupling reactions and the steric effect from *ortho*-substituents is minimal, albeit for groups that are able to chelate the metal to slow the rate of reductive elimination. Finally, this process is general for five- through seven-membered vinylsiloxanes (n = 0–2), albeit the α-substituted derivatives (R^1) undergo cross-coupling with significantly reduced reaction rates.

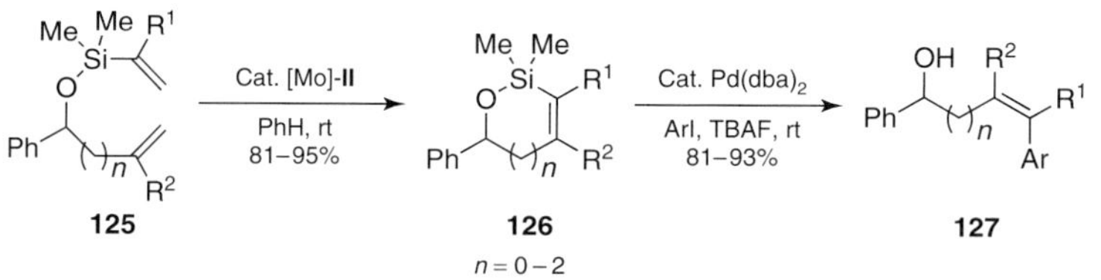

Scheme 8.25 Palladium-mediated cross-coupling of vinylsiloxanes.

Denmark and Yang also reported the intramolecular variant, which provides a novel approach to the construction of a variety of medium ring carbocycles that contain a homoannular 1,3-diene (Equation 8.4) [49b,c]. This is a particularly interesting transformation given that cyclization reactions that afford medium-sized rings are generally disfavored entropically. The cyclic vinylsiloxanes **128** were prepared in a manner analogous to that described earlier using [Mo]-**I** catalyst, albeit with a pendant vinyl iodide. Preliminary studies demonstrated that conditions that generally favor the intra- over the intermolecular reaction manifold, namely, high dilution or slow addition, were not optimal for this process. Additional studies demonstrated that the quantity of TBAF and the palladium catalyst were critical for optimal yields. The intramolecular palladium-catalyzed cross-coupling of the vinylsiloxanes **128** ($n = 1$, $m = 0–6$) in the presence of excess TBAF (10 equiv.) furnished the carbocyclic alcohols **129** in 55–72% yield.

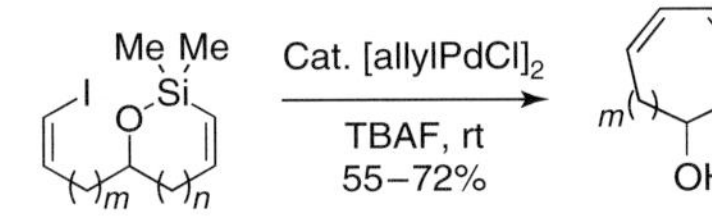

128
$m = 0–6$, $n = 1$

129

(8.4)

Additional studies demonstrated that the TST-RCM in conjunction with the intramolecular silicon-assisted metal-catalyzed cross-coupling could be readily adapted to prepare cyclic ethers, which have proven extremely challenging to prepare using conventional methods. The process provided the key transformation in the total synthesis of the antifeedant (+)-brasilenyne, which is a nine-membered homodiene containing cyclic ether that was isolated from the digestive gland of the sea hare *Aplysia brasiliana* (Scheme 8.26) [50]. Silylation of homoallylic alcohol **130** followed by RCM with [Mo]-**I** catalyst furnished the cyclic vinylsiloxane **131** in 84% overall yield. Intramolecular silicon-assisted palladium-catalyzed cross-coupling of the vinylsiloxane **131** in the presence of excess TBAF afforded the cyclic ether **132** in 61% yield, which was converted to (+)-brasilenyne (**133**) in an additional six steps.

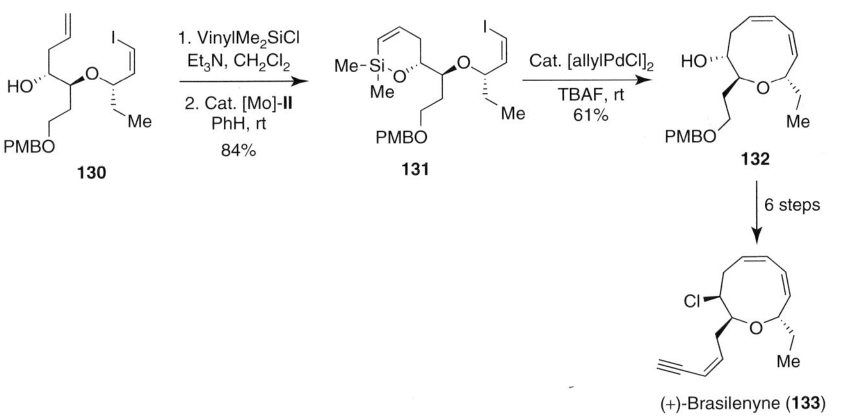

Scheme 8.26 Total synthesis of (+)-brasilenyne.

Vilarrasa *et al.* combined the aforementioned methodologies to install the *E*-trisubstituted alkene present in the cytotoxic macrolide amphidinolide X, which was isolated from *Amphidinium dinoflagellate* and possesses an unusual asymmetric diolide structure (Scheme 8.27) [51]. Although it was envisioned that the challenging trisubstituted C12–C13 olefin could be introduced *via* RCM, the attempted cyclization failed to directly provide the 16-membered macrocycle with a variety of

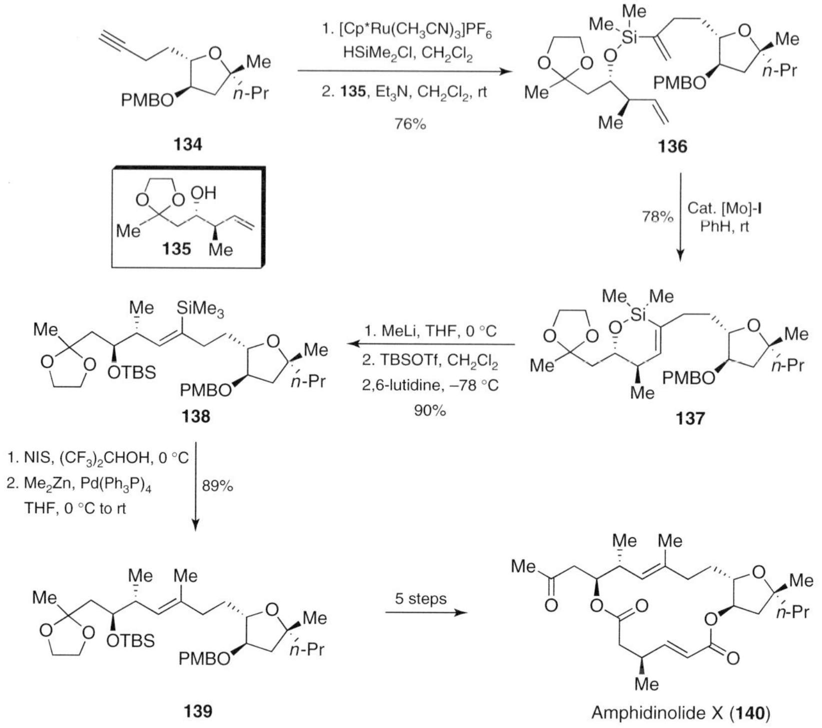

Scheme 8.27 Total synthesis of amphidinolide X.

catalysts and substrates with stereochemical variations. This prompted the introduction of the challenging trisubstituted alkene using the TST-RCM strategy in a fragment-coupling step. Hydrosilylation of the alkyne **134** with chlorodimethylsilane and Trost catalyst $[Cp^*Ru(CH_3CN)_3]PF_6$ afforded the triorganochlorosilane, which was coupled with the secondary alcohol **135** to provide the acyclic vinylsiloxane **136** in 76% overall yield. Treatment of the acyclic vinylsiloxane **136** with Schrock catalyst [Mo]-**I** provided the cyclic variant **137** in 78% yield. The direct conversion of the vinylsiloxane **137** into the trisubstituted alkene **139** proved more problematic than anticipated, which prompted the development of a novel reaction sequence. Ring opening of the vinylsiloxane **137** with methyllithium followed by the protection of the secondary alcohol as the *tert*-butyldimethylsilyl ether afforded the vinylsilane **138** in excellent yield. Iododesilylation of the vinylsilane **138** with *N*-iodosuccinimide in $(CF_3)_2CHOH$ followed by a Negishi coupling with dimethylzinc afforded the *E*-trisubstituted alkene **139** in 89% overall yield. The elaboration of this fragment using a five-step sequence completed the total synthesis of amphidinolide X (**140**). Overall, this provides another excellent example of the successful manipulation of a cyclic vinylsiloxane to afford substituted alkenes in a highly selective manner.

8.2.4.4 Related Applications and Developments

In related studies, the TST-RCM of allylsiloxanes was featured as a homologation step in a novel approach to *trans*-fused oxepane polyethers [52]. Additionally, the enantioselective TST-RCM of prochiral alkenes has been developed using molybdenum and ruthenium chiral complexes with excellent enantioselectivities [53]. Finally, Yao developed a polymer-bound ruthenium carbene complex, which was both robust and recyclable enough to be utilized for the TST-RCM of eight-membered allylsiloxanes after being employed in an alternative RCM reaction [54]. Despite the advent of several new catalysts, Schrock catalyst remains the optimal metal-complex for vinylsiloxane RCM.

8.2.5
Enyne TST-RCM: Tri- and Tetrasubstituted Acyclic Dienes

Yao described the stereoselective construction of acyclic tri- and tetrasubstituted 1,3-dienes using the enyne TST-RCM to circumvent the problems with controlling chemo-, regio- and stereoselectivity (Scheme 8.28) [55]. Treatment of the enyne siloxane **141** with catalyst [Ru]-**I** furnished the diene **142**, which was subjected to an oxidative cleavage to afford the diene **143** in excellent overall yield. Interestingly, this study demonstrates that a variety of substituents are tolerated, particularly on the alkyne (R^2), in which the most challenging examples utilize ethylene gas to overcome low reactivity that extends the reaction times. Additional studies demonstrated that the method was also applicable to the preparation of seven-membered siloxanes and a tandem dienyne metathesis under high dilution conditions to provide a tetrasubstituted diene with Grubbs's second-generation catalyst [Ru]-**II**. Although Lee *et al.* have continued to extend this methodology, it has still not been applied to natural product synthesis [56].

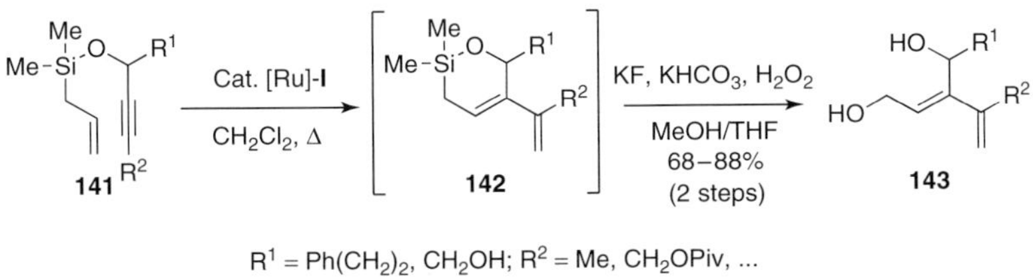

141 **142** **143**

Scheme 8.28 Enyne TST-RCM of enynesiloxanes.

8.2.5.1 Illudins

Movassaghi *et al.* employed this strategy to prepare the cyclohexadiene moiety present in the semisynthetic analogs of illudin S, namely, (−)-acylfulvene and (−)-irofulven (Figure 8.14) [57]. The semisynthetic analogs were developed to overcome the severe toxicity exhibited by the naturally occurring agents. For example, (−)-irofulven is currently in phase II clinical trials for the treatment of ovarian, prostrate, and thyroid cancer. Although these targets were ultimately prepared using a *bis*-alkoxyalkynylsilane enyne TST-RCM, this example serves

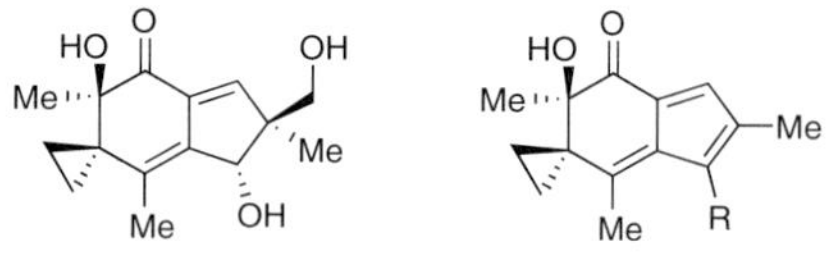

Figure 8.14 Illudin S, (−)-acylfulvene (R = H), and (−)-irofulven (R = CH_2OH).

to demonstrate the potential for total synthesis despite the inherent limitations associated with the functionalization of the siloxanes in complex molecules.

Treatment of the enynesiloxane **144**, which had been prepared through the silylation of the corresponding propargylic alcohol, with catalyst [Ru]-**II** afforded the cyclic diene **145** in 52–76% yield (Scheme 8.29). A critical feature of this approach is that it provides the 1,3-diene motif in good yield, and it is tolerant to a variety of substituents: for example, alkenes, hydroxyesters, and alkyl iodides. Unfortunately, the dienes **145** were particularly sensitive to chromatography on silica gel and alumina, which is evident from the attempted protodesilylation, which led to significant decomposition. Although the oxidation of the dienylsiloxane **145** provides the triol **146**, the poor yield of this step in addition to the sensitivity of the substrates prompted the examination of the *bis*-alkoxyalkynylsilane to circumvent the oxidation reaction and thus provide a more viable route for the synthesis of these agents.

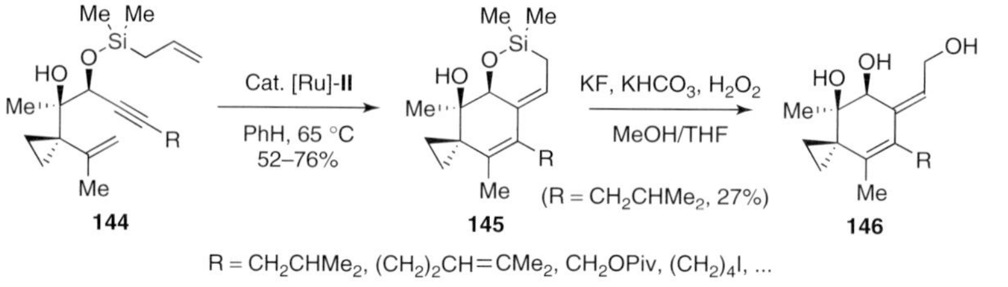

Scheme 8.29 Enyne TST-RCM for the synthesis of (−)-acylfulvene and (−)-irofulven.

8.3
Conclusions and Outlook

The TST-RCM provides a versatile synthetic strategy for target-directed synthesis as exemplified by the examples outlined herein. The TST-RCM provides homo- and cross-coupling reactions in both a stereospecific and stereoselective fashion, which is particularly useful for the construction of *E*- and *Z*-substituted alkenes, in addition to dienes. The advent of more reactive RCM catalysts that have exquisite functional group tolerance has increased the versatility of this process, which will dramatically increase the number of applications to the challenging motifs present in total synthesis applications. Furthermore the recent development of the enantioselective TST-RCM is likely to provide a very useful method for the asymmetric synthesis of challenging alkene-containing substrates for target-directed synthesis. The ability to circumvent the limitations associated with the assembly and functionalization of the unsymmetrical *bis*-alkoxyalkynylsilanes and the organosiloxanes, respectively, will inspire the development of more creative solutions to these problems. Finally, in the absence of catalysts that can control the formation of *E*- and *Z*-alkenes, particularly *Z*-di- and *E*/*Z*-trisubstituted versions, the TST-RCM will provide a versatile alternative that will continue to gain favor in natural product synthesis.

Acknowledgments

The author would like to acknowledge the intellectual and experimental contributions of his former coworkers in the metathesis subgroup. In particular, he would like to highlight the contributions of Jian Cui, Srinivasa Murthy, and Alexei Polosukhin. He is also grateful to Alen Cusak for his assistance in the preparation of this chapter. Finally, the National Institutes of Health (GM54623) is thanked for generous financial support, in addition to the numerous pharmaceutical companies for unrestricted funds.

References

1 For a recent review on disposable tethered transformations, see: Gauthier, D.R. Jr., Zandi, K.S., and Shea, K.J. (1998) *Tetrahedron*, **54**, 2289–2338.

2 (a) Bender, M.L. and Neveu, M.C. (1958) *J. Am. Chem. Soc.*, **80**, 5388–5391; (b) Page, M.I. (1973) *Chem. Soc. Rev.*, **2**, 295–323; (c) Kirby, A. (1980) *J. Adv. Phys. Org. Chem.*, **17**, 183–278.

3 (a) Nishiyama, H., Kitajima, T., Matsumoto, M., and Itoh, K. (1984) *J. Org. Chem.*, **49**, 2298–2300; (b) Stork, G. and Kahn, M. (1985) *J. Am. Chem. Soc.*, **107**, 500–501.

4 For a review on temporary silicon-tethered reactions, see: (a) Bols, M. and Skrydstrup, T. (1995) *Chem. Rev.*, **95**, 1253–1277; (b) Fensterbank, L., Malacria, M., and Sieburth, S.McN. (1997) *Synthesis*, 813–854.

5 For a direct comparison of the temporary silicon tethered strategy to enzyme catalyzed reaction, see: Xin, Y.C., Mallet, J.-M., and Sinaÿ, P. (1993) *J. Chem. Soc., Chem. Commun.*, 864–865.

6 Fersht, A. (1999) *Structure and Mechanism in Protein Science*, W. H. Freeman and Co., New York.

7 For selective deprotection of silyl ethers see: (a) Nelson, T.D. and Crouch, R.D. (1996) *Synthesis*, **9**, 1031–1069; (b) Crouch, R.D. (2004) *Tetrahedron*, **60**, 5833–5871.

8 Brook, M.A. (2000) *Silicon in Organic, Organometallic and Polymer Chemistry*, Wiley-Interscience, New York.

9 For review on olefin metathesis see: (a) Grubbs, R.H., Miller, S.J., and Fu, G.C. (1995) *Acc. Chem. Res.*, **28**, 446–452; (b) Schuster, M. and Blechert, S. (1997) *Angew. Chem. Int. Ed.*, **36**, 2036–2056; (c) Grubbs, R.H. and Chang, S. (1998) *Tetrahedron*, **54**, 4413–4450; (d) Armstrong, S.K. (1998) *J. Chem. Soc., Perkin Trans. 1*, 371–388; (e) Blechert, S. (1999) *Pure Appl. Chem.*, **71**, 1393–1399; (f) Phillips, A.J. and Abell, A.D. (1999) *Aldrichim. Acta*, **32**, 75–89; (g) Fürstner, A. (2000) *Angew. Chem. Int. Ed.*, **39**, 3012–3043; (h) Schrock, R.R. and Hoveyda, A.H. (2003) *Angew. Chem. Int. Ed.*, **42**, 4592–4633; (i) Deiters, A. and Martin, S.F. (2004) *Chem. Rev.*, **104**, 2199–2238; (j) Nicolaou, K.C., Bulger, P.G., and Sarlah, D. (2005) *Angew. Chem. Int. Ed.*, **44**, 4490–4527; and pertinent references therein.

10 Chatterjee, A.K., Choi, T.-L., Sanders, D.P., and Grubbs, R.H. (2003) *J. Am. Chem. Soc.*, **125**, 11360–11370.

11 Fu, G.C. and Grubbs, R.H. (1992) *J. Am. Chem. Soc.*, **114**, 5426–5427.

12 Levisalles, J., Rudler, H., and Villemin, D. (1979) *J. Organomet. Chem.*, **164**, 251–257.
13 Evans, P.A. and Murthy, V.S. (1998) *J. Org. Chem.*, **63**, 6768–6769.
14 Amador, M., Ariza, X., Garcia, J., and Ortiz, J. (2004) *J. Org. Chem.*, **69**, 8172–8175.
15 Hoye, T.R. and Promo, M.A. (1999) *Tetrahedron Lett.*, **40**, 1429–1432.
16 Lobbel, M. and Köll, P. (2000) *Tetrahedron: Asymmetry*, **11**, 393–396.
17 (a) Huang, J., Stevens, E.D., Nolan, S.P., and Petersen, J.L. (1999) *J. Am. Chem. Soc.*, **121**, 2674–2678; (b) Scholl, M., Trnka, T.M., Morgan, J.P., and Grubbs, R.H. (1999) *Tetrahedron Lett.*, **40**, 2247–2250; (c) Briot, A., Bujard, M., Gouverneur, V., Nolan, S.P., and Mioskowski, C. (2000) *Org. Lett.*, **2**, 1517–1519.
18 Scholl, M., Ding, S., Lee, C.W., and Grubbs, R.H. (1999) *Org. Lett.*, **1**, 953–956.
19 Evans, P.A. and Murthy, V.S. (1999) *Tetrahedron Lett.*, **40**, 1253–1256.
20 Boiteau, J.-G., Van de Weghe, P., and Eustache, J. (2001) *Tetrahedron Lett.*, **42**, 239–242.
21 (a) Gierasch, T.M., Chytil, M., Didiuk, M.T., Park, J.Y., Urban, J.J., Nolan, S.P., and Verdine, G.L. (2000) *Org. Lett.*, **2**, 3999–4002; (b) Harrison, B.A. and Verdine, G.L. (2001) *Org. Lett.*, **3**, 2157–2159; (c) Harrison, B.A., Gierasch, T.M., Neilan, C., Pasternak, G.W., and Verdine, G.L. (2002) *J. Am. Chem. Soc.*, **124**, 13352–13353.
22 Evans, P.A., Cui, J., and Buffone, G.P. (2003) *Angew. Chem. Int. Ed.*, **42**, 1734–1737.
23 Van de Weghe, P., Aoun, D., Boiteau, J.-G., and Eustache, J. (2002) *Org. Lett.*, **4**, 4105–4108.
24 Marjanovic, J. and Kozmin, S.A. (2007) *Angew. Chem. Int. Ed.*, **46**, 8854–8857.
25 Hooper, A.M., Dufour, S., Willaert, S., Pouvreau, S., and Pickett, J.A. (2007) *Tetrahedron Lett.*, **48**, 5991–5994.
26 Casey, E.M., Teesdale-Spittle, P., and Harvey, J.E. (2008) *Tetrahedron Lett.*, **49**, 7021–7023.
27 Evans, P.A., Cui, J., Gharpure, S.J., Polosukhin, A., and Zhang, H.-R. (2003) *J. Am. Chem. Soc.*, **125**, 14702–14703.
28 Brown, L.J., Spurr, I.B., Kemp, S.C., Camp, N.P., Gibson, K.R., and Brown, R.C.D. (2008) *Org. Lett.*, **10**, 2489–2492.
29 Hoye, T.R., Eklov, B.M., Jeon, J., and Khoroosi, M. (2006) *Org. Lett.*, **8**, 3383–3386.
30 Gaich, T. and Mulzer, J. (2005) *Org. Lett.*, **7**, 1311–1313.
31 Van de Weghe, P., Bourg, S., and Eustache, J. (2003) *Tetrahedron*, **59**, 7365–7376.
32 Hara, O., Sugimoto, K., and Hamada, Y. (2004) *Tetrahedron*, **60**, 9381–9390.
33 Postema, M.H.D. and Piper, J.L. (2002) *Tetrahedron Lett.*, **43**, 7095–7099.
34 (a) Diver, S.T. and Giessert, A.J. (2004) *Chem. Rev.*, **104**, 1317–1382; (b) Maifeld, S.V. and Lee, D. (2005) *Chem. Eur. J.*, **11**, 6118–6126; (c) Van de Weghe, P., Bisseret, P., Blanchard, N., and Eustache, J. (2006) *J. Organomet. Chem.*, **691**, 5078–5108; (d) Chattopadhyay, S.K., Karmakar, S., Biswas, T., Majumdar, K.C., Rahaman, H., and Roy, B. (2007) *Tetrahedron*, **63**, 3919–3952.
35 Hansen, E.C. and Lee, D. (2006) *Acc. Chem. Res.*, **39**, 509–519.
36 (a) Grimm, J.B. and Lee, D. (2004) *J. Org. Chem.*, **69**, 8967–8970; (b) Grimm, J.B., Otte, R.D., and Lee, D. (2005) *J. Organomet. Chem.*, **690**, 5508–5516; (c) Kim, Y.J., Grimm, J.B., and Lee, D. (2007) *Tetrahedron Lett.*, **48**, 7961–7964.
37 (a) Kim, Y.J. and Lee, D. (2006) *Org. Lett.*, **8**, 5219–5222; (b) Kim, Y.J. and Lee, D. (2007) *Org. Lett.*, **9**, 3859.
38 Mukherjee, S. and Lee, D. (2009) *Org. Lett.*, **11**, 2916–2919.
39 Chang, S. and Grubbs, R.H. (1997) *Tetrahedron Lett.*, **38**, 4757–4760.
40 Meyer, C. and Cossy, J. (1997) *Tetrahedron Lett.*, **38**, 7861–7864.
41 (a) Cassidy, J.H., Marsden, S.P., and Stemp, G. (1997) *Synlett*, 1411–1413; (b) Miles, S.M., Marsden, S.P., Leatherbarrow, R.J., and Coates, W.J. (2004) *J. Org. Chem.*, **69**, 6874–6882.
42 (a) Miles, S.M., Marsden, S.P., Leatherbarrow, R.J., and Coates, W.J.

(2004) *Chem. Commun.*, 2292–2293; (b) Akindele, T., Marsden, S.P., and Cumming, J.G. (2005) *Org. Lett.*, **7**, 3685–3688.

43 (a) Jiménez-González, L., Álvarez-Corral, M., Muñoz-Dorado, M., and Rodríguez-García, I. (2005) *Chem. Commun.*, 2689–2691; (b) Jiménez-González, L., García-Muñoz, S., Álvarez-Corral, M., Muñoz-Dorado, M., and Rodríguez-García, I. (2007) *Chem. Eur. J.*, **13**, 557–568.

44 Ahmed, M., Atkinson, C.E., Barrett, A.G.M., Malagu, K., and Procopiou, P.A. (2003) *Org. Lett.*, **5**, 669–672.

45 Li, F. and Miller, M.J. (2006) *J. Org. Chem.*, **71**, 5221–5227.

46 Tung, J.C., Chen, W., Noll, B.C., Taylor, R.E., Fields, S.C., Dent, W.H. III, and Green, F.R. III (2007) *Synthesis*, 2388–2396.

47 Ahmed, M., Barrett, A.G.M., Beall, J.C., Braddock, D.C., Flack, K., Gibson, V.C., Procopiou, P.A., and Salter, M.M. (1999) *Tetrahedron*, **55**, 3219–3232.

48 Barrett, A.G.M., Beall, J.C., Braddock, D.C., Flack, K., Gibson, V.C., and Salter, M.M. (2000) *J. Org. Chem.*, **65**, 6508–6514.

49 (a) Denmark, S.E. and Yang, S.-M. (2001) *Org. Lett.*, **3**, 1749–1752; (b) Denmark, S.E. and Yang, S.-M. (2002) *J. Am. Chem. Soc.*, **124**, 2102–2103; (c) Denmark, S.E. and Yang, S.-M. (2004) *Tetrahedron*, **60**, 9695–9708.

50 (a) Denmark, S.E. and Yang, S.-M. (2002) *J. Am. Chem. Soc.*, **124**, 15196–15197; (b) Denmark, S.E. and Yang, S.-M. (2004) *J. Am. Chem. Soc.*, **126**, 12432–12440.

51 Rodríguez-Escrich, C., Urpí, F., and Vilarrasa, J. (2008) *Org. Lett.*, **10**, 5191–5194.

52 Brouard, I., Hanxing, L., and Martin, J.D. (2000) *Synthesis*, **6**, 883–892.

53 (a) Zhu, S.S., Cefalo, D.R., La, D.S., Jamieson, J.Y., Davis, W.M., Hoveyda, A.H., and Schrock, R.R. (1999) *J. Am. Chem. Soc.*, **121**, 8251–8259; (b) Kiely, A.F., Jernelius, J.A., Schrock, R.R., and Hoveyda, A.H. (2002) *J. Am. Chem. Soc.*, **124**, 2868–2869; (c) Funk, T.W., Berlin, J.M., and Grubbs, R.H. (2006) *J. Am. Chem. Soc.*, **128**, 1840–1846.

54 Yao, Q. (2000) *Angew. Chem. Int. Ed.*, **39**, 3896–3898.

55 Yao, Q. (2001) *Org. Lett.*, **3**, 2069–2072.

56 (a) Miller, R.L., Maifeld, S.V., and Lee, D. (2004) *Org. Lett.*, **6**, 2773–2776; (b) Maifeld, S.V., Miller, R.L., and Lee, D. (2004) *J. Am. Chem. Soc.*, **126**, 12228–12229; (c) Park, S., Kim, M., and Lee, D. (2005) *J. Am. Chem. Soc.*, **126**, 9410–9415; (d) Park, S. and Lee, D. (2007) *Synthesis*, 2313–2316.

57 Movassaghi, M., Piizzi, G., Siegel, D.S., and Piersanti, G. (2006) *Angew. Chem. Int. Ed.*, **45**, 5859–5863.

9
Metathesis Involving a Relay and Applications in Natural Product Synthesis

Thomas R. Hoye and Junha Jeon

9.1
Introduction

As each of the chapters in this monograph makes clearly evident, alkene (and alkyne) metathesis is one of the most powerful and versatile synthetic methodologies in all of organic chemistry. In this chapter discussion is focused on (ruthenium–alkylidene-mediated) ring-closing metathesis (RCM) reactions involving a *relay* process that is initiated at a remote alkene site within the substrate. The chapter is organized as follows: the concept is introduced in Section 9.1; the earliest discoveries and reports that established the viability of the strategy are presented in Section 9.2; uses of relay metathesis in settings not directly motivated by natural product synthesis are presented in Section 9.3; applications in natural product synthesis are discussed in Section 9.4; instructive examples in which the desired relay event is thwarted are presented in Section 9.5 and concluding comments are offered in Section 9.6.

9.1.1
The Relay Concept

RCM (**1** to **2**, Scheme 9.1) is one of the most common and important types of olefin metathesis reaction. Accommodation of considerable structural complexity within "A" in RCM substrates **1** is one of the hallmarks of this powerful and broadly applicable reaction. Nonetheless and despite significant advantages available by modification of external factors (for example, catalyst selection, additive use, solvent choice, or ethylene concentration), not every diene substrate **1** can be successfully cyclized to cycloalkene **2**. Because of the advent of *relay ring-closing metathesis* (RRCM, **4a** to **4b** to **3** to **2**, Scheme 9.1), chemists now have an alternative for resuscitating/redirecting these otherwise lifeless/recalcitrant systems when such a limitation is encountered.

More specifically, altering the parent diene substrate **1** by installation of an extender arm that includes a tether element "T" provides the opportunity to

Metathesis in Natural Product Synthesis: Strategies, Substrates and Catalysts. Edited by Janine Cossy, Stellios Arseniyadis, and Christophe Meyer

ISBN: 978-3-527-32440-8

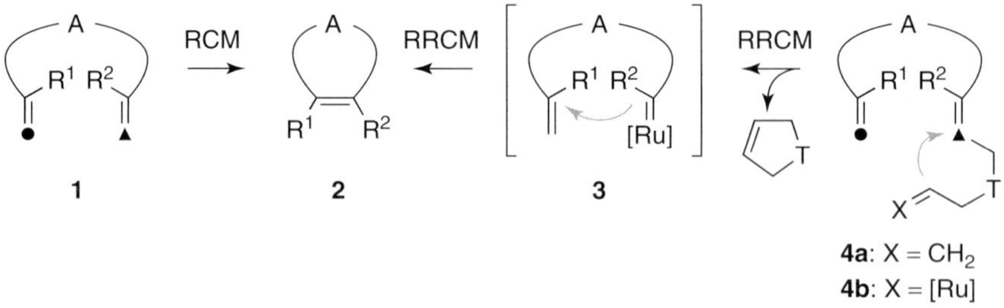

Scheme 9.1 Generic RCM (**1** to **2**) versus relay RCM (RRCM, **4a** to **2** via **4b** and **3**) processes.

1) load ruthenium metal and its associated ligands (designated herein as [Ru]) to a reactive terminal alkene (**4a** to **4b**);
2) relay [Ru] by an initial RCM event that ejects a disposable cycloalkene (generally five-membered) to give **3**; and
3) complete the cyclization to the desired product **2**. In practice, the relay tether (T) has been $C(CO_2R)_2$, O, CH_2, $OSiR_2$, or C(=O).

As the examples discussed in this chapter show, there is a variety of reasons why a relay-modified substrate might be used and/or why it might lead to a successful cyclization outcome.

9.1.2 Basic Tenets of RCM

A few fundamental principles that directly impact the level of efficiency associated with every specific metathesis reaction are worth stating at the outset of this discussion. As with any catalytic process, the rate of the primary reaction versus the lifetime of the working catalyst system – in effect, the turnover number (TON) – is a critical parameter. In the field of alkene/alkyne metathesis, there has been considerable evolution in the design and implementation of the catalysts (or precatalysts) that are used. Often, these advances have been driven by the desire to increase the TON by improving the longevity of the catalytic intermediates (catalyst robustness toward destructive side reactions) or speeding up the rate of the slow step(s) in each metathesis cycle (or both).

The rate-limiting event in substrate turnover is often the rate of loading of the metal onto a substrate alkene, although the subsequent rate of cyclization of the initially loaded species is sometimes the slowest stage. An alternative to catalyst (and reaction condition) alteration for improving either the TON and/or the selectivity of a given metathesis reaction is substrate modification. Of course, requirements dictated by the structure of the desired product place obvious constraints on this approach. On the other hand, two of the four alkene carbon atoms in a simple RCM substrate are expendable (see • and ▲ in **1**, Scheme 9.1), since they are absent in the product. This means that, in principle, those two terminal atoms can bear substituents other than hydrogen (and, correspondingly, that the by-product of the reaction can be a second alkene other than ethylene). In retrospect, it was this seemingly simple notion that opened our eyes to the possibility of using a relay fragment to advantage as we set out to demonstrate RRCM and some of its strategic benefits.

A few comments about yields of reaction product(s) are in order. A reported chemical yield can represent several different things; it often, necessarily, has a substantial error bar, especially when the experiment has been performed on a small scale. This is often the case with complex and precious substrates: for example, those in hand at the closing stage of a total synthesis effort. Even when the yield value is accurate, it is often a combination of inherent imperfections in both the reaction chemistry and the purification/handling routines. Similar issues affect the way one interprets the catalyst loading in metal-mediated reactions. Although we have provided the yields and catalyst loadings reported for the metathesis reactions described here, we encourage readers to be mindful of the above points and not to permit yield vagaries to overshadow the more important conceptual and strategic lessons that are revealed and taught by each of the case studies described. Importantly, in every example the product shown represents the major (and often sole) reaction path that was reported by the original investigators.

9.2
Early Relay Metathesis Discoveries

The earliest demonstration of a RRCM reaction in our research group at the University of Minnesota was carried out by Hongyu Zhao. He took up the challenge presented by the low reactivity of 1,1-disubstituted ethylenes toward the early generation ruthenium alkylidenes [1], most commonly represented by diene **5**, and its failure to cyclize to **6** (Scheme 9.2) [2]. The problem was successfully solved by preparing and then cyclizing the relay-armed triene substrate **8** [3–5]. In this RRCM experiment, initiator [Ru]-**I** was used, since the experiment predated the report of

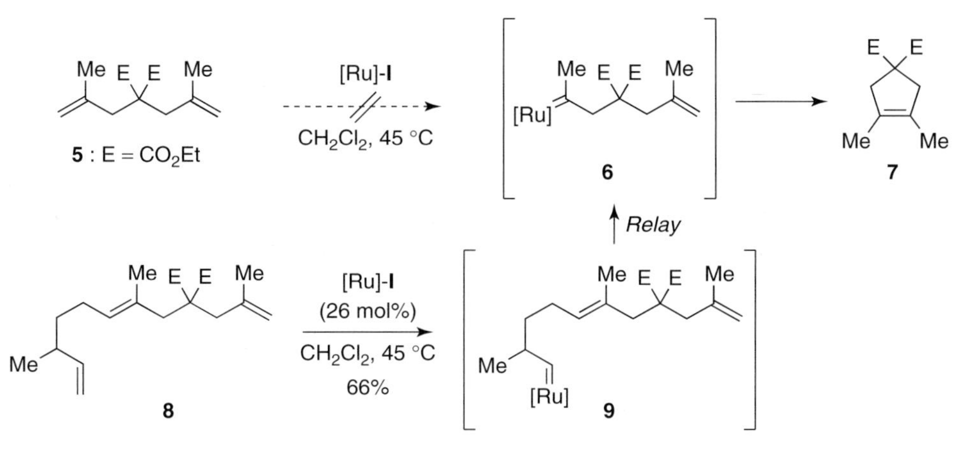

Scheme 9.2 Our first example of RRCM: closure of **8** to **7**.

the discovery of [Ru]-**II**. (Although these ruthenium complexes are commonly referred to as *catalysts*, it is more correct to refer to them as metathesis initiators or precatalysts, since they are not recovered unchanged once they have performed their duty.) This showed that by use of a relay strategy, net loading of the metal to a substrate that is otherwise inert, in this case to produce **6** (via **9**), is achievable.

Independent efforts by the Piscopio, Lee, and Porco groups at Array Biopharma, the University of Wisconsin, Madison, and Boston University, respectively, also laid the base for further development of relay metathesis. In the first of these reports, the RRCM concept was presented as shown in Scheme 9.3 [6].

The example shown in Scheme 9.4 constitutes the first case of a relay cross-metathesis (relay-CM) process [7]. Reaction to give **16** is faster and more efficient for the relay-activated substrate **17** than for the "parent" CM partner **14**.

In the synthesis of oximidine III (**23**), Porco *et al.* faced the need to cyclize a polyene like **18** to access macrolide **19**, but this process failed for either (*E*)- or (*Z*)-**18** (Scheme 9.5) [8]. Use of the relay-activated substrate **20** and the second-generation

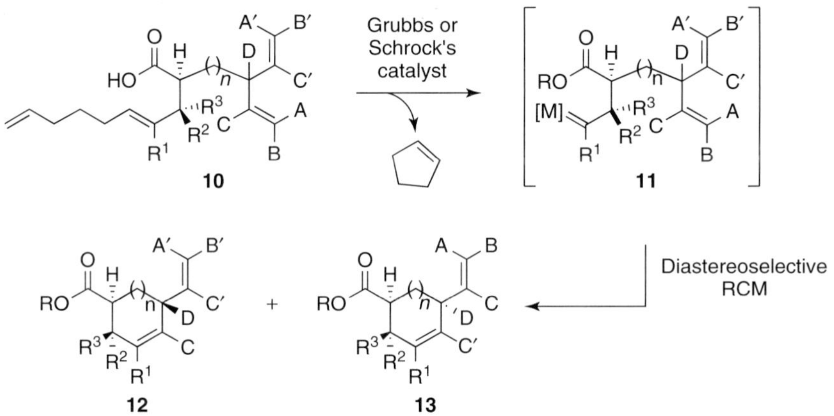

Scheme 9.3 Generic representation of prospects for RRCM processes.

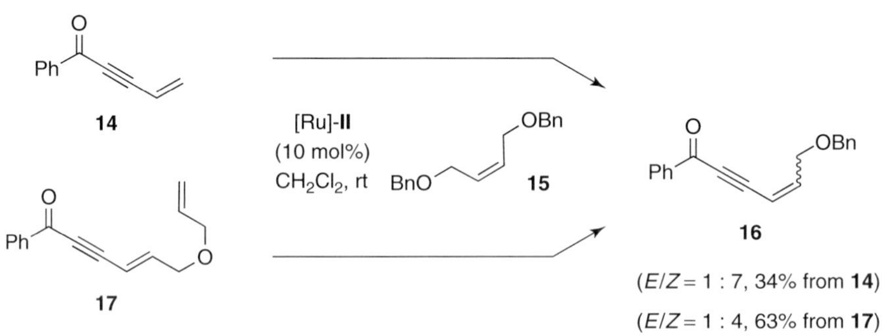

Scheme 9.4 The first relay cross-metathesis (relay-CM) reaction: cross-coupling of donor ene–yne **17** with acceptor alkene **15**.

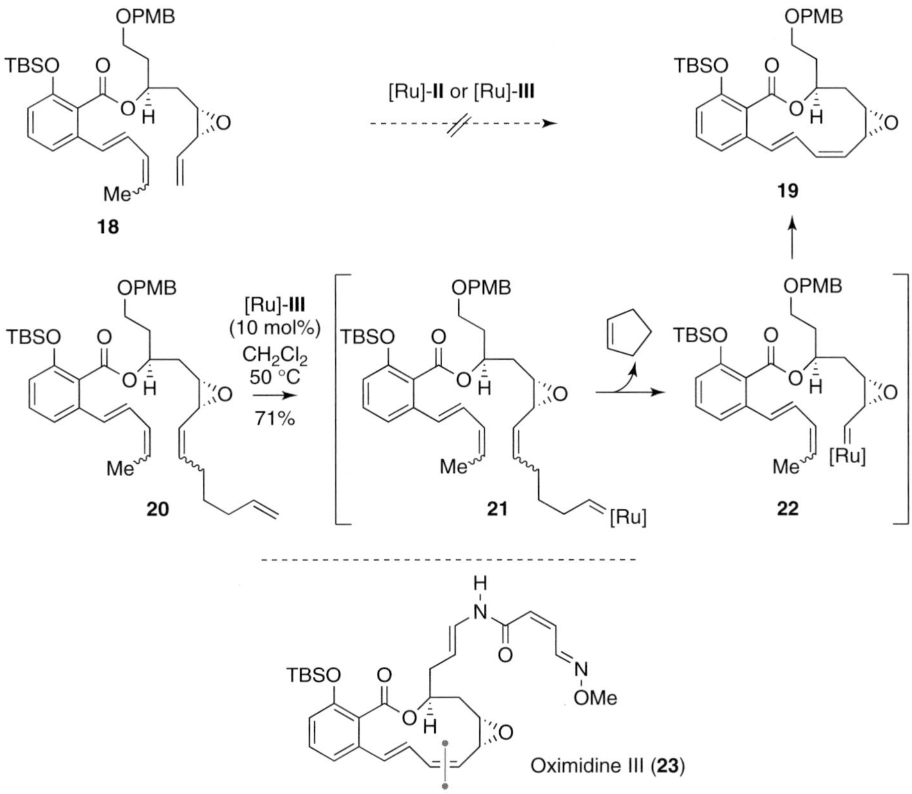

Scheme 9.5 The first successful application of RRCM in a natural product synthesis.

Hoveyda–Grubbs initiator [Ru]-**III** nicely solved this problem, presumably because ruthenium loading adjacent to the oxirane was now viable via the relay of **21** to **22**. This constituted the first report of a successful application of an RRCM reaction in the context of a natural product total synthesis.

Other early studies of RRCM applications in our labs were carried out by Danielson [9], Tennakoon [4, 10], Wang [4, 11], and Jeffrey [4, 12]. RCM of the silicon-tethered substrates **25** and **28** to give the hindered (*Z*)-trisubstituted alkene, an intermediate for use in our synthesis of peloruside A (**29**), was successful in both cases (Scheme 9.6) [5, 10]. However, it was difficult to prepare **25** in high diastereomeric purity. More specifically, the two enantiomers of the secondary alcohol precursor (±)-**24** were barely discriminated by several lipases that we examined. By contrast, citronellene-derived alcohol **27**, a 1 : 1 mixture of epimers at C15, could be kinetically resolved quite efficiently. Thus, the relay tether served an auxiliary but nonetheless important role, merely because of its size.

In a model study of a potential application of RRCM for formation of the C4–C5-trisubstituted alkene in oocydin A (**33**), the substrates **30a** and **30b** were each successfully cyclized to the 13-membered lactones **32a** and **32b**, respectively

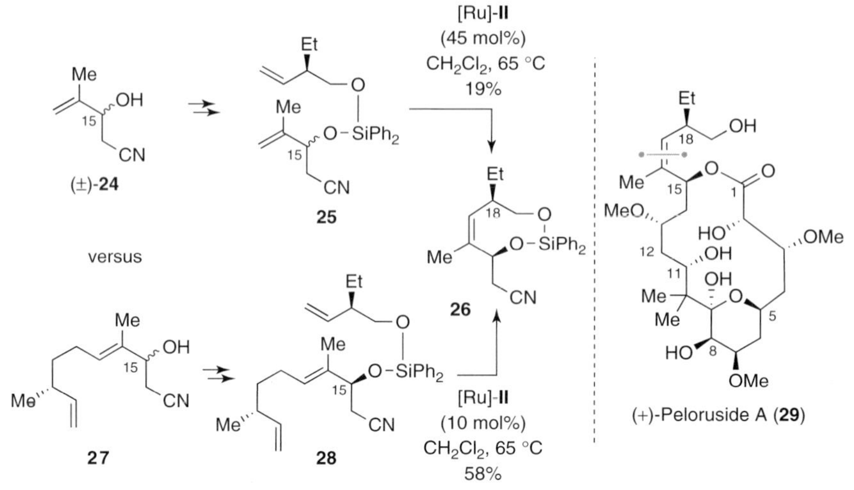

Scheme 9.6 RRCM permitting the use of an otherwise inefficient enzymatic resolution.

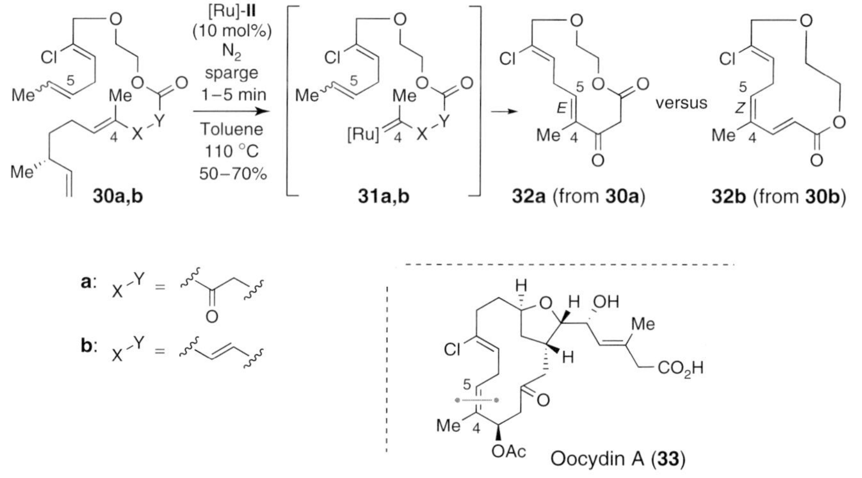

Scheme 9.7 RRCM through alkylidenes that are otherwise difficult to access (**31a** and **31b**).

(Scheme 9.7) [5, 11]. This demonstrated for the first time the ability to initiate a key cyclization through either an electron-deficient β-ketoalkylidene (**31a**) or an allylidene (**31b**) complex by the strategic design of the relay substrate. Interestingly, the E/Z selectivities with respect to the newly formed $\Delta^{4,5}$-bond were absolute and opposite for the two substrates.

Another example of a successful cyclization via an alkylidene adjacent to an electron-deficient functional group is the cyclization of **34** to **35** with [Ru]-**I** (Scheme 9.8) [4, 5]. The complementary closure of **36** to the same product

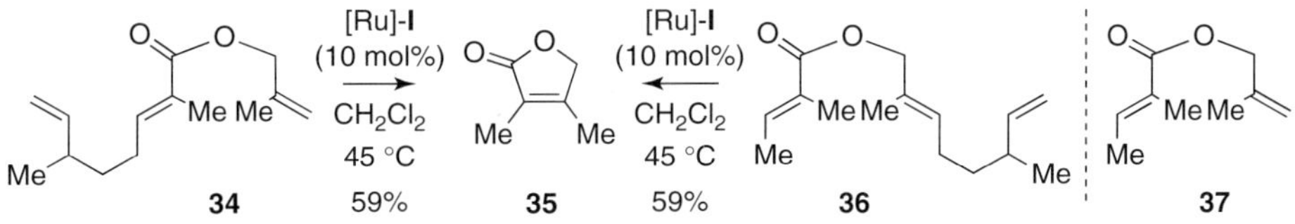

Scheme 9.8 Access to a carboalkoxyalkylidene complex (from **34**).

demonstrates versatility, and it is noteworthy that the parent, non-relay-activated diene **37** was unreactive under the conditions employed. Presumably, in the absence of a relay tether, ruthenium loading to either alkene in **37** was simply too slow relative to the initiator/catalyst lifetime.

Finally, a different type of complementarity is clearly evident from the RRCM results summarized in Scheme 9.9. Ene–yne–ene cascade cyclization (for a non-relay, directed example see [13]) of either of the isomeric, relay-activated substrates **38** or **41** gave either isomeric bicyclic ether **39** or **40**, respectively, each with very high regiochemical selectivity [5, 12]. This example shows the ability to use RRCM to dictate the formation of different constitutional isomers merely by relocation of the relay tether. That is, the directionality ("endedness") of the process from "right-to-left" versus "left-to-right" can be controlled at will. Interestingly, the parent, non-relay-activated ene–yne–ene substrate (not shown) gave a nearly 1 : 1 product ratio, and the use of [Ru]-**I** rather than [Ru]-**II** gave greater regioselectivity for each of **38** and **41**.

Although the above earliest examples were described by various terminologies ("RRCM" [4], "metathesis trigger" [6], "RCM–RCM" [8], or "catalyst delivery vehicle" [7]), these words all converge to the same concept. Relay metathesis "involves the design of substrates that, in effect, permit[s] one to dictate the sequence of metathesis events by choreographing the metal atom . . . through the individual steps" [5] of the metathesis cascade. The above, first-generation relay metathesis experiments were summarized in the earliest review of the subject [14].

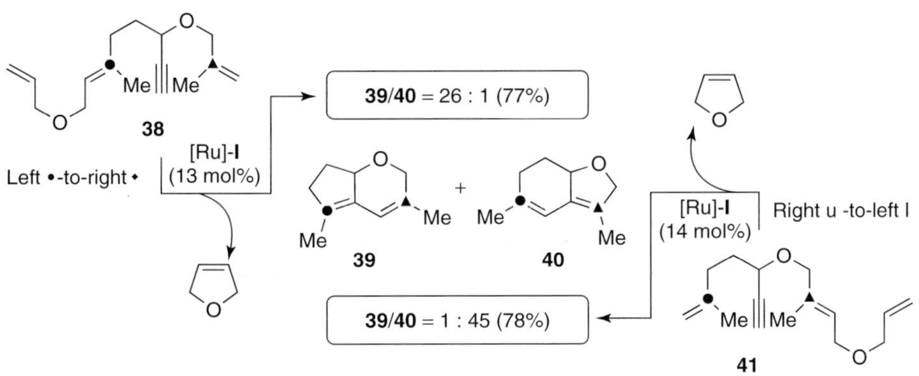

Scheme 9.9 Complementary product constitution by controlling the "directionality" of ene–yne–ene RRCM.

9.3
Examples of Relay Metathesis Directed at Targets Other than Natural Products

In their efforts to implement a strategy by which the chirality in the menthol-derived auxiliary in **42** could be transmitted into nonracemic heterocycles like **43**, Spino *et al.* sought to cyclize the very hindered parent substrate **42a** via RCM (Scheme 9.10), but the latter compound proved to be unreactive [15]. Substrate modification to **42b** and its subsequent RRCM closure using the [Ru]-**VI** initiator resulted in formation of lactam **43** (via an electron-deficient alkylidene intermediate); the remainder of the product was the parent diene **42a**. We refer to this type of process as *truncation* [5], and it results when the RCM cyclization rate of the relayed ruthenium intermediate is slow relative to the competing bimolecular process by which another molecule of substrate, **42b** here, cross-metathesizes with net transfer of a methylene group. That obstacle was overcome here by an insightful modification of the relay tether involving the addition of one more substituent. Specifically, the use of the substrate analog **42c** resulted in the formation of **43**, with impressively high efficiency. Arguably, substrates **42b** and **42c**, whose cyclization was absolutely enabled by the relay strategy, might be the most demanding ever to have been successfully ring-closed by metathesis.

Fujiwara *et al.* conceived a plan for the synthesis of nonracemic oxacycles like **47** that capitalized on sequential chirality transfer and then RRCM (Scheme 9.11) [16]. In the first stage, an Ireland–Claisen rearrangement of the serine-derived allylic ester **44** produced **45**, thereby simultaneously repositioning both alkene and stereogenic carbon centers. This RRCM substrate was cyclized using [Ru]-**II** to give **47** (via **46**).

Grandbois and Collins have used the asymmetric initiator (*R*,*R*)-[Ru]-**XIV** to kinetically resolve a racemic sample of the doubly armed, C_2-symmetric tetraene (±)-**48** (Scheme 9.12) [17]. Interestingly, the enantiomeric excess of the [7]helicene

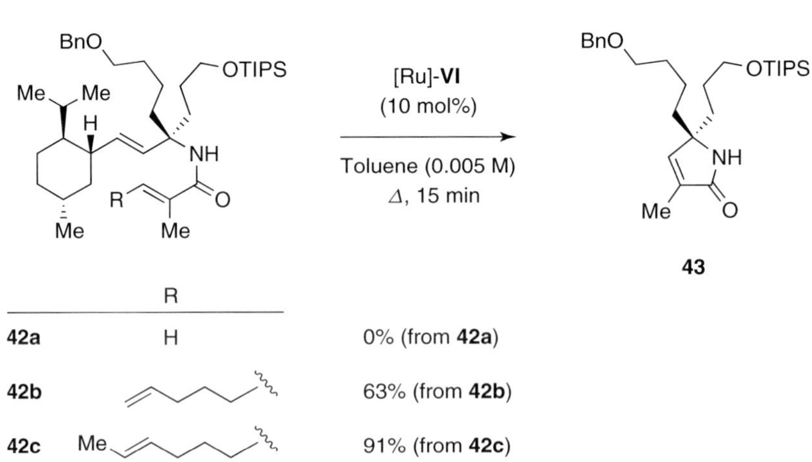

Scheme 9.10 RRCM activation to allow closure of the remarkably hindered substrates **42**.

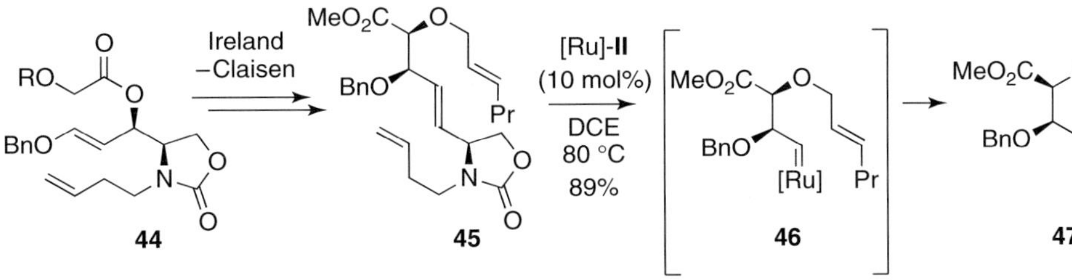

Scheme 9.11 Sequential chirality transfer and RRCM of a serine derivative.

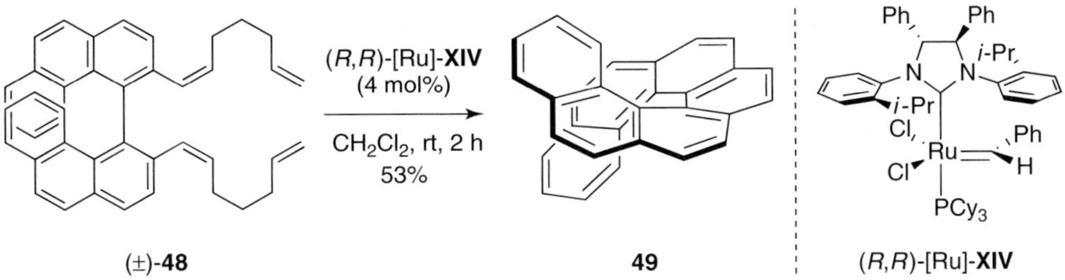

Scheme 9.12 Use of RRCM and an asymmetric initiator to kinetically resolve the helically chiral substrate (±)-**48**.

49 that was produced was higher when the RRCM cyclization was used compared to the case of the parent, non-relay-activated, divinylbiphenanthrene substrate.

9.4 Examples of Relay Metathesis Motivated by Natural Product Synthesis

In the course of studying the effects of noncovalent interactions on various RCM reactions to produce macrocycles, Collins *et al.* showed that the relay-armed substrate **50** closed efficiently to the macrocycle **52**, whereas the parent diene **53** gave mostly dimer **54** (Scheme 9.13) [18]. The ability to direct the main cyclization event via alkylidene **51** (vs that derived by loading the metal onto the more reactive monosubstituted alkene in **53**) substantially affected the outcome.

Motivated by the structures of various monomeric and dimeric members of the longithorone family [for example, longithorone C (**57**) and longithorone I (**60**)], these investigators expanded their studies (Scheme 9.14) [19]. Hence, the RRCM closure of pentaene **55** by use of the [Ru]-**VIII** initiator gave the triene **56** [19b], which contains many of the core structural elements of longithorone C (**57**). The ene–yne RRCM of substrate **58a** or **58b** using [Ru]-**III** gave diene **59a** or **59b**, respectively [19c].

Trienomycin A (**64**) and the related geldanamycin (**65**) are ansamycin antibiotics containing metacyclophane–macrolactams. With an eye toward preparation of structurally simpler analogs, Peng and Blagg synthesized trienes **61a** and **61b** and studied their RCM reactivities (Scheme 9.15) [20]. Upon being exposed to [Ru]-**I**,

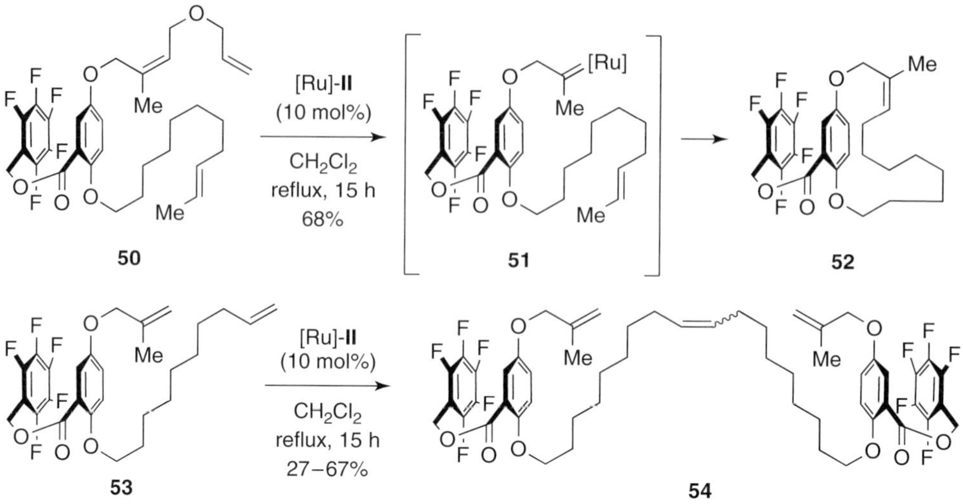

Scheme 9.13 Use of RRCM to divert dimerization (**53** to **54**) to cyclization (**50** to **52**).

the former gave only the RRCM product **62**, whereas the methylated analog **61b** gave a preponderance of the macrocyclic diene **63b**. A combination of factors likely contributes to this different outcome: first, the metal relay event is expected to be slower in **61b** versus **61a** and, second, since either of these two products can also arise from initial alkylidene formation at C8 in the nonenyl amide moiety, the non-relay mode of cyclization to give **62** should also be slower from **61b** than from **61a**.

In the course of their total synthesis of (−)-mucocin (**71**), Crimmins *et al.* targeted intermediate dihydrofurans like **67** (Scheme 9.16) [21]. Earlier, they had established that trienes like **66** gave poor regioselectivity for formation of the desired **67** versus the isomeric six-membered dihydro-2*H*-pyran **68** [22]. This result is consistent with the competitive cyclizations, in either direction, suggested by arrows *a* versus *b* in **66**. This problem was solved by the cleverly conceived use of substrate **69**, which underwent RRCM with high efficiency (via *c*) to give the desired dihydrofuran **70**.

The key step in the total synthesis of dihydro-*epi*-deoxyarteannuin B (**73**) reported by Dudley *et al.* involved metathesis closure of the hindered alkene in **72** [23]. This was first attempted, unsuccessfully, with the parent diene **72** (Scheme 9.17). CM, presumably with benzylidene fragments from the initiator, was observed. It is likely that obligate loading of the metal to either of the hindered alkene carbons (• or ▲) in **72** was simply too difficult. This problem was nicely and efficiently solved through the RRCM of analog **74**. Here, the relay event provided for insertion of the metal onto carbon ▲ and the subsequent RCM.

In their total synthesis of (−)-archazolid B (**76**), Trauner *et al.* successfully ring-closed the relay-activated polyene **75** at a late stage via an intermediate ruthenium allylidene complex (at carbon ▲) (Scheme 9.18) [24]. They designed this RRCM event into the synthesis plan from the outset, in part to minimize the

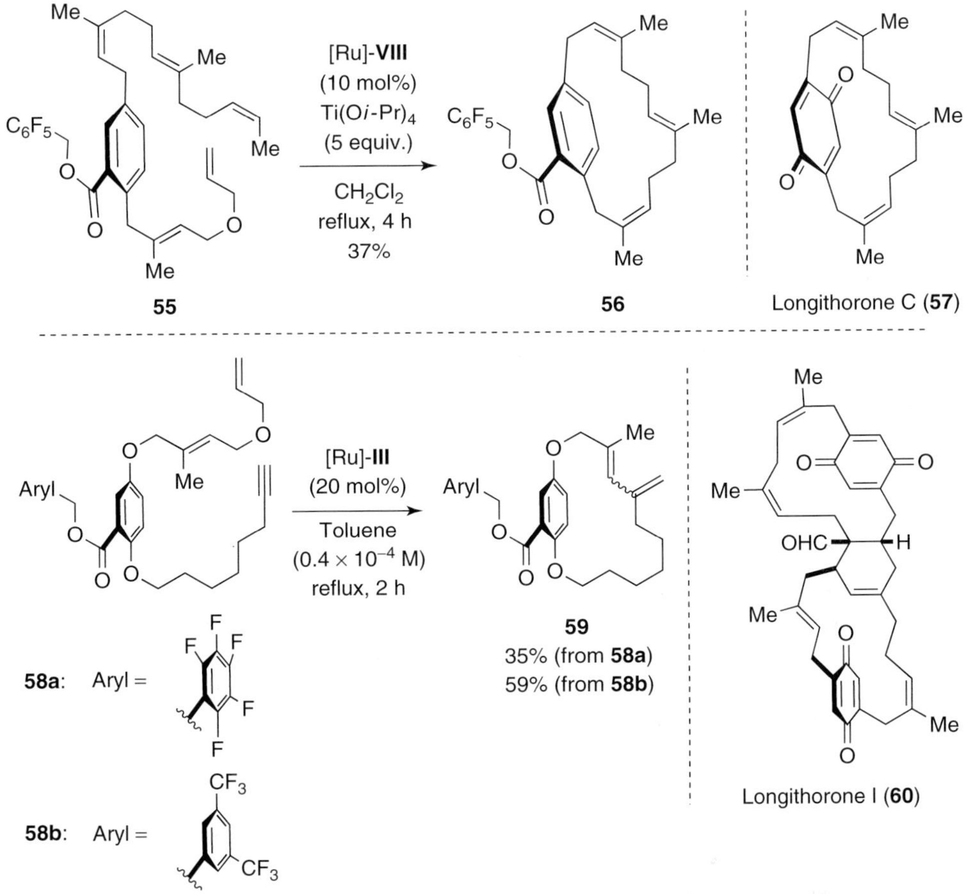

Scheme 9.14 RRCM cyclizations relevant to longithorones **57** and **60**.

possible complication of initial activation at C21 (carbon •) in **75**, from which they speculated that undesired cyclization to excise a δ-valerolactone might occur.

Kim *et al.* used the relay-CM strategy as a key step in the construction of the conjugated ene–yne in (+)-scanloenyne (**79**) (Scheme 9.19) [25]. The relay-activated partner **77**, under the action of [Ru]-**XII**, was successfully cross-metathesized with the terminal alkene in **78** to give the ene–yne in **79** with high *Z*-selectivity, a type of stereocontrol first demonstrated by Lee [7a].

In a ploy to distinguish the pair of diastereotopic vinyl groups in substrate **80**, Burke *et al.* used RCM to convert that triene to the bicyclic ketal **81**, a key intermediate in their synthesis of (+)-didemniserinolipid B (**85**) (Scheme 9.20) [26]. They were badgered by the secondary formation of **82**, via subsequent CM of the primary product **81** with the stoichiometric amount of styrene that was coproduced by the initial RCM event. By insightful incorporation of the designer relay auxiliary seen in substrate **83**, they changed the profile of the metathesis

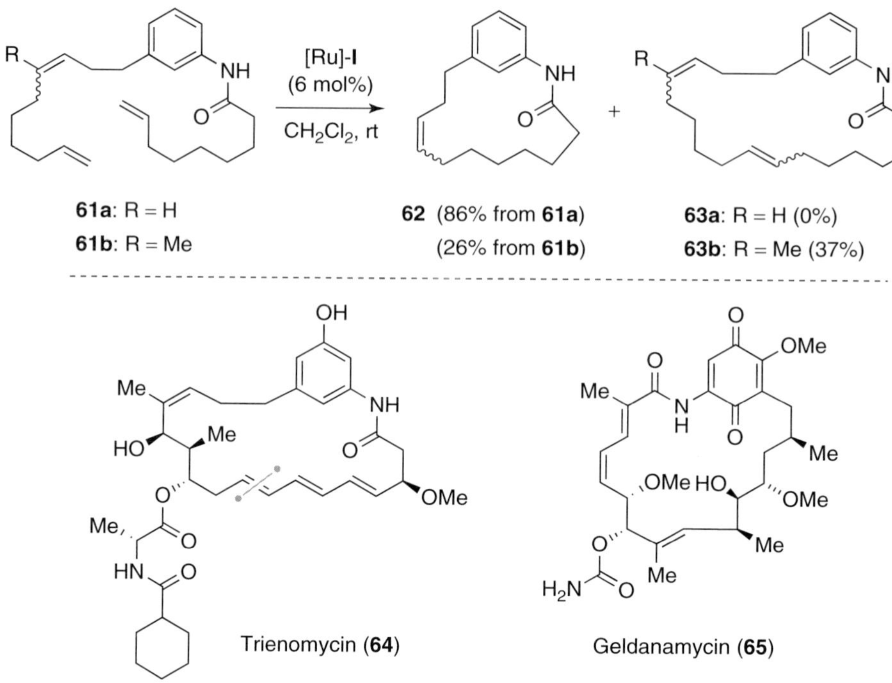

Scheme 9.15 Competitive RCM and RRCM cyclizations to trienomycin A (**64**) analogs.

by-products to ethylene and indene **84**, neither of which eroded the vinyl group in **81**. The efficiency of this RRCM reaction was excellent.

Njardarson *et al.* recently reported the RRCM cyclization of **86** to the inside–outside bicyclic triene **87** (Scheme 9.21), an analog of **89** that, in turn, is an atropisomer of the natural product hypoestoxide (**88**) [27]. This is another example in which the workers wisely designed the RRCM into their *de novo* plans, presumably envisioning that closure of the disubstituted ruthenium alkylidene at C6 to the less substituted alkene carbon at C5 (product numbering) had a greater likelihood of success than the alternative.

Parker *et al.* developed an effective method for the stereocontrolled synthesis of a key intermediate in a projected discodermolide (**94**) synthesis that involved RCM closure to the dihydrooxasiline **91** and its subsequent conversion to the acyclic vinyl iodide **93**, the C9–C14 subunit of **94** (Scheme 9.22) [28]. Attempts to cyclize the hindered triene **90** with either [Mo]-**I** or [Ru]-**II** failed. The steric hindrance in **90** is a direct consequence of the strategy used for the otherwise efficient synthesis of this nonracemic substrate, which involved a [2,3]-Wittig rearrangement. A solution to the metathesis cyclization was discovered in the form of the [Ru]-**II**-promoted RRCM of tetraene **92**, which was also formed via a [2,3]-Wittig rearrangement and which produced **91** with extremely high efficiency. The success of this RRCM was due at least in part to the fact that the site of initiation (carbon •) strongly favored formation of only the six-membered oxasilacycle. Initiation at the vinyl

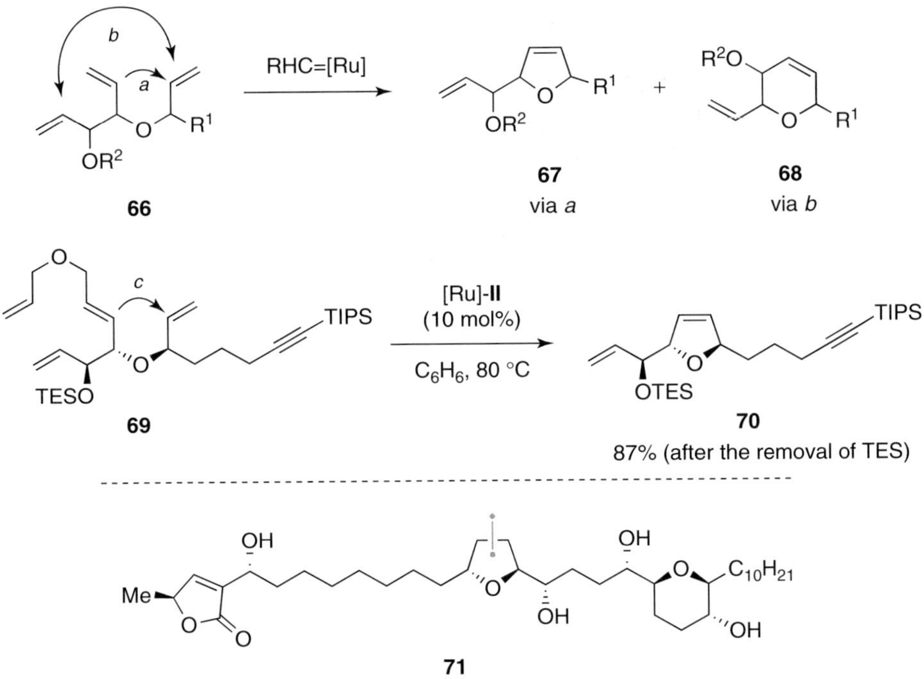

Scheme 9.16 RRCM as a key step in the synthesis of (−)-mucocin (**71**).

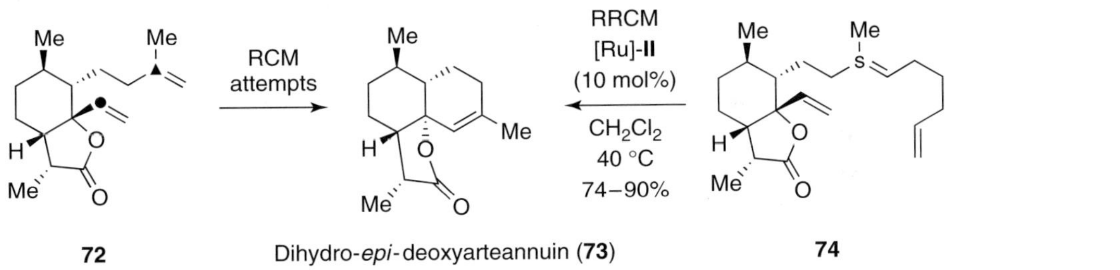

Scheme 9.17 RRCM in the synthesis of dihydro-*epi*-deoxyarteannuin B (**73**).

silane (carbon ▲) would provide the opportunity for competitive formation of five- versus six-membered cyclization products.

Meyer, Cossy, *et al.* recently described the use of an RRCM to close the dihydropyranone ring present in phoslactomycin B (**97**) [29]. Specifically, the tetraene **95** was efficiently converted to **96** using [Ru]-**II** (Scheme 9.23), a key step in their formal total synthesis of **97**. In this case, cyclization of the non-relay armed parent triene could be achieved, albeit much more slowly than the closure of **95**. As a consequence, a substantially higher initiator loading [Ru]-**II** was required to achieve comparable conversion efficiency in the absence of the relay activation unit. This represents another type of advantage that can accrue from use of RRCM.

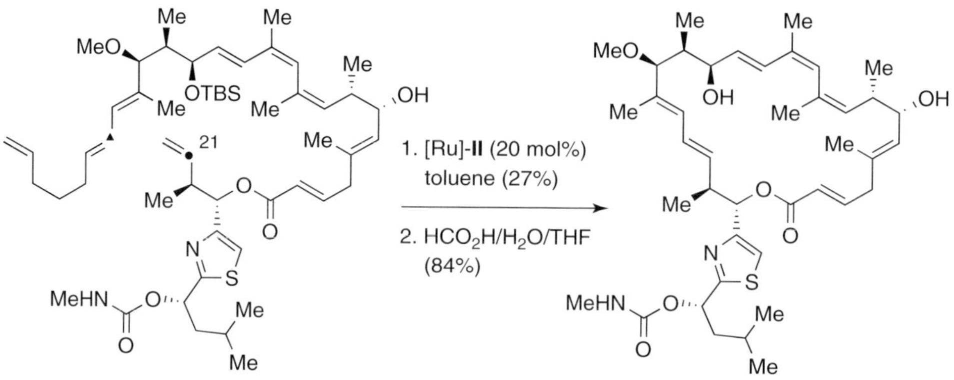

75

(−)-Archazolid B (**76**)

Scheme 9.18 RRCM closure via an allylidene in the synthesis of (−)-archazolid B (**76**).

In parallel with the Parker example shown in Scheme 9.22, the stereocontrolled construction of **95** also capitalized on a [2,3]-Wittig rearrangement to create the C4–C5 bond.

Lee *et al.* have used a relay metathesis trigger to further enable their studies of metallatropic [1,3]-shifts. For example, both of the relay-activated diene–diynes **98** underwent CM with (*Z*)-1,4-diacetoxy-2-butene (**99**) to give ene–diyne **100** (Scheme 9.24) [30]. This process was designed to involve insertion of ruthenium onto the relay subunit in **98** to give **101**, followed by rapid migration of [Ru] to the isomeric species **102** and **103** and drainage of the most reactive of these, **103**, from the equilibrium manifold by final CM with **99**.

More recently, the same investigators have reported the successful implementation of this process in the total synthesis of three structurally related natural products tricholomenyn A (**106**), asperpentyn (**109**), and harveynone (**110**) [30, 31]. The tandem relay ene–yne RCM/metallatropic shift reactions of **104** gave **105**, which lacks the dihydrofuran relay by-product but is otherwise isomeric with **104** (Scheme 9.25). Product **105** was then efficiently elaborated to tricholomenyn A (**106**). Interestingly, the pair of related, non-relay-activated substrates **107a** and **107b** had different fates upon treatment with [Ru]-**II**. The former did not undergo a tandem

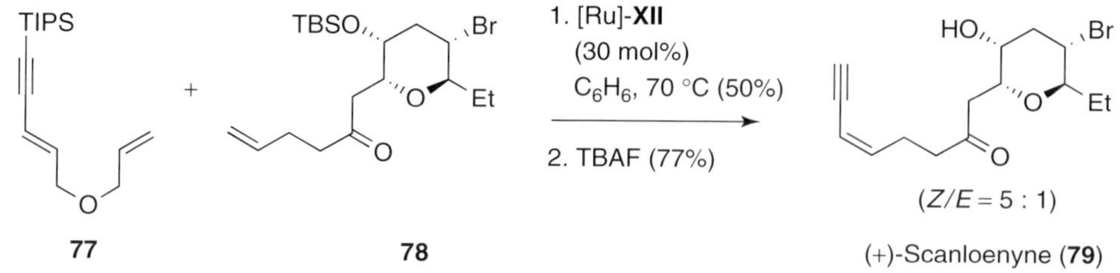

77 78 (+)-Scanloenyne (**79**)

Scheme 9.19 *Z*-Selective relay-CM in the synthesis of (+)-scanloenyne (**79**).

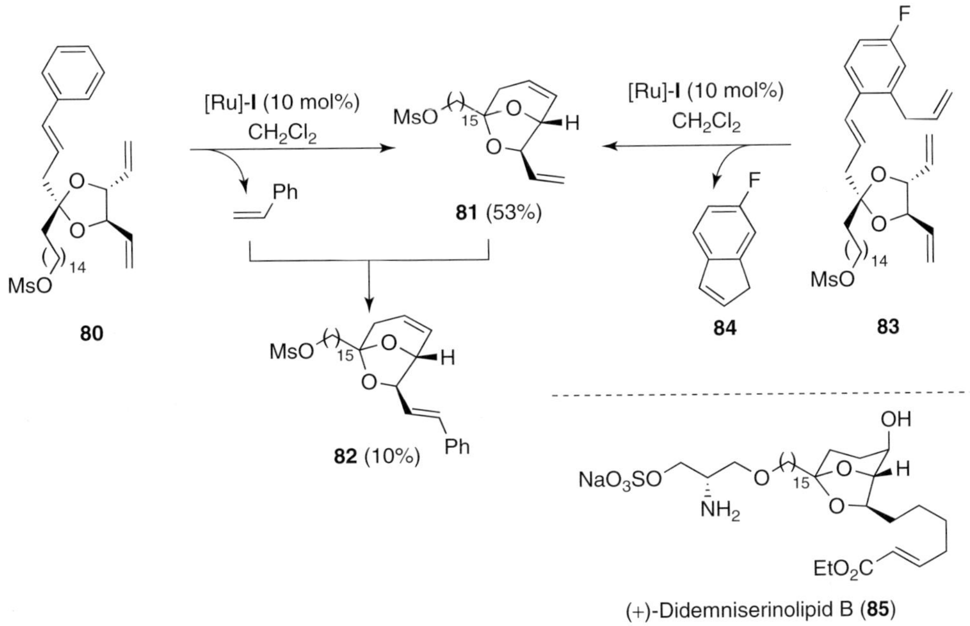

Scheme 9.20 RRCM substrate **83** altering the profile of RCM by-products in the synthesis of (+)-didemniserinolipid B (**85**).

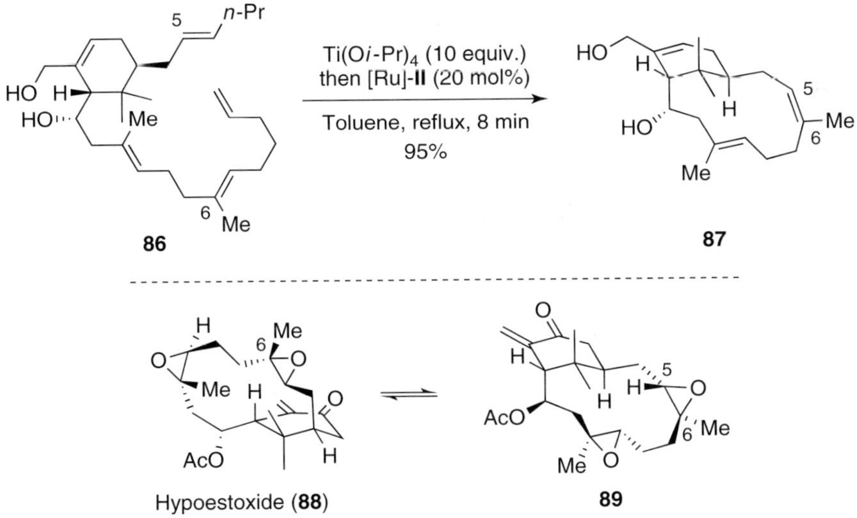

Scheme 9.21 RRCM to control the directionality of ring closure in the synthesis of macrocycle **87**.

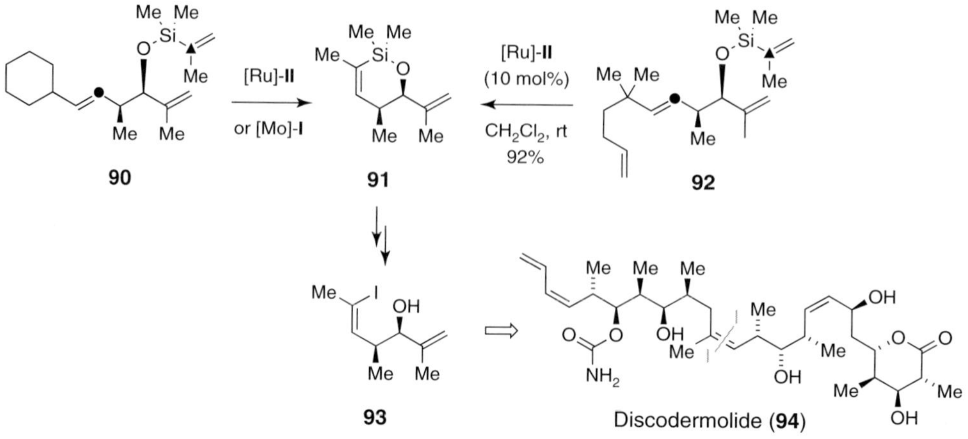

Scheme 9.22 Fully regiocontrolled RRCM closure of **92** to **91**.

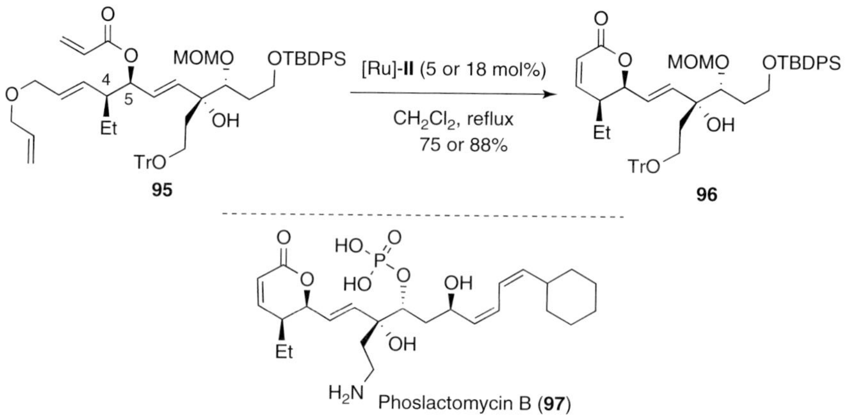

Scheme 9.23 RRCM enhancing the efficiency of cyclization, permitting a lower level of initiator loading.

cyclization analogous to that of **104**, but the latter did to give **108**. This intermediate was readily carried forward to both asperpentyn (**109**) and harveynone (**110**).

In a related and more mechanistically motivated study, Lee *et al.* were able to structurally characterize an intermediate product of metallatropic rearrangement. Specifically, upon treatment of **112** with 1 equiv. of [Ru]-**I** they isolated crystalline **111** as the sole product (Scheme 9.26) [32]. Additionally, they observed a divergence of reactivity depending on the choice of the initiator. The action of [Ru]-**II** on substrate **112** gave the truncation product **113** exclusively. In light of additional strong evidence they accumulated for the reversible nature of the metal migration, the authors rationalized that the more hindered nature of the heterocyclic carbene ligand in [Ru]-**II** sterically destabilized ruthenium complex **114**.

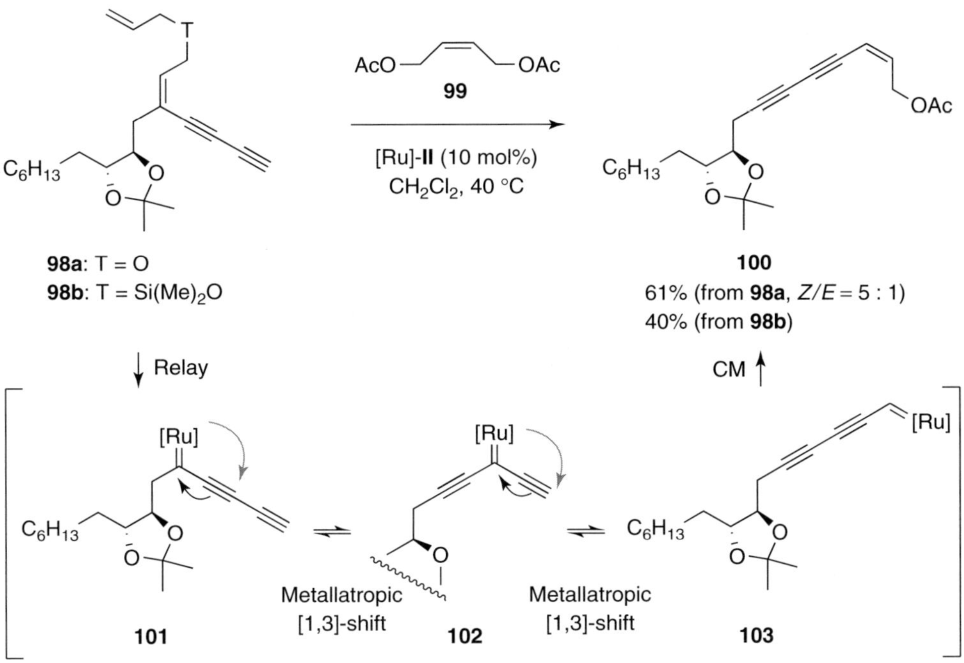

Scheme 9.24 Cascade of relay initiation, metallatropic rearrangement, and CM.

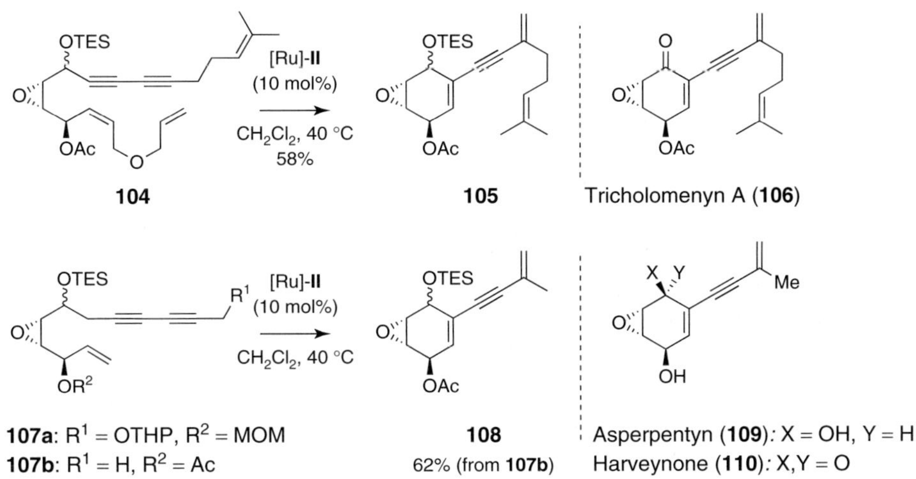

Scheme 9.25 Integration of RRCM with metallatropic shift in the syntheses of tricholomenyn A (**106**), asperpentyn (**109**), and harveynone (**110**).

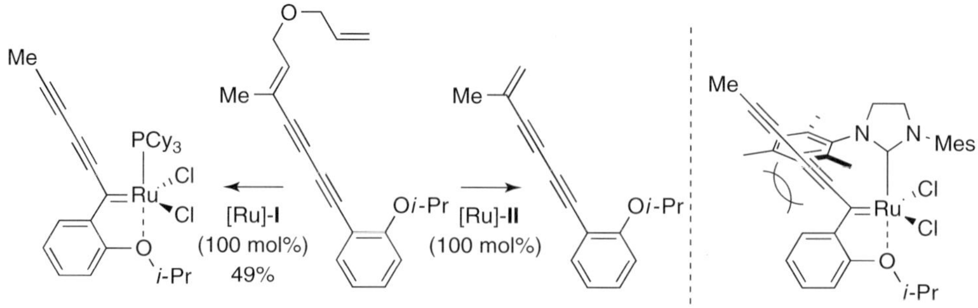

111 112 113 114

Scheme 9.26 Relay activation to access pathways that lend mechanistic insight.

9.5 Examples of Relay Metatheses Thwarted in Achieving the Desired Outcome

Not every attempt to implement a relay metathesis strategy has resulted in the intended cyclization to a desired cyclic alkene. The obstructions fall mostly into one of two categories:

1) interference from a truncation event,
2) interference from premature macrocyclization.

9.5.1 Interference from a Truncation Event

As briefly introduced earlier (see the transformation of **42b** to **42a** in Scheme 9.10), in some instances the main RCM event turns out to be slower than that of bimolecular CM with other alkene-containing, methylene-donor molecules in the reaction medium. These donors are nearly always either the terminal alkene in an additional molecule of the substrate or the stoichiometric by-product of metathesis cyclization – most often ethylene.

This truncation event was observed in early studies of RRCM with the substrate **8** (Scheme 9.2). In that experiment, approximately 2% of the parent (and unreactive) **5** was obtained in competition with closure to the RRCM product **7**. As with all cases of truncation, this is most likely the result of competitive reaction processes of the metal-relayed intermediate **6**, which can evolve to **7** by a unimolecular cyclization or to **5** by a bimolecular reaction with **8** and/or ethylene.

This problem was again encountered soon thereafter in the attempted RRCM cyclization of substrates **115a–d** to give **117** (Scheme 9.27), a key intermediate in the synthesis of callipeltoside A (**118**) [9]. In every instance, the outcome was the formation of the truncation product and parent diene **116**, a substrate type we already knew to be unresponsive to the RCM protocols available at the time [33]. In this hindered and heavily branched system, truncation was always faster than cyclization, even when the experiment was performed at high dilution and with

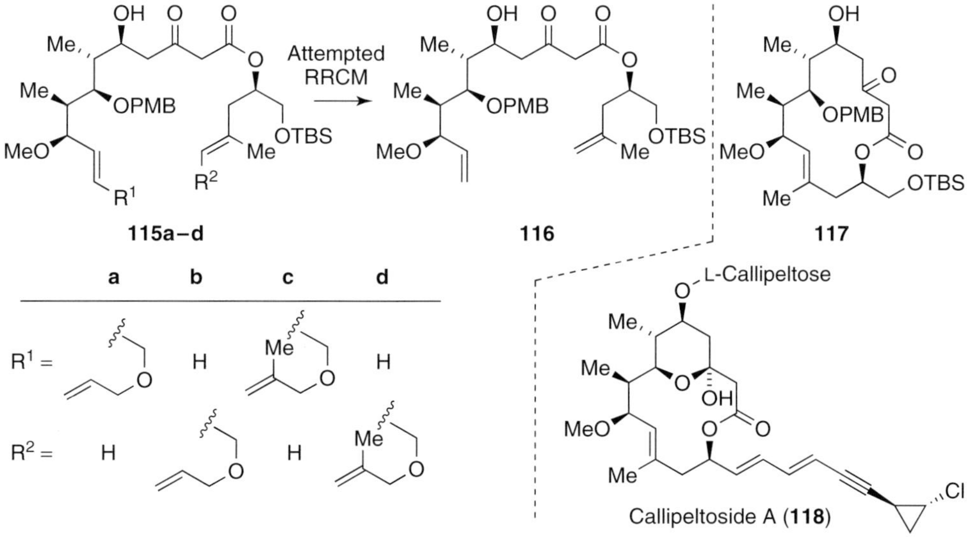

Scheme 9.27 Truncation: a problem when the rate constant for the final RCM step is small.

the relay tethers present in **115c** and **115d**, which are less reactive CM partners compared to **115a** and **115b**, respectively.

There are a few additional reports of truncation reaction products [11c, 20]. Rychnovsky *et al.* envisioned a possible double RCM within the coordination sphere of the Ca(II) biscarboxylate salt of the diene substrate **120** or of its relay-tethered analog **119** as an innovative approach to a synthesis of the dimeric C_2-symmetric macrodiolide natural product SCH 351448 (**121**) (Scheme 9.28) [34]. Again, metal loading and relay within **119** were effective, but the internal cyclization was sufficiently slow that truncated **120** was the only product observed. Under slightly more forcing conditions, the parent diene **120** produced a monomeric macrocyclic lactone RCM product.

In studies aimed at a synthesis of characiol (**124**), Helmboldt and Hiersemann observed that it was possible to close triene **123b** by RCM to give 17-norcharaciol, but that the methyl-substituted analog **123a** was unreactive (Scheme 9.29) [35]. They then attempted to cyclize the relay-activated analog **122**, but observed that it was converted to **123a** rather than the macrocyclic dienedione present in **124**. This reinforces the tenet that the rate of the primary unimolecular cyclization event of interest must be greater than that of the competing bimolecular CM reaction with another of the substrate molecules in order to RRCM to suceed.

9.5.2
Interference from Premature Macrocyclization

Another potential pathway that can interfere with the intended RRCM outcome is the internal macrocyclization of the initial ruthenium alkylidene. This becomes

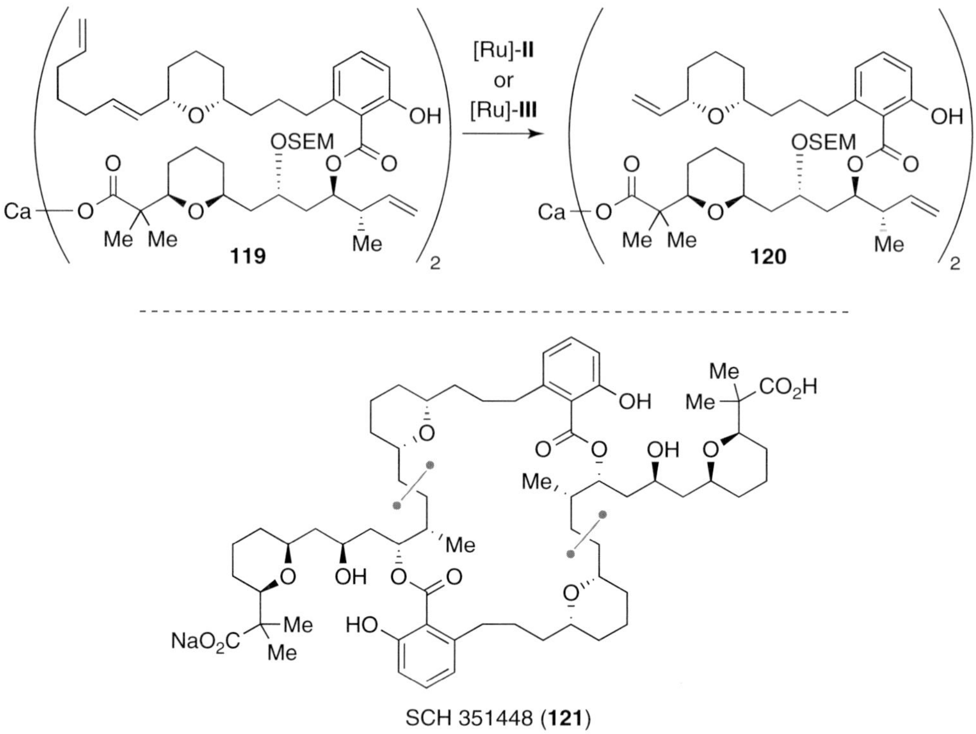

Scheme 9.28 Truncation: a problem when the rate constant for the final RCM step is small.

a serious problem when the large-ring cyclization process is faster than the relay event (for example, in the generic RRCM process depicted in Scheme 9.1, closure of the alkylidene in **4b** directly onto the R^1-bearing alkene prior to relay to the R^2-bearing alkene).

Such a process was first observed as a competitive event during the RRCM cyclization of **28** to give **26**, which was described earlier (Scheme 9.6); that is, the macrocyclic diene **127** was formed along with the desired **26** (Scheme 9.30). It is not clear whether this macrocyclization product arises via **126a** and its reaction through path *a* in competition with relay contraction, or from competitive initial formation of **126b** and its cyclization via path *b*.

In the course of their synthesis of a ring-expanded bryostatin analog **131**, Trost *et al.* studied the RCM cyclization of substrate **128**, which was armed for a potential RRCM reaction (Scheme 9.31) [36]. This complex polyene cyclized very cleanly under the action of [Ru]-**III** to produce the cyclic diene **129**. There was no evidence of formation of the smaller RRCM product. The rate of the relay event here is presumably lowered by the pair of quaternized carbons allylic (•) and homoallylic (▲) to C17 in the alkylidene derived from **128**.

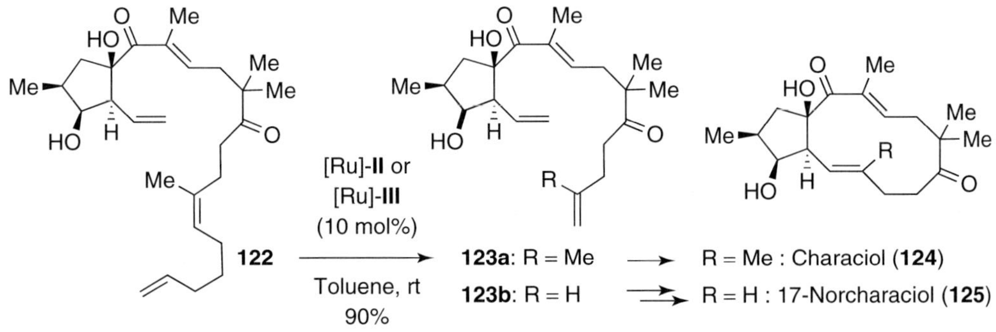

Scheme 9.29 Substitution on the alkylidene intermediate slowing cyclization en route to **124**.

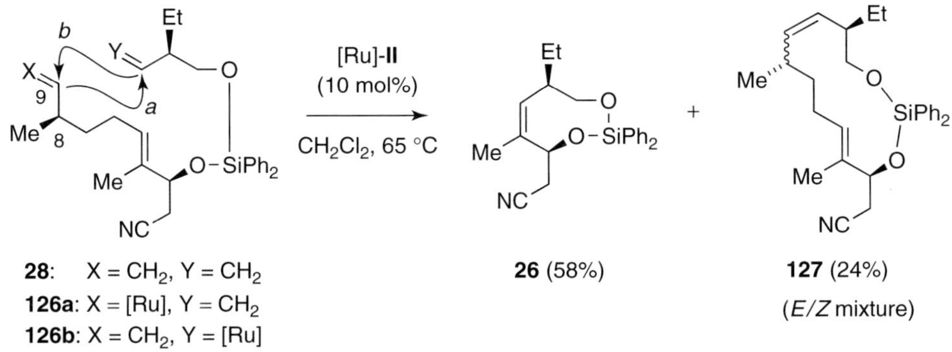

Scheme 9.30 Competitive rates of relay versus direct cyclization resulting in premature ring closing.

Fürstner *et al.* attempted to close **132a** to afford the skeleton of spirastrellolide A (**134**) by an RRCM reaction [the corresponding parent, non-relay-activated diene (not shown) was unreactive]. When exposed to initiator [Ru]-**II**, triene **132a** gave, instead, the non-relay-contracted macrocyclic diene **133** (Scheme 9.32) [37]. The cumulative steric hindrance from several branched sites on the distal end of the initially formed ruthenium alkylidene **132b** again presumably slowed the rate of the internal relay event to the point where cyclization within **132b** to C25 of the remote vinyl group superceded.

9.6 Conclusion

We have reviewed here the discovery, evolution, and applications of "metathesis involving a relay." While most of the examples fall into the category of RRCM, the handful comprising relay-CM processes further reinforces both the power and

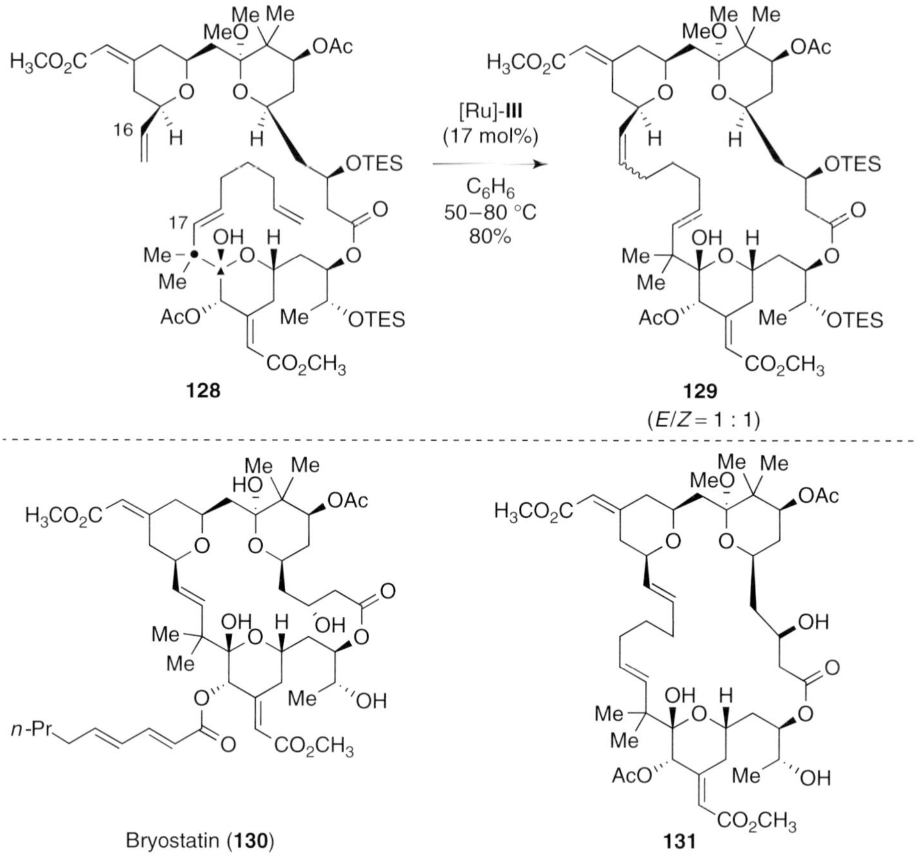

Scheme 9.31 Macrocyclization of the relay-armed RRCM substrate **128** occurring via direct RCM (to **129**) en route to the bryostatin analog **131**.

potential of relay metathesis. As this collection of examples demonstrates, relay metathesis has been implemented at the strategic level for a number of different reasons and to address a variety of different types of reactivity and selectivity problems and issues. The feature common to each application is the opportunity that a relay metathesis strategy provides to capitalize on a redirecting effect that can alter, and thereby potentially complement, the outcome of the parent (or non-relay) metathesis reaction. Given the huge (and still growing) impact that olefin metathesis has had on the science of organic synthesis, it is reasonable to assume that limitations in classical metathesis reactions will continue to be encountered, especially so in the often more complex settings of natural product synthesis. Hopefully, the relay metathesis results archived here will serve to stimulate readers' new ideas not only for solutions to their specific problems but also for new and innovative relay-based conceptual advances as well. For relevant

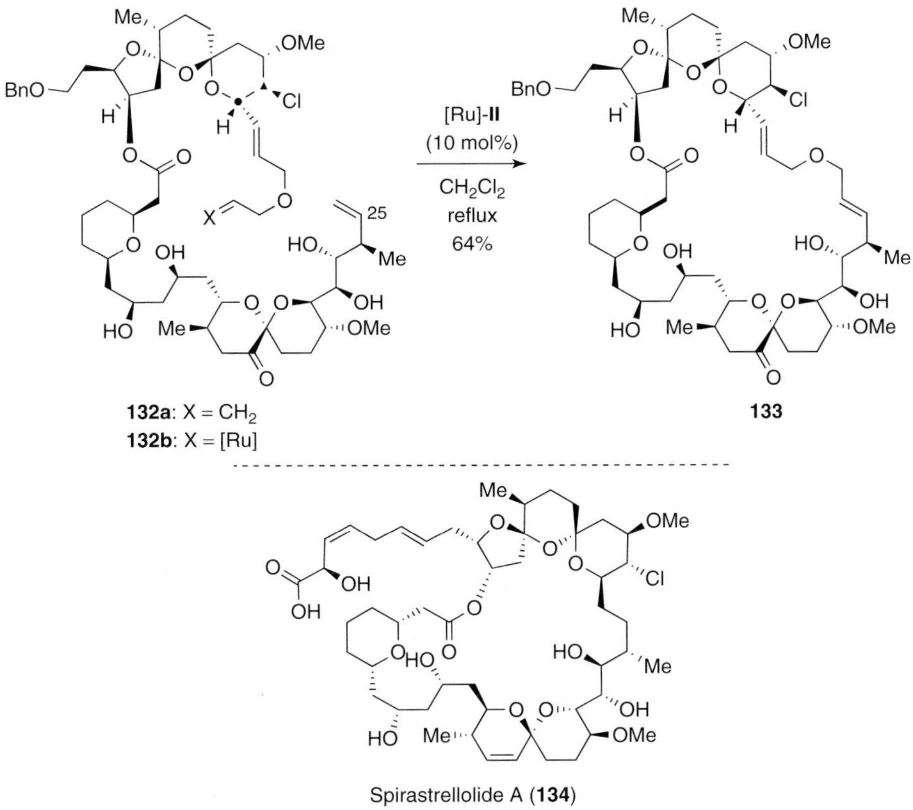

132a: X = CH_2
132b: X = [Ru]

133

Scheme 9.32 Premature macrocyclization, which is faster than the relay event, precluding RRCM.

recent studies appearing after submission of this chapter to the publisher and added ca. five months later at the stage of review of pageproof see Refs 38–43.

Acknowledgments

The authors thank Matthew J. Jansma and the editors of this book for their critical reviews of the manuscript, and the National Institutes of Health of the Department of Health and Human Services for research grants (GM65597 and CA76497) that have supported various aspects of our studies in this area here at the University of Minnesota.

References

1 Ulman, M. and Grubbs, R.H. (1998) *Organometallics*, **17**, 2484–2489.

2 (a) Kirkland, T.A. and Grubbs, R.H. (1997) *J. Org. Chem.*, **62**, 7310–7318; (b) refs in Stewart, I.C., Ung, T., Pletnev, A.A., Berlin, J.M., Grubbs, R.H., and Schrodi, Y. (2007) *Org. Lett.*, **9**, 1589–1592.

3 (a) Hoye, T.R. and Zhao, H. (1999) *Abstracts of Papers, 218th National Meeting of the American Chemical Society, August 22–26, 1999, New Orleans, LA*, American Chemical Society, Washington, DC, ORGN-620; (b) Hoye, T.R. and Zhao, H. (1999) *Abstracts of Papers, 218th National Meeting of the American Chemical Society, August 22–26, 1999, New Orleans, LA*, American Chemical Society, Washington, DC, ORGN-303.

4 Zhao, H. (2000) PhD Thesis, Studies towards a total synthesis of callipeltoside A, University of Minnesota, Minneapolis, MN.

5 Hoye, T.R., Jeffrey, C.S., Tennakoon, M.A., Wang, J., and Zhao, H. (2004) *J. Am. Chem. Soc.*, **126**, 10210–10211.

6 Robinson, J., Piscopio, A.D., and Zhu, L. (2003) *Abstracts of Papers, 226th National Meeting of the American Chemical Society, September 7–11, 2003, New York*, American Chemical Society, Washington, DC, ORGN-118.

7 (a) Hansen, E.C. and Lee, D. (2004) *Org. Lett.*, **6**, 2035–2038; (b) Hansen, E.C. (2006) PhD Thesis, Ruthenium-catalyzed enyne couplings: reactivity, selectivity and synthetic utility, University of Wisconsin, Madison, WI.

8 (a) Wang, X., Bowman, E.J., Bowman, B.J., and Porco, J.A. (2004) *Angew. Chem. Int. Ed. Engl.*, **43**, 3601–3605; (b) Wang, X. (2005) PhD Thesis, Studies toward the synthesis of chondriamides A and C, oximidines II and III, and the tetrapetalon A core, Boston University, Boston, MA.

9 Danielson, M.E. (2003) PhD Thesis, A total synthesis of callipeltoside A, University of Minnesota, Minneapolis, MN.

10 Tennakoon, M.A. (2001) PhD Thesis, efforts toward the total synthesis of scyphostatin; synthetic studies towards peloriside A, University of Minnesota, Minneapolis, MN.

11 (a) Hoye, T.R. and Wang, J. (2003) *Abstracts of Papers, 226th National Meeting of the American Chemical Society, September 7–11, 2003, New York*, American Chemical Society, Washington, DC, ORGN-670; (b) Wang, J. (2005) PhD Thesis, 1. The first total synthesis of haterumalide NA (oocydin A); 2. Deciphering the coupling constants of spiruchostatins by NMR spectroscopy; 3. Design and synthesis of analogs of latminculin A and B, Actin polymerization inhibitors, University of Minnesota, Minneapolis, MN; (c) Hoye, T.R. and Wang, J. (2005) *J. Am. Chem. Soc.*, **127**, 6950–6951.

12 Jeffrey, C.S. (2004) Master thesis, Part I: The epoxidation of 1.0 mg of geraniol. Part II: Relay ring closing metathesis (RRCM): Selectivity in tandem enyne synthesis, University of Minnesota, Minneapolis, MN.

13 Maifeld, S.V. and Lee, D. (2005) *Chem. Eur. J.*, **11**, 6118–6126.

14 Wallace, D.J. (2005) *Angew. Chem. Int. Ed. Engl.*, **44**, 1912–1915.

15 (a) Spino, C., Bolsvert, L., Douville, J., Roy, S., Lauzon, S., Minville, J., Gagnon, D., Beaumier, F., and Chabot, C. (2006) *J. Organomet. Chem.*, **691**, 5336–5355; (b) Lauzon, S., Tremblay, F., Gagnon, D., Godbout, C., Chabot, C., Mercier-Shanks, C., Perreault, S., DeSeve, H., and Spino, C. (2008) *J. Org. Chem.*, **73**, 6239–6250.

16 Sato, D., Fujiwara, K., Kawal, H., and Suzuki, T. (2008) *Tetrahedron Lett.*, **49**, 1514–1517.

17 Grandbois, A. and Collins, S.K. (2008) *Chem. Eur. J.*, **14**, 9323–9329.

18 El-Azizi, Y., Schmitzer, A., and Collins, S.K. (2006) *Angew. Chem. Int. Ed. Engl.*, **45**, 968–973.

19 (a) El-Azizi, Y., Zakarian, J.E., Bouillerand, L., Schmitzer, A.R., and Collins, S.K. (2008) *Adv. Synth. Catal.*, **350**, 2219–2225; (b) Zakarian, J.E., El-Azizi, Y., and Collins, S.K. (2008) *Org. Lett.*, **10**, 2927–2930; (c) Collins, S.K., El-Azizi, Y., and Schmitzer, A.R. (2007) *J. Org. Chem.*, **72**, 6397–6408.

20 Peng, W. and Blagg, B.S.J. (2006) *Org. Lett.*, **8**, 975–978.

21 Crimmins, M.T., Zhang, Y., and Diaz, F.A. (2006) *Org. Lett.*, **8**, 2369–2372.

22 She, J. (2004) PhD Thesis, Cost efficient asymmetric aldol reactions: application towards the total synthesis of (+)-gigantecin, University of North Carolina, Chapel Hill, NC.

23 Dudley, G.B., Engel, D.A., Ghiviriga, I., Lam, H., Poon, K.W., and Singletary, J.A. (2007) *Org. Lett.*, **9**, 2839–2842.

24 Roethle, P.A., Chen, I.T., and Trauner, D. (2007) *J. Am. Chem. Soc.*, **129**, 8960–8961.

25 Lee, H., Kim, K.W., Park, J., Kim, H., Kim, S., Kim, D., Hu, X., Yang, W., and Hong, J. (2008) *Angew. Chem. Int. Ed. Engl.*, **47**, 4200–4203.

26 Marvin, C.C., Voight, E.A., Suh, J.M., Paradise, C.L., and Burke, S.D. (2008) *J. Org. Chem.*, **73**, 8452–8457.

27 McGrath, N.A., Lee, C.A., Araki, H., Brichacek, M., and Njardarson, J.T. (2008) *Angew. Chem. Int. Ed. Engl.*, **47**, 1–5.

28 Xie, Q., Denton, R.W., and Parker, K.A. (2008) *Org. Lett.*, **10**, 5345–5348.

29 Druais, V., Hall, M.J., Corsi, C., Wendeborn, S.V., Meyer, C., and Cossy, J. (2009) *Org. Lett.*, **11**, 935–938.

30 Cho, E.J. and Lee, D. (2008) *Org. Lett.*, **10**, 257–259.

31 Li, J., Park, S., Miller, R.L., and Lee, D. (2009) *Org. Lett.*, **11**, 571–574.

32 Yun, S.Y., Kim, M., Lee, D., and Wink, D.J. (2009) *J. Am. Chem. Soc.*, **131**, 24–25.

33 Hoye, T.R. and Zhao, H. (1999) *Org. Lett.*, **1**, 169–171.

34 Cheung, L.L., Marumoto, S., Anderson, C.D., and Rychnovsky, S.D. (2008) *Org. Lett.*, **10**, 3101–3104.

35 Helmboldt, H. and Hiersemann, M. (2009) *J. Org. Chem.*, **74**, 1698–1708.

36 Trost, B.M., Yang, H., Thiel, O.R., Frontier, A.J., and Brindle, C.S. (2007) *J. Am. Chem. Soc.*, **129**, 2206–2207.

37 Fürstner, A., Fasching, B., O'Neil, G., Fenster, M.D.B., Godbout, C., and Ceccon, J. (2007) *Chem. Commun.*, 3045–3047.

38 Tannert, R., Hu, T.-S., Arndt, H.-D., and Waldmann, H. (2009) *Chem. Commun.*, 1493–1495.

39 Donohoe, T.J., fishlock, L.P., Basutto, J.A., Bower, J.F, Procopiou, P.A. and Thompson, A.l (2009) *Chem. Commun.*, 3008–3010.

40 Crimmins, M.T, Jacobs, D.L. (2009) *Org. lett.*, **11**, 2695–2698.

41 Ashworth, I.W., Miles, J.A.L., Nelson, D.J., Percy, J.M., singh, K. (2009) *Tetrahedron*, **65**, 9637–9646.

42 Donohoe, T.J., Bower, J.F., Basutto, J.A., fishlock, L.P., procopiou, P.A, and callens, C.K.A. (2009) *Tetrahedron*, **65**, 8969–8980.

43 Siegel, D.S., Piizzi, G., Piersanti, G., and Movassaghi, M. (2009) *J. Org. Chem.*, **2009**, ASAP.

10
Cross-metathesis in Natural Products Synthesis

Joëlle Prunet and Laurence Grimaud

10.1
Introduction

This chapter is divided into six sections. The first three sections loosely reflect the role of alkene cross-metathesis (CM) in the general plan of the synthesis of natural products, which can be the functionalization of terminal olefins appending a side chain to the core of a complex compound, or coupling two fragments to build the entire skeleton of the target molecule. Afterwards, tandem processes involving CM will be presented, followed by a few examples of ene–yne and alkyne CM in natural product synthesis [1].

This review is not comprehensive, but we have tried to select significant examples in each of the categories of CM reactions cited above, while at the same time emphasizing to which types [2] the olefins involved in the metathesis reactions belong.

10.2
Functionalization of Olefins

10.2.1
Cross-metathesis with Acrylate Derivatives

Electron-deficient olefins such as acrylate derivatives fall into the type II category. Cross-metatheses of these substrates (used in excess) with type I terminal olefins are usually easy and diastereoselective in favor of the *E*-isomer, except for acrylonitrile, which leads to the *Z*-products.

10.2.1.1 Acrylonitrile

Stockman, Fuchs, and coworkers employed a two-directional CM reaction between dienone **1** and acrylonitrile to synthesize the core of histrionicotoxin (Scheme 10.1), a noncompetitive inhibitor of nicotinic acetylcholine receptors [3]. Thermal conditions for this metathesis reaction required refluxing in tetrahydrofuran

Metathesis in Natural Product Synthesis: Strategies, Substrates and Catalysts. Edited by Janine Cossy, Stellios Arseniyadis, and Christophe Meyer

ISBN: 978-3-527-32440-8

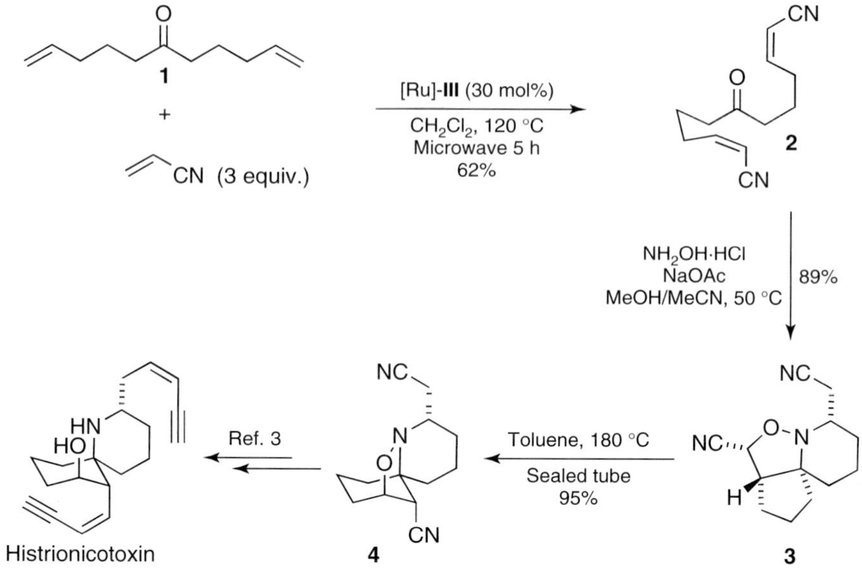

Scheme 10.1 CM with acrylonitrile in the synthesis of histrionicotoxin.

(THF) for two weeks and adding 5 mol% of catalyst every two days. Fortunately, CM could be accelerated by microwave irradiation and compound **2** could be obtained in 62% yield after only 5 hours at 120 °C. Kinetic tricyclic product **3** was then obtained by an oxime formation/Michael addition/1,4-prototropic shift/[3+2]-cycloaddition cascade, before being isomerized to the thermodynamic tricycle **4** via a retro-[3+2]/[3+2]-cycloaddition sequence.

10.2.1.2 **Thioacrylates**

The synthesis of faranal, the trail pheromone of the common ant *Monomorium pharaonis*, was recently reported by Minnaard, Feringa, and coworkers [4]. Hence, CM of **5** with ethyl thioacrylate furnished the unsaturated thioester **6** in good yield along with 5% of the homodimer of **5** (Scheme 10.2). The second stereocenter was installed by an enantioselective conjugate addition, which led to **7** in both high yield and high selectivity.

10.2.1.3 **Acrylic Acid**

Ruthenium catalysts are tolerant to many functional groups including carboxylic acids. BouzBouz and Cossy took advantage of this property for the construction of the *seco* acid of colletol, a bis-lactone isolated from the fermentation broth of *Collectotrichum capsici* [5]. Conjugated hexenoic acid derivative **9** was obtained in 78% from olefin **8** and acrylic acid, while the C9–C10 double bond was formed with an *E*/*Z* selectivity better than 20 : 1 (Scheme 10.3). Macrolactonization under Yamaguchi's conditions followed by the cleavage of the methoxymethyl (MOM)

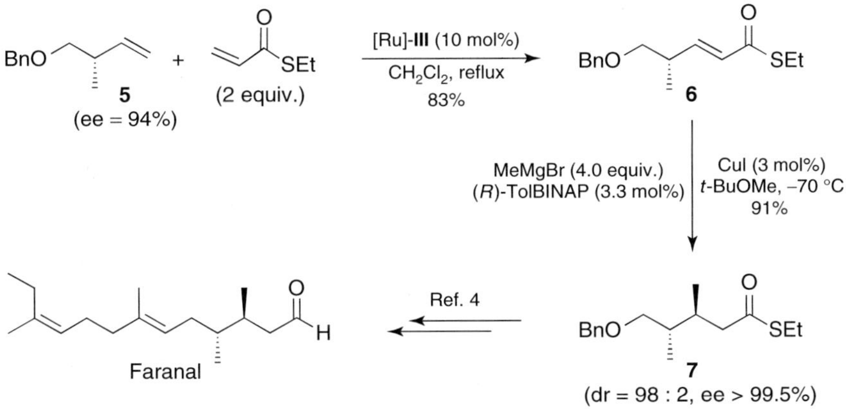

Scheme 10.2 Synthesis of faranal involving CM with ethyl thioacrylate.

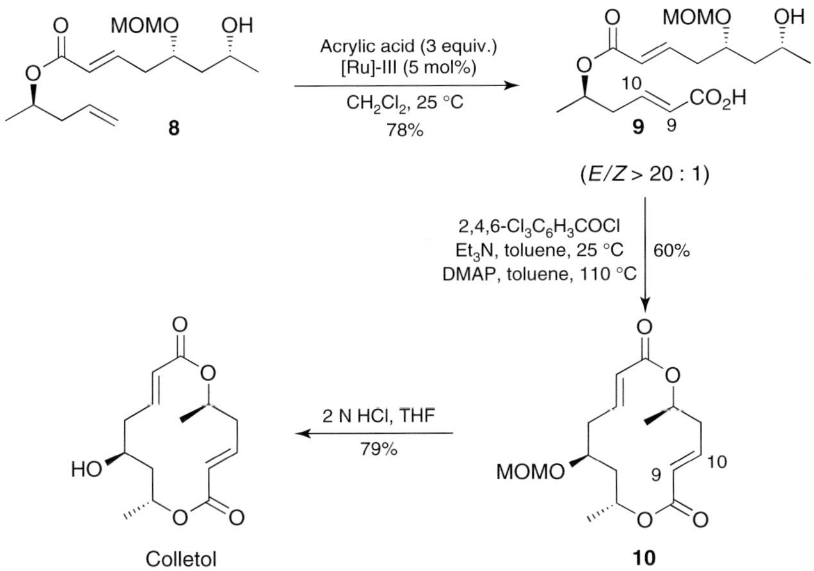

Scheme 10.3 CM with acrylic acid toward colletol.

ether led to colletol in good overall yield. It is interesting to note that an alternative route involving acylation of **8** with acrylic acid followed by ring-closing metathesis (RCM) led to macrocycle **10** as a 2.8 : 1 mixture of *E*- and *Z*-isomers.

10.2.1.4 **Acrylimides**

During their total synthesis of cylindramide A [6], a macrocyclic tetramic acid isolated from *Halicondria cylindrata*, which is cytotoxic to B16 melanoma cells, Hart and Phillips performed a highly diastereoselective Diels–Alder reaction between acrylimide **13** and cyclopentadiene, applying the methodology developed

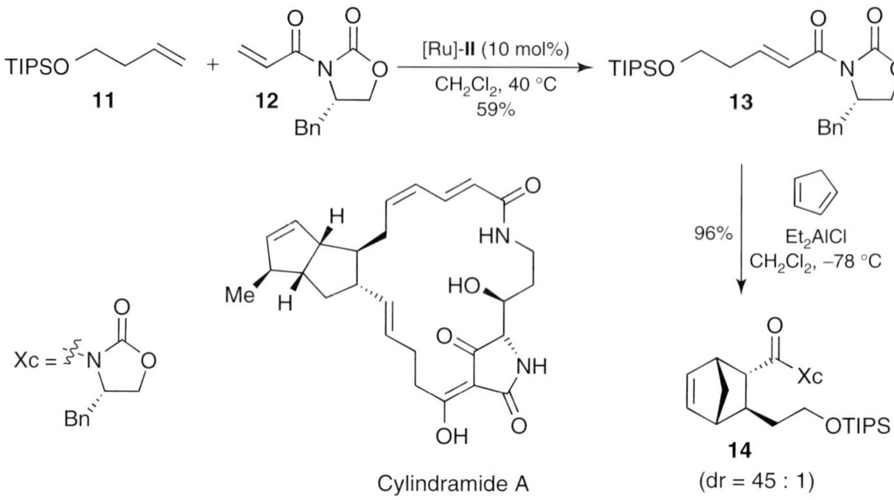

Scheme 10.4 CM with an acrylimide derivative en route to cylindramide A.

by Evans *et al.* [7]. Dienophile **13** was prepared via a CM between olefin **11** and acryloyl oxazolidinone **12** using [Ru]-**II** (Scheme 10.4).

10.2.1.5 Acrylates

There are numerous examples of CM reactions with ethyl or methyl acrylate. An interesting case was uncovered by Trost *et al.* during their formal synthesis of isoaltholactone, a cytotoxic agent selective for HT-29 colon cancer cells [8]. CM of **15** was regioselective, with metathesis only occurring on the side of the allylic alcohol to give unsaturated ester **16** in good yield (Scheme 10.5). The authors invoke the

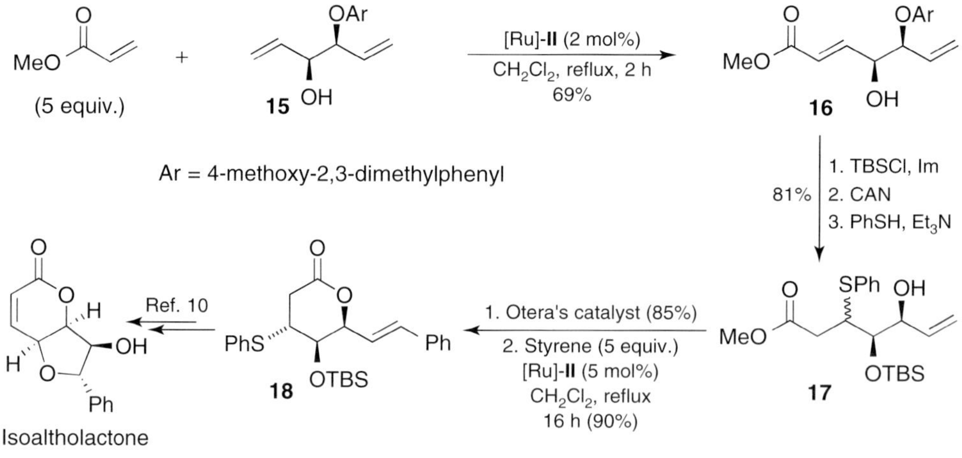

Scheme 10.5 Regioselective CM with methyl acrylate for the formal synthesis of isoaltholactone.

deactivating effect of the electron-rich arylether to explain this selectivity. Further functionalization of compound **16** led to ester **17**, which was lactonized in the presence of Otera's catalyst [9]. CM of the resulting lactone with styrene then furnished lactone **18**, while elimination of the benzenethiol moiety completed the formal synthesis of isoaltholactone [10].

10.2.1.6 **Acrolein**

Cossy and BouzBouz designed a retrosynthesis of the C1–C14 fragment of amphidinol 3 with three CM reactions between homoallylic alcohols and acrylic derivatives [11]. The first metathesis between **19** and acrolein was straightforward, giving the expected product **20** in 79% yield (Scheme 10.6). Aldehyde **20** was then transformed into bis-acetate **21** by treatment with the allyltitanium complex (*S,S*)-**24**, first reported by Hafner, Duthaler, and coworkers [12], followed by acetylation. Interestingly, the second CM with acrolein leading to **22** was very chemoselective and only the terminal olefin reacted. The internal double bond did not undergo metathesis, probably because the corresponding metallacyclobutane was deactivated by complexation of the ruthenium by the carbonyl moiety of the acetate group. It had been shown on model systems that the same reaction on the corresponding alcohols was not selective [11]. Aldehyde **22** was then transformed into homoallylic acetate **23** as previously, while the C1 and C2 carbons were installed by CM with ethyl acrylate. In each case, the *E*/*Z* ratio obtained after metathesis was better than 50 : 1.

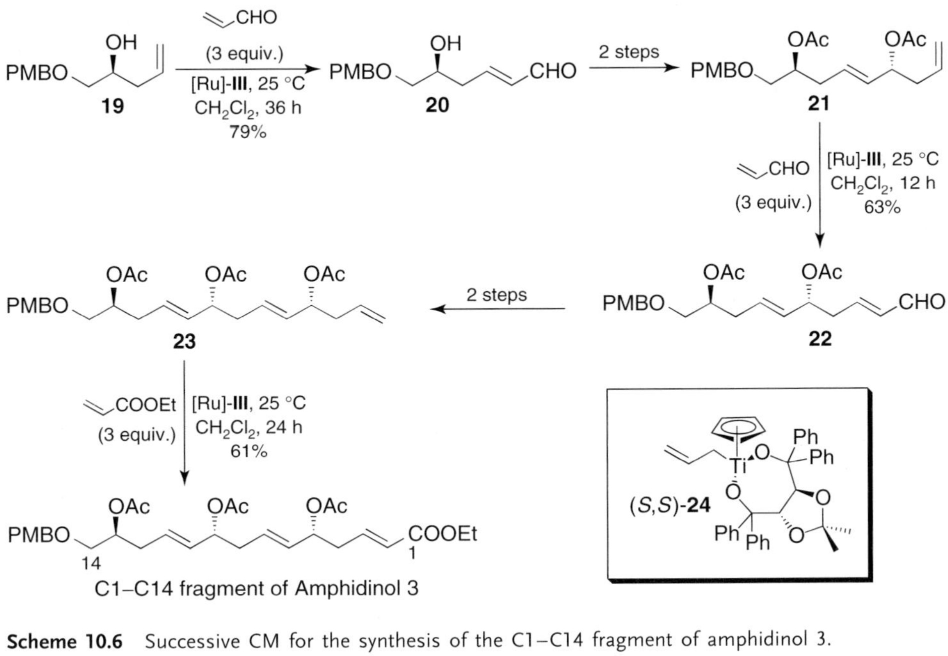

Scheme 10.6 Successive CM for the synthesis of the C1–C14 fragment of amphidinol 3.

10.2.1.7 Vinyl Ketones

In their synthesis of nupharamine [13], isolated from the rhizome of the water lily *Nuphar japonica*, Gebauer and Blechert effected a CM of homoallylic protected amine **26** (prepared in three steps from the corresponding phthalimide **25**) with furyl vinyl ketone, leading to the desired compound **27** in good yield (Scheme 10.7). Hydrogenation of the double bond followed by deprotection of the amine and reductive amination then furnished nupharamine.

The vinyl ketone partner used in the CM reaction can be of a size comparable to that of the other partner. For example, Cossy *et al.* recently described the synthesis of the monomeric counterpart of marinomycin A, isolated from the saline culture of *Marinospora* strain CNQ-140, which shows significant antibiotic activity against *Staphylococcus aureus* or *Enterococcus faecium*, using a CM between a vinyl ketone and a type I olefin [14]. Another case of CM between a complex vinyl ketone and a type I olefin was reported by Rychnovsky and coworkers for the construction of the C1–C52 segment of amphidinol 3 [15].

10.2.2 Cross-metathesis with Vinyl Derivatives

CM reactions can also be used to prepare partners for cross-coupling reactions such as vinyl boronates, vinyl silanes, or vinyl halides, which are type II olefins.

10.2.2.1 Vinyl Boronates

CM reactions with vinyl boronates were first reported by Morril and Grubbs [16]. They are usually diastereoselective in favor of the *E*-isomer. Vincent and Prunet used this type of CM for the synthesis of the C1–C15 fragment of dolabelide C, a cytotoxic macrolactone isolated from the sea hare *Dolabella auricularia* [17]. Hence, 1,1-disubstituted olefin **28** was reacted with 4 equiv. of vinyl boronate **29**, and the resulting product was converted to the corresponding vinyl iodide in 89% yield for the two steps (Scheme 10.8). As it is often the case with 1,1-disubstituted

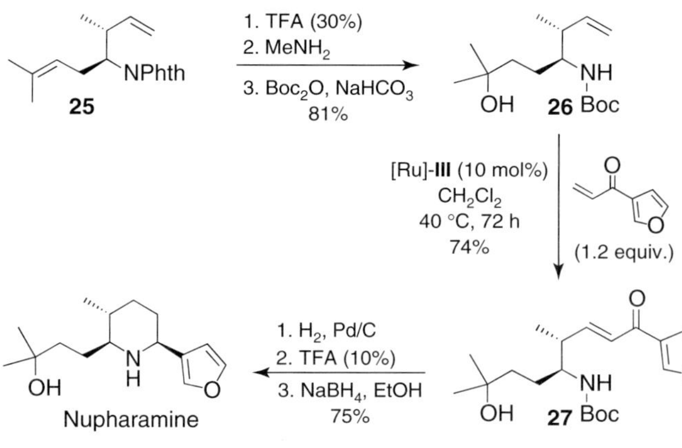

Scheme 10.7 Synthesis of nupharamine by a CM/reductive amination sequence.

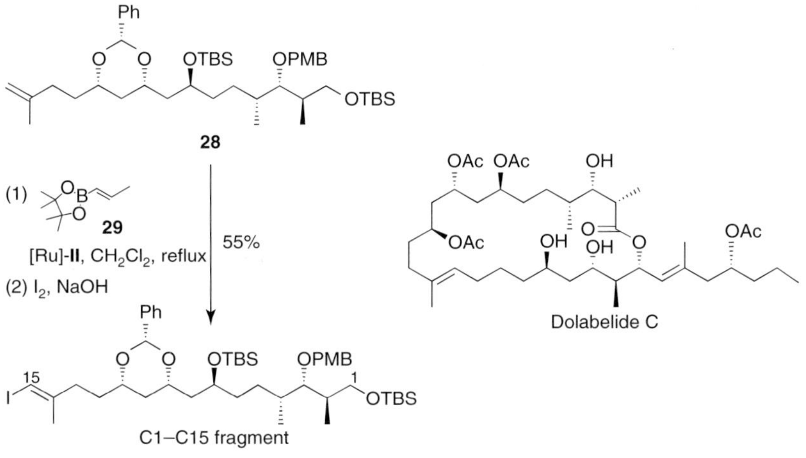

Scheme 10.8 CM with a vinyl boronate for the synthesis of the C1–C15 fragment of dolabelide C.

olefins with no strong steric bias, the selectivity of the metathesis reaction was modest ($E/Z = 2:1$) [2]; however, the desired *E*-olefin could be obtained in 55% overall yield after separation of the two diastereomers.

A CM reaction between vinyl boronate **29** and a terminal olefin was also employed by Rahn and Kalesse during their synthesis of chlorotonil A (see Scheme 10.11 for the structure of this compound), which has been isolated from the myxobacterium *Sorangium cellulosium* [18]. The resulting product was then engaged in a Suzuki coupling.

10.2.2.2 **Vinyl Silanes**

The same kind of strategy was employed by Kirschning and coworkers for their synthesis of thuggacin B, a polyketide isolated from the myxobacterium *Sorangium cellulosium*, which shows promising activity against *Mycobacterium tuberculosis* [19]. CM of terminal olefin **30** (encompassing a diol motif) with vinyl triethoxysilane proceeded in satisfactory yield (Scheme 10.9). When the two hydroxy groups present in **30** were protected as TBS ethers, the yield of the metathesis was poor, probably due to steric hindrance. After protection of the diol moiety as an acetonide, vinyl silane **31** was transformed into vinyl iodide **32**, which was later engaged in a Sonogashira coupling.

10.2.3 **Cross-metathesis with Allylic Derivatives**

This category of coupling usually involves two type I olefins, and proceeds in good yield but with variable stereoselectivity.

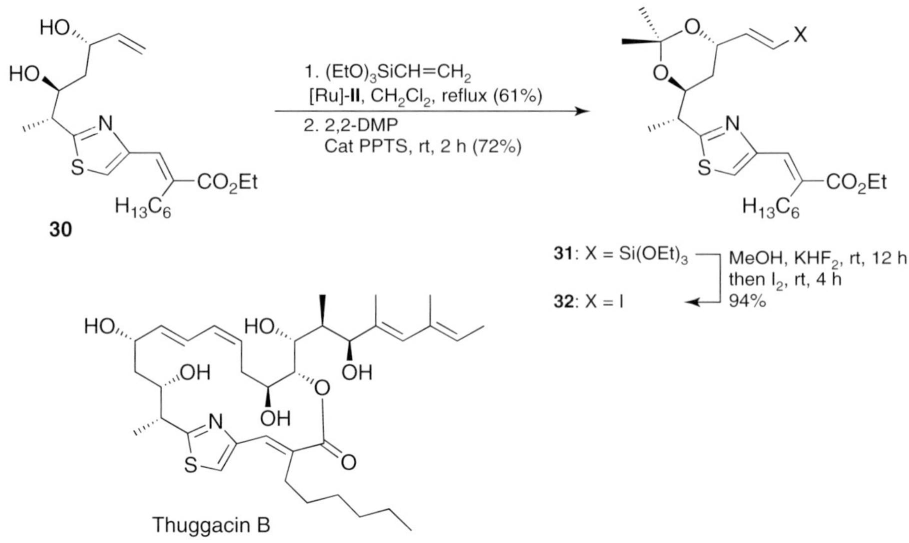

Scheme 10.9 CM with a vinyl silane en route to thuggacin B.

10.2.3.1 Allyl Silanes

Allyl trimethylsilane has been used as a CM partner by Fukuyama and coworkers in the synthesis of the 3,8-diazabicyclo[3.2.1] core of lemonomycin, a tetrahydroisoquinoline alkaloid isolated from *Streptomyces candidus* with interesting antibiotic activity against *Staphylococcus aureus* or *Enterococcus faecium*, as well as cytotoxicity against HCT-116 human colon cancer cell line [20]. CM between disubstituted olefin **33** and allyl trimethylsilane led to the desired product **34** in 51% yield as a mixture of *E*- and *Z*-isomers (Scheme 10.10). This compound, when treated with $BF_3{\cdot}OEt_2$, yielded bicycle **35** via the corresponding acyl iminium. When the metathesis reaction was performed with the allyl analog of **33** without the terminal methyl group, a low yield was obtained because of partial decomposition.

10.2.3.2 Allyl Phosphonates

Rahn and Kalesse employed a diastereoselective CM reaction during their synthesis of chlorotonil A between **36** and allyl phosphonate **37**, which led to phosphonate **38** as a single isomer (Scheme 10.11) [18]. This phosphonate was then used in an olefination with aldehyde **39** which occurred without elimination of the allylic OPMB group to afford diene **40**.

10.2.3.3 Allylic Alcohol Derivatives

Roulland recently described the total synthesis of Oocydin A, a chlorinated macrolide isolated from the bacterium *Serratia marcescens* [21]. CM of diol **41** with allyl acetate proceeded to give compound **42** in good yield and excellent selectivity (E/Z = 95 : 5) (Scheme 10.12). This compound was then engaged in a palladium-catalyzed allylic substitution that led to THF **43** in a highly diastereoselective fashion.

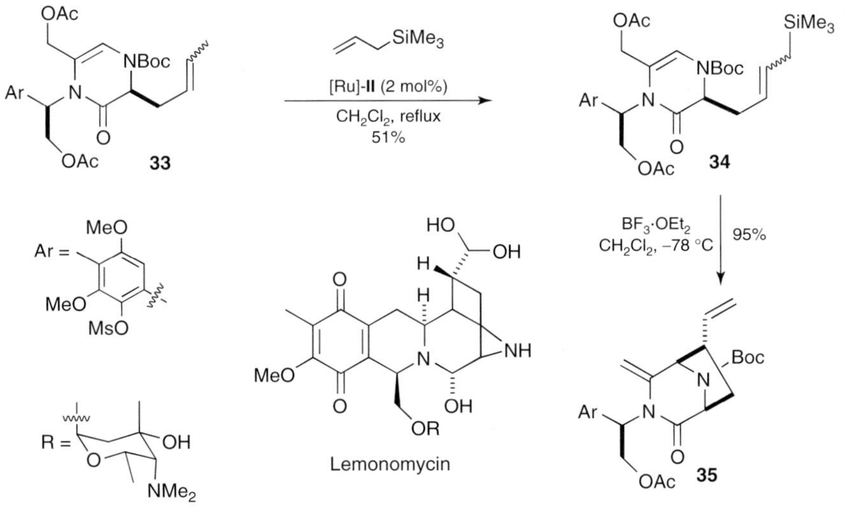

Scheme 10.10 CM with allyl silane for the construction of the lemonomycin core.

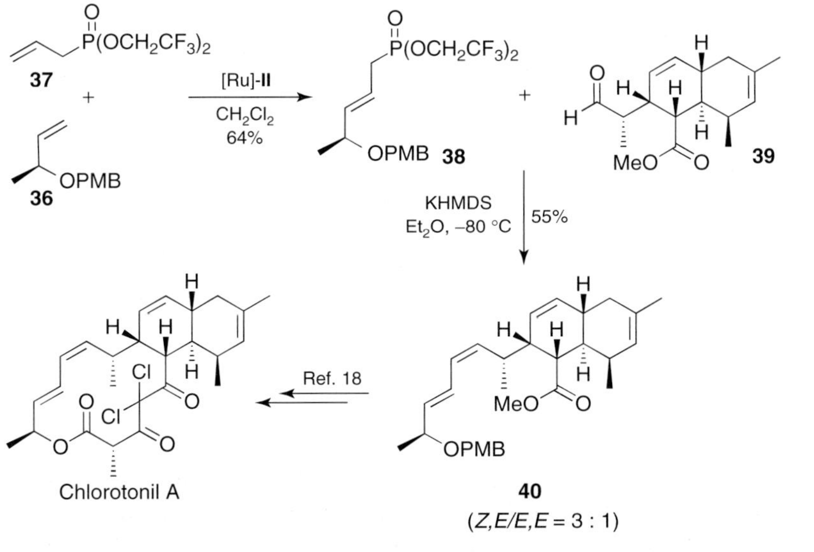

Scheme 10.11 Formation of an allyl phosphonate by CM in the synthesis of chlorotonil A.

10.2.3.4 Miscellaneous

Lasonolide A, a macrolide isolated form the Caribbean marine sponge *Forcepia* sp., exhibits potent antitumor activity against A-549 human lung carcinoma and Panc-1 human pancreatic carcinoma. The synthesis of this molecule described by Ghosh and Gong involves a CM between olefin **44** and homoallyltetrazolylsulfone **45**, which proceeds in good yield (Scheme 10.13) [22]. After protection of the hydroxy group present in **46** as a methylthiomethyl (MTM) ether, the resulting sulfone was

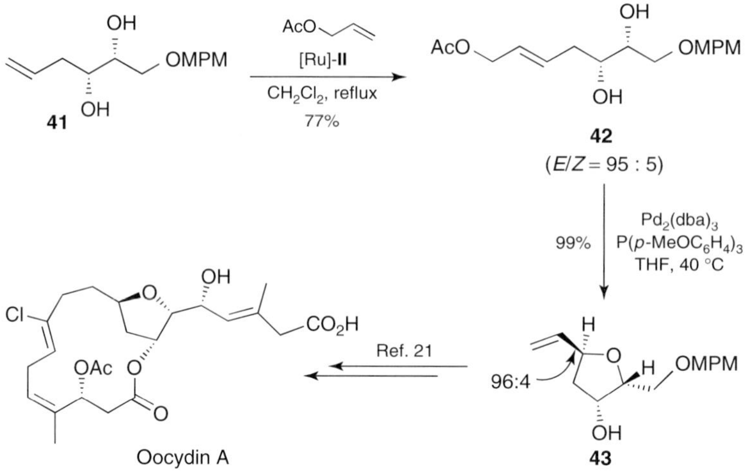

Scheme 10.12 Synthesis of oocydin involving a CM with allyl acetate.

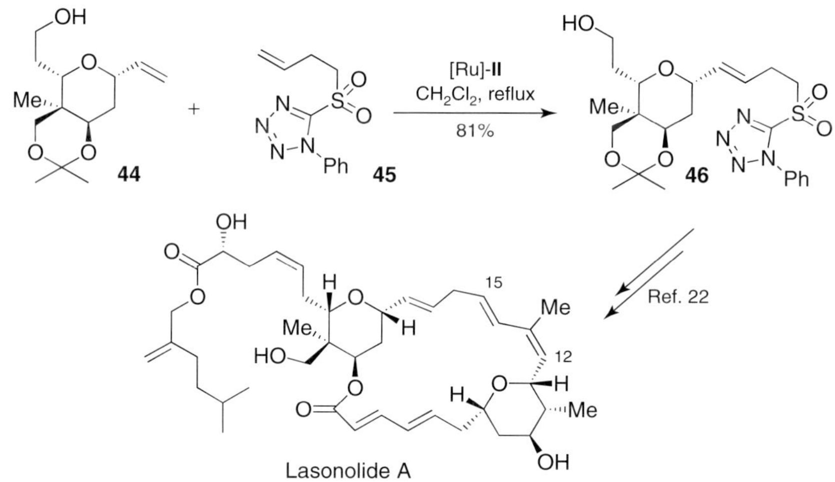

Scheme 10.13 CM with homoallyl tetrazolyl sulfone for the construction of lasonolide A.

engaged in a Julia–Kocienski coupling with an unsaturated aldehyde to form the C12–C15 dienic system of lasonolide A.

10.3
Appending a Side Chain

Appending a side chain to the core of a complex molecule very often involves a type II olefin (the core) and a type I olefin (the side chain). However, in some cases the side chain can be of type II or III as exemplified in this section.

10.3.1
With No Functional Group

10.3.1.1 A Simple Case

Reiser and coworkers reported the enantioselective synthesis of five paraconic acids with interesting antibiotic and antitumor properties [23]. These 3-carboxylic acid-substituted γ-butyrolactones differ *inter alia* by their side chain at C4, which were attached by CM (Scheme 10.14). For the synthesis of roccellaric acid, lactone **47a** was treated with 1.5 equiv. of 1-dodecene and catalyst [Ru]-**I**. Compound **48a** was obtained in 57% yield as a 3.5 : 1 mixture of *E*- and *Z*-isomers. In an analogous way, CM of lactone **47b** with 1-decene furnished **48b**, a precursor of nephrosteranic acid, in 53% yield. Even though a large excess of 1-decene (4 equiv.) and the more active catalyst [Ru]-**II** were used, the conversion was not complete (27% recovered **47b**). However, the selectivity was much higher in the latter case ($E/Z = 7:1$).

10.3.1.2 The Specific Case of Isopropylidene

1,1-Disubstituted alkenes do not usually dimerize in the presence of catalyst [Ru]-**II** but when isobutylene is used as solvent, a small amount of background dimerization is observed. Grubbs and coworkers reported the superior behavior of 2-methyl-2-butene for this kind of reactions [24], while Porco and Qi employed this reagent for the synthesis of clusianone, a potential anti-HIV therapeutic agent isolated from the floral resin of *Clusia* species [25]. The isoprenyl side chain was installed in good yield from **49** by allylic acetate reduction followed by CM in neat 2-methyl-2-butene in the presence of catalyst [Ru]-**II**. Compound **50** was isolated and then transformed into the target molecule by nucleophilic demethylation (Scheme 10.15).

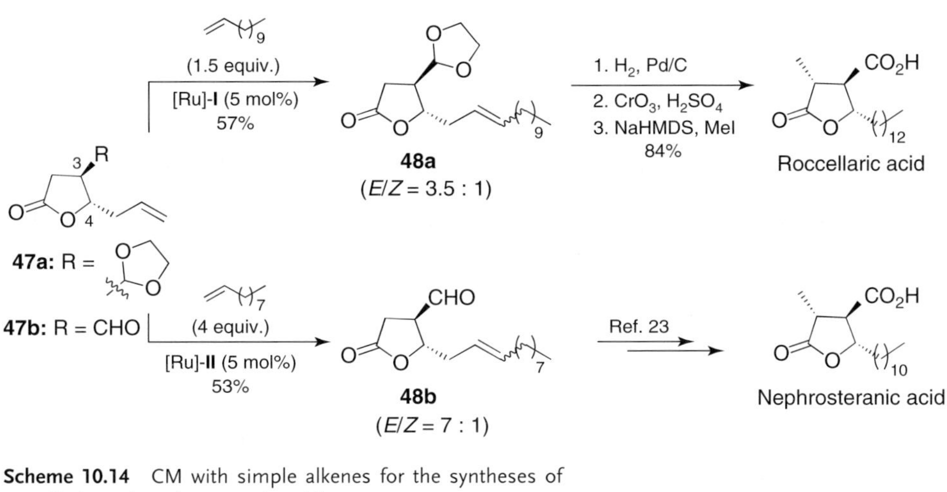

Scheme 10.14 CM with simple alkenes for the syntheses of roccellaric and nephrosteranic acids.

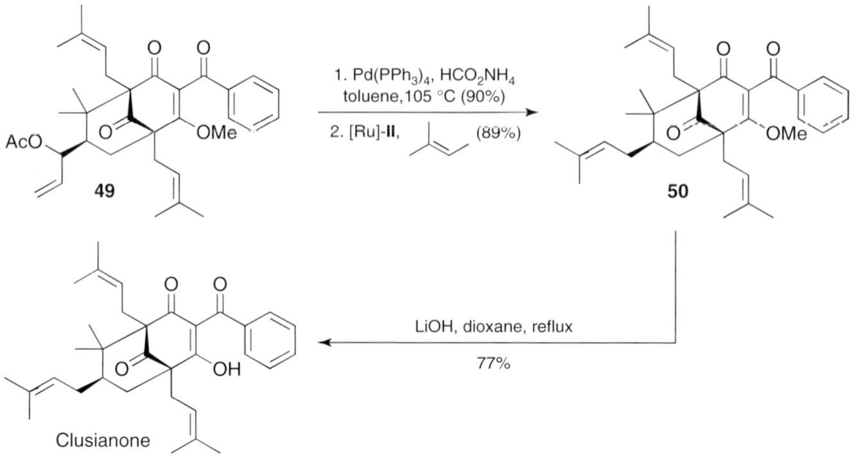

Scheme 10.15 Transforming an allyl into an isoprenyl side chain using CM with 2-methyl-2-butene.

10.3.1.3 Removing Part of a Side Chain

Wipf *et al.* converted an allyl substituent into an ethyl group during their synthesis of tuberostemonine (Scheme 10.16) [26], a compound from the family of the *Stemona* alkaloids, which have been used against tuberculosis and bronchitis in Eastern folk medicine. Isomerization of the terminal double bond in **51** was

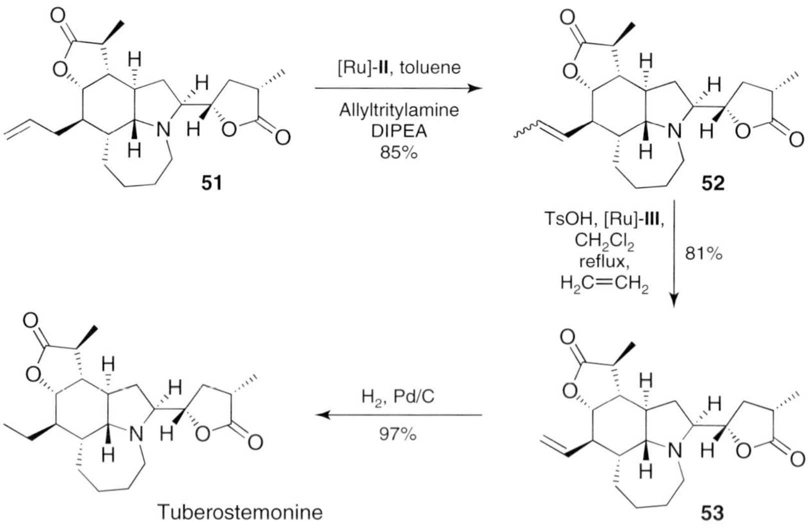

Scheme 10.16 Isomerization/CM sequence for the synthesis of the side chain of tuberostemonine.

accomplished with catalyst [Ru]-**II** in refluxing toluene in the presence of allyltritylamine and diisopropylethylamine (DIPEA) [27]. CM reaction of **52** with ethylene in the presence of catalyst [Ru]-**III** and *para*-toluenesulfonic acid (TsOH) in refluxing CH_2Cl_2 furnished compound **53** in 81% yield. Interestingly, the use of the phosphine-free catalyst for this step was important in order to avoid lengthy chromatographic separation that led to decomposition of the product. Hydrogenation of the vinyl substituent finally provided tuberostemonine in excellent yield.

Franck and coworkers had previously employed a similar isomerization/CM sequence for the preparation of sugar derivatives used in the synthesis of α-*C*-galactosyl ceramide immunostimulants, but in their case the isomerization reaction was catalyzed with $PdCl_2$ [28].

10.3.2
With Functional Groups

During their synthesis of prostaglandins $F_{2\alpha}$ and J_2, Mulzer and coworkers studied the effect of allylic and homoallylic substituents on CM [29]. Hence, reaction of lactone **54a** (R = Bz) with allylic silylether **55**, in the presence of complex [Ru]-**II** added via syringe pump, furnished compound **56a** in only 40% yield and 7 : 1 *E*/*Z* selectivity along with 25% of the corresponding debenzoylated product **56c** (R = H) (Scheme 10.17). With catalyst [Ru]-**III**, the yield dropped to 25%. These poor results are probably due to the deactivation of the catalyst by complexation of the ruthenium by the carbonyl moiety of the benzoate. Indeed, when lactone **54b** (R = TBS) was submitted to CM, the desired product **56b** was obtained in 84% yield and 17 : 1 *E*/*Z* selectivity. The result with lactone **54c** (R = H) was also satisfactory, as the corresponding coupled product was isolated in 70% yield and 17 : 1 *E*/*Z* selectivity (in all cases, the catalyst was added over a period of

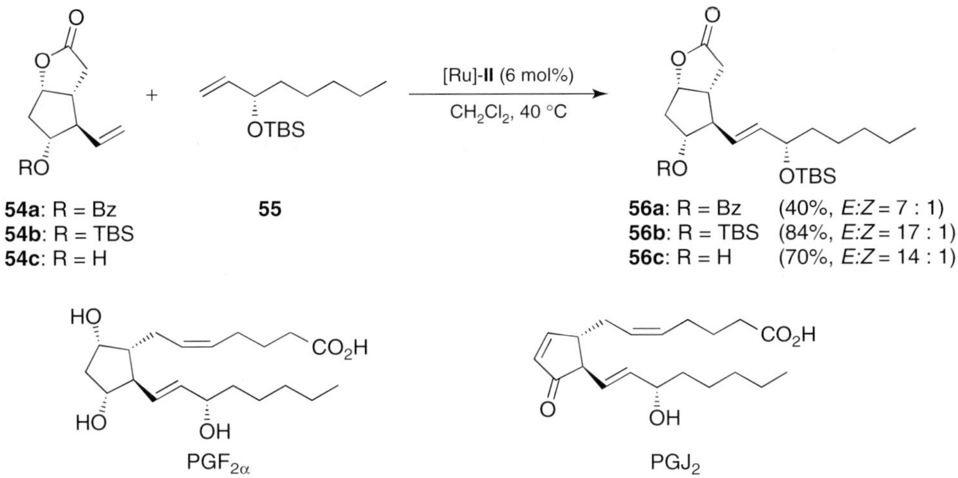

Scheme 10.17 Effect of homoallylic substituents on CM en route to prostaglandins.

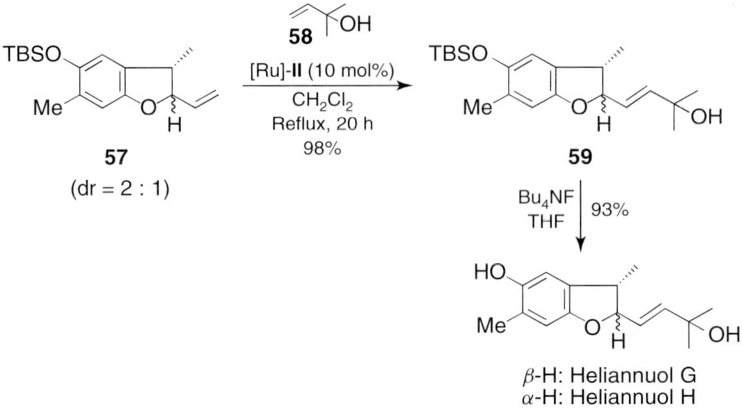

Scheme 10.18 Installing the side chain of heliannuols G and H using CM with a hindered allylic alcohol.

8–12 hours). Compounds **56b** and **56c** were then converted into prostaglandins $F_{2\alpha}$ and J_2, respectively. It is noteworthy that when catalyst [Ru]-**III** was used with **54b** or **54c**, only the homodimers of each partner were obtained.

Shishido and coworkers have used CM with a hindered allylic alcohol for appending the side chain of heliannuols G and H, two sesquiterpenes isolated from the leaves of *Helianthus annus* L. SH-22, and YPP, which exhibit interesting phytotoxic allelopathic activity [30]. Reaction of TBS ether **57** with 2-methyl-3-buten-2-ol **58** led to the direct precursor of heliannuols G and H in excellent yield (Scheme 10.18). However, when the phenolic analog of **57** was submitted to similar reaction conditions, a mixture of heliannuols G and H was obtained in only 32% yield, thereby showing the importance of the phenol protecting group on the CM.

10.4 Couplings

There are numerous examples of CM being used to assemble the skeleton of natural products.

For example, in the last step of the synthesis of melithiazole C, a fungicide extracted from myxobacterium *Melittangium lichenicola*, Cossy and coworkers optimized a CM between vinyl thiazole **60** and allylic methylether **61** (Scheme 10.19) [31]. While catalyst [Ru]-**I** was ineffective, CM of an equimolar amount of **60** and **61**

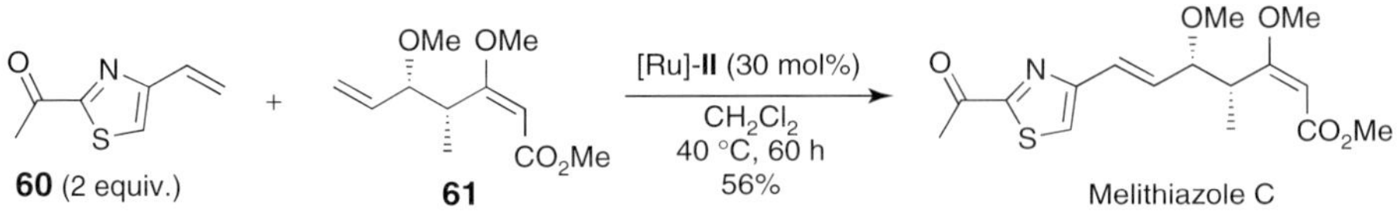

Scheme 10.19 CM with a vinyl thiazole for the last step of melithiazole C synthesis.

with [Ru]-**II** or [Ru]-**III** resulted in the formation of the homodimer of thiazole **60**. Since compound **61** is a type II olefin that dimerizes slowly, more forcing conditions were employed (30 mol% of catalyst) and the target compound was obtained in 56% yield (68% conversion) with an excellent *E*/*Z* ratio. Higher catalyst loading or heating the reaction in benzene at 60 °C did not improve the yield.

Marshall and Sabatini recently reported the synthesis of asimicin, an *Annonaceus* acetogenin exhibiting a broad range of biological activities [32]. Two successive CM reactions were envisaged for the construction of **64** (Scheme 10.20). When C_2-symmetric bis-alkene **62a** was reacted with 1-decene, the product resulting from a bis-metathesis reaction **62b** was obtained in 80% yield. No selective mono CM was possible since **62a** is a type II olefin as opposed to a type I partner. However, further metathesis of type III olefin **62b** with type I olefin **63** furnished the desired product **64** in 38% yield along with 28% of recovered **62b** and 9% of the bis-metathesis product. A single recycle of the side products furnished an additional 25–30% yield in **64**. Hydrogenation of the two alkene moieties and acidic removal of the protecting groups led to asimicin in good yield. Bis-metathesis of **62a** with **63** followed by CM with 1-decene was also performed, but the overall process was slightly less efficient.

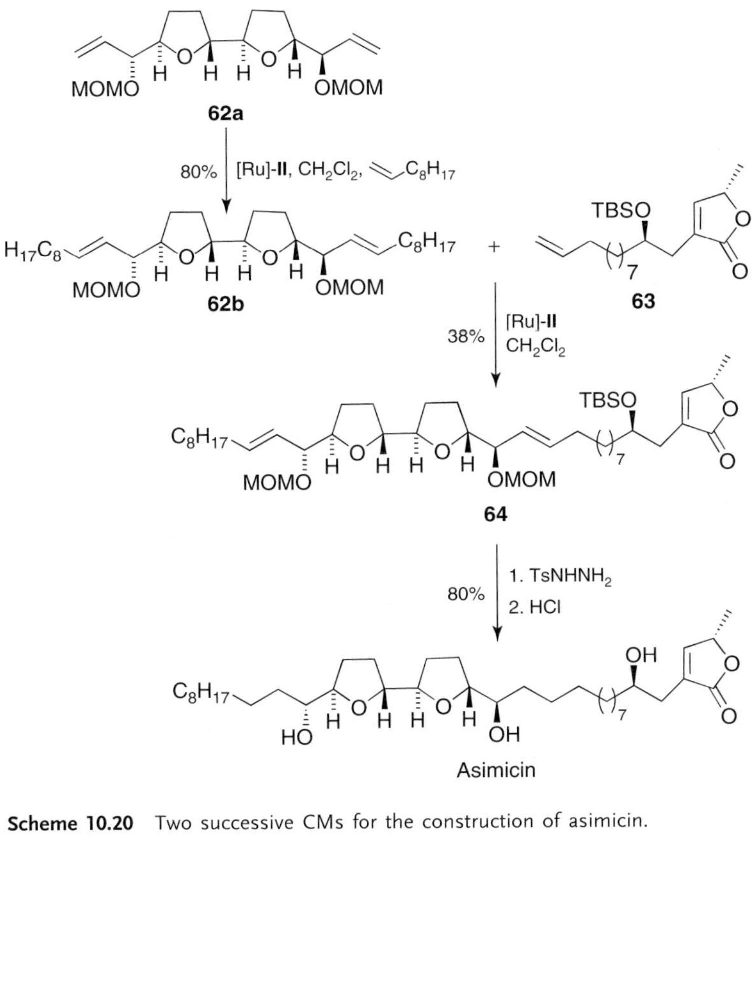

Scheme 10.20 Two successive CMs for the construction of asimicin.

Two other syntheses of acetogenins involving CM were described in the same year by Crimmins *et al.* (mucocin) [33] and Hoye *et al.* (gigantecin) [34].

CM involving a diene partner can be difficult when the other olefin is hindered. Morita and Kuwahara had to employ an allylic pivalate as a diene precursor for the preparation of the dienoic fragment of elaiolide [35]. Koide and coworkers resorted to the use of the nitro-substituted [Ru]-**VII** [36] for the metathesis of a diene leading to FR901464, which is a potent anticancer product isolated from *Pseudomonas* sp. No. 2663 [37]. However, the CM of diene **66** with 2 equiv. of allylic alcohol **65** performed by Crimmins *et al.* led to diene **67** in 63% yield along with 31% of recovered **66** and the homodimer of **65**, which could be recycled (Scheme 10.21) [38]. The *E*-selectivity on the other hand was total. Compound **67** was then transformed into apoptolidinone, the aglycone of apoptolidine, which induces cell death by inhibiting the mitochondrial F_0F_1-ATPase.

Quinn *et al.* designed a tandem RCM/CM process for the synthesis of three members of the *Annonaceous* acetogenin family, muricatacin [39], rollicosin [40], and

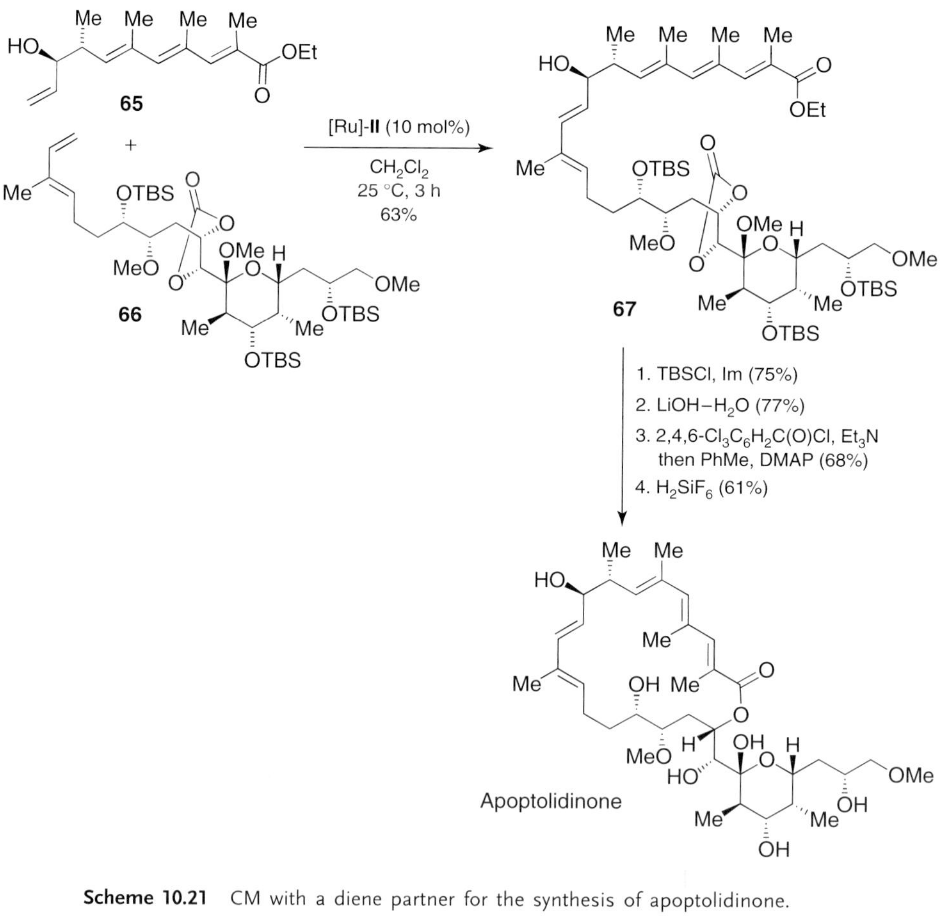

Scheme 10.21 CM with a diene partner for the synthesis of apoptolidinone.

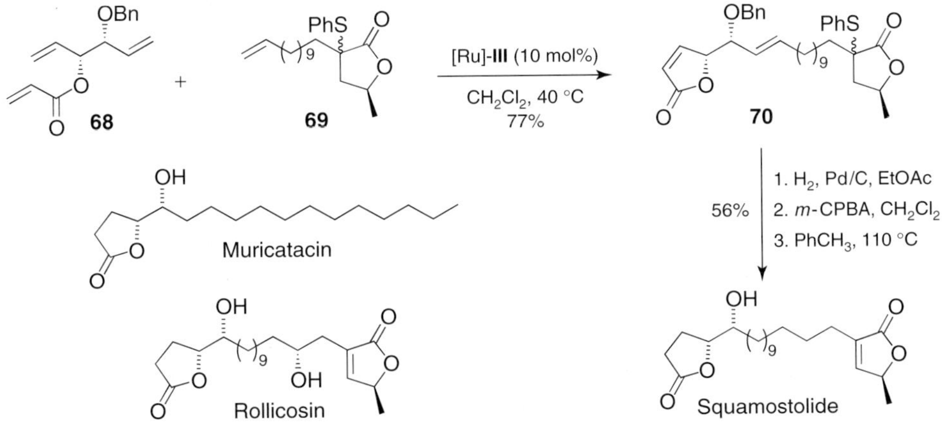

Scheme 10.22 Tandem RCM/CM in the synthesis of squamostolide.

squamostolide [41] (Scheme 10.22). Formation of the unsaturated five-membered ring lactone by RCM and appending the side chain by CM of the other alkene occurred in one pot (Scheme 10.22). The optimized conditions necessitated slow addition of the catalyst by syringe pump to a 0.01 M solution of **68** and **69** in refluxing CH_2Cl_2 over 6 hours and further heating for an additional 14 hours. The simple RCM product was observed as an intermediate when catalyst [Ru]-**II** was used (confirming that the RCM process is faster than the CM reaction), but not with complex [Ru]-**III** which proved to be superior in this case. Since homodimerization of type I olefin **69** was very fast, the CM leading to **70** likely involved this dimer. Interestingly, the other possible RCM product (the six-membered ring unsaturated lactone) was never observed. Simple transformations ultimately led to squamostolide in good overall yield.

10.5 Cascade Processes Involving CM

Since olefin metathesis reactions involve alkenes both as starting materials and products, they are potential candidates for cascade processes.

10.5.1 ROM/CM

Parrain and coworkers described an efficient cascade involving the ring opening of cyclopropene ketal of type **71** followed by CM with a variety of alkenes in the presence of catalyst [Ru]-**I** for the synthesis of protected divinyl ketones [42]. This sequence was employed by Kozmin and coworkers for the preparation of the linear center portion of bistramide A, a marine metabolite isolated from *Lissoclinum bistratum* which exhibits potent cytotoxicity against several cell lines [43], as well as for the synthesis of spirofungin A, an inhibitor of isoleucyl-tRNA synthase isolated from *Streptomyces violaceusniger* Tü [44], and routiennocin, an antibiotic agent extracted from *Streptomyces routienii* Huang sp. nov. (ATCC39446) [45]. For the latter case, ring opening of cyclopropene acetal **71** with catalyst [Ru]-**II** followed by CM with alkene **72** produced the elongated acetal, which proved inert to subsequent metathesis (Scheme 10.23). The corresponding dienone obtained by *in situ* acidic hydrolysis was subjected to a second CM with alkene **73** to furnish **74** in 59% yield. Subsequent hydrogenation and saponification of the methyl ester led to the target molecule.

Asymmetric ring-opening metathesis (AROM) followed by CM proved very efficient for the synthesis of baconipyrone C realized by Hoveyda and Gillingham (Scheme 10.24) [46]. Hence, desymmetrization of compound **75** was best accomplished by ROM/CM with styrene in the presence of [Ru]-**XV** generated

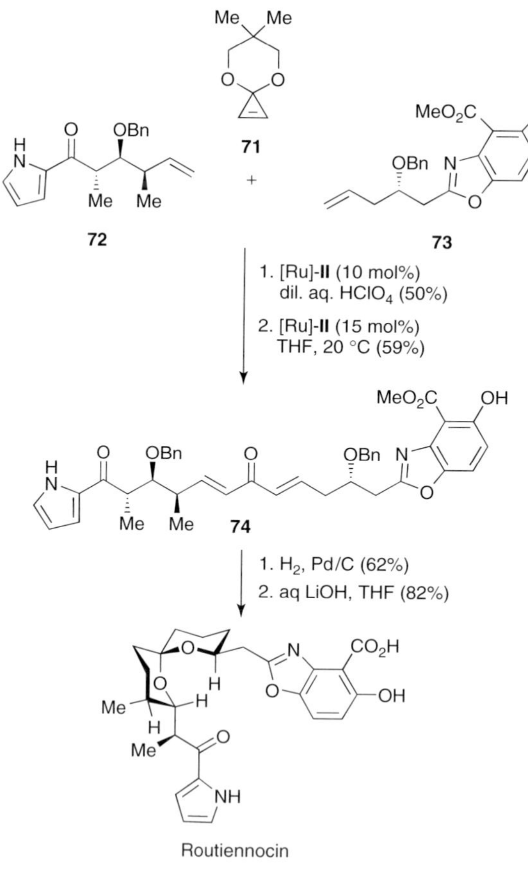

Scheme 10.23 Cascade ROM/CM of cyclopropene ketal for the synthesis of routiennocin.

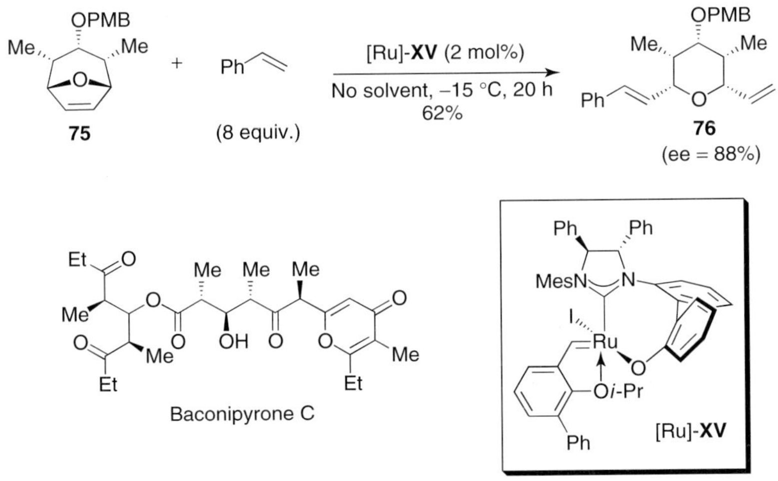

Scheme 10.24 AROM/CM in the synthesis of baconipyrone C.

in situ. Interestingly, when the corresponding isolated catalyst was used [47], higher catalyst loading was required for a lower yield.

10.5.2
ROM/CM/RCM

A three-step process was reported by Tadano and coworkers for the construction of the 9-oxabicyclo[4.2.1]nona-2,4-diene core of mycoepoxydiene [48]. The reactions were first examined in a stepwise manner. Ring opening of **77** followed by CM of the carbene intermediates with 1,3-butadiene furnished a mixture of regioisomeric products **78** and **79** (Scheme 10.25). Since only the *Z*-stereomers can undergo subsequent ring closure, this reaction was optimized for the formation of the *Z*-olefins. Catalyst [Ru]-**I** gave the best ratio, while complex [Ru]-**II** always favored the thermodynamic *E*-isomers. Formation of cyclooctadiene **80** necessitated the more active catalyst [Ru]-**II**; therefore, for the one-pot procedure this carbene was added to the reaction after consumption of **77** and 1,3-butadiene degassing, and the mixture was heated at reflux. Desilylation was then effected to enable separation of **80** from the by-products probably resulting from the unproductive *E*-isomers of **78** and **79**.

10.5.3
ROM/RCM/CM

Hart and Phillips have used a CM for the preparation of a direct precursor of vinyl ketone **81**, see Section 2.1 (Scheme 10.4) [6]. Their synthesis of cylindramide A then involved an ROM/RCM cascade of norbornene derivative **81**, followed by CM with alkene **82**, which led to the fused 5,5-bicyclic compound **83** in 59% yield as a 2 : 1 mixture of separable *E*- and *Z*-diastereomers (Scheme 10.26).

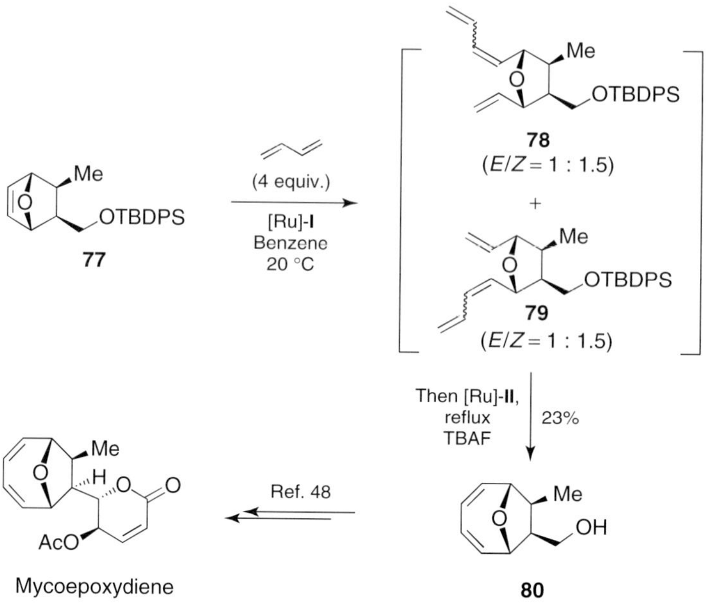

Scheme 10.25 ROM/CM/RCM process for the preparation of mycoepoxydiene.

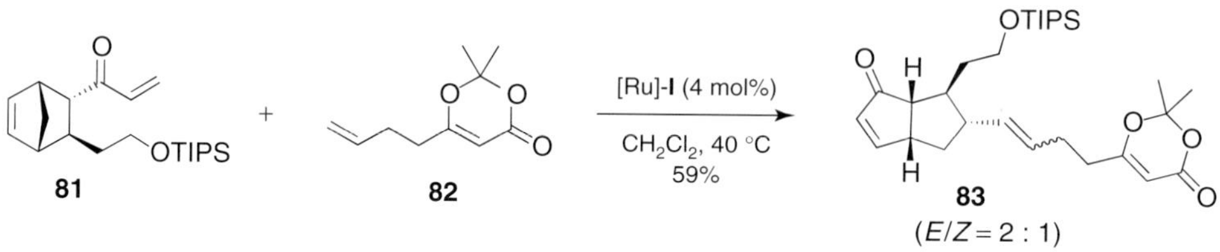

Scheme 10.26 ROM/RCM/CM cascade en route to cylindramide A.

10.5.4 CM/RCM

Smith *et al.* reported an elegant synthesis of cytotoxic agents cylindrocyclophanes A and F, which were isolated from the terrestrial blue-green algae *Cylindrospermum lichenforme* [49]. Dimerization of bis-olefin **84** with [Ru]-**I**, followed by RCM of the resulting dimer furnished exclusively the [7,7]-paracyclophane **85** in 61% yield as a single *E,E*-isomer (Scheme 10.27). The yield could not be improved with catalyst [Ru]-**II**; however, the use of [Mo]-**I** gave better results (72% yield). In addition, none of the "head-to-head" dimerization products was observed during this reaction. In an interesting experiment, these dimers, prepared by another route, were independently submitted to [Mo]-**I** and compound **85** was obtained as

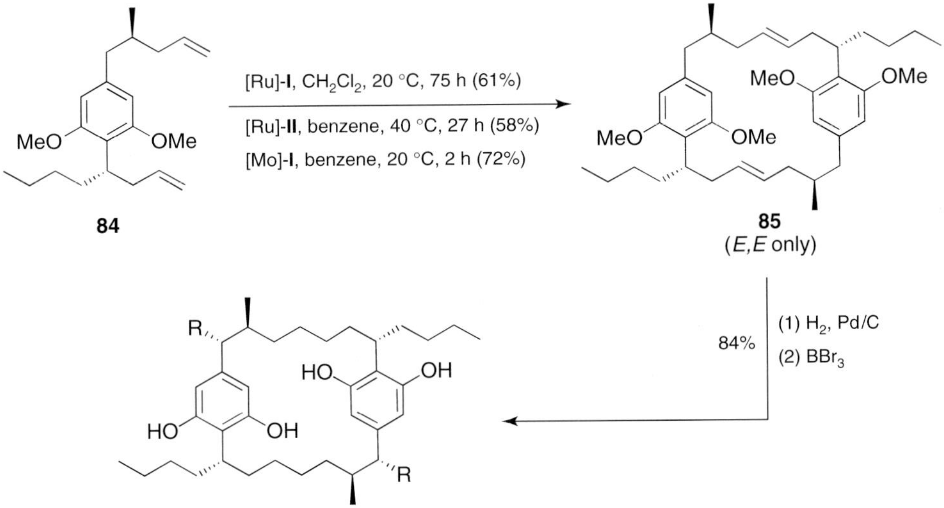

84

85
(*E*,*E* only)

Cylindrocyclophane F: R = H
Cylindrocyclophane A: R = OH

Scheme 10.27 Synthesis of cylindrocyclophane F using a CM/RCM sequence.

the exclusive product in each case, as the [8,6]-cyclophanes were not detected. The dimers probably undergo a cascade of metathesis reactions leading to the most stable [7,7]-paracyclophane **85**. This compound was then converted in two steps into cylindrocyclophane F.

10.5.5
RCEYM/CM

Finally, metathesis between an alkene and an alkyne, first reported by Kinoshita and Mori [50], is a very efficient way of producing 1,3-dienes [51]. The ring-closing version of this reaction (ring-closing ene–yne metathesis (RCEYM)) leads to semicyclic 1,3-dienes, which can undergo CM with another alkene. Clark and coworkers utilized this sequence of reactions for the construction of polyether substructures found in the dinoflagellates toxins gambierol and hemibrevetoxin B [52]. Compound **86** was thus transformed into diene **87** by RCEYM, followed by CM with the dimer of allyl acetate in 62% overall yield (Scheme 10.28). When they used the one-pot process first reported by Grimaud, El Kaïm, and coworkers [53], **87** was produced in 54% yield along with 25% of the ene–yne RCM product.

Hoveyda and coworkers also employed this sequence of reactions for the synthesis of the antimycobacterial agent erogorgiaene [54]. The stepwise operation was efficient but the yield for the one-pot operation was never over 20%.

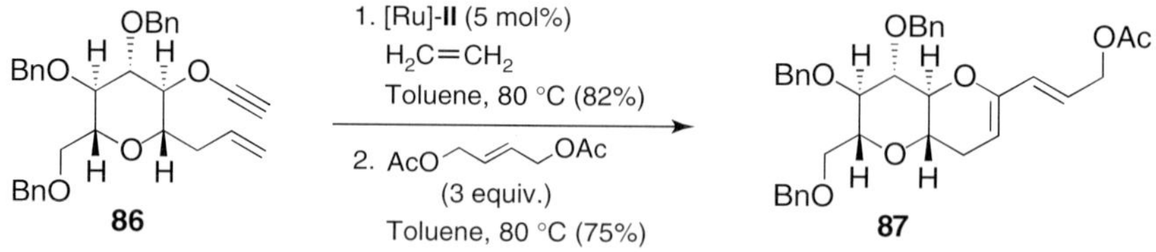

Scheme 10.28 Synthesis of a polycyclic ether subunit by an ene–yne RCM/CM sequence.

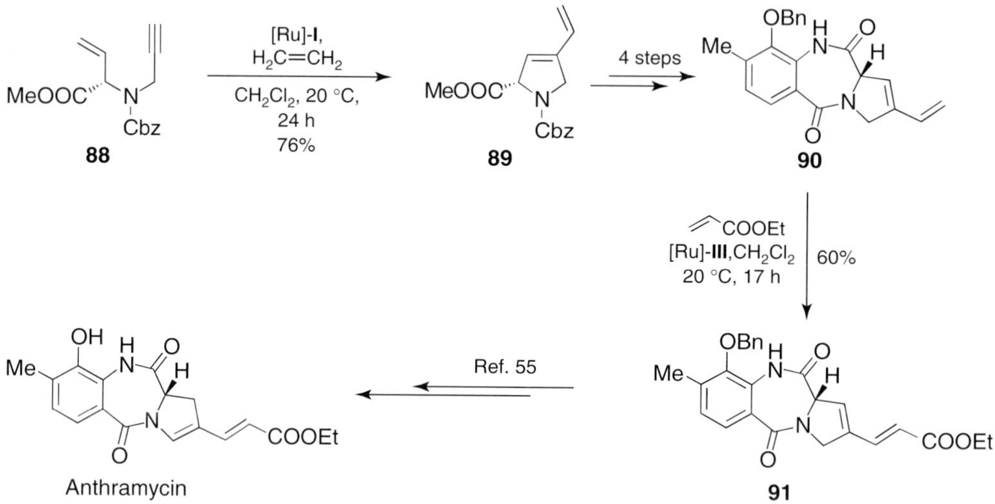

Scheme 10.29 Synthesis of an anthramycin derivative.

A derivative of the antitumor agent anthramycin was synthesized by Mori and coworkers by performing a CM reaction on a diene that was formed by ene–yne metathesis, but several steps separated the two operations [55]. [Ru]-**I** efficiently catalyzed the formation of five-membered diene **89** (Scheme 10.29). After four steps, the resulting product **90** was submitted to CM in the presence of ethyl acrylate and [Ru]-**III** to furnish compound **91**. The dienamine system could be isomerized into the one present in anthramycin by heating **91** with rhodium trichloride ($RhCl_3{\cdot}3H_2O$) in ethanol at 110 °C.

10.6
Ene–yne CM

Very few examples of ene–yne CM have been reported in the field of natural products synthesis. Mori *et al.* employed this reaction for the synthesis of anolignans A and B, two HIV-1 reverse transcriptor inhibitors extracted form *Anogeissus acuminate* [56]. CM of internal alkyne **92** with ethylene in the presence of [Ru]-**II**

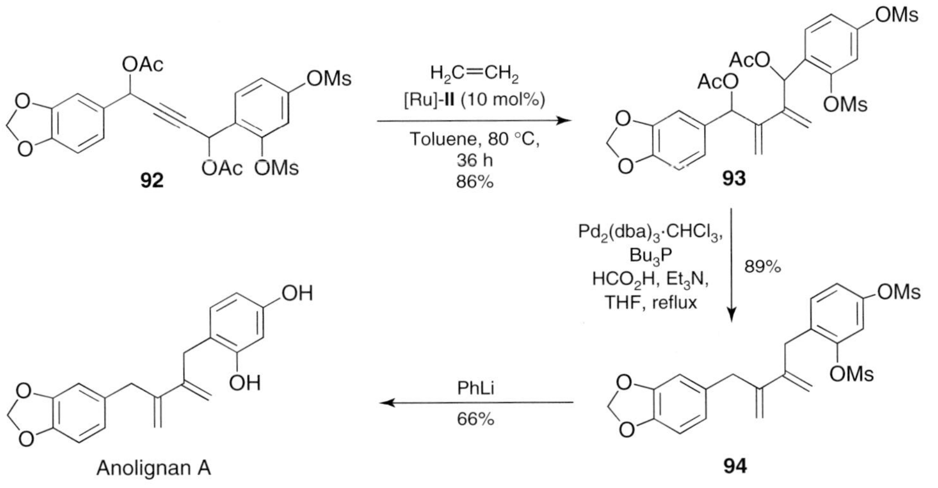

Scheme 10.30 Ene–yne CM with ethylene toward anolignan A.

led to the desired 2,3-substituted diene **93** in 86% yield (Scheme 10.30). The same reaction with [Ru]-**I** afforded **93** in 65% yield, along with 15% of recovered **92**. Hydrogenolysis under palladium catalysis then furnished **94**, which was converted to anolignan A by removal of the mesylates.

A CM between a terminal alkyne and an allylsilylether was successfully performed by Rodriguez and coworkers for the preparation of the dienic system of mycothiazole [57]. Diver and coworkers also described an elegant ene–yne CM with 1,5-hexadiene followed by RCM en route to a cyclohexadienic intermediate of scabrosins [58].

10.7 Alkyne CM

Alkyne metathesis has only been recently applied to the synthesis of natural products [59]. The development of a well-defined molybdenum catalyst by Fürstner *et al.* largely contributed to this success. Up to now, few examples of the CM version have been reported. However, Fürstner *et al.* described the syntheses of dehydrohomoancepsenolide [60] and PGE_2 methyl ester [61] using this strategy. The triaryl amido molybdenum complex **[Mo]-V** did not induce metathesis, but when it was activated by CH_2Cl_2, the corresponding chloro complex was formed, and CM of alkyne **95** with internal alkyne **96** led to compound **97** in 51% yield (Scheme 10.31). This alkyne was then hydrogenated and the silylether hydrolyzed to furnish PGE_2 methyl ester in good overall yield.

Another example of alkyne CM can be found in the synthesis of terpestacin reported by Chan and Jamison [62].

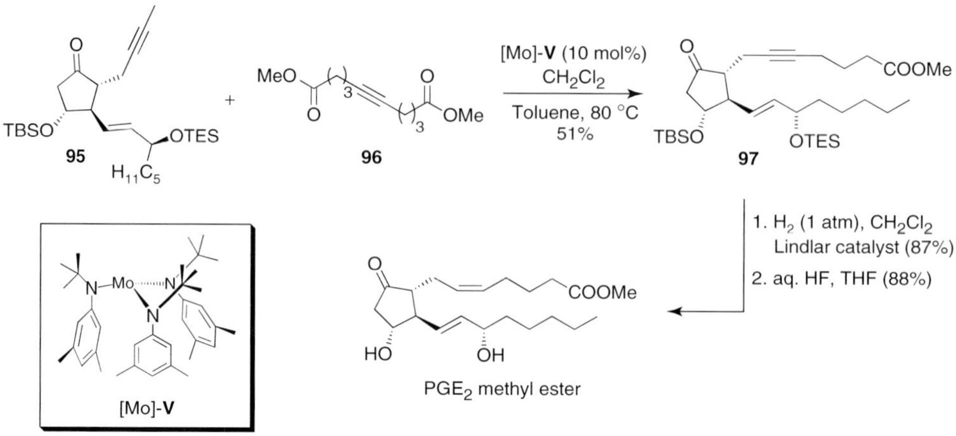

Scheme 10.31 Synthesis of PGE_2 methyl ester using alkyne CM.

10.8 Conclusion and Perspectives

CM was developed later than RCM, but can now be found in numerous syntheses of natural products. This reaction has largely benefited from the discovery of second-generation ruthenium catalysts, and in most cases, high yielding chemo- and *E*-stereoselectivities are obtained. More importantly, the outcome of CM can be predicted using the classification of olefins according to their ability to dimerize proposed by Grubbs and coworkers [2], making it a reliable reaction when planning the synthesis of complex targets.

However, CM involving hindered olefinic partners can be difficult, and in many cases require addition of fresh catalyst after a certain reaction time or syringe pump addition of the catalyst. The construction of trisubstituted double bonds [63] and *Z* double bounds [64] by CM is also not an easy task, but progress is being made in this direction with the development of new Ru or Mo catalysts.

Acknowledgments

Financial support for this work was provided by the CNRS, the Ecole Polytechnique, and the Ecole Nationale Supérieure des Techniques Avancées.

References

1 For a previous review on the application of olefin CM to the synthesis of natural products, see: Prunet, J. (2005) *Curr. Top. Med. Chem.*, **5**, 1559–1577.

2 (a) Chatterjee, A.K., Choi, T.-L., Sanders, D.P., and Grubbs, R.H. (2003) *J. Am. Chem. Soc.*, **125**, 11360–11370; (b) Blackwell, H.E., O'Leary, D.J.,

Chatterjee, A.K., Washenfelder, R.A., Bussmann, D.A., and Grubbs, R.H. (2000) *J. Am. Chem. Soc.*, **122**, 58–71.

3 Karatholuvhu, M.S., Sinclair, A., Newton, A.F., Alcaraz, M.-L., Stockman, R.A., and Fuchs, P.L. (2006) *J. Am. Chem. Soc.*, **128**, 12656–12657.

4 Van Zijl, A.W., Szymanski, W., Lopez, F., Minnaard, A.J., and Feringa, B.L. (2008) *J. Org. Chem.*, **73**, 6994–7002.

5 BouzBouz, S. and Cossy, J. (2006) *Tetrahedron Lett.*, **47**, 901–904.

6 Hart, A.C. and Phillips, A.J. (2006) *J. Am. Chem. Soc.*, **128**, 1094–1095.

7 Evans, D.A., Chapman, K.T., and Bisaha, J. (1984) *J. Am. Chem. Soc.*, **106**, 4261–4263.

8 Trost, B.M., Aponick, A., and Stanzl, B.N. (2007) *Chem. Eur. J.*, **13**, 9547–9560.

9 Orita, A., Sakamoto, K., Hamada, Y., and Otera, J. (1999) *Tetrahedron*, **55**, 2899–2910.

10 For the synthesis of isoaltholactone, see: Harris, J.M. and O'Doherty, G.A. (2001) *Tetrahedron*, **57**, 5161–5171.

11 BouzBouz, S. and Cossy, J. (2001) *Org. Lett.*, **3**, 1451–1454.

12 Hafner, A., Duthaler, R.O., Marti, R., Rihs, G., Rothe-Streit, P., and Schwarzenbach, F. (1992) *J. Am. Chem. Soc.*, **114**, 2321–2336.

13 Gebauer, J. and Blechert, S. (2005) *Synlett*, **18**, 2826–2828.

14 Amans, D., Bellosta, V., and Cossy, J. (2007) *Org. Lett.*, **9**, 1453–1456.

15 Huckins, J.R., de Vicente, J., and Rychnovsky, S.D. (2007) *Org. Lett.*, **9**, 4757–4760.

16 Morril, C. and Grubbs, R.H. (2003) *J. Org. Chem.*, **68**, 6031–6034.

17 Vincent, A. and Prunet, J. (2006) *Synlett*, **14**, 2269–2271.

18 Rahn, N. and Kalesse, M. (2008) *Angew. Chem. Int. Ed.*, **47**, 597–599.

19 Bock, M., Dehn, R., and Kirschning, A. (2008) *Angew. Chem. Int. Ed.*, **47**, 9134–9137.

20 Rikimaru, K., Mori, K., Kan, T., and Fukuyama, T. (2005) *Chem. Commun.*, 394–396.

21 Roulland, E. (2008) *Angew. Chem. Int. Ed.*, **47**, 3762–3765.

22 Ghosh, A.K. and Gong, G. (2007) *Org. Lett.*, **9**, 1437–1440.

23 Chhor, R.B., Nosse, B., Sörgel, S., Böhm, C., Seitz, M., and Reiser, O. (2003) *Chem. Eur. J.*, **9**, 260–270.

24 Chaterjee, A.K., Sanders, D.P., and Grubbs, R.H. (2002) *Org. Lett.*, **4**, 1939–1942.

25 Qi, J. and Porco, J.A. (2007) *J. Am. Chem. Soc.*, **129**, 12682–12683.

26 (a) Wipf, P., Rector, S.R., and Takahashi, H. (2002) *J. Am. Chem. Soc.*, **124**, 14848–14849; (b) Wipf, P. and Spencer, S.R. (2005) *J. Am. Chem. Soc.*, **127**, 225–235.

27 Hu, Y.-J., Dominique, R., Das, K.S., and Roy, R. (2000) *Can. J. Chem.*, **78**, 838–845.

28 Chen, G., Schmieg, J., Tsuji, M., and Franck, R.W. (2004) *Org. Lett.*, **6**, 4077–4080.

29 Sheddan, N., Arion, V., and Mulzer, J. (2006) *Tetrahedron Lett.*, **47**, 6689–6693.

30 Morimoto, S., Shindo, V., Yoshida, M., and Shishido, K. (2006) *Tetrahedron Lett.*, **47**, 7353–7356.

31 Gebauer, J., Arseniyadis, S., and Cossy, J. (2007) *Org. Lett.*, **9**, 3425–3427.

32 Marshall, J.A. and Sabatini, J.J. (2006) *Org. Lett.*, **8**, 3557–3560.

33 Crimmins, M.T., Zhang, Y., and Diaz, F.A. (2006) *Org. Lett.*, **8**, 2369–2372.

34 Hoye, T.R., Eklov, B.M., Jeon, J., and Khoroosi, M. (2006) *Org. Lett.*, **8**, 3383–3386.

35 Morita, A. and Kuwahara, S. (2007) *Tetrahedron Lett.*, **48**, 3163–3166.

36 Michrowska, A., Bujok, R., Harutyunyan, S., Sashuk, V., Dolgonos, G., and Grela, K. (2004) *J. Am. Chem. Soc.*, **126**, 9318–9325.

37 Albert, B.J., Sivaramakrishnan, A., Naka, T., Czaicki, N.L., and Koide, K. (2007) *J. Am. Chem. Soc.*, **129**, 2648–2659.

38 Crimmins, M.T., Christie, H.S., Chaudhary, K., and Long, A. (2005) *J. Am. Chem. Soc.*, **127**, 13810–13812.

39 Quinn, K.J., Isaacs, A.K., and Arvary, R.A. (2004) *Org. Lett.*, **6**, 4143–4145.

40 Quinn, K.J., Isaacs, A.K., De Christopher, B.A., Szklarz, S.C., and Arvary, R.A. (2005) *Org. Lett.*, **7**, 1243–1245.

41 Quinn, K.J., Smith, A.G., and Cammarano, C.M. (2007) *Tetrahedron*, **63**, 4881–4886.
42 Michaut, M., Parrain, J.-L., and Santelli, M. (1998) *Chem. Commun.*, 2567–2568.
43 Statsuk, A.V., Liu, D., and Kozmin, S.A. (2004) *J. Am. Chem. Soc.*, **126**, 9546–9547.
44 Marjanovic, J. and Kozmin, S.A. (2007) *Angew. Chem. Int. Ed.*, **46**, 8854–8857.
45 Matsumoto, K. and Kozmin, S.A. (2008) *Adv. Synth. Catal.*, **350**, 557–560.
46 Gillingham, D.G. and Hoveyda, A.H. (2007) *Angew. Chem. Int. Ed.*, **46**, 3860–3864.
47 Van Veldhuizen, J.J., Campbell, J.E., Giudici, R.E., and Hoveyda, A.H. (2005) *J. Am. Chem. Soc.*, **127**, 6877–6882.
48 Takao, K.-I., Yasui, H., Yamamoto, S., Sasaki, D., Kawasaki, S., Watanabe, G., and Tadano, K.-I. (2004) *J. Org. Chem.*, **69**, 8789–8795.
49 Smith, A.B.III, Adams, C.M., Kozmin, S.A., and Paone, D.V. (2001) *J. Am. Chem. Soc.*, **123**, 5925–5937.
50 Kinoshita, M. and Mori, M. (1994) *Synlett*, 1020–1022.
51 For reviews on ene-yne metathesis, see: (a) Poulsen, C.S., Madsen, R. (2003) *Synthesis*, **1**, 1–18; (b) Diver, S.T., Giessert, A.J. (2004) *Chem. Rev.*, **104**, 1317–1382; (c) Mori, M. (2007) *Adv. Synth. Catal.*, **349**, 121–135.
52 (a) Clark, J.S. (2006) *Chem. Commun.*, 3571–3581; (b) Clark, J.S., Elustondo, F., and Kimber, M.C. (2004) *Chem. Commun.*, 2470–2471.
53 Royer, F., Vilain, C., Elkaïm, L., and Grimaud, L. (2003) *Org. Lett.*, **5**, 2007–2009.
54 Cesati, R.R.III, de Armas, J., and Hoveyda, A.H. (2004) *J. Am. Chem. Soc.*, **126**, 96–101.
55 Kitamura, T., Sato, Y., and Mori, M. (2004) *Tetrahedron*, **60**, 9649–9657.
56 Mori, M., Tonogaki, K., and Nishiguchi, N. (2002) *J. Org. Chem.*, **67**, 224–226.
57 Rodriguez-Conesa, S., Candal, P., Jimenez, C., and Rodriguez, J. (2001) *Tetrahedron Lett.*, **42**, 6699–6702.
58 Middleton, M.D., Peppers, B.P., and Diver, S.T. (2006) *Tetrahedron*, **62**, 10528–10540.
59 Fürstner, A. and Davies, P.W. (2005) *Chem. Commun.*, 2307–2320.
60 Fürstner, A. and Dierkes, T. (2000) *Org. Lett.*, **2**, 2463–2465.
61 Fürstner, A. and Mathes, T. (2001) *Org. Lett.*, **3**, 221–223.
62 Chan, J. and Jamison, T.F. (2004) *J. Am. Chem. Soc.*, **126**, 10682–10691.
63 Stewart, I.C., Douglas, C.J., and Grubbs, R.H. (2008) *Org. Lett.*, **10**, 441–444.
64 Ibrahem, I., Yu, M., Schrock, R.R., and Hoveyda, A.H. (2009) *J. Am. Chem. Soc.*, **131**, 3844–3845.

11
Cascade Metathesis in Natural Product Synthesis

Marta Porta and Siegfried Blechert

11.1
Introduction

Alkene metathesis is well established as a valuable synthetic tool in organic chemistry, and has proven to be a unique technique for the construction of complex structures in a rapid and effective way [1]. Since all metathesis transformations (except diyne metathesis) can be promoted by the same carbene catalysts, it is possible to combine them in a sequential way and enable a metathesis cascade in one pot. A *domino reaction* has been defined as "a process involving two or more bond-forming transformations which take place under the same reaction conditions without additional reagents and catalysts, and in which the subsequent reactions result as a consequence of the functionality formed in the previous step" [2]. These domino processes, also called *tandem*, provide a tremendous increase in molecular complexity within the use of one single catalyst in one pot. By carefully adjusting the conditions and judicious choice of the catalytic complex, it is often possible to conduct the process with excellent regio-, chemo-, diastereo-, and even enantioselectivity. Domino processes are especially attractive not only because they provide elegance to the synthetic route but also for being ecologically and economically favorable since they minimize the amount of waste, solvents, reagents, and energy compared to stepwise reactions. Much effort has been invested in developing new methodologies involving metathesis domino processes that have been successfully applied to the preparation of biologically active and naturally occurring molecules. In this chapter, a selection of synthetic approaches with ruthenium-catalyzed domino reactions that include exclusively metathesis steps will be discussed, whereby syntheses involving tandem reactions connecting a metathesis reaction with a reaction of a different type [3] or several metathesis transformations occurring independently at the same time will not be described [4]. Through this collection of natural product syntheses, we would like to highlight the high efficiency and utility of metathesis domino reactions in organic synthesis, discuss its driving forces, limits, and scope, and ultimately illustrate the enormous synthetic potential of this transformation.

Metathesis in Natural Product Synthesis: Strategies, Substrates and Catalysts. Edited by Janine Cossy, Stellios Arseniyadis, and Christophe Meyer

ISBN: 978-3-527-32440-8

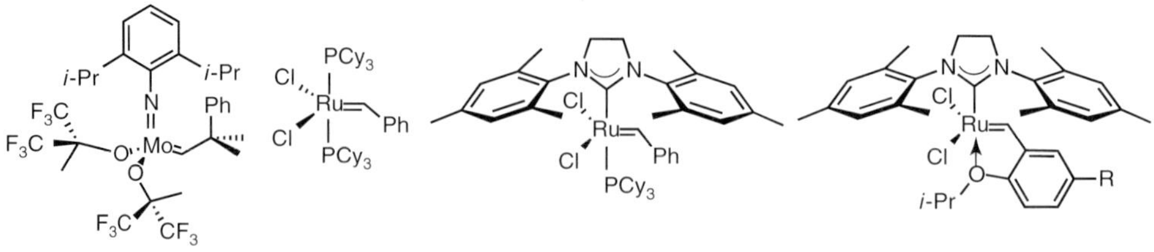

Scheme 11.1 Carbene complexes used in metathesis cascades.

At the heart of advancement in metathesis and its establishment as a unique methodology for carbon–carbon bond formation lies the development of effective catalysts. A great variety of olefin metathesis catalysts have been developed over the past 20 years. Currently, there are up to 20 ruthenium catalysts commercially available; several reviews have been published with an overview of the most significant ones, where their reactivity, stability, and selectivity have been also addressed [5]. Despite the large number of these carbene complexes, catalysts [Mo]-**I** [6], [Ru]-**I** [7], [Ru]-**II** [8], and [Ru]-**III** [9] are the most widely applied in domino metathesis with just a few exceptions (Scheme 11.1). Catalyst [Mo]-**I** exhibits high olefin metathesis activity but lacks general functional group tolerance and suffers from strong air and moisture sensitivity. Functional group compatibility is crucial for the synthesis of complex molecules, and therefore commercially available catalysts [Ru]-**I**, [Ru]-**II**, and [Ru]-**III** have been utilized in synthesis to a greater extent than [Mo]-**I**. Catalysts of the first generation tend to furnish preferentially kinetic compounds, while [Ru]-**II** and [Ru]-**III** usually drive the reaction toward thermodynamically stable products.

11.2 RCM–CM Sequences

There are many combinations possible of metathesis reactions that can be used in domino metatheses, increasing process complexity with the number of transformations included. Natural product syntheses with ring-closing metathesis (RCM)–cross-metathesis (CM) sequences are described in this section. Ene–ene and ene–yne–ene cascade reactions in a two-metathesis step process will also be discussed.

11.2.1 Ene–ene RCM–CM

11.2.1.1 Synthesis of (3*R*,9*R*,10*R*)-Panaxytriol

This type of metathesis cascade can be considered as an example of relay ring-closing metathesis (RRCM) (Chapter 9)[10]. This methodology is classically used in RCM of

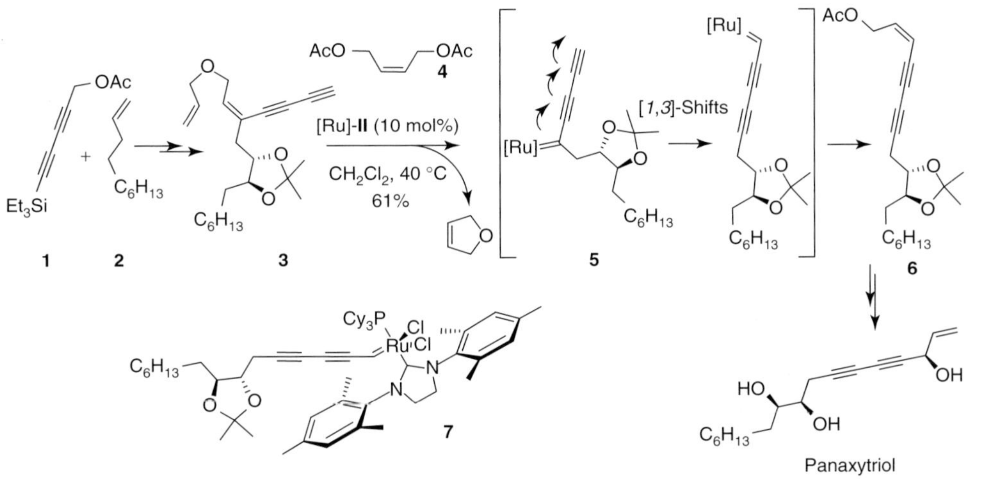

Scheme 11.2 Synthesis of panaxytriol.

sterically hindered olefins, which remains unreactive to standard metathesis conditions because of the difficulty of initiating the catalytic process. The relay approach involves incorporation of a temporary tether, containing a sterically accessible olefin, where the catalytic cycle is initiated. The purpose of this strategy is the formation of a kinetically favorable five-membered ring with concomitant delivery of a ruthenium carbene onto the originally less accessible position to facilitate the following metathesis step. In 2005, Lee *et al.* published an innovative tandem methodology combining metathesis and metallotropy for the construction of enediynes and oligoenynes [11a] and have recently applied this strategy in the synthesis of panaxytriol [11b], a component of Red Ginseng. Panaxytriol was first isolated in 1983 and exhibits inhibitory activity against a range of tumor cell types, such as human gastric adenocarcinoma, human breast carcinoma, and mouse lymphoma.

Diyne **3** was synthesized in eight steps from **1** and **2** using a ruthenium-catalyzed Alder-ene reaction and a Cadiot–Chodkiewicz reaction as key transformations (Scheme 11.2). By treatment of **3** with catalyst [Ru]-**II** (10 mol%), propagating carbene **5** is generated, followed by two consecutive metallotropic [1,3]-shifts to form a new ruthenium species that undergoes CM with alkene **4**, to finally complete the catalytic cycle. The desired product is obtained in 61% yield as a mixture of isomers with an unusual high *E*/*Z* selectivity (1 : 5). Interestingly, under this reaction conditions, it was possible to isolate the ruthenium complex **7** in 10% yield, suggesting that the carbene species derived from **7** is a viable intermediate in the catalytic cycle.

11.2.2
Ene–yne–ene RCM–CM

Enyne metathesis is widely used for its capacity to carry out multiple bond formations in tandem fashion, which is due to the ability of the alkyne moiety to

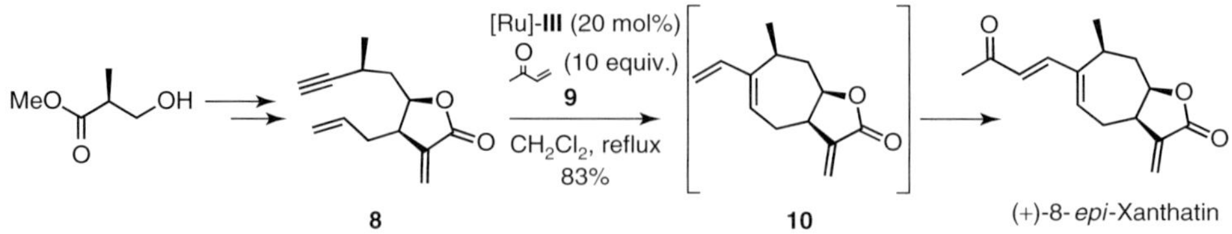

Scheme 11.3 Synthesis of (+)-8-*epi*-xanthatin through a RCM–CM cascade.

form a conjugated ruthenium carbene as result of the reaction with an olefin, which ultimately can react with a second double bond to form a substituted 1,3-diene [12].

11.2.2.1 Synthesis of (+)-8-*epi*-Xanthatin

There is only one example in the literature where an enyne RCM coupled to an intermolecular CM in a sequence process has been applied in the synthesis of a natural product. The synthesis of (+)-8-*epi*-xanthatin, isolated from the leaves extracts from *Xanthium canadense*, was proposed by Martin *et al.* through an enyne RCM–CM cascade combining ring closing of **8** with a CM with **9** [13]. The special challenge for this transformation consists in the use of an electron-poor olefin for the final metathesis reaction. The use of [Ru]-**III** (20 mol%), the catalyst of choice for this type of transformations [14], and an excess (10 equiv.) of the methyl vinyl ketone **9** resulted in the formation of the desired product in 83% yield (Scheme 11.3).

11.3
Ene–yne–ene RCM–RCM

When an olefin happens to be attached to the ring that has been formed after an initial ene–yne RCM, a new bicyclic moiety can be formed by a second RCM. Such metathetical transformation is highly useful for the construction of polycyclic ring systems and it has been applied in the synthesis of natural products and derivatives to a great extent. Herein, we report a selection of the most significant examples (Scheme 11.4).

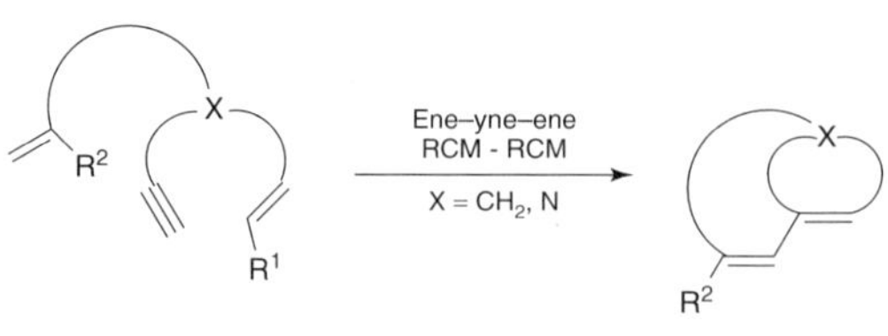

Scheme 11.4 General principle of ene–yne–ene RCM–RCM domino reactions.

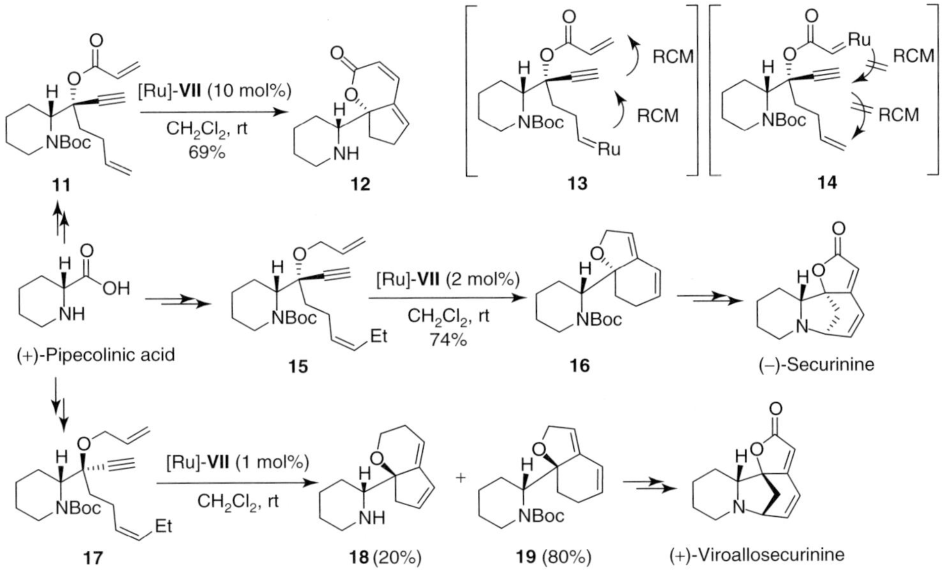

Scheme 11.5 Synthesis of (−)-securinine and (+)-viroallosecurinine through an ene–yne–ene metathesis cascade.

11.3.1
Synthesis of Bicyclic Structures

11.3.1.1 Synthesis of (−)-Securinine and (+)-Viroallosecurinine

A suitable example of how to steer selectivity in an ene–yne–ene metathesis cascade is illustrated in the synthesis of securinine [15] and viroallosecurinine [16] published by Honda *et al.* Naturally occurring *securinega* alkaloids comprise about 20 tetracyclic compounds, isolated from *Securinega suffruticasa* in 1956, which show a great scope of biological activity. Securinine, the most abundant alkaloid of this group, has been used clinically since 1968 as a central nervous system (CNS) stimulating drug, acting as a stereospecific antagonist at the gamma aminobutyric acid (GABA) binding site of the GABA–receptor complex.

First synthetic attempts with compound **11** using catalyst [Ru]-**VII** [17] delivered the undesired bicyclic structure **12**, which results from the initial attack of the ruthenium carbene complex at the terminal olefin via intermediate **13** (Scheme 11.5). To direct the initiation of the catalytic cycle into the other double bond, through intermediate **14**, Honda *et al.* decided to introduce steric hindrance using a disubstituted olefin in the alkyl chain. Their results showed, however, that the resulting compound was not reactive enough to undergo metathesis under the chosen reaction conditions. With the use of a more electron-rich allyl ether moiety such as in **15**, it was possible to isolate the desired product in 74% yield and successfully complete the synthesis of securinine. Ruthenium complexes [Ru]-**I** and [Ru]-**II** were also tested, leading to lower yields and requiring higher reaction temperatures as well as longer reaction times [18]. The same synthetic strategy was applied in the synthesis of its analog viroallosecurinine with the epimeric compound **17**. Under the same catalytic conditions, the metathesis domino reaction delivered the desired product **19** in 80% yield together with compound **18** in 20% yield.

11.3.1.2 Total Synthesis of *ent*-Lepadin F and G

Lepadins are members of the decahydroquinoline alkaloids, which are found in different marine sources, with a normally pentasubstituted *cis*-fused ring. Blechert *et al.* have recently reported the total synthesis of *ent*-lepadin F and G by a well-directed dienyne RCM cascade where aforethought functionalization of the corresponding olefins has permitted obtaining selectively the desired product [19]. When substrate **20** (Scheme 11.6) is subjected to metathesis conditions, two different reaction pathways can be envisioned: metathesis may initiate at the terminal double bond to produce ruthenium carbene intermediate **23**, which then reacts with the alkyne moiety to furnish the desired bicyclic compound **21**. Metathesis may also initiate on the disubstituted alkenyl moiety to produce ruthenium carbene species **24**, which would then lead to the bicyclic structure **22**. Selectivity in the metathesis cascade was achieved by a coordinative effect of the unprotected alcohol, which favors the initial catalyst attack of the allylic alcohol olefin, together with the use of a disubstituted alkene as a final cross-partner, which is a common strategy for the selective catalytic cycle initiation at the mono-substituted alkene. For the metathesis domino reaction, several conditions were tested with different catalysts at different temperatures. When the metathesis precursor **20** was exposed to catalyst [Ru]-**II** in CH_2Cl_2 both at room temperature and at 40 °C, no reaction occurred. The use of the more thermally stable

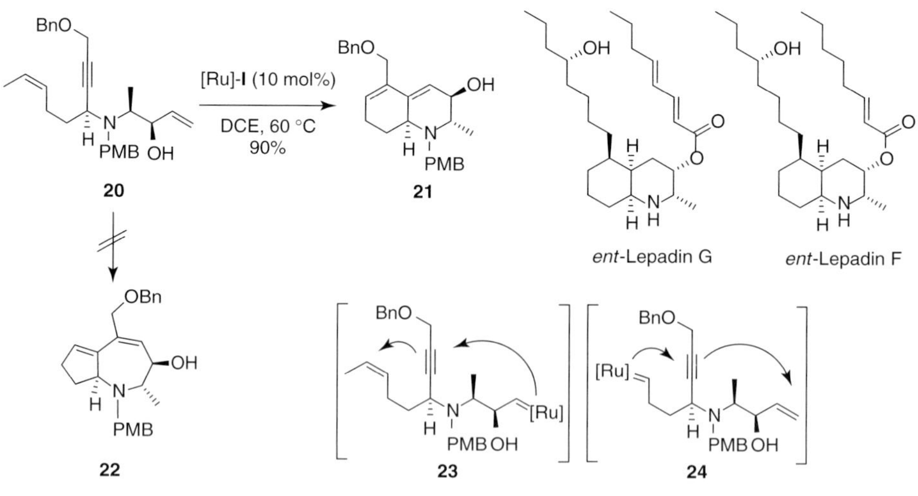

Scheme 11.6 Synthesis of lepadine alkaloids.

catalyst [Ru]-**III** in toluene at 80 °C resulted in complete conversion of **20** after 3 hours and the desired product **21** was obtained in 62% yield. Optimized conditions were found with the use of the more reactive but less stable catalyst [Ru]-**I** at 60 °C in 1,2-dichloroethane (DCE). Under these conditions, **21** was isolated without significant formation of by-products and with an increased yield of 90%.

11.3.2
Synthesis of Tricyclic Compounds

11.3.2.1 Synthesis of (±)-Guanacastepene A

The first application of an ene–yne–ene RCM–CM cascade in natural product synthesis was published by Hanna and Boyer in 2002 with their studies on the tricyclic ring systems related to guanacastepene A [20a]. Originally isolated from a fungus collected in the Guanacaste Conservation Area in Costa Rica, this diterpene became an attractive synthetic target because of its novel highly functionalized 5,7,6-ring fused structure and an interesting antibacterial activity revealed by biological studies. Hanna *et al.*'s synthetic approach started with the construction of trienes with different substitution patterns on the alkyne moiety (Scheme 11.7). In a first attempt, the use of catalyst [Ru]-**II** (10 mol%) in refluxing CH_2Cl_2 resulted in a mixture of products. By lowering the reaction temperature, the desired product **29** was obtained in 40–60% yield along with remaining starting material. In their efforts to optimize the reaction conditions, the authors found that by stirring the

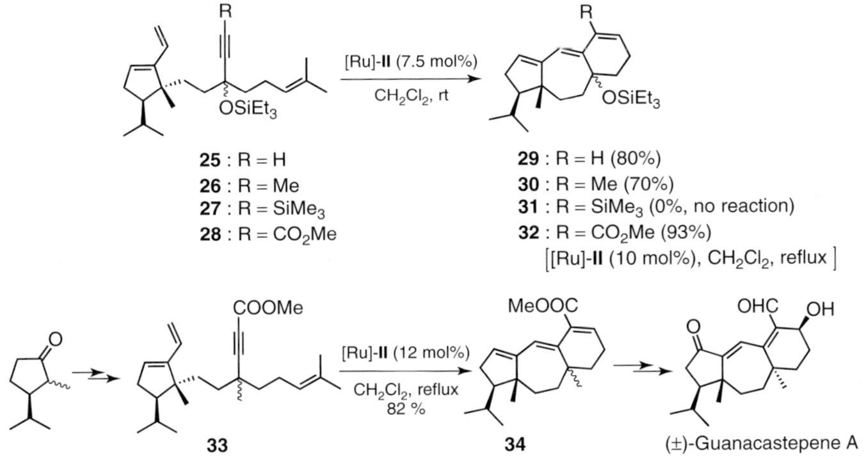

Scheme 11.7 Metathesis studies on (±)-guanacastepene A tricyclic core and its total synthesis.

reaction mixture under a stream of nitrogen, so that the solvent slowly evaporated, a complete conversion of the starting material was achieved and **29** could be isolated in 80% yield. The use of other ruthenium complexes did not show any improvement and the starting diene–ynes were recovered unchanged. Methyl substitution on the alkyne moiety was also tolerated, but a bulkier substituent such as trimethylsilane (TMS) impeded the diene–yne metathesis cascade. In the presence of an electron-withdrawing group on the alkyne (compound **28**), conversion to the desired product **32** was achieved in 93% yield, although harsher conditions were required. The total synthesis of guanacastepene A with a diene–yne RCM domino reaction as the key step was published two years later by the same group [20]. In this case, triene–yne **33** was efficiently converted to **34** in 82% yield by the second-generation catalyst [Ru]-**II**. In this unprecedented synthetic approach to this terpene skeleton, it is quite remarkable that the tandem metathesis reaction initiates at the relatively unreactive conjugated diene and also that the second RCM, despite steric hindrance at the trisubstituted olefin, can occur even at room temperature, probably because of the formation of a thermodynamically favored conjugated triene.

11.3.2.2 **Approach to Taxane Analogs**

As shown above, domino RCM–CM can be successfully applied in the synthesis of tri- and tetracyclic structures. Granja *et al.* have recently reported a diene–yne domino metathesis to construct the tricyclic carbon framework characteristic for taxanes [21]. Their approach evidences that success of the metathesis cascade largely depends upon structural strain, and thus the sterical alignment of the reactant olefins, implying selective transformations. Their studies are based on an array of different metathesis substrates that differ in the grade of functionalization and in the configuration of each chiral center. Tactically, and in the same way as in **33**, the terminal isopropylidene favors the initial incorporation of the ruthenium to the less-substituted double bond (Scheme 11.8). Their preliminary results showed that only the isomer **35** permitted obtaining the desired tricyclic core, while all other diastereoisomers yielded bicycle **39** resulting from the RCM between the alkyne and the terminal olefin. When the structural complexity of the skeleton was increased with the introduction of some relevant functional groups for the biological activity of taxanes, unfortunately, neither the hydroxyketones (R^1 = OH and Y = CO) (structure **36**) nor the corresponding 1,3-diols (R^1 = OH and Y = CH–OH) were transformed into the desired tricyclic compound under the applied metathesis conditions. Instead, complete decomposition of the dienynes was observed. With the corresponding protected alcohols in compound **37** (R^1 = OMEM and Y = CH–OMEM), it was possible to obtain the desired tricyclic structure, whereas all other stereoisomers at C2, C3, and C8 yielded again bicyclic structures of type **39** as a result of the interrupted domino reactions. When these trienes were subjected to metathesis conditions, they could not be converted to the tricyclic product, presumably because of steric hindrance of both olefins and structural strain in the metallacyclic intermediate.

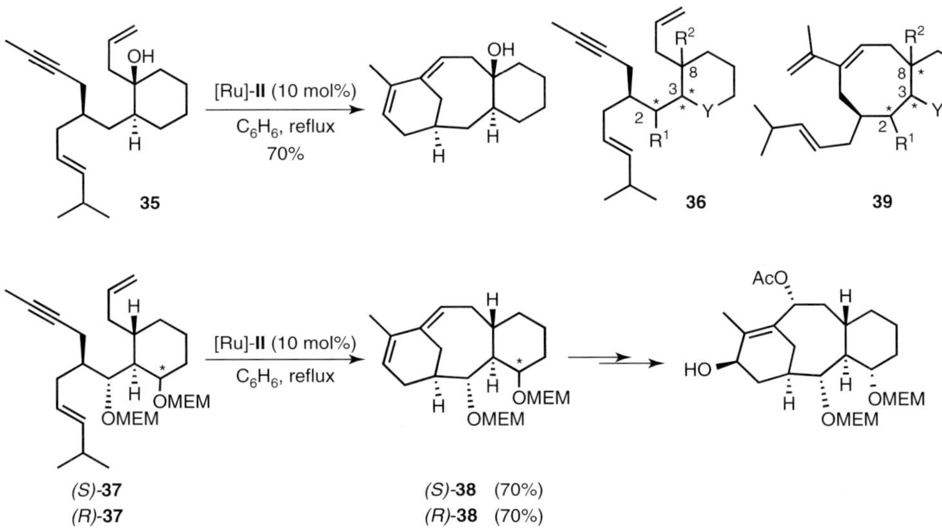

Scheme 11.8 Metathesis results on the synthetic approach to taxane analogs.

11.3.3
Synthesis of Natural Products Containing Tetracycles

11.3.3.1 Synthesis of Erythrina Alkaloids

Erythrina alkaloids were isolated from several species of the *Erythrina* genus and display a variety of pharmacological effects, such as sedative, hypotensive, neuromuscular-blocking, and CNS activity. Hatakeyama *et al.* [22] and Mori *et al.* [23] published almost simultaneously a synthetic route to erythravine and erythrocarine, respectively, using a diene–yne RCM domino metathesis as the key step. For the synthesis of erythravine, protection of the alcohol functionality was necessary, whereby the choice of the protecting group had a very significant influence on the stereoisomer distribution, even though a 1 : 1 epimeric mixture was used as a starting material (Scheme 11.9). With the use of an ester as protecting group, the desired product was obtained as a 63 : 37 mixture of α and β epimers in 78% yield. The discrepancy between the initial and the obtained epimeric ratio may arise from different rates of cyclization and/or decomposition rates under the metathesis conditions.

In Mori *et al.*'s approach, metathesis precursor **41** was synthesized with a nickel-mediated alkyne carboxylation as a key step. A better yield was achieved by converting the amino group into its corresponding hydrochloride. First-generation Grubbs catalyst [Ru]-**I** was again the catalyst of choice, which afforded the desired product **42** quantitatively as a 1 : 1 mixture of α and β isomers after 18 hours at room temperature.

More challenging was the synthesis of the nonaromatic derivate erythroidine published by Hatakeyama *et al.* in 2006 (Scheme 11.10) [24]. RCM of **43**, **46**,

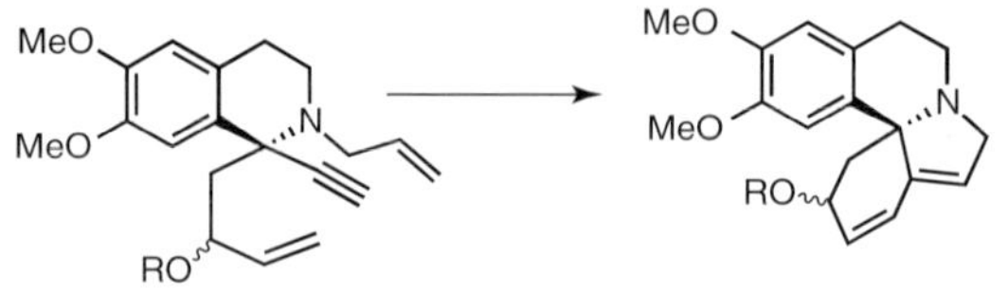

Substrate	Conditions	Yield (%)	α	β
R = H	[Ru]-**I** or [Ru]-**II** (10–20 mol%) CH_2Cl_2 or toluene, rt to reflux	0	–	–
R = TES	[Ru]-**II** (10 mol%), CH_2Cl_2, reflux	63	27	73
R = Ac	[Ru]-**II** (10 mol%), CH_2Cl_2, reflux	78	63	37

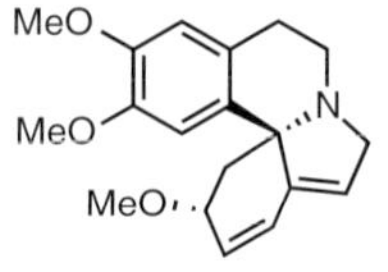

Erythravine

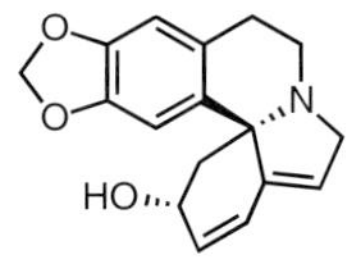

Erythrocarine

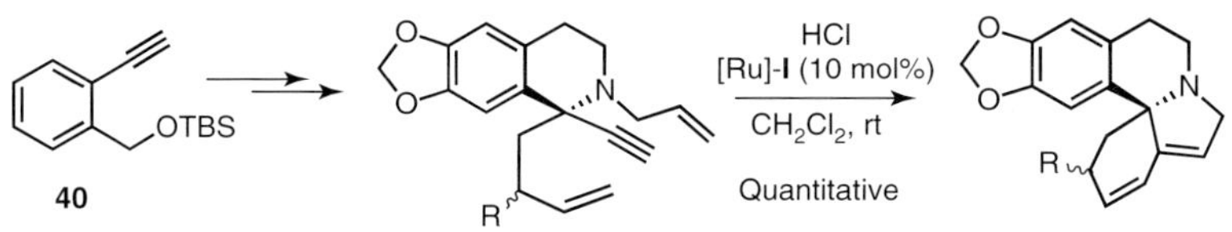

40 **41** : R = α,β-OAc **42** : R = α,β-OAc

Scheme 11.9 Metathesis domino reaction toward erythravine and erythrocarine.

(*S*)- or (*R*)-**48**, and **51** using catalysts [Ru]-**I** and [Ru]-**II**, the last being generally more effective, afforded product mixtures containing starting material, bicyclic structures resulting from the RCM of the double bonds, and only **43** and (*S*)-**48** delivered the desired tricyclic products **44** and **49**, respectively, in moderate yields. In the case of **43**, compound **45** resulting from the initiation of the metathesis cascade on the allylic ether olefin was also generated. Unfortunately, attempts toward the synthesis of erythroidine with **44** and **49** failed, presumably because of their instability under basic conditions. With compound **53** as substrate, which is structurally closer to the target molecule, erythroidine was furnished in 42% yield with [Ru]-**I** (10 mol%) (CH_2Cl_2, rt) together with 4% of **54**, along with unconverted starting material (31%). In this case, [Ru]-**II** turned out to be less effective and gave erythroidine in less than 30% yield.

11.4 ROM–CM Sequences

Because of the reversible character of the metathesis itself, it is possible to perform ring-opening metathesis (ROM) of endocyclic olefins if the process is thermodynamically favored. This particular type of tandem process is especially atom-economical (as no ethylene is released), although the negative entropy

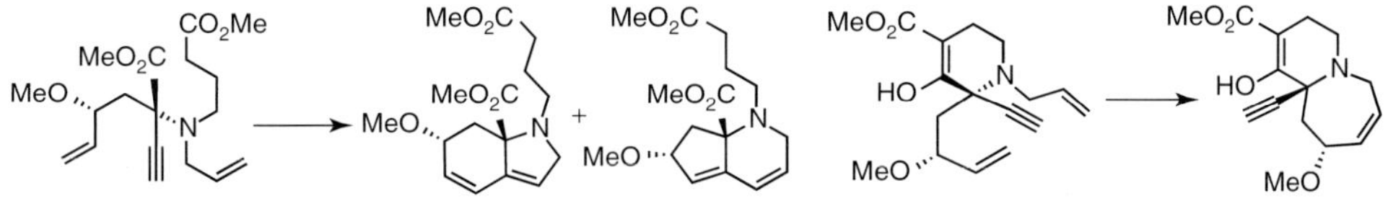

43 **44** **45** **46** **47**

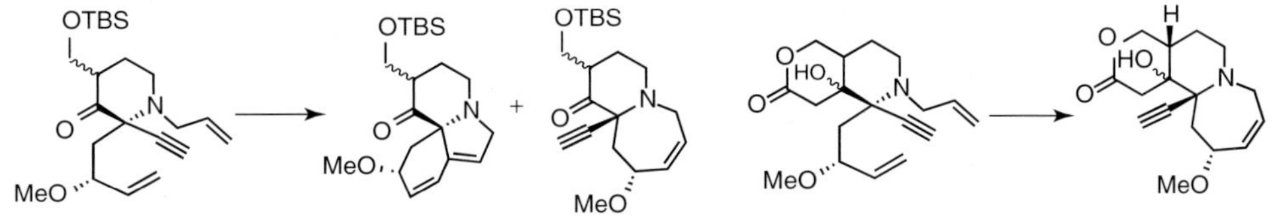

48 **49** **50** **51** **52**

Substrate	Conditions	Products (Yield %)		
43	[Ru]-**I** (10 mol%), CH_2Cl_2, reflux, 14 h	**43** (27)	**44** (0)	**45** (26)
43	[Ru]-**II** (10 mol%), CH_2Cl_2, reflux, 1 h	**43** (57)	**44** (15)	**45** (0)
46	[Ru]-**I** (10 mol%), CH_2Cl_2, reflux, 12.5 h	**46** (55)	**44** (7)	—
(*S*)-**48**	[Ru]-**I** (20 mol%), CH_2Cl_2, reflux, 14 h	**48** (0)	**49** (27)	**50** (9)
(*R*)-**48**	[Ru]-**II** (20 mol%), CH_2Cl_2, rt, 4 h	**48** (66)	**49** (0)	**50** (0)
51	[Ru]-**II** (20 mol%), CH_2Cl_2, reflux, 14 h	**51** (0)	**52** (55)	—

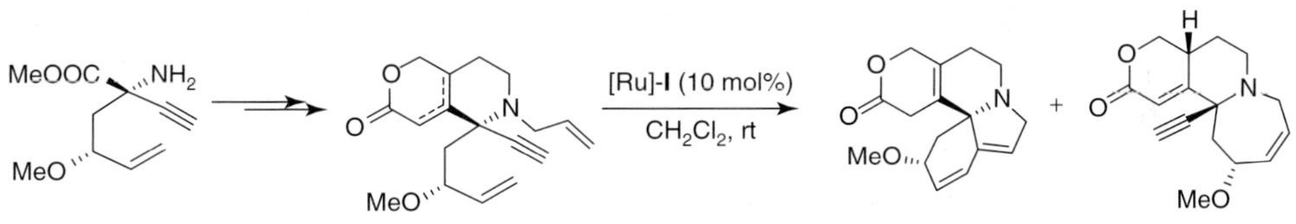

53 Erythroidine (42%) **54** (4%)

Scheme 11.10 Metathesis studies on the synthesis of erythroidine.

resulting from two molecules being combined into one must be overcome. Despite that, natural product syntheses has been reported using ROM–CM sequences where ethylene is used as cross-partner [25]; herein the focus will be on the use of more complex olefins as cross-partners (Scheme 11.11) [26].

11.4.1
Synthesis of Bistramide A

Bistramide A is a novel marine metabolite of *Lissoclinum bistratum* which possesses a profound effect on cell cycle regulation, with highly selective

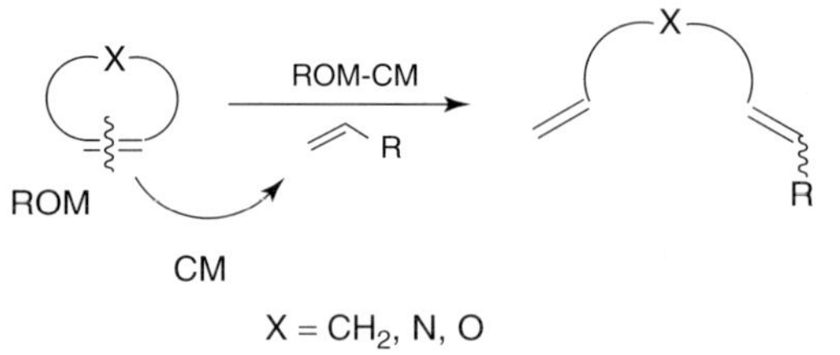

Scheme 11.11 General principle of ROM–CM domino reactions.

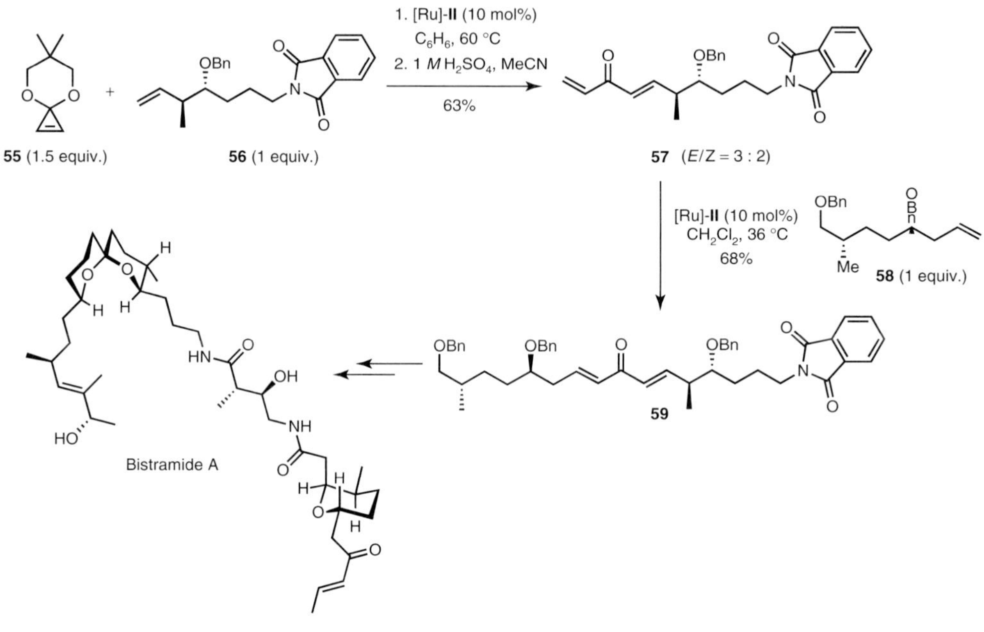

Scheme 11.12 Synthesis of bistramide A.

activation of a unique isotype from protein kinase C. Kozmin *et al.*'s inventive approach for the synthesis of the spiroketal domain involves a ROM–CM of cyclopropenone acetal **55** with alkene **56** (Scheme 11.12) [27]. This transformation was catalyzed by catalyst [Ru]-**II** (10 mol%), and only a slight excess (1.5 equiv.) of **55** was needed to afford dienone **57** in 63% overall yield after acidic hydrolysis. Interestingly, the corresponding intermediate acetal proved to be inert toward subsequent metathesis, which also justifies the absence of the formation of double-cross-coupled product and also impedes the performance of a ROM-double CM domino process. The next envisaged CM step had to be performed separately, and the desired product **59** was isolated in 68% yield after treating **57** with 10 mol% catalyst [Ru]-**II** and 1 equiv. of olefin **58**.

11.5
RCM–ROM Sequences – Ring-rearrangement Metathesis (RRM)

So far the synthetic potential of RCM and ROM has been exemplified when combined with CM in a domino process to construct demanding structures, in terms of complexity and functionalization, in a rapid manner starting from relatively easily accessible substrates. When RCM and ROM themselves are sequentially combined, this domino process is classified as ring-rearrangement metatheses (RRM). Among all the possible combinations of metathesis reactions, this methodology has proven its extraordinary synthetic potential when it comes to constructing carbocycles and heterocycles, serving as a key step in an increasing number of natural product syntheses since 1999, when the first application of ruthenium-catalyzed RRM in the synthesis of (−)-halosaline was published [28]. By definition, in these domino reactions a strained carbocyclic alkene is transformed into a new cycle by intramolecular ROM–RCM with an exocyclic double bond. Since stereocenters remain unchanged during the metathesis reaction, probably the most important feature of RRM is the capacity to transfer any stereochemical information from the starting material into the product. While the application of RCM can be limited by the accessibility of the chiral acyclic precursor, RRM offers the possibility of transferring stereocenters, easily accessible in a ring system, to the side chain or vice versa, and thus avoids complicated multistep syntheses of chiral structures which are otherwise difficult to access (Scheme 11.13).

Ring-rearrangement domino processes can be applied to mono- and polycyclic structures of different sizes. Because of the reversibility of this metathesis cascade, driven by thermodynamic factors (loss of ring strain) as well as kinetic effects (such as the formation of a less reactive carbene complex), the strategic selection of protecting groups, reaction conditions, and electronic properties of the reacting parts is crucial in order to shift the equilibrium to the desired product. Bicyclic systems deserve special attention among the different types of substrates suitable for metathesis reactions. Owing to the highly strained ring system, these substrates are highly reactive, and the driving force for the reaction can be clearly attributed to the release of energy. The introduction of an external olefin, usually ethylene, is necessary in most of the cases to avoid oligomerization, a common side reaction in RRM.

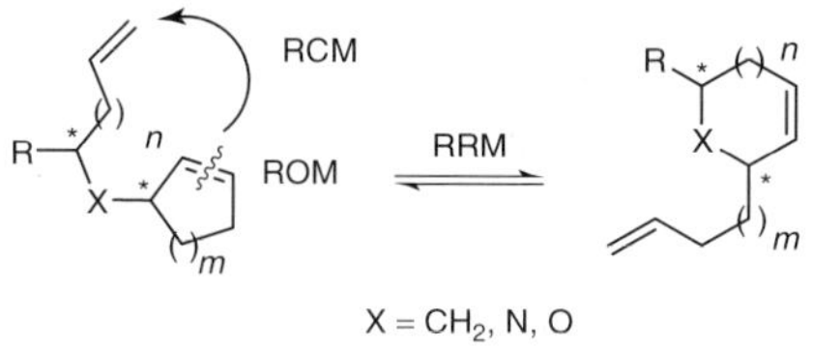

Scheme 11.13 General principle of the RRM cascade.

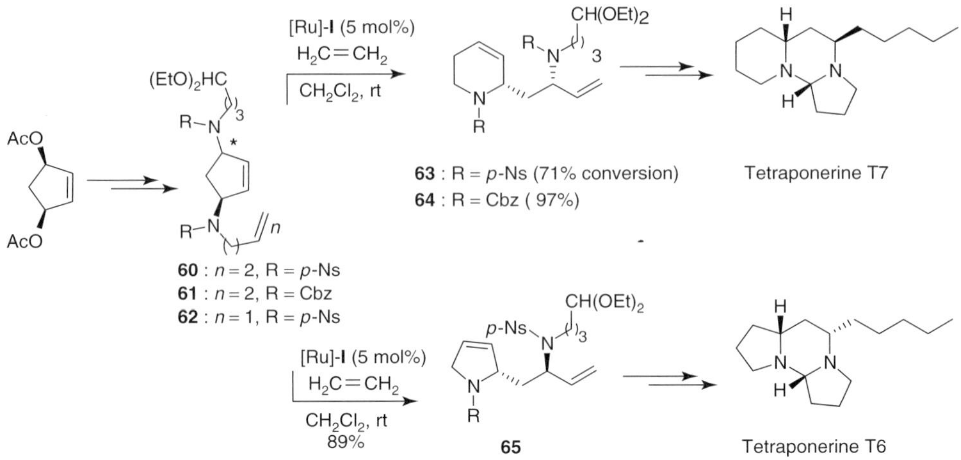

Scheme 11.14 Synthesis of tetraponerines.

11.5.1
RRM of Monocyclic Substrates

11.5.1.1 Synthesis of Tetraponerines

An early application of RRM in target-oriented synthesis consisted in a skeletal reorganization of enantiomerically pure five-membered carbocycles (Scheme 11.14), which were readily accessible with a palladium-catalyzed allylic substitution of cyclopentene derivatives. Four tetraponerine alkaloids T4, T6, T7, and T8 isolated from the New Guinean ant *Tetraponera* were successfully synthesized by Blechert *et al.* with a flexible approach that permitted constructing both five- and six-membered heterocycles by varying the length of the *N*-alkenyl substituent on the cyclopentene [29]. As a representative example, subjecting metathesis precursors **60** and **62** to catalyst [Ru]-**I** (5 mol%) resulted in 71 and 91% conversion, respectively, after two days under ethylene atmosphere. By changing the N-protecting group to benzyloxycarbonyl (Cbz), the transformation proceeded quantitatively in 12 hours at room temperature and tetrahydropyridine **64** was isolated in 97% isolated yield. A plausible explanation might be attributed to a change in the complexation behavior with the ruthenium carbene during the course of the reaction.

11.5.1.2 Synthesis of (−)-Swainsonine and (+)-Castanospermine

Members of the polyhydroxylated indolizidine family have received continuous, intense scrutiny from synthetic chemists owing to their densely functionalized, stereochemically rich structures. Swainsonine, isolated from the fungus *Rhizoctonia leguminicolain*, has especially attracted great attention mainly because of its important antimetastatic, antitumor-proliferative, anticancer, and immunoregulating activities, being the first compound to be selected for testing as an anticancer drug, reaching Phase I in clinical trials. Blechert *et al.* presented a synthetic approach

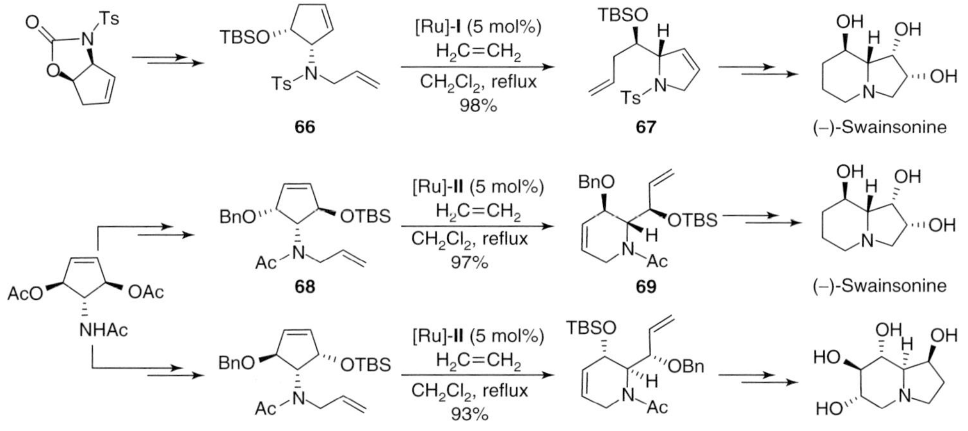

Scheme 11.15 Synthesis of (−)-swainsonine and (+)-castanospermine by RRM of cyclopentene derivatives.

to this polyhydroxylated indolizidine alkaloid based on RRM of a cyclopentene derivative obtained by palladium-catalyzed desymmetrization of cyclopentenediol [30]. When the metathesis precursor **66** (Scheme 11.15) was treated with commercially available [Ru]-**I** (5 mol%) under ethylene atmosphere, the desired rearranged product could be isolated in a yield of up to 98%, whereas rearrangement of its benzyl ether analog afforded a starting material/product ratio of 18 : 1. The sterically demanding *tert*-butyldimethylsilyl (TBS) ether proved to be essential to shift the rearrangement equilibrium completely toward the product, apparently due to an increase of the steric interactions between the substituents, thus facilitating ring opening and also preventing product dimerization.

Mariano *et al.* focused on the RRM reaction of unsymmetrically substituted aminocyclopentenediol derivatives to investigate the regiochemical and stereochemical control of the process [31]. Tetrahydropyridine **68** rearranged to **69**, in which the benzyloxy group is positioned next to the *N*-acyl group, whereas reaction of diastereomer **70** selectively furnished the regioisomer **71** in 97% yield. The high selectivity of this process was ascribed to an energy difference between the ruthenacyclobutane intermediates, thus suggesting the influence of kinetic factors. Tetrahydropyridine **69** was further converted into (−)-swainsonine, whereas **71** served as starting material in the synthesis of (+)-castanospermine (Scheme 11.15) [32].

11.5.1.3 Synthesis of (+)-*trans*-195A

RRM is not limited to cyclopentene derivatives, although applications of such metathesis domino process in six-membered cyclo-olefins have seldom been reported. Even though ring rearrangement of cyclohexenes to five-membered heterocycles is thermodynamically disfavored, synthesis of tetrahydropyridine

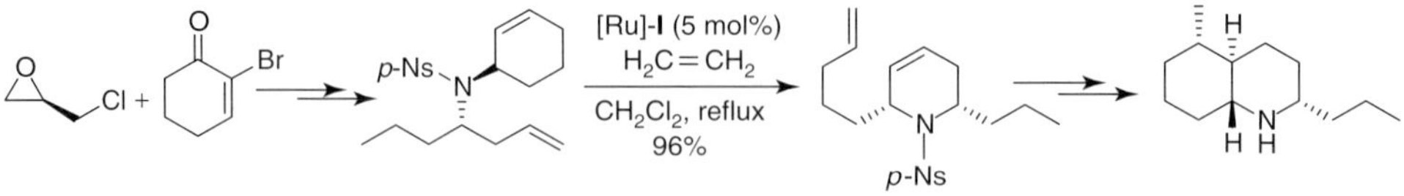

72 73 (+)-*trans*-195A

Scheme 11.16 Application of RRM of cyclohexenes in the synthesis of (+)-*trans*-195A.

derivatives can be achieved, thereby indicating that the resulting heterocycle is thermodynamically more favored than the corresponding cyclohexene. This concept was used in the synthesis of the poison frog alkaloid (+)-*trans*-195A, where **72** was selectively converted to the *cis*-2,6-disubstituted tetrahydropyridine **73** in 96% yield, with catalyst [Ru]-**I** (5 mol%) under an ethylene atmosphere (Scheme 11.16) [33].

11.5.1.4 Synthesis of (−)-Centrolobine – Diastereoselective RRM (*d*-RRM)

There are two possible pathways for the generation of a new chiral center by means of metathesis: asymmetric induction by a chiral complex itself (asymmetric metathesis, Chapter 12), or diastereoselective induction where the chiral information present in the substrate influences the sterics of the newly formed chiral center. Metathesis domino processes can also be performed in a diastereoselective manner generating new stereocenters with high selectivity. The challenge for such a process is to shift the reaction to the desired isomer, whereby the reaction can be kinetically or thermodynamically controlled. The selectivity is mainly catalyst dependent; thus electronic and, specially, steric effects of the remaining ligands in the initiated catalytic complex are decisive for the reaction outcome. Blechert and Böhrsch presented in 2006 the single application in natural product synthesis of this methodology with the construction of (−)-centrolobine [34], a tetrahydropyranic antibiotic that has shown activity against leishmania amazonensis promastigotes, a major health problem in Brazil. The key features of its synthetic approach involve not only a *d*-RRM but also three consecutive transformations catalyzed by complex [Ru]-**II**. The enantiopure metathesis precursor **74**, obtained by an asymmetric iridium-catalyzed allylic etherification, undergoes a ROM–RCM domino reaction in which the stereocenter at C5 is established diastereoselectively. Different conditions and precatalysts were tested: [Ru]-**I** (5 mol%) led to poor diastereoselectivity (*cis* : *trans* = 2 : 1), whereas the employment of [Ru]-**II** (3 mol%) in benzene provided a significant improvement of the diastereoisomeric ratio to 4 : 1 (*cis* : *trans*). Optimized reaction conditions were found to be under an ethylene atmosphere in saturated CH_2Cl_2 at 50 °C in a pressure vessel. A selective double bond isomerization of the terminal alkene with *in situ* generated ruthenium hydride species followed the metathesis cascade. A one-pot CM-hydrogenation procedure completed the synthesis of centrolobine in five steps, with 22% overall yield and 98% ee (Scheme 11.17).

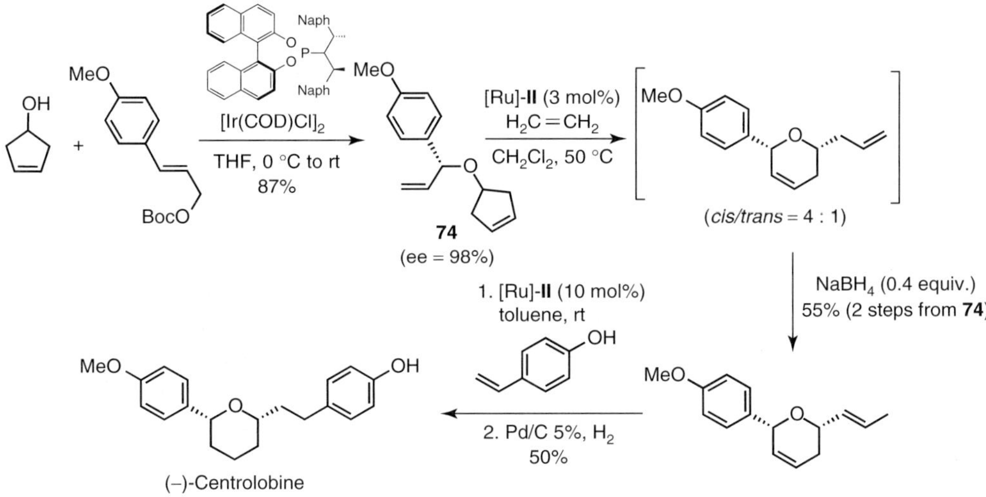

Scheme 11.17 *d*-RRM cascade in the synthesis of (–)-centrolobine.

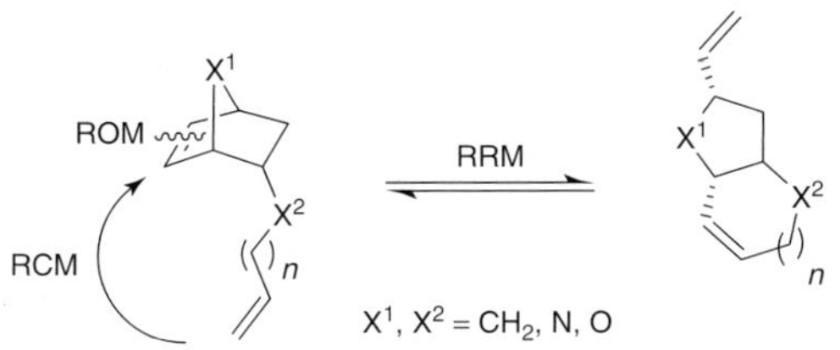

Scheme 11.18 General principle of RRM of norbornene derivatives.

11.5.2
RRM of Bicyclic Substrates

Norbornene derivatives are common substrates for ring-rearrangement domino processes. Because of their highly strained structure, equilibrium is strongly shifted to the product and, owing to the highly stereocontrolled norbornene synthesis, it is granted easy access to new architectures which otherwise would be difficult to synthesize (Scheme 11.18).

In general, different mechanistic pathways can be taken into account for these domino reactions (Scheme 11.19). The first would involve initial attack at the terminal exocyclic double bond, which would then undergo metallocycloaddition onto the endocyclic olefin. Alternatively, ring opening may occur first leading to two possible intermediates **A** and **D**. As observed in early studies [35], favored formation of **A** or **D**, and therefore **B** or **E**, depends on the ring size of the final product as well as on the presence of ethylene in the reaction mixture. Under these conditions, it is possible to convert **A** into **C** and shift the equilibrium toward

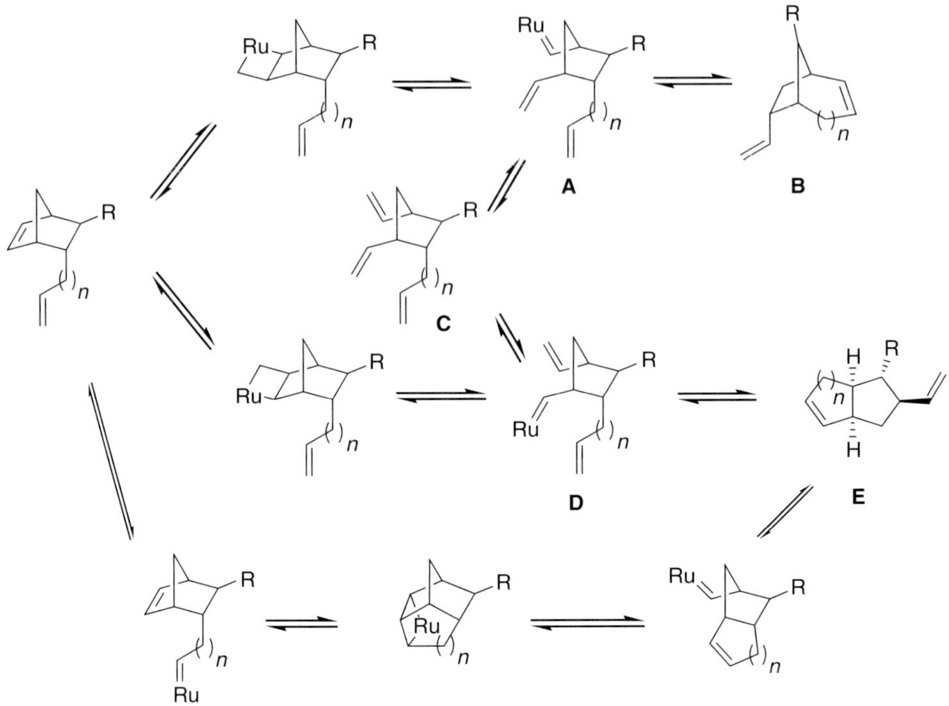

Scheme 11.19 Mechanistic aspects of RRM domino processes.

the most stable compound **E**. Moreover, dimeric or oligomeric materials can be reverted back to monomeric ruthenium-alkylidene, thereby avoiding the formation of the major by-products of these type of domino metathesis. Attack at the less hindered and kinetically favored olefin cannot be excluded, although it seems more plausible that the metathesis cascade starts with ROM, with loss of ring strain being the main driving force.

11.5.2.1 Synthesis of Indolizidine 251F, (±)-*trans*-Lumausyne and Aburatubolactam A

Following the pioneering work on the synthesis of [x.3.0] bicyclic systems by RRM from norbornene derivatives [35], Aubé *et al.* applied this strategy to synthesize indolizidine alkaloid 251F, where enone **76** was obtained from the rearrangement of norbornene **75** (Scheme 11.20) [36]. Early attempts afforded the desired product in only poor yield (30%) mainly because of olefin oligomerization. Under an atmosphere of ethylene, it was possible to optimize the metathesis cascade and the bicyclic compound **76** was isolated in 93% yield with the use of catalyst [Ru]-**I** (5 mol%).

An equivalent approach was used by Chandler and Phillips in the synthesis of *trans*-kumausyne, isolated from red algae of the genus *Laurencia* in Japan [37].

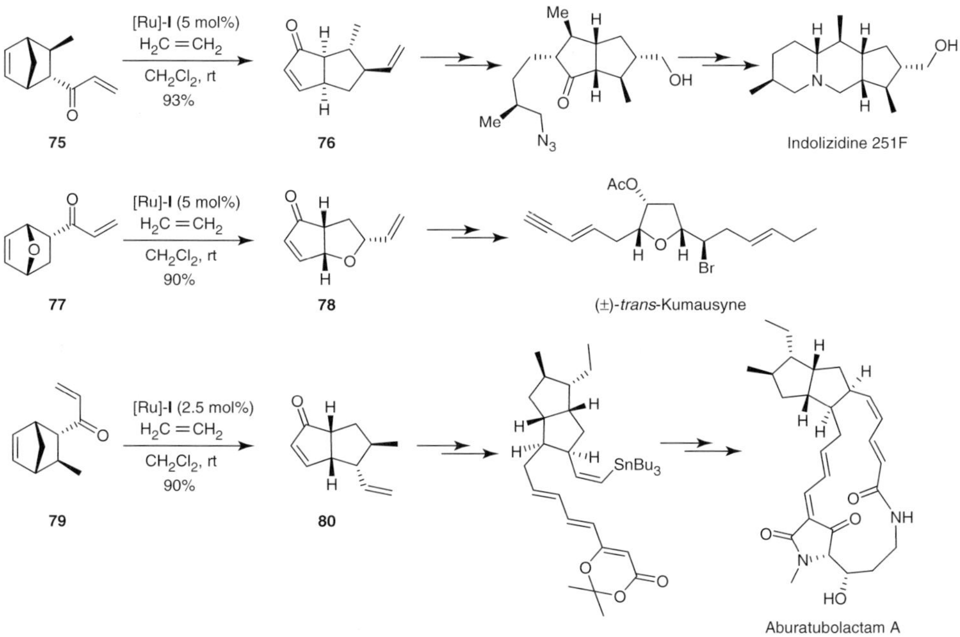

Scheme 11.20 Rearrangement of norbornene derivatives in the synthesis of indolizidine 251F, (±)-*trans*-kumausyne and aburatubolactam A.

Subjecting the oxanorbornene derivative **77** to catalyst [Ru]-**I** (5 mol%) in CH_2Cl_2 under an atmosphere of ethylene resulted in smooth ROM–RCM cascade, providing tetrahydrofuran **78** in 83% yield.

The latest of the applications of this type of RRM cascade has been recently published by the same group in the synthesis of aburatubolactam A [38]. This macrolactam, isolated from bacteria of a marine mollusc collected in Japan, is a member of a growing class of mixed polyketide amino acid metabolites that displays a diverse range of biological activities including cytotoxicity, antimicrobial activity, and the inhibition of superoxide generation. Phillip *et al.*'s approach, which included Stille coupling and Lacey–Dieckmann cyclization as key features, started with the synthesis of **80** through a ROM–RCM domino process. When treated with the first-generation catalyst [Ru]-**I** under an ethylene atmosphere, enone **79** was rearranged to the desired bicyclo[3.3.0]octenone **80** in 90% yield. The absence of an ethylene atmosphere decreased the efficiency of the transformation and led to **80** in 41% yield.

11.5.2.2 Synthesis of (+)-*ent*-Lepadin B

Also very recently, Barbe and Charette published another domino metathesis in the synthesis of *ent*-lepadin B [39]. This lepadine was isolated from the tunicate of *Clavelina lepadiformis* and was shown to exhibit significant *in vitro* cytotoxicity

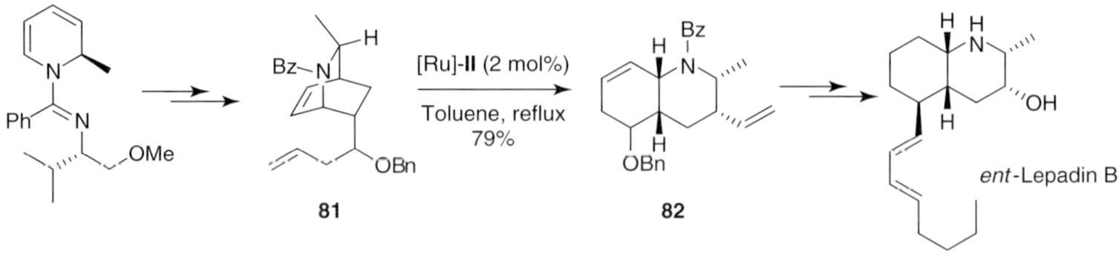

Scheme 11.21 Synthesis of *ent*-lepadin B by RRM cascade.

against several human cancer cell lines. Lepadin B has been recently identified as a potent blocker of neuronal nicotinic acetylcholine receptors, which are implicated in several neurological disorders including nicotinic addiction, epilepsy, as well as Parkinson and Alzheimer diseases. Subjecting **81** to catalyst [Ru]-**II** (2 mol%) (toluene, 80 °C) afforded the desired rearranged product **82** in 79% yield after only 2 minutes (Scheme 11.21). Interestingly, performing the reaction under an ethylene atmosphere decreased the reaction rate and did not provide any yield improvement.

11.6
RCM–ROM Sequences Combined with Other Metathesis Reactions

The synthetic potential of RRM can be amplified with the introduction of a third olefin, permitting the access to much more complex structures in one single transformation, by a further intermolecular CM or a second RCM. As discussed above, two mechanistic pathways can be envisaged. The metathesis cascade may involve an initial formation of a ruthenium–alkylidene complex at the terminal olefin, or alternatively, the sequence may begin with [2+2]-cycloaddition of the ruthenium carbene onto endocyclic double bond, followed by ROM–RCM or CM. Neither of these possibilities can be ruled out, and the occurrence of one or the other may be influenced according to ring strain, steric and electronic properties of the corresponding olefins, and also the reaction conditions employed (Scheme 11.22).

11.6.1
RCM–ROM–RCM

These types of transformations were first explored by Grubbs *et al.* [40] and have often been applied in the synthesis of many natural products. From a mechanistic point of view, these metathesis cascades possess an additional driving force, apart from the release of ring strain, which consists in an entropy gain with the release of a molecule of ethylene as volatile by-product. Like other RRM reactions discussed, oligomerization is an important side reaction, which may be suppressed by the use of an ethylene atmosphere and by performing the reaction at higher dilution.

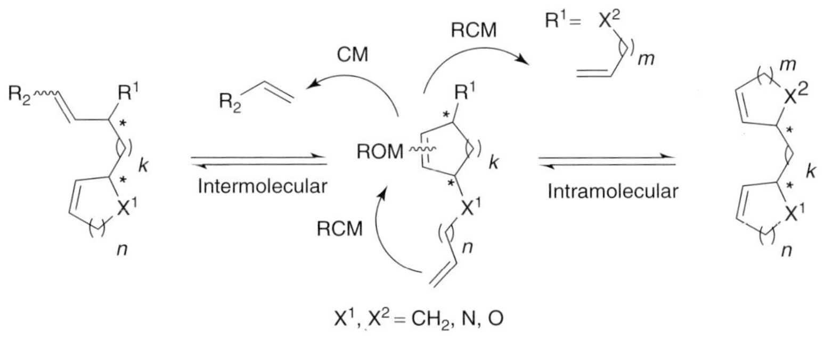

Scheme 11.22 General concept of the RCM–ROM–RCM and RCM–ROM–CM sequences.

11.6.1.1 RCM–ROM–RCM Cascades of Monocyclic Structures

Synthesis of (−)-Anaferine, (+)-Dihydrocuscohygrine, Cuscohygrine Synthetic applications of RCM–ROM–RCM cascades have been mainly championed in Blechert *et al.* in the elegant synthesis of a variety of structurally diverse alkaloids. With a flexible synthetic approach, ring rearrangement of the seven-membered rings **83** and **84** led to the core of (−)-anaferine [41], (+)-dihydrocuscohygrine, and cuscohygrine [42] by just modifying the *N*-alkenyl chain length (Scheme 11.23). Tandem RRM was carried out under an inert atmosphere with catalyst [Ru]-**I** in refluxing CH_2Cl_2 and provided bis-tetrahydropyridine derivative **85** after 48 hours in 87% yield. However, in the case of **83**, a one-pot metathesis cascade-hydrogenation was performed in order to avoid ruthenium-promoted decomposition of the rearranged products, and the dipyrrolidine derivative **86** was isolated in a satisfactory 72% yield.

Synthesis of (+)-Dumetorine A and (+)-Astrophylline The first enantioselective synthesis of (+)-dumetorine was also achieved through an RCM–ROM–RCM domino reaction as a key transformation in only six steps and a 27% overall yield from the functionalized cyclopentene **87** [43]. Preliminary investigation with catalyst [Ru]-**I** (10 mol%) allowed obtaining the desired product **89** in only 7% yield. The low yield was attributed to the preferential formation of a stable cyclic carbene complex **88** (Scheme 11.24) which inhibited further reaction and interrupted the catalytic cycle. With the use of $Ti(Oi\text{-}Pr)_4$ (30 mol%) as a co-additive, the formation of the chelated intermediate was inhibited and the desired ring-rearranged product **89** was obtained in good yield (80%) with only 5 mol% of catalyst [Ru]-**I**.

The first synthesis of (+)-astrophylline was performed in a similar manner by Blechert and Schaudt [44]. The functionalized cyclopentene **91**, in which the stereocontrolled generation of the 1,2-*trans* relative configuration was achieved via a *n*-BuLi-mediated [2,3]-Wittig–Still rearrangement of **90**, was successfully transformed to **92** with the use of the second-generation catalyst [Ru]-**II** (1 mol%).

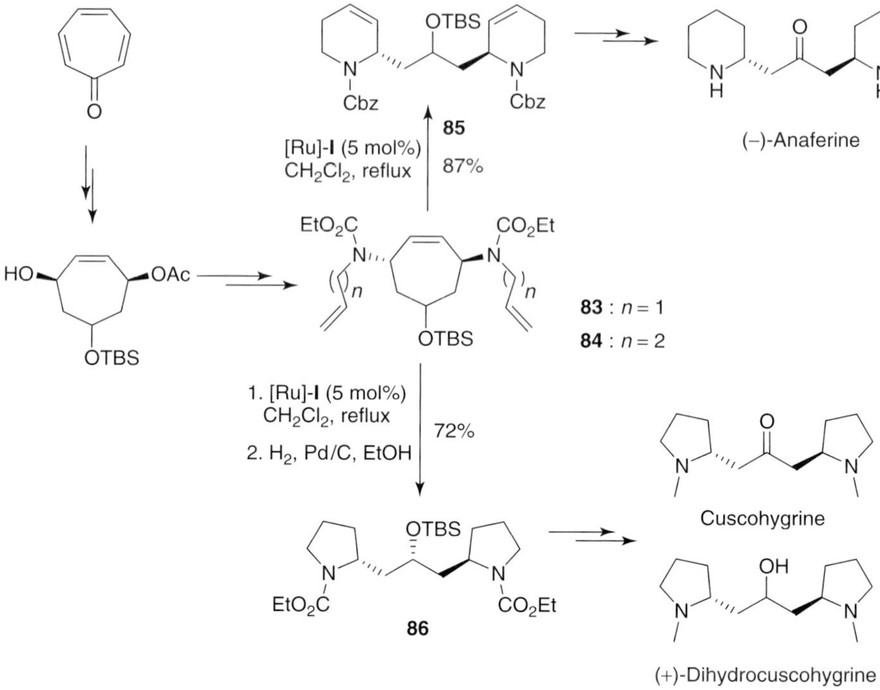

Scheme 11.23 RRM domino processes of seven-membered rings in the synthesis of naturally occurring alkaloids.

C22–C34 Subunit of Halichondrin B A more sterically challenging rearrangement of this type was employed by Burke *et al.* in the synthesis of a known synthetic precursor of halichondrin B [45]. This polyether macrolide, isolated from marine sponges, is a tubulin-interactive antimitotic agent that has been recommended by the National Cancer Institute for stage A preclinical development because of potent activity against various chemoresistant human solid tumor xenografts. In initial investigations, treatment of triene **93** with [Ru]-**I** (25 mol%) led to the formation of a mixture of diastereomeric dihydropyrans **95** and **96** in 21% yield (Scheme 11.25) as a result of CM at the catalytic cycle initiation. The corresponding propagating carbene species failed to undergo intramolecular ring closing, presumably because of steric hindrance in the [2+2]-cycloaddition, resulting in interruption of the metathesis sequence. While efforts to optimize the transformation with an ethylene atmosphere yielded only oligomerized products, the use of the more reactive, but also more sensitive, Schrock catalyst [Mo]-**I** effected the desired transformation and provided the C22–C34 halichondrin B precursor **94** in 79% yield.

Synthesis of (−)-Halosaline and (−)-Indolizidine 167B The synthesis of halosaline presents an early application of domino metathesis in natural product synthesis

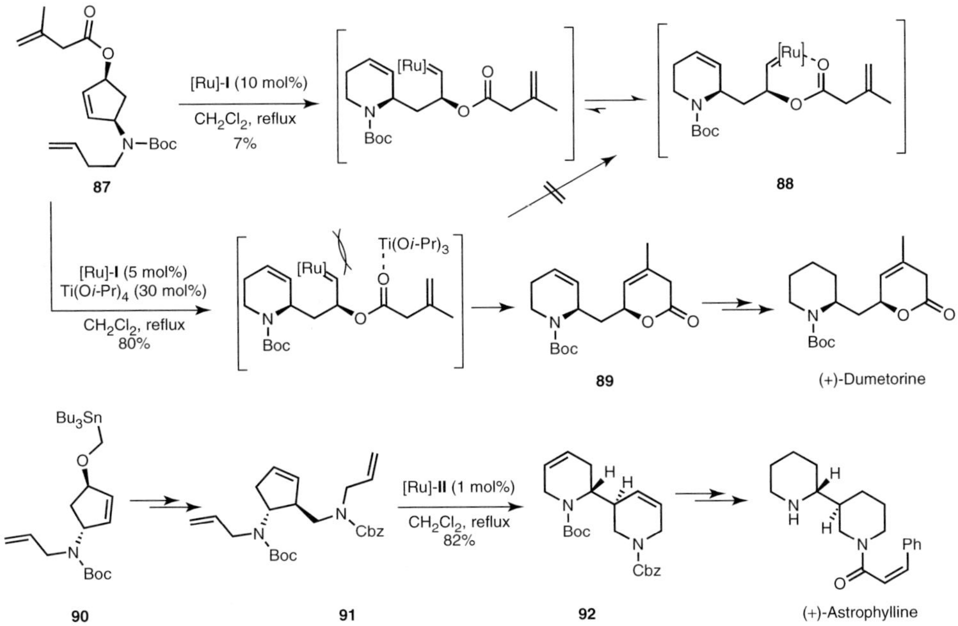

Scheme 11.24 RRM of cyclopentene derivatives in the synthesis of (+)-dumetorine A and (+)-astrophylline.

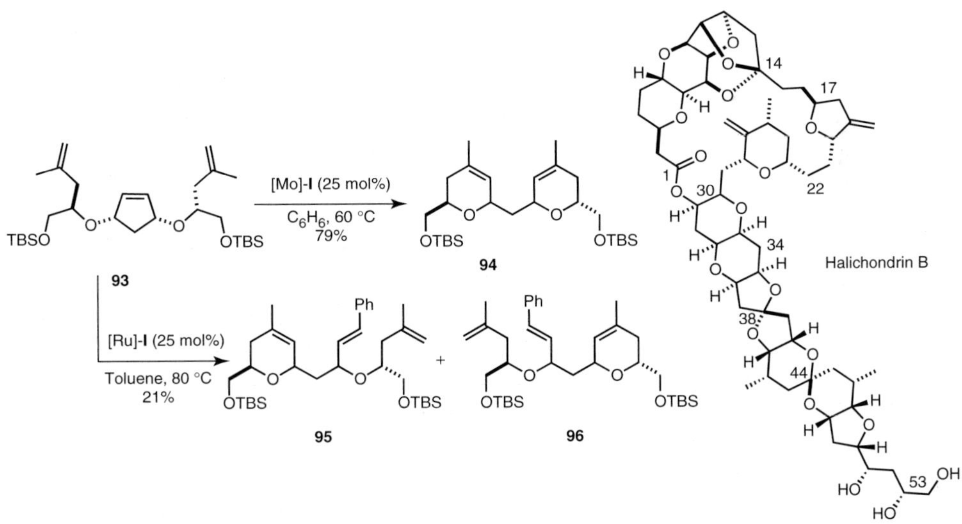

Scheme 11.25 Synthesis of a halichondrin B precursor via ring-opening double ring-closing metathesis.

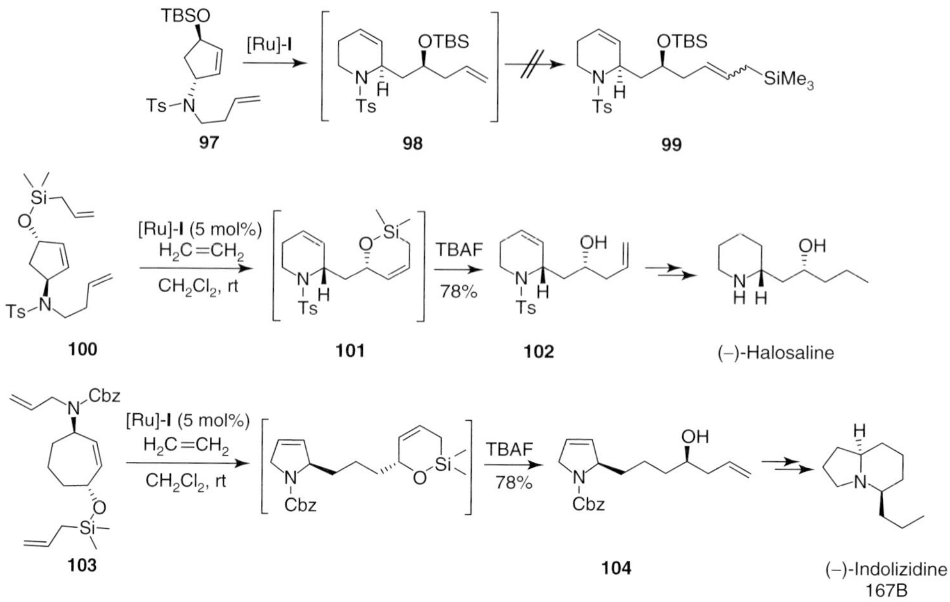

Scheme 11.26 RCM–ROM–RCM cascades of silylethers in the synthesis of alkaloids.

by Blechert *et al.* [28]. Synthetic studies toward this alkaloid began with a tandem RCM–ROM–CM sequence using allyltrimethylsilane as cross-partner to generate **99**. Performing the transformation with [Ru]-**I** as the catalyst led to the desired metathesis product **99** in less than 10% yield. The main product obtained was **98**, suggesting that steric hindrance hampered the CM step. The use of catalyst [Mo]-**I**, which is known to give higher yields for sterically hindered substrates (as seen in the synthesis of **94**), only resulted in CM of the terminal double bond of cyclopentene **97** with allyltrimethylsilane. A clever modification of the metathesis substrate was the addition of a tethered functionality, which enabled the synthesis of **102** by a one-pot RRM desilylation procedure [46]. Replacement of the CM step by an additional RCM overcame the problem of steric hindrance and allowed the use of a cheaper and less sensitive ruthenium catalyst. Compound **102** was finally obtained in an increased overall yield of 78% with catalyst [Ru]-**I** (5 mol%) (Scheme 11.26).

A similar approach served the same group in the synthesis of (−)-indolizidine 167B, an alkaloid isolated from a neotropical frog species. RRM of **103** with [Ru]-**I** (5 mol%), followed by cleavage of the silyl ether functionality, furnished **104** in 78% yield. With this procedure, Blechert *et al.* also exemplified that RCM–ROM–RCM domino processes are not restricted to the formation of same sized bicyclic systems.

11.6.1.2 RCM–ROM–RCM Cascades of Bicyclic Structures

Synthesis of (+)-Cyanthiwigin U Phillips and Pfeiffer have also combined the RRM metathesis of bicyclic compounds described in Section 4.2 with another metathesis

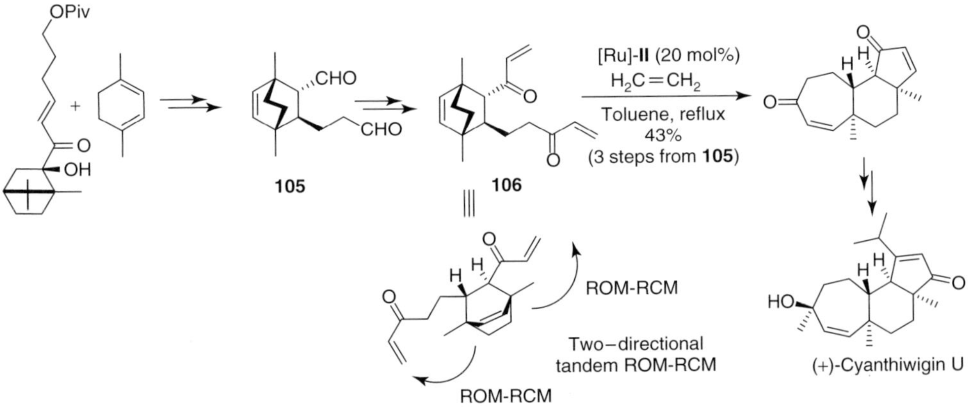

Scheme 11.27 RCM–ROM–RCM cascade in the synthesis of (+)-cyanthiwigin U.

step [47]. In this case, the substituted bicyclo[2.2.2]octane **106** served as a precursor for the synthesis of cyanthiwigin U, a diterpenoid isolated from both fungal and marine sources.

An elegant directional domino ROM–RCM sequence enabled the construction of the desired tricyclic core in the presence of catalyst [Ru]-**I** (20 mol%) and CH_2Cl_2 atmosphere. Four further transformations completed a concise synthesis of this complex compound in 12 steps and 17% overall yield (Scheme 11.27).

11.6.2 RCM–ROM–CM

The domino processes discussed so far concern intramolecular reactions. Extension of the intramolecular domino metathesis processes described in previous sections to an intermolecular reaction, namely, CM, was also successfully applied in organic synthesis (Scheme 11.22). Although CM is afflicted with difficulties in E/Z selectivity and chemoselectivity (three different products are possible in a reaction with two different olefins), the choice of suitable substrates and catalysts alleviates the limitation of these problems. Both mono- and bicycles are suitable substrates. In this case, however, the loss of ring strain, which usually shifts the domino metathesis process toward the desired product in RRM of bicyclic structures, does not guarantee a successful reaction, as CM displays the main origin of side reactions.

11.6.2.1 **Synthesis of (−)-Lasubine II**

Following the studies on the synthesis of halosaline, Blechert and Zaja presented the synthesis of (−)-lasubine II [48]. Cyclopentenone **107** was chosen as the metathesis substrate in order to avoid the failure of the CM step observed in **98** with the incorporation of the bulky TBS group. Rearrangement of **107** in the presence

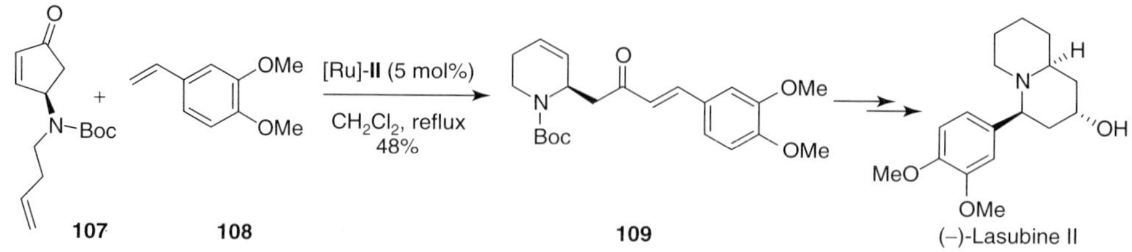

Scheme 11.28 Synthesis of (−)-lasubine II by RCM–ROM–CM.

of dimethoxystyrene and [Ru]-**I** (5 mol%) resulted only in traces of CM of the terminal double bond of **107** with **108**. Fortunately, utilization of 5 mol% of either [Ru]-**II** or [Ru]-**III** furnished the desired *E*-configured product **109** in a gratifying yield of 48 and 44%, respectively. The main by-product isolated resulted, again, from the CM between **107** with **108**, which demonstrates the main concurrent reaction and, therefore, the main difficulty of these types of transformations (Scheme 11.28).

11.6.2.2 **Synthesis of (+)-Cylindramide A and Bicyclic Core of Geodin A**

Phillips *et al.* have also extended the domino metathesis of norbornene derivatives with a CM step and employed this strategy in the construction of the bicyclic fragment of geodin A [49] and the synthesis of (+)-cylindramide A [50]. In both cases, the bicyclic compound **110** served as the metathesis precursor, using different olefins as cross-partners (Scheme 11.29). Metathesis cascade with **111** led to the formation of 54% of product **113** (E/Z = 1.5 : 1) along with 30% of **112**, indicating the partial failure of the final CM step. Interestingly, when olefin **114** was employed, none of the rearranged product lacking CM was found and the desired bicyclic structure **115** was formed in 59% yield with a selectivity of E/Z = 2 : 1.

11.6.2.3 **Total Synthesis of (+)-Mycoepoxydiene**

Although the domino metathesis performed in the synthesis of the fungal metabolite (+)-mycoepoxydiene [51] does not really involve a ring rearrangement by definition, it demonstrates the synthetic utility of ROM–CM–RCM processes for the expansion (or contraction) of macrocyclic ring systems. This type of sequence, previously explored by the groups of Diver and Grubbs among others, was first studied on the oxabicyclic compound **116** separating the first ROM–CM from the RCM step. ROM and CM with 1,3-butadiene (4 equiv.) furnished a mixture of regioisomeric trienes **117** and **118** in 75% yield when catalyzed by of [Ru]-**I** (5 mol%). The ring closure of these two structures was found to be poorly effective, and the desired product **119** was obtained in only 29% yield. The one-pot procedure, however, permitted obtaining **120** in a better overall yield than in a stepwise manner with the sequential use of catalysts [Ru]-**I** and [Ru]-**II** (Scheme 11.30).

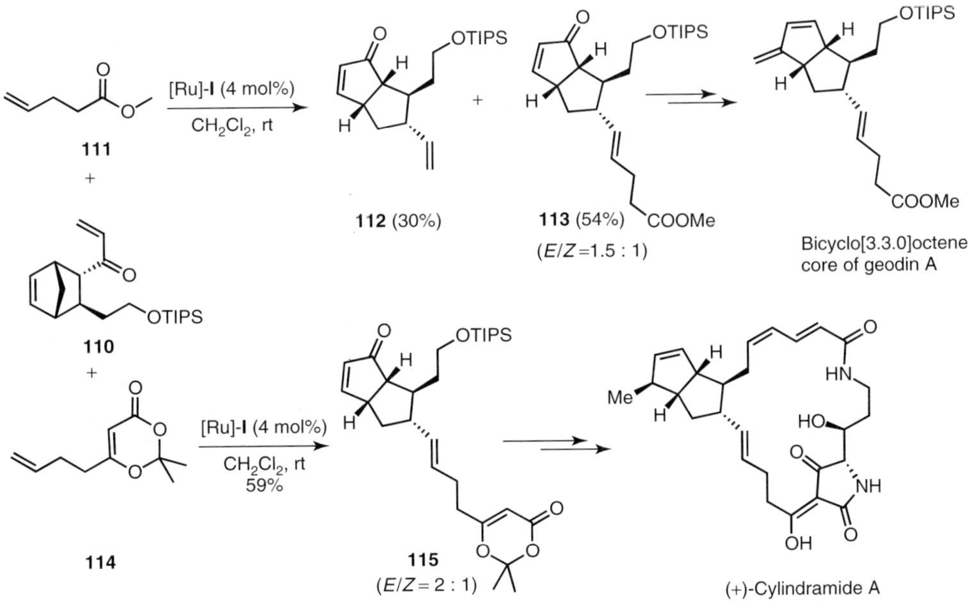

Scheme 11.29 RCM–ROM–CM cascades on bicyclic systems.

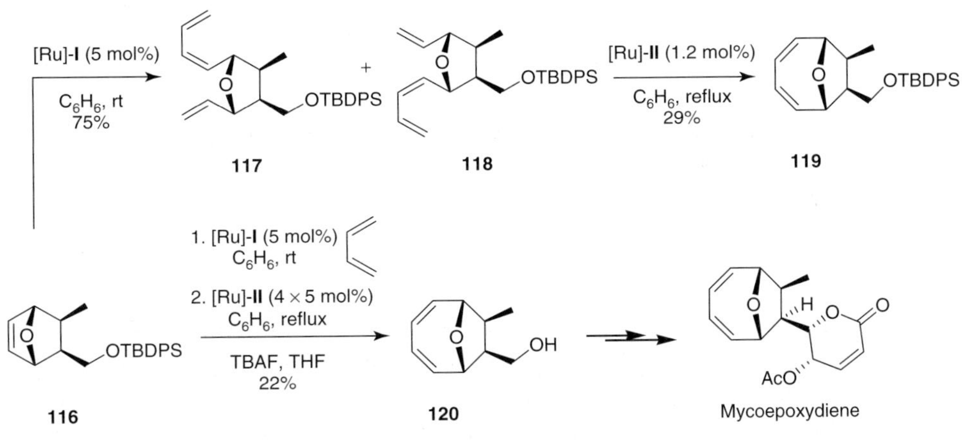

Scheme 11.30 Synthesis of mycoepoxydiene by ROM–CM–RCM.

11.7
Conclusions and Outlook

Alkene metathesis has revolutionized modern approaches to the synthesis of a wide variety of useful organic molecules with far-reaching applications in natural product synthesis. The aforementioned examples are testimony to the utility of catalytic olefin metathesis in domino reactions and show how these domino processes permit multi C–C bond-forming–bond-breaking reactions allowing rapid access to structures of impressive complexity which have proved otherwise very challenging. ROM–CM, RCM–CM, ene–yne RCM–RCM, and RCM–RCM cascades have been described, which have proven to be very efficient and versatile methods for the construction of carbocycles, heterocycles, and polycyclic structures. The extension of the RCM–RCM sequence by further metathesis steps has been also established in organic synthesis and has led to even more complex structures, although the catalysts with higher activities are still required. The transfer of stereoinformation in RRM processes is an attractive feature, but chiral starting materials are needed and further development of asymmetric processes is therefore essential. Given the advantages of metathesis cascades in terms of atom economy and catalytic efficiency, we can only expect the interest in these processes to grow further.

References

1 (a) Grubbs, R.H. (ed.) (2003) *Handbook of Metathesis*, Vols. 1, 2, 3, Wiley-VCH Verlag GmbH, Weinheim; For reviews of the alkene-metathesis reaction, see: (b) Holub, N. and Blechert, S. (2007) *Chem. Asian J.*, **2** (9), 1064–1082; (c) Schmidt, B. and Hermanns, J. (2004) *Top. Organomet. Chem.*, **7**, 223–267; (d) Connon, S.J. and Blechert, S. (2004) *Top. Organomet. Chem.*, **7**, 93–124; (e) Fürstner, A. (2000) *Angew. Chem. Int. Ed.*, **39**, 3012–3043; (f) Connon, S.J. and Blechert, S. (2003) *Angew. Chem. Int. Ed.*, **42**, 1900–1923; (g) Grubbs, R.H. and Chang, S. (1998) *Tetrahedron*, **54**, 4413–4450; (h) Schuster, M. and Blechert, S. (1997) *Angew. Chem. Int. Ed.*, **36**, 2037–2056; For its application in total synthesis, see: (i) Nicolaou, K.C., Bulger, P.G., and Sarlah, D. (2005) *Angew. Chem. Int. Ed.*, **44**, 4490–4527.

2 Tietze, L.F. (1996) *Chem. Rev.*, **96**, 115–136.

3 For examples on natural product synthesis see: (a) Moreno-Manas, M., Pleixats, R., and Santamaria, A. (2001) *Synlett*, 1784–1786; (b) Limanto, J. and Snapper, M.L. (2000) *J. Am. Chem. Soc.*, **122**, 8071–8072.

4 For examples see: (a) Piper, J.L. and Postema, M.H.D. (2004) *J. Org. Chem.*, **69**, 7395–7398; (b) Wallace, D.J., Bulger, P.G., Kennedy, D.J., Ashwood, M.S., Cottrell, I.F., and Dolling, U.-H. (2001) *Synlett*, 357–360.

5 For a recent review, see: Deshmukh, P.H. and Blechert, S. (2007) *Dalton Trans.*, **24**, 2479–2491.

6 Schrock, R.R., Murdzek, J.S., Bazan, G.C., Robbins, J., DiMare, M., and O'Regan, M. (1990) *J. Am. Chem. Soc.*, **112**, 3875–3886.

7 Nguyen, S.T., Johnson, L.K., Grubbs, R.H., and Ziller, J.W. (1992) *J. Am. Chem. Soc.*, **114**, 3974–3975.

8 Scholl, M., Ding, S., Lee, C.W., and Grubbs, R.H. (1999) *Org. Lett.*, **1**, 953–956.

9 (a) Garber, S.B., Kingsbury, J.S., Gray, B.L., and Hoveyda, H. (2000) *J. Am. Chem. Soc.*, **122**, 8168–8179; (b) Gessler, S. and Blechert, S. (2000) *Tetrahedron Lett.*, **41**, 9973–9976.

10 (a) Hoye, T.R., Jeffrey, C.S., Tennakoon, M.A., Wang, J., and Zhao, H. (2004) *J. Am. Chem. Soc.*, **126**, 10210–10211; (b) Wallace, D. (2005) *Angew. Chem. Int. Ed.*, **44**, 1912–1915.

11 (a) Kim, M. and Lee, D. (2005) *J. Am. Chem. Soc.*, **127**, 18024–18025; (b) Cho, E.J. and Lee, D. (2008) *Org. Lett.*, **10**, 257–259.

12 For specific reviews of the enyne-metathesis reaction, see: (a) Villar, H., Fringsa, M., and Bolm, C. (2007) *Chem. Soc. Rev.*, **36**, 55–66;

(b) Diver, S.T. and Giessert, A.J. (2004) *Chem. Rev.*, **104**, 1317–1382.

13 (a) Kummer, D., Brenneman, J., and Martin, S.F. (2005) *Org. Lett.*, 7, 4621–4623; (b) Kummer, D.A., Brenneman, J.B., and Martin, S.F. (2006) *Tetrahedron*, **62**, 11437–11449.

14 Dewi, P., Randl, S., and Blechert, S. (2005) *Tetrahedron Lett.*, **46**, 577–580.

15 Honda, T., Namiki, H., Kaneda, K., and Mizutani, H. (2004) *Org. Lett.*, **6**, 87–89.

16 Honda, T., Namiki, H., Watanabe, M., and Mizutani, H. (2004) *Tetrahedron Lett.*, **45**, 5211–5213.

17 Michrowska, A., Bujok, R., Harutyunyan, S., Sashuk, V., Dolgonos, G., and Grela, K. (2004) *J. Am. Chem. Soc.*, **126**, 9318–9325.

18 Honda *et al.*, unpublished results.

19 Niethe, A., Fischer, D., and Blechert, S. (2008) *J. Org. Chem.*, **73**, 3088–3093.

20 (a) Boyer, F.-D. and Hanna, I. (2002) *Tetrahedron Lett.*, **43**, 7469–7472; (b) Boyer, F.D., Hanna, I., and Ricard, L. (2004) *Org. Lett.*, **6**, 1817–1820.

21 Aldegunde, M.J., Castedo, L., and Granja, J.R. (2008) *Org. Lett.*, **10**, 3789–3792.

22 Fukumoto, H., Takahashi, K., Ishihara, J., and Hatakeyama, S. (2003) *Tetrahedron Lett.*, **44**, 8047–8049.

23 Shimizu, K., Takimoto, M., and Mori, M. (2003) *Org. Lett.*, **5**, 2323–2325.

24 Fukumoto, H., Takahashi, K., Ishihara, J., and Hatakeyama, S. (2006) *Angew. Chem. Int. Ed.*, **45**, 2731–2734.

25 For examples see: (a) Limanto, J. and Snapper, M.L. (2000) *J. Am. Chem. Soc.*, **122**, 8071–8072; (b) Bassindale, M.J., Hamley, P., and Harrity, J.P.A. (2001) *Tetrahedron Lett.*, **42**, 9055–9057.

26 For another application of a ROM-CM cascade see: Schrader, T.O. and Snapper, M.L. (2002) *J. Am. Chem. Soc.*, **124**, 10998–11000.

27 Alexander, A., Statsuk, V., Liu, D., and Kozmin, S.A. (2004) *J. Am. Chem. Soc.*, **126**, 9546–9547.

28 Stragies, R. and Blechert, S. (1999) *Tetrahedron*, **55**, 8179–8188.

29 Stragies, R. and Blechert, S. (2000) *J. Am. Chem. Soc.*, **122**, 9584–9591.

30 Buschmann, N., Rückert, A., and Blechert, S. (2002) *J. Org. Chem.*, **67**, 4325–4329.

31 Song, L., Duesler, E.N., and Mariano, P.S. (2004) *J. Org. Chem.*, **69**, 7284–7293.

32 Zhao, Z., Song, L., and Mariano, P.S. (2005) *Tetrahedron*, **61**, 8888–8894.

33 Holub, N., Neidhöfer, J., and Blechert, S. (2005) *Org. Lett.*, **7**, 1227–1229.

34 Böhrsch, V. and Blechert, S. (2006) *Chem. Commun.*, 1968–1970.

35 Stragies, R. and Blechert, S. (1998) *Synlett*, 169–170.

36 Wrobleski, A., Sahasrabudhe, K., and Aubé, J. (2004) *J. Am. Chem. Soc.*, **126**, 5475–5481.

37 Chandler, C.L. and Phillips, A.J. (2005) *Org. Lett.*, **7**, 3493–3495.

38 Henderson, J.A. and Phillips, A.J. (2008) *Angew. Chem. Int. Ed.*, **47**, 1–4.

39 Barbe, G. and Charette, A.B. (2008) *J. Am. Chem. Soc.*, **130**, 13873–13875.

40 Zuercher, W.J., Hashimoto, M., and Grubbs, R.H. (1996) *J. Am. Chem. Soc.*, **118**, 6634–6640.

41 Blechert, S. and Stapper, C. (2002) *Eur. J. Org. Chem.*, **16**, 2855–2858.

42 Stapper, C. and Blechert, S. (2002) *J. Org. Chem.*, **67**, 6456–6460.

43 Rückert, A., Deshmukh, P.H., and Blechert, S. (2006) *Tetrahedron Lett.*, **47**, 7977–7981.

44 Schaudt, M. and Blechert, S. (2003) *J. Org. Chem.*, **68**, 2913–2920.

45 Burke, S.D., Quinn, K.J., and Chen, V.J. (1998) *J. Org. Chem.*, **63**, 8626–8627.

46 Zaminer, J., Stapper, C., and Blechert, S. (2002) *Tetrahedron Lett.*, **43**, 6739–6741.

47 Pfeiffer, M.W.B. and Phillips, A.J. (2005) *J. Am. Chem. Soc.*, **127**, 5334–5335.

48 Zaja, M. and Blechert, S. (2004) *Tetrahedron*, **60**, 9629–9634.

49 Phillips, A.J., Hart, A.C., and Henderson, J.A. (2006) *Tetrahedron Lett.*, **47**, 3743–3745.

50 Hart, A.C. and Phillips, A.J. (2006) *J. Am. Chem. Soc.*, **128**, 1094–1095.

51 Takao, K., Yasui, H., Yamamoto, S., Sasaki, D., Kawasaki, S., Watanabe, G., and Tadano, K. (2004) *J. Org. Chem.*, **69**, 8789–8795.

12
Catalytic Enantioselective Olefin Metathesis and Natural Product Synthesis

Amir H. Hoveyda, Steven J. Malcolmson, Simon J. Meek, and Adil R. Zhugralin

12.1
Introduction

In this chapter, a brief account of various applications of catalytic enantioselective olefin metathesis [1] to the synthesis of biologically active molecules is presented. The availability of chiral catalysts for olefin metathesis offers schemes for enantioselective synthesis of natural products that are more concise than those entailing the use of achiral Ru- or Mo-based complexes and enantiomerically pure substrates [2]. Enantioselective olefin metathesis plays a critical role in the development of more effective catalysts. A less appreciated role of investigations directed toward chiral catalyst development is that such efforts can lead to the identification of better catalysts. In designing a chiral catalyst, the structure – the steric and electronic characteristics – of the promoter molecule might be altered to the extent that a fundamentally new type of catalyst, one that likely has different reactivity and/or selectivity profiles, emerges.

12.2
Total Synthesis of Natural Products with Enantiomerically Pure Chiral Olefin Metathesis Catalysts Bearing a C_2-symmetric Diolate Ligand

12.2.1
Total Synthesis of Coniine through Enantioselective RCM with Substrates Bearing a Tertiary Amine

In 2005, the first examples of enantioselective olefin metathesis involving substrates that bear Lewis basic amines were reported [3]. Catalytic activity of number of chiral Mo-based diolates (such as **2**, Scheme 12.1) were probed; enantioselective ring-closing metathesis (RCM) reactions deliver various azacyclic structures in up to 99/1 enantiomeric ratio (er). It should be noted that Ru-based catalysts can be largely inhibited in the presence of the same class of amine substrates [4].

Metathesis in Natural Product Synthesis: Strategies, Substrates and Catalysts. Edited by Janine Cossy, Stellios Arseniyadis, and Christophe Meyer

ISBN: 978-3-527-32440-8

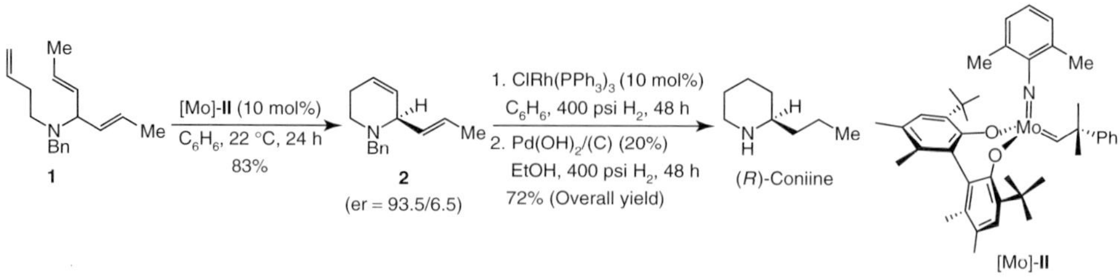

Scheme 12.1 Total synthesis of (*R*)-coniine through enantioselective RCM.

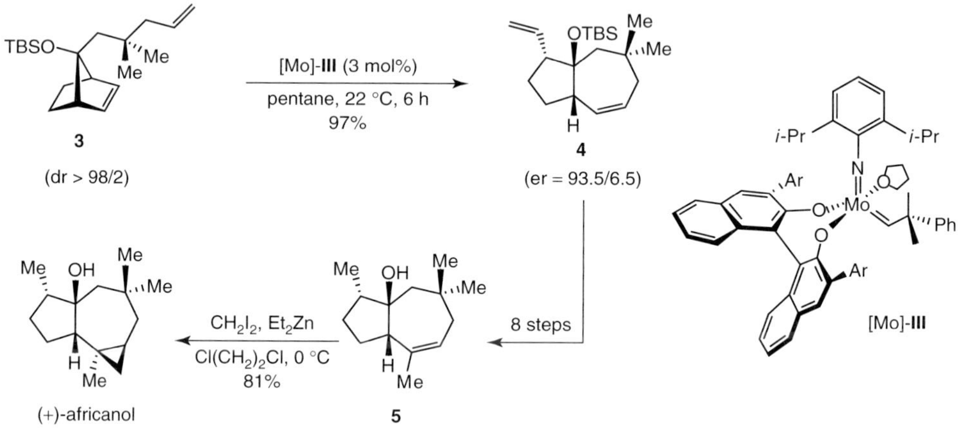

Scheme 12.2 Synthesis of Africanol through an enantioselective ring-opening/ring-closing metathesis (ROM/RCM) reaction. Ar = 2,4,6-$Me_3C_6H_2$.

The utility of the above-mentioned class of catalytic transformations has been demonstrated in the context of preparation of enantiomerically enriched coniine, the neurotoxin found in the hemlock plant. Thus, Mo-catalyzed RCM of benzylamine **1** in the presence of 10 mol% of chiral [Mo]-catalyst ([Mo]-**II**) gives rise to the formation of unsaturated piperidine **2** in 93.5/6.5 er and 83% yield (Scheme 12.1). A two-step sequence furnishes the poisonous alkaloid.

12.2.2
Enantioselective Synthesis of Africanol by a Ring-opening/Ring-closing Metathesis Reaction

As illustrated in Scheme 12.2, treatment of *meso* tertiary silyl ether **3**, prepared diastereoselectively through alkylation of the corresponding norbornenone, with 3 mol% of chiral alkylidene [Mo]-**III** gives rise to **4** in high enantiomeric purity (er = 93.5/6.5) and 97% yield. A bicyclic structure containing the desired stereochemistry is thus accessed through a single, efficient catalytic process. Transformation to **5**,

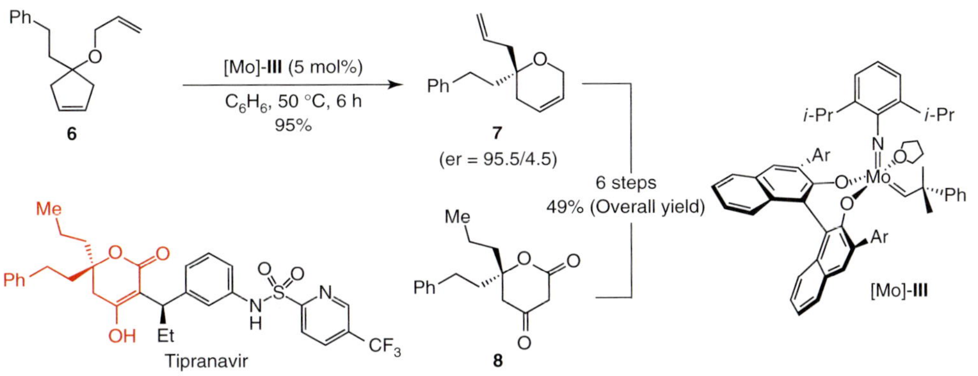

Scheme 12.3 Enantioselective synthesis of the lactone fragment of tipranavir through a ring-opening/ring-closing metathesis (ROM/RCM) reaction. Ar = 2,4,6-$Me_3C_6H_2$.

bearing the trisubstituted cyclic alkene and conversion of the vinyl unit to the desired methyl substituent require eight additional steps. The natural product is then accessed through a hydroxyl-directed cyclopropanation reaction [5].

12.2.3
Enantioselective Synthesis of the Lactone Fragment of Anti-HIV Agent Tipranivir

Unsaturated pyran **6** (Scheme 12.3) has been utilized (via **8**) in the synthesis of the lactone fragment of tipranavir, a molecule discovered to possess potent activity against the HIV-protease enzyme [6]. Synthesis of enantiomerically enriched **7** (er = 95.5/4.5), a chiral product that would otherwise be difficult to access in the nonracemic form, is therefore achieved through a catalytic reaction of an easily available achiral substrate. Enantioselective preparation of **7** by an RCM process that involves the use of an achiral catalyst demands an effective protocol for preparation of the requisite O-substituted quaternary carbon stereogenic center.

12.3
Enantioselective Synthesis of Quebrachamine through an Exceptionally Challenging RCM Reaction

An enantioselective synthesis of *Aspidosperma* alkaloid quebrachamine, summarized in Scheme 12.4, has been recently accomplished [7]. The late-stage RCM requires ring-closure onto one of two sterically hindered vinyl groups at a congested all-carbon quaternary center [8] in the presence of a Lewis basic tertiary amine. [Mo]–diolate complexes (e.g., [Mo]-**II**, Scheme 12.1) do not promote any reaction even under forcing conditions (up to 50 mol% catalyst, up to 80 °C). When

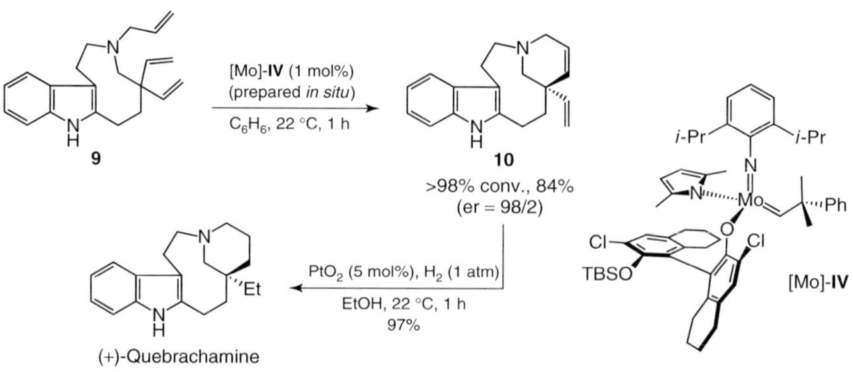

Scheme 12.4 A stereogenic-at-Mo complex as chiral catalyst for enantioselective RCM in a stereoselective synthesis of quebrachamine.

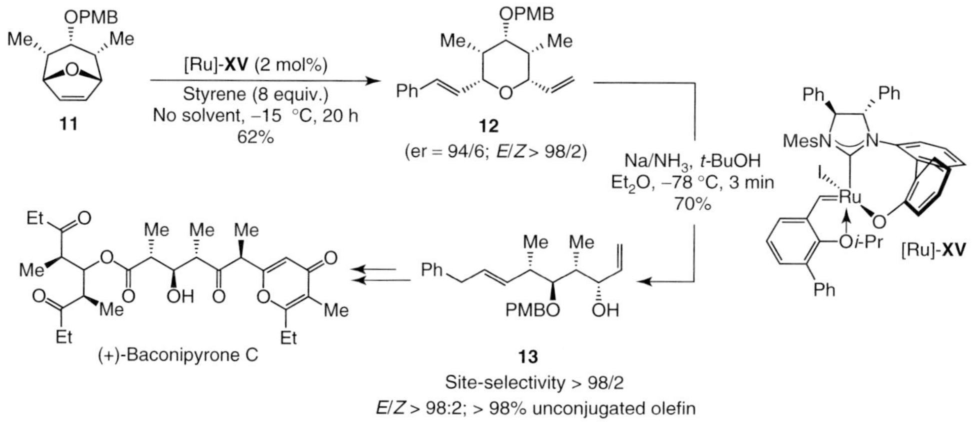

Scheme 12.5 Total synthesis of baconipyrone C by an enantioselective Ru-catalyzed ring-opening/cross-metathesis (ROCM) reaction. PMB = *p*-methoxybenzyl; Mes = 2,4,6-$Me_3C_6H_2$.

a chiral [Ru]-catalyst (e.g., the chiral complex illustrated in Scheme 12.5) is used, *rac*-**10** is obtained.

To address the above reactivity and selectivity problems, a new class of stereogenic-at-[Mo] complexes, represented by monopyrrolide [Mo]-**IV**, was designed [7a]. The stereogenic-at-[Mo] complex [Mo]-**IV**, prepared and used *in situ* from the corresponding achiral Mo-bispyrrolide and a chiral aryl alcohol, readily promotes the difficult RCM. Tetracyclic **10** is obtained efficiently in 84% yield and an enantiomeric ratio of 98/2[9].

12.4
Synthesis of Baconipyrone C by Ru-catalyzed Enantioselective ROCM

Total synthesis of baconipyrone C, summarized in Scheme 12.5 [10], is the first and only application of Ru-catalyzed enantioselective olefin metathesis to natural product synthesis. Treatment of oxabicycle **11** with styrene and 2 mol% chiral Ru complex [Ru]-**XV** [11] leads to the formation of pyran **12** in 62% yield and with an enantiomeric ratio of 94 : 6. Ru-carbene [Ru]-**XV** is generated *in situ* by subjecting the corresponding Ag-based *N*-heterocyclic carbene (NHC) to an achiral Ru–PCy_3 complex and NaI. It is also worthy of note that the diketone fragment of baconipyrone C was synthesized through a tandem double-allylic alkylation process promoted by a chiral NHC–Cu complex that is structurally related to carbene [Ru]-**XV**.

12.5
Conclusions and Future Outlook

Discovery and development of chiral catalysts has an important relationship with natural product synthesis; it is not only that the utility of such catalysts are best demonstrated in a total synthesis setting, but also that efforts to prepare natural products with maximum efficiency and selectivity illustrates the need for more effective catalysts. Therefore, catalyst development driven by demands presented in a total synthesis (problem in search of a solution) may prove to be a more valuable approach than the strategy of catalyst discovery followed by search for an appropriate application (solution in search of a problem).

There is little doubt that more active and robust olefin metathesis catalysts need to be discovered and developed for this important class of transformations to achieve their maximum potential [12]. One avenue of research that can lead to identification of such superior catalysts likely involves natural product synthesis, since challenges faced in the context of such efforts often demand innovation in catalyst design.

Acknowledgments

We are grateful to the National Science Foundation (Grants CHE–0213009 and CHE–0715138) and the National Institutes of Health (Grant GM-59426) for generous support of our efforts toward the development of effective chiral olefin metathesis catalysts and applications of the derived methods to target-oriented synthesis.

References

1 Schrock, R.R. and Hoveyda, A.H. (2003) *Angew. Chem. Int. Ed.*, **42**, 4592–4633.

2 For reviews on applications of catalytic olefin metathesis reactions (largely promoted by achiral catalysts) to natural product synthesis, see: (a) Deiters, A. and Martin, S.F. (2004) *Chem. Rev.*, **104**, 2199–2238; (b) Nicolaou, K.C., Bulger, P.G., and Sarlah, D. (2005) *Angew. Chem. Int. Ed.*, **44**, 4490–4527.

3 Sattely, E.S., Cortez, G.A., Moebius, D.C., Schrock, R.R., and Hoveyda, A.H. (2005) *J. Am. Chem. Soc.*, **127**, 8526–8533.

4 For representative examples, see: (a) Lee, K.L., Goh, J.B., and Martin, S.F. (2001) *Tetrahedron Lett.*, **42**, 1635–1638; (b) Wipf, P., Rector, S.R., and Takahashi, H. (2002) *J. Am. Chem. Soc.*, **124**, 14848–14849; (c) Wipf, P. and Spencer, S.R. (2005) *J. Am. Chem. Soc.*, **127**, 225–235; (d) Cortez, G.A., Schrock, R.R., and Hoveyda, A.H. (2007) *Angew. Chem. Int. Ed.*, **46**, 4534–4538.

5 Hoveyda, A.H., Evans, D.A., and Fu, G.C. (1993) *Chem. Rev.*, **93**, 1307–1370.

6 Cefalo, D.R., Kiely, A.F., Wuchrer, M., Jamieson, J.Y., Schrock, R.R., and Hoveyda, A.H. (2001) *J. Am. Chem. Soc.*, **123**, 3139–3140.

7 (a) Malcolmson, S.J., Meek, S.J., Sattely, E.S., Schrock, R.R., and Hoveyda, A.H. (2008) *Nature*, **456**, 933–937; (b) Sattely, E.S., Meek, S.J., Malcolmson, S.J., Schrock, R.R., and Hoveyda, A.H. (2009) *J. Am. Chem. Soc.*, **131**, 943–953.

8 (a) Christophers., J. and Baro, A. (eds) (2006) *Quaternary Stereocenters: Challenges and Solutions for Organic Synthesis*, Wiley-VCH Verlag GmbH, Weinheim; (b) Denissova, I. and Barriault, L. (2003) *Tetrahedron*, **59**, 10105–10146.

9 Formation of an all-carbon quaternary stereogenic center through enantioselective olefin metathesis is relatively rare. For the first example of such a transformation, see: Lee, A.-L., Malcolmson, S.J., Puglisi, A., Schrock, R.R., and Hoveyda, A.H. (2006) *J. Am. Chem. Soc.*, **128**, 5153–5157.

10 Gillingham, D.G. and Hoveyda, A.H. (2007) *Angew. Chem. Int. Ed.*, **46**, 3860–3864.

11 Van Veldhuizen, J.J., Campbell, J.E., Giudici, R.E., and Hoveyda, A.H. (2005) *J. Am. Chem. Soc.*, **127**, 6877–6882.

12 Zhugralin, A.R. and Hoveyda, A.H. (2007) *Nature*, **450**, 243–251.

13
Metathesis Reactions in Solid-phase Organic Synthesis

Sofia Barluenga, Pierre-Yves Dakas, Rajamalleswaramma Jogireddy, Gaële Valot, and Nicolas Winssinger

13.1
Introduction

As is evident from the different chapters of this book, the mildness and versatility of the metathesis reaction [1–3] have had a tremendous impact on organic chemistry, which enabled it to be rapidly embraced by the community of synthetic chemists as it offered a new manifold of disconnections in retrosynthetic analysis [4] and new opportunities in diversity-oriented synthesis [5]. The discovery of the first efficient and tolerant soluble catalyst was reported at a time of high activity in the field of combinatorial chemistry. The potential of this enabling reaction was therefore rapidly evaluated in the context of solid-phase organic synthesis (see introduction for the most relevant catalysts). Importantly, it was swiftly established that the most popular support for solid-phase organic synthesis (polystyrene-based resins) is compatible with the ruthenium-based metathesis catalysts, and the past 10 years have shown that conditions developed for solution-phase chemistry generally translate well to solid-phase synthesis. An important consideration for carrying out a ring-closing metathesis (RCM) is the rate of cyclization versus oligomerization which will dictate the concentration at which the reaction should be performed to maximize the desired outcome. Analogously, the outcome of metathesis reactions on polymer-bound alkenes will depend on the density of the alkenes (loading of the resin) and mobility or conformational flexibility of the alkenes (cross-linking of the resin and nature of the linker). From the first reports in 1996/1997 to the early 2000s, all reactions on solid phase were carried out with the first-generation ruthenium catalyst [Ru]-**II** bearing phosphine ligands. Following the commercial availability of second-generation catalysts with a carbene ligand, its superior reactivity was frequently favored. As for solution-phase chemistry, subtle differences in the outcome of the reaction were sometimes observed depending on the choice of catalyst ([Ru]-**II** versus [Ru]-**III**). While both catalysts lead to the same intermediate following the first step of the catalytic cycle, differences in the rate and reversibility of this first step have been found to lead to significantly improved performances in some circumstances. This is particularly true for cross-metathesis

Metathesis in Natural Product Synthesis: Strategies, Substrates and Catalysts. Edited by Janine Cossy, Stellios Arseniyadis, and Christophe Meyer

ISBN: 978-3-527-32440-8

(CM) reaction with α,β-conjugated alkenes for which [Ru]-**III** often performs best. A welcomed addition to this family is the development of a fluorous-tagged catalyst that dramatically facilitates catalyst isolation, a critical point if the metathesis is used in the cleavage of high-throughput synthesis programs.

This chapter is divided into six subsections which cover the most prominent applications of the metathesis reaction in solid-phase organic chemistry: RCM for cleavages from the resin; synthesis of small rings to macrocycles on the resin via RCM; the use of the metathesis reaction for dimerization of polymer-bound small molecules; RCM to constrict peptide conformations (while this is formally an RCM, it is treated separately because of the unique considerations of peptide chemistry); CM between a polymer-bound alkene and an alkene in solution; and ene–yne metathesis. As the field evolved relatively rapidly with major advances in the catalysts, each subsection is organized in chronological order for the sake of clarity.

13.2
Metathesis-based Cyclorelease Reaction

Following the first publications reporting the broad functional group compatibility of the new ruthenium-based metathesis catalyst, it became apparent that an alkene functionality could serve as a linker to immobilize the small molecules of interest. The first applications of this principle exploited the metathesis reaction to perform an RCM with concomitant cleavage of the substrate from the resin, the so-called cyclorelease reaction, to afford an endocyclic alkene product (Figure 13.1, Equation 1). Conversely, if a terminal olefin is sought rather than an endocyclic alkene, the linker could be designed to release the terminal alkene while leaving the ring-closing product polymer bound (Figure 13.1, Equation 2. An important consideration in the former case is that the catalyst remains bound to the resin and an alkene must be within reacting distance for the catalyst to turn over. The efficiency of the turnover is strongly dependent on the substrate, the type of resin, and the conditions used. Nonetheless, it has been shown that external alkenes can be used to promote efficient catalyst turnover in some cases [6, 7]. As the reaction is reversible and under thermodynamic control, whether the external alkene will also compete for CM and ring-opening metathesis (ROM) depends on the strain of the cyclized product. In the second case, release of a terminal alkene upon RCM on the polymer, the catalyst remains associated with the alkene in solution and

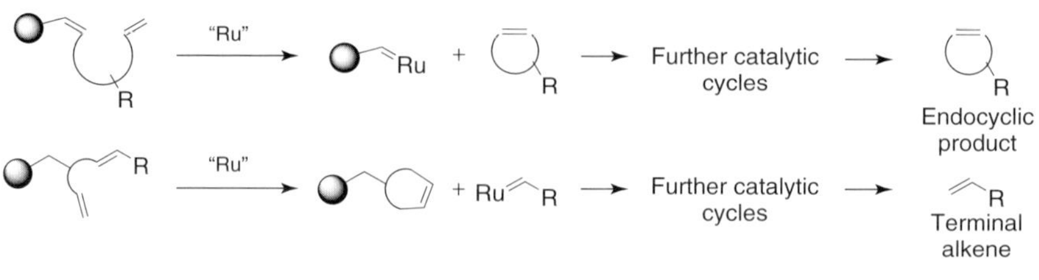

Figure 13.1 Metathesis-based cyclorelease reaction.

can readily turn over. An important consideration in these reactions is that the product will inevitably be contaminated with the catalyst, which will require some form of purification prior to biological screenings. The more recent development of practicable methods to remove the catalyst such as oxidation with dimethyl sulfoxide (DMSO) and Ph_3PO [8], which converts the catalyst to easily removable polar by-products, are important advances. Alternatively, the light fluorous-tagged Grubbs–Hoveyda catalyst reported by Curran (f-[Ru]-**III**) [9] which can be removed by fluorous solid-phase extraction offers another solution which is compatible with automation.

Pioneering examples of the cyclorelease cleavage to obtain biologically relevant products include the preparation of tetrapeptide macrocycles reported by Blechert *et al.* [10], the synthesis of epothilones by Nicolaou *et al.* [11], and lactams [12] as well as β-turn mimetics by Piscopio *et al.* [13] (Figure 13.2). In the case of the

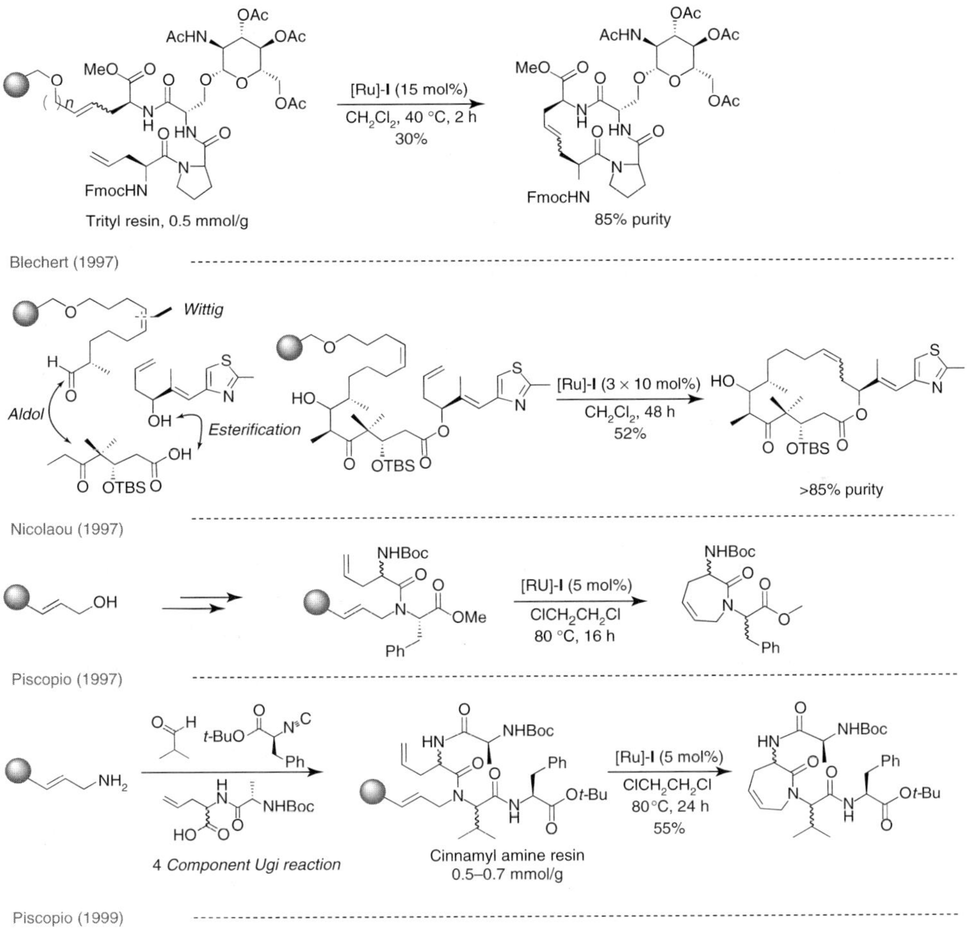

Figure 13.2 Pioneering examples of the RCM cyclorelease.

epothilones, the chemistry was used to make over 100 analogs, shedding important light on the structure activity of this pharmacophore [14].

More recently, this strategy was used by Overkleeft *et al.* to prepare a pilot library of nine pyranofurans from mannitol using [Ru]-**II** for the cyclorelease (Figure 13.3). The reaction was found to be extremely efficient, affording over 90% yield in several examples with just 5 mol% catalyst using a moderately high loading resin (0.55 mmol g^{-1}) [15]. The effect of an external alkene in the cyclization to facilitate catalyst turnover was revisited by Brown *et al.* in the context of the cyclorelease of seven-membered sulfonamide on a fairly high loading resin (0.7–1.2 mmol g^{-1}). The cyclorelease proceeded with 30% yield using as little as 1 mol% catalyst and afforded nearly quantitative yield with 5% catalyst; however, the addition of octene did not improve the outcome [16]. A cyclorelease strategy was also used recently in a medicinal chemistry effort following the identification of a potent modulator of the human Motilin receptor (*h*MOT-R) from a high-throughput program [17]. To this end, Marsault *et al.* prepared a focused library of macrocycles containing

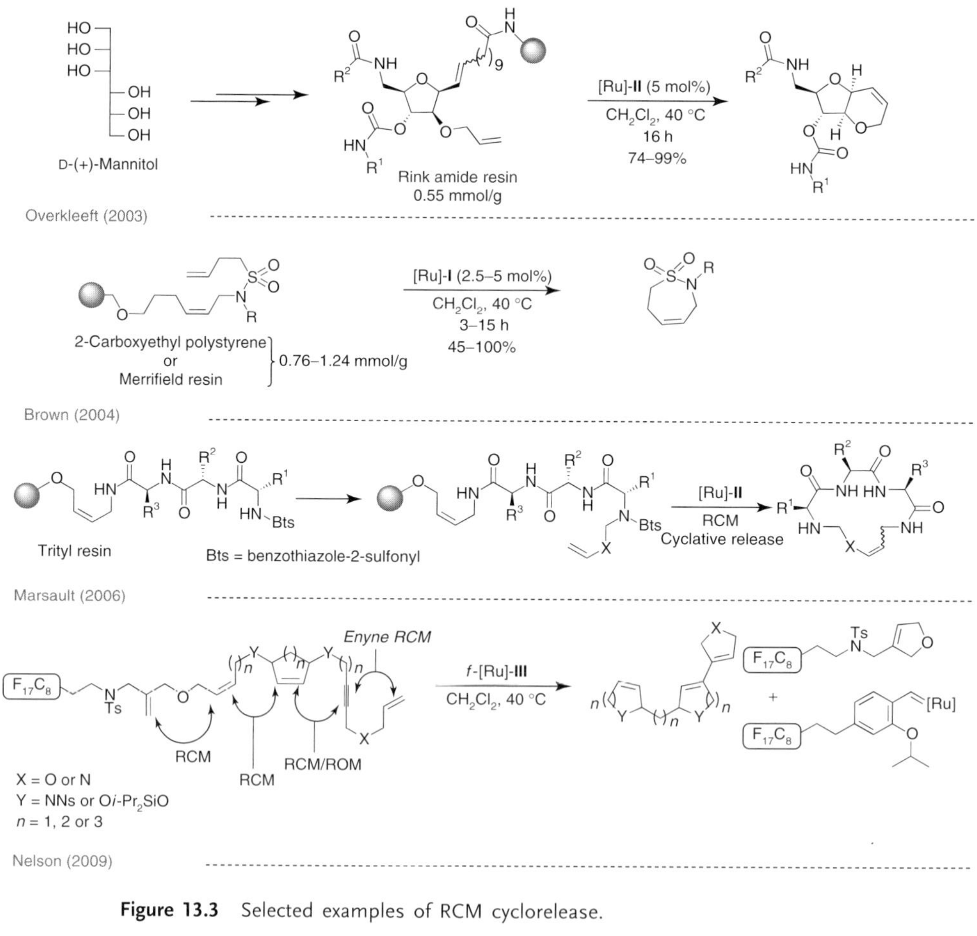

Figure 13.3 Selected examples of RCM cyclorelease.

variations of the tripeptide motif and the alkene linkage. Over 30 analogs were prepared on a fairly high loading resin (1.2 mmol g^{-1}) to afford the macrocycles which were individually purified. Nelson *et al.* reported a very elegant extension of the metathesis-based cyclorelease reaction leveraged on combinations of RCM and ROM in conjunction with ene–yne metathesis to afford over 80 different polycyclic scaffolds [18]. Rather than using a polymer matrix to immobilize the substrates, Nelson opted for the light fluorous tag technology developed by Curran *et al.* [19–21], wherein molecules covalently linked to a perfluorinated chain of three to nine carbons can be adsorbed on fluorinated silica gel in solid-phase extractions (F-SPE). Though the metathesis was performed in solution phase, this elegant example has been included, as the logic of linker connectivity and the underlying motivation (ease of automation/purification) are the same as for solid-phase chemistry. The final metathesis cascade was performed with a fluorous-tagged metathesis catalyst [9] (Figure 13.1) such that the final F-SPE not only removed the linker by-product but also the catalyst. This work represents a significant landmark in the degree of skeletal diversity achieved by a sequence involving only three bond-building steps prior to the metathesis cascade and attests to the power of the olefin metathesis to generate molecular diversity in so-called diversity-oriented synthesis.

Regarding the second mode of RCM-based cleavage affording terminal alkenes (Figure 13.1, Equation 2), Schmidt and Knerr investigated the release of polymer-bound oligosaccharides by RCM to afford *O*-allyl glycosides which represented versatile protected intermediates for further manipulation (Figure 13.4) [22, 23]. An alkene linker is particularly attractive for oligosaccharide synthesis, as it is orthogonal to the acidic nature of most glycosylation methods and ester type protecting groups which are required to control β-glycosylations. Alternatively, Seeberger has used a CM to release *O*-alkenyl glycosides (Figure 13.11, Section 13.6). While a direct comparison of both systems has not been reported, it can be anticipated that the linker reported by Schmidt and Knerr would cleave faster on the basis of the fact that the kinetics of the metathesis reaction is strongly influenced by the level of substitution of the alkene and that RCM for five-membered rings are extremely fast. In 2002, Waldmann *et al.* published the solid-phase synthesis of a library of dysidiolide analogs (a protein phosphatase inhibitor) leveraging a similar cleavage strategy that yielded compounds with improved biological activity [24, 25]. In this case, the final step of the synthesis is a cyclorelease reaction yielding a highly substituted alkene. It is interesting to note that the use of cyclorelease cleavage kinetically relayed the catalyst to only one of two trisubstituted alkenes. In this case, the use of [Ru]-**II** did not improve the yield of the reaction relative to [Ru]-**I**. The synthesis of the natural product was achieved in 11 steps in an overall yield of 14% representing an average yield of 84% for each step. The analogs were synthesized in a total of 8–11 steps on solid phase with an overall yield of 6–27% based on a Merrifield resin with a loading of 1.1 mmol g^{-1}. The breadth of reaction employed in the synthesis (organometallic additions, Wittig and Diels–Alder reactions) testifies to the versatility and robustness of the alkene linker. This concept was further extended by van der Marel *et al.* to a

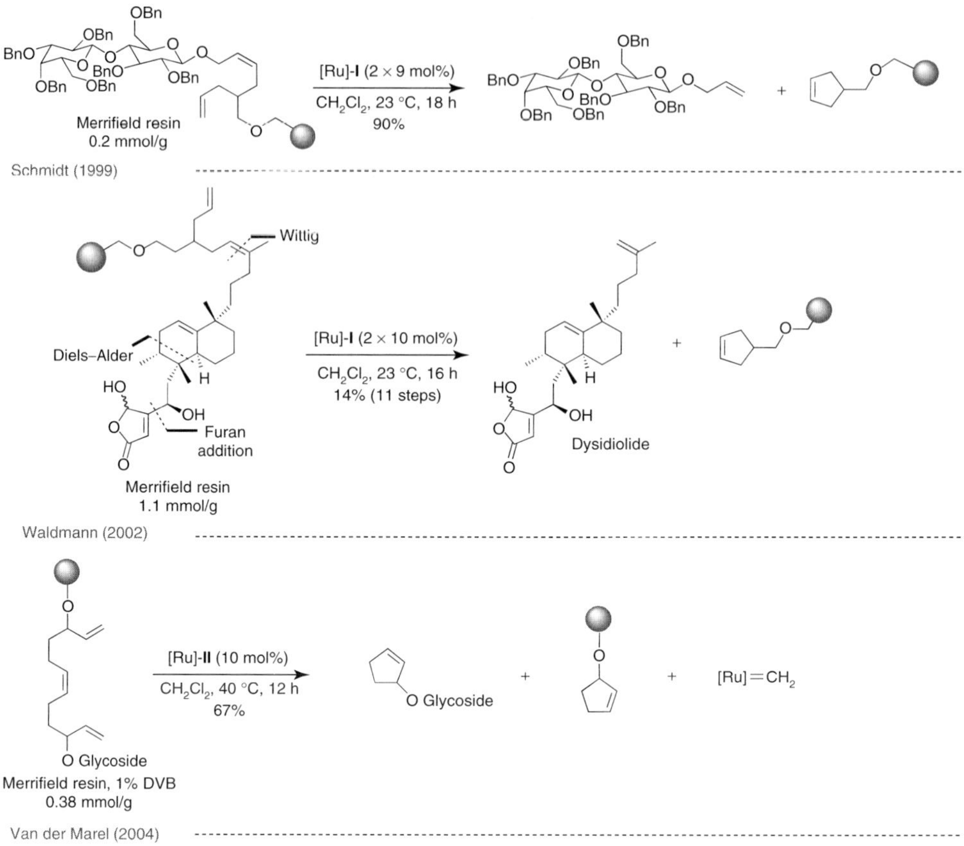

Figure 13.4 Selected examples of cyclorelease.

tandem RCM reaction in the context of a linker for carbohydrate synthesis. In the cleavage modes previously described (Figure 13.1, equation 2), the catalyst is regenerated after a catalytic cycle in the form of a polymer-bound carbene or a carbene linked to the product of interest. This latter strategy bypasses this issue, as the tandem metathesis produces a very reactive methylene carbene. Indeed, good yields (>68%) were achieved in just 2 hours for the cleavage of a disaccharide [26].

13.3 Ring-closing Metathesis (RCM)

While the first examples of RCM on solid phase exploited the opportunity to simultaneously release the substrate from the resin, RCM without concomitant cleavage has been even more extensively used in solid-phase synthesis and has the advantage that the catalyst is recovered in the solution at the end of the catalytic cycle and thus unhindered to turn over. However, the proportion of cyclization versus CM with another polymer-bound alkene will still depend on the nature of the reaction and the loading of the resin. Early work on solid-phase organic synthesis had already investigated the possibility of using polymeric support to "isolate" reactive intermediates and create a quasi-infinite dilution environment. It was concluded that even at low loading (>0.1 mmol g^{-1}), interaction between polymer-bound intermediates is still possible [27]. Nevertheless, the likelihood and frequency of such intraresin interactions will decrease at lower loading and is dependent not only on the loading but also on the level of cross-linking, the solvent, and the swelling properties of the polymer as well as the linker. The number of parameters and the heterogeneous nature of all polymeric supports make a general conclusion difficult. As will be discussed in the next section, the intraresin interaction can be favored with highly loaded resins (typically higher than 1 mmol g^{-1}) if there is no competition for thermodynamically more favorable RCM. One of the first examples of RCM in solid-phase organic synthesis is the modular assembly of functionalized azacycles described by Blechert *et al.* (Figure 13.5) [28].

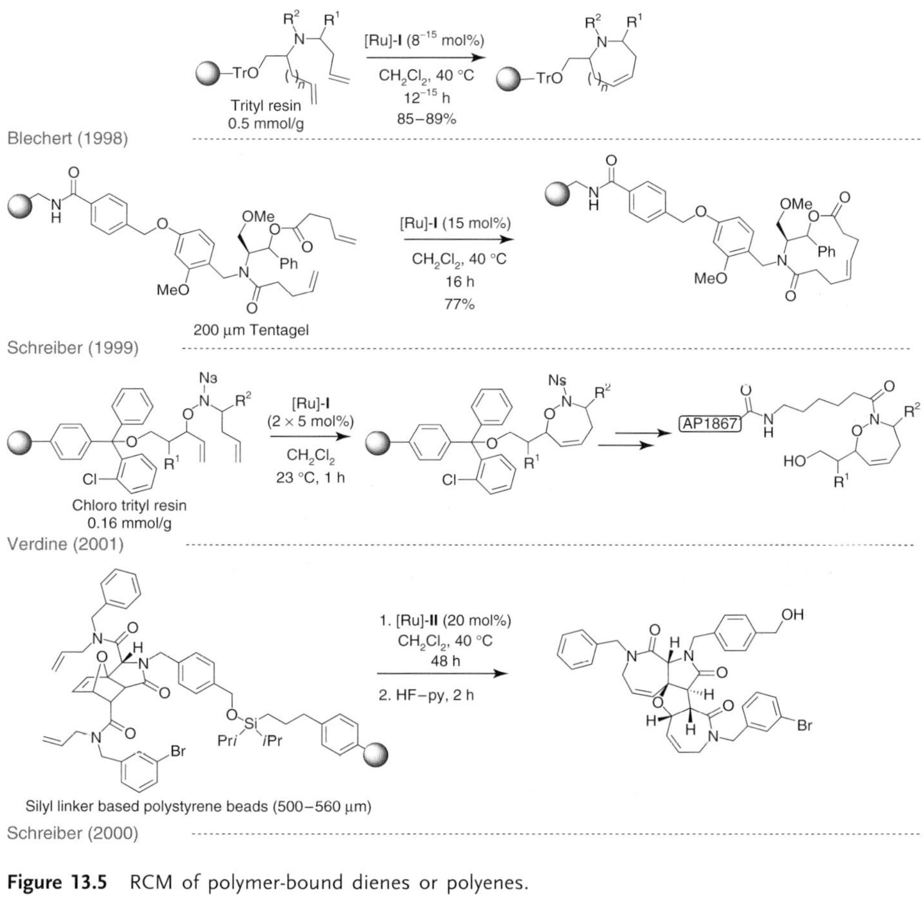

Figure 13.5 RCM of polymer-bound dienes or polyenes.

The approach was based on a RCM to form heterocycles of different ring sizes. On the basis of the recognition that many bioactive secondary metabolites have macrocyclic structure, Schreiber *et al.* evaluated the possibility of using a RCM to access diverse macrocycles. As only [Ru]-**I** was available at the time, the study was performed with the first-generation catalyst and as a proof of concept. Though only one example of a macrocycle was reported on solid phase, it suggests that this approach should be compatible with split-pool synthesis of larger libraries [29]. Indeed, a small molecule emanating from this chemistry was recently found to bind to the Sonic hedgehog protein and inhibit its signaling pathway [30]. Beyond these early proofs of principle regarding the efficacy of the RCM on solid phase, Verdine's lab developed the solid-phase synthesis of tetrahydrooxazepines in which the critical step of the synthetic scheme was a RCM to form the *cis*-endocyclic olefin (Figure 13.5). This was also the first example of a RCM being performed on a molecule containing an N–O bond. The resin of choice was a polystyrene-based trityl resin. The metathesis reaction was performed at room temperature for 1 hour using [Ru]-**I** as the catalyst. For all the substrates examined, the metathesis reaction was found to produce a single compound with complete conversion [31], thereby enabling the preparation of a library of 320 compounds. In a very elegant example that stands as an early milestone in the development of diversity-oriented synthesis, Schreiber *et al.* reported a synthetic pathway leverage on one of the most powerful reactions in combinatorial chemistry, the Ugi multicomponent reaction, which delivers a scaffold bearing two terminal alkenes and a strained internal alkene. Exposure of this polymer-bound intermediate to [Ru]-**II** (20 mol%, CH_2Cl_2, 40 °C, 60 hours) afforded the thermodynamically more favorable rearranged product having a seven-five-five-seven fused polycyclic structure (Figure 13.5) [32]. On the basis of their investment in HF-mediated cleavages for library preparation, the work was performed using a silyl linker.

Arya *et al.* have made extensive use of RCM for the synthesis of natural product-like polycyclic compounds on solid phase. They have prepared a wide variety of tetrahydroquinoline derivatives and polycyclic compounds including 6-, 8-, and 10-membered rings as shown in Figure 13.6. The phenolic hydroxyl group was utilized as an anchorage point to a Wang resin which can be conveniently cleaved with triflouroacetic acid (TFA) [33–35]. An extension of this work was published more recently in which an aminoindoline-based library was prepared. In this case 6-, 7- and 10-membered rings were generated via RCM using a silyl-based resin with macrobeads [36, 37]. Inspired by the plethora of biological activities found in polyketides bearing recurring 1,3-diols with α,β-unsaturated lactones, Waldmann *et al.* reported the synthesis of natural product-derived libraries via iterative stereoselective allylations and ozonolyses. Esterification of the final allylation product and RCM afforded α,β-unsaturated δ-lactones with multiple oxygenated side chains (Figure 13.6). The stereoselective synthesis on a Wang resin was achieved in a 40–60% overall yield in 10 steps and with a 63–80% diastereoselectivity. The metathesis reaction was carried out using [Ru]-**III** as the catalyst of choice. After the RCM, the resin turned brown as a result of the absorption of the metal; however, after cleavage the ruthenium content of the product was still significant and a

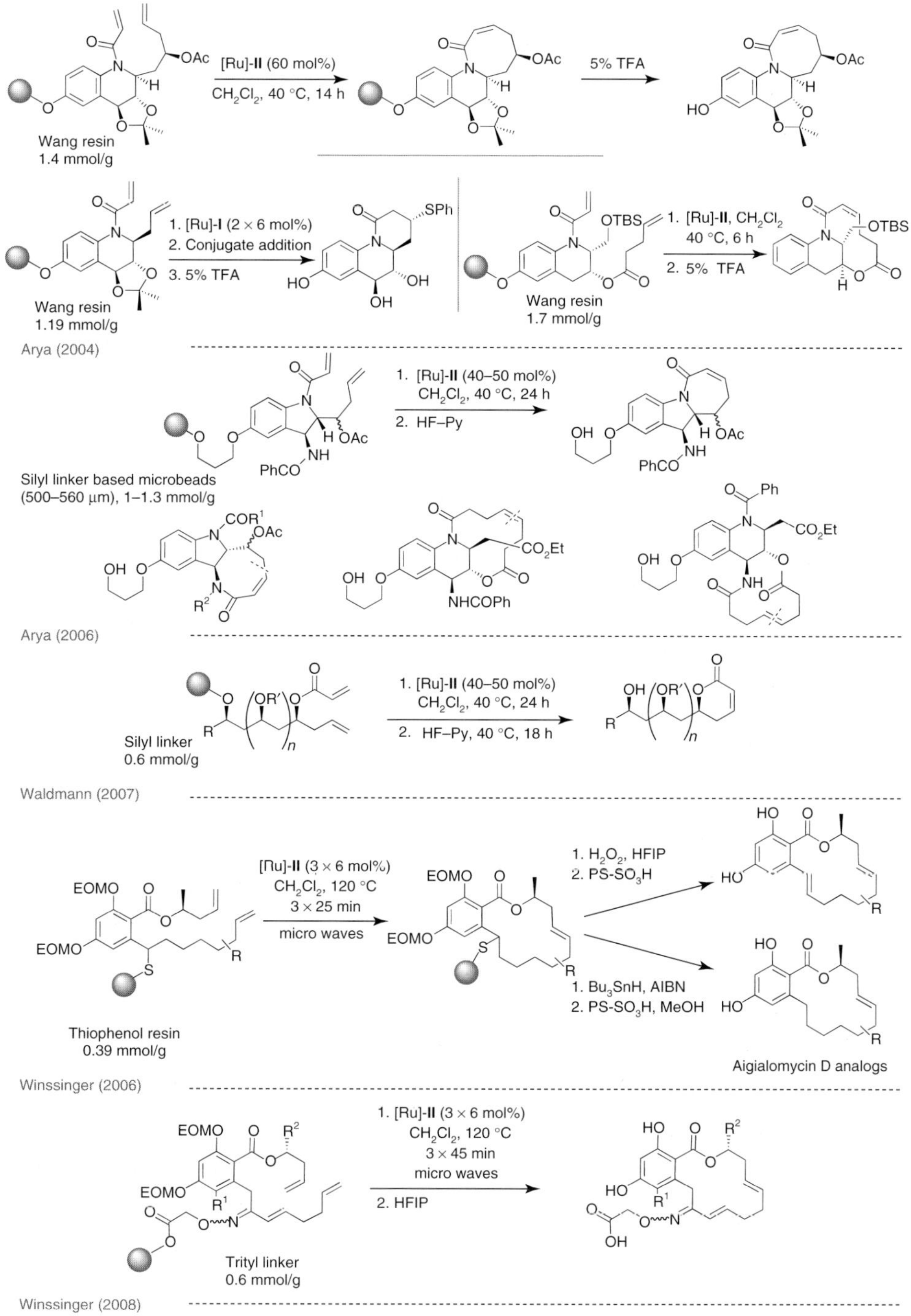

Figure 13.6 Selected examples of RCM.

purification step by high-performance liquid chromatography (HPLC) was required [38]. Our group has also relied on the use of RCM for the solid-phase synthesis of natural products and libraries thereof. We have been attracted to the resorcylic acid lactones (RALs) on the basis of the observation that the small library available from natural sources contains several potent kinase and ATPase inhibitors [39]. For the solid-phase synthesis of aigialomycin D and analogs, a RCM with [Ru]-**II** provided the 14-membered macrocycle in excellent yield. However, we had noted that at lower temperatures and with shorter reaction times, mixtures of (*E*)- and (*Z*)-isomers were obtained [40]. To drive the reaction to the desired (*E*)-isomer, the reaction was carried out at elevated temperatures (120 °C), which was most conveniently achieved under microwave irradiation. As the catalyst is unstable at that temperature, it was added in three portions over 75 minutes. It is noteworthy that the presence of the thioether linker did not impede the reaction. More recently, RCM was used to refine the biological activity of Hsp90 (heat shock protein 90) inhibitors through the synthesis of a focused pochoxime library. The reaction conditions were the same as for aigialomycins and also proceeded in excellent yields (HPLC analysis of crude cleavage product showed a complete and clean conversion) [41].

13.4 Intraresin Dimerization

As previously discussed, the fact that polymer-bound intermediates can react with each other had been observed since the first investigations into solid-phase organic synthesis in the early 1970s [27]. Motivated by the knowledge that many signal transduction pathways are triggered by receptor dimerization, Schreiber *et al.* reasoned that the metathesis could be harnessed to dimerize libraries of polymer-bound small molecules [42]. Indeed, using a high-capacity (1–2 mmol g^{-1}), lightly cross-linked (1% DVB) polystyrene with a silyl linker, they achieved over 90% yield in a number of examples suggesting that this methodology could be general to obtain libraries of monomers and dimers from the same synthetic pathways. As anticipated, the metathesis afforded a mixture of (*E*) and (*Z*) isomers. This methodology was used by Liao *et al.* for the synthesis of benzofuran-based dimeric molecules synthesized by a Sonogashira reaction followed by a palladium-carbonylative annulations. Using the same resin and conditions as previously reported by Schreiber, the dimers were obtained in 70–80% yield as *E*/*Z* mixtures [43]. Dervan and coworkers also ingeniously capitalized on the ability to dimerize polymer-bound intermediates to generate a library of H-pin polyamides via on-resin cross-linking (Figure 13.7), as the dimeric nature of these polyamides is important for their DNA binding properties [44]. The reaction was achieved on highly loaded PAM resin (1.2 mmol g^{-1}) using [Ru]-**I**. To avoid ambiguities steaming from alkene isomers, the double bond was hydrogenated post cleavage. DNA binding affinities were analyzed for

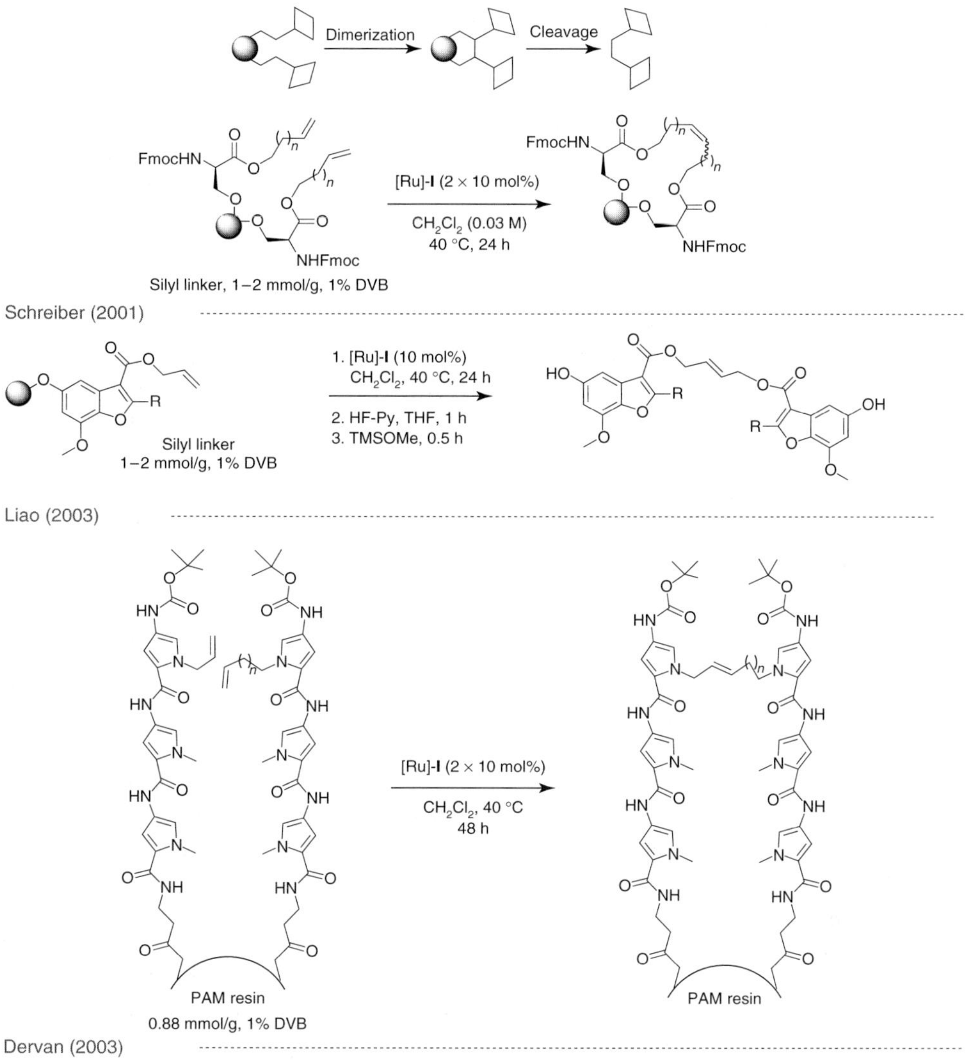

Figure 13.7 Selected examples of intraresin dimerization.

each member of the library, thereby enabling a rapid optimization of the optimal linker length.

13.5
Restricting Peptide Conformation through Cyclization

While proteinic ligands are typically endowed with high affinity and specificity for their target, attempts to reconstitute these activities with the peptidic fragment involved in the interaction are usually disappointing because of the fact that short peptides do not form the stable secondary structures which are involved in the interactions. To this end, a number of efforts have been reported to stabilize secondary structures and in particular α-helical structures. One of the most used technologies is the formation of disulfide bonds that are readily formed between natural cysteine residues but which suffer from being not pharmacologically stable. Grubbs *et al.* were the first to explore the potential of the metathesis reaction to form a stable covalent bond between residues on an adjacent helical turn of an α-helix. One helical turn being on average 3.6 amino acid residues, a serine or a homoserine bearing an allyl group was introduced in a heptameric peptide selected for its capacity to form a helix and therefore position the terminal residues within reaction distance from the second residue and the sixth residue ($i + 4$). Treatment of this peptide with [Ru]-**I** afforded the cross-linked peptide in high yield. While it was obtained as a mixture of olefin geometry (about $E/Z = 5:1$), a single compound was readily obtained after hydrogenation of the alkene. Circular dichroism spectra showed that the cross-linked peptides were indeed helical and demonstrated that the metathesis reaction with tolerant catalysts was uniquely suited to rigidify peptide motifs [45]. This first example carried out in solution phase after cleavage of the peptide from the resin inspired a number of other research groups to harness the robustness of this chemistry with a variety of unnatural amino acids bearing terminal alkenes. As the side chain of the residue in a helix points in a parallel direction rather than toward each other, Verdine *et al.* investigated the cross-linking of helices using residues having opposite stereochemistry. To neutralize the detrimental impact on helical formation of a D-amino acid residue, quaternary amino acids were used (Figure 13.8). The metathesis proceeded in high yield in just 2 hours using a 10 mM concentration of catalyst. As expected, high yields were obtained only for alkenes with sufficient distance to reach over and react with the second alkene. For the "stapling" of a single turn using homochiral amino acids (pointing in a parallel direction), 8 atoms were sufficient while 11 were required to "staple" two helical turns bearing opposite stereochemistry. It is noteworthy that the stapling of a double turn with an 11-carbon spacer represents a 33-membered ring. The "stapled" peptides were shown to have enhanced helical conformation and were over 40-fold more resistant to proteolytic degradation relatively to their unstapled counterpart [46].

This methodology has been elegantly applied to two therapeutically relevant problems. In the first one, a 23-mer peptide with a single helical turn staple was used to agonize the BH3 domain, which is an important regulator to apoptosis and is generally disabled in cancer cell lines. Importantly, the stable peptide had improved pharmacological properties (nearly 10-fold extension in half-life in serum) and induced apoptosis in a series of cancer lines. This peptide was shown to be effective in prolonging the life span of leukemic mice [47]. In the second application, a 12-mer peptide containing a double helical turn staple was used to agonize p53, a transcription factor playing an important role in protecting cells from malignant transformations. The designed peptide was shown to agonize the p53 pathway in cells [48]. The success of these examples suggests that this stapling methodology by RCM may enable the development of a wide array of peptidic compounds to probe protein interactions and has a good therapeutic potential. In an alternative approach to helical stabilization by side chain stapling, Arora *et al.* have shown that artificial α-helical turns with a short peptide can be obtained by mimicking the key C=O:H–N hydrogen bond with a carbon–carbon bond accessed by metathesis, thereby affording conformationally stable structures at physiological conditions (Figure 13.8) [49]. They later showed that microwave irradiation provided significant improvements in the efficiency of this transformation [50]. As a general principle, the biological activity of linear peptides can be enhanced by cyclization if such cyclized products favor the bioactive conformation. To this end, Kessler and Schmiedeberg investigated the efficacy of a metathesis on peptides containing two allyl glycine residues or homoallyglycine residues at various distances and

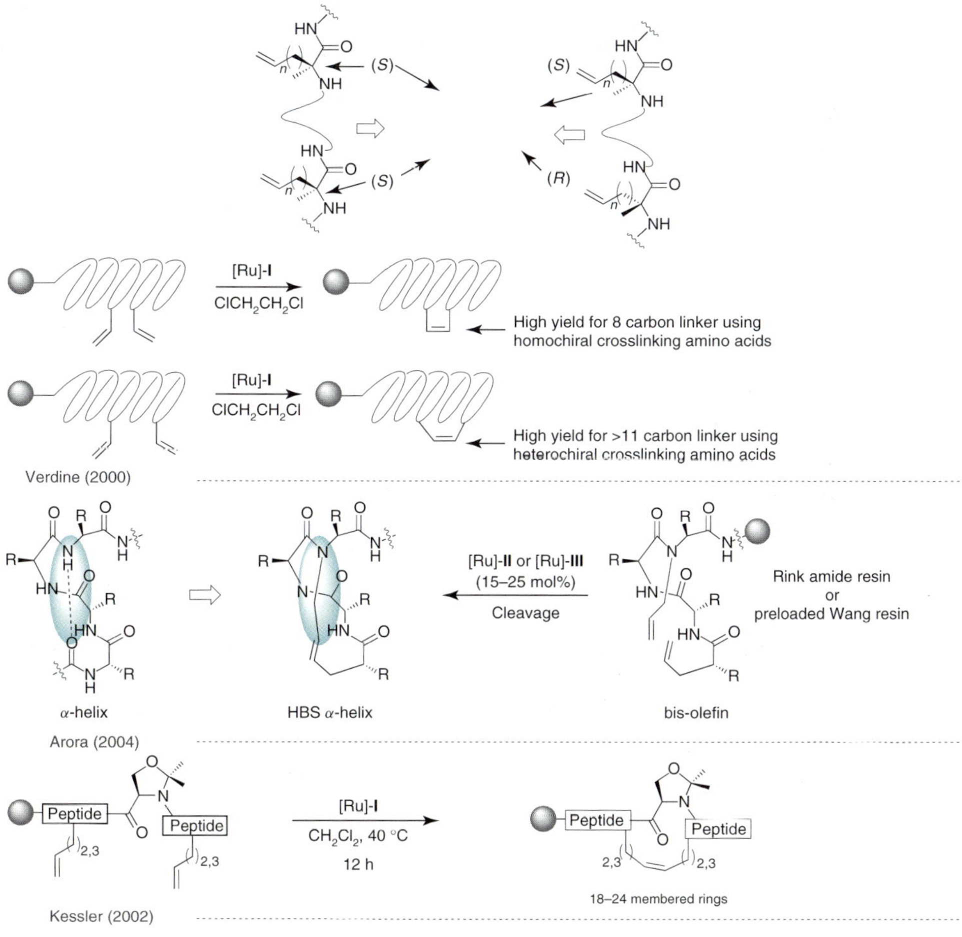

Figure 13.8 Restricting peptide conformation through cyclization.

noted that the cyclization efficiency in longer peptides was significantly reduced [51]. However, this could be overcome if a pseudoproline (acetonide-protected serine) was incorporated in the sequence, thereby providing the necessary level of preorganization for efficient cyclization. Nature has relied extensively on disulfide bridges between cysteine residues to restrict peptide conformation, and a number of naturally occurring biologically active peptides contain this motif. For the aforementioned reasons, an alkene surrogate is an attractive alternative. Vederas *et al.* have relied on this approach to prepare analogs of the antibiotic leucosin A, which has nanomolar activity against pathogens such as *Listeria monocytogenes* (Figure 13.9) [52]. In this case, treatment of the polymer-bound fully protected peptide (37-mer) bearing two allyl glycine residues did not afford any cyclization. It was found that precomplexation of the peptide by brief treatment with LiCl in dimethylformamide (DMF) prior to the metathesis afforded the desired cyclization product. Interestingly, cyclization of shorter peptides did not require similar pretreatment [53]. The corresponding carbocyclic analog still displayed nanomolar activity but was about 10-fold less potent than the parent compound. They applied the same methodology to the synthesis of a carbocyclic antibiotic analog bearing three macrocycles, using sequential on-resin RCM [54]. It is interesting to note that the second and third RCM did not result in any scrambling in the precedent ring-closure products, attesting to the lower reactivity of endocyclic alkenes and the

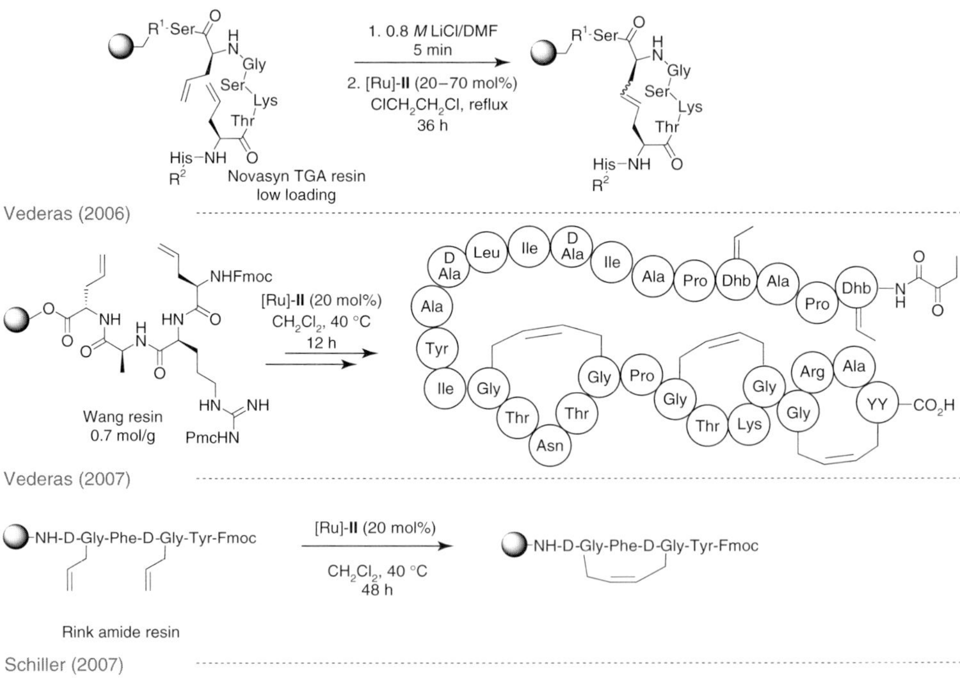

Figure 13.9 Restricting peptide conformation through cyclization.

thermodynamic stability of the first and second macrocycles. The replacement of the sulfide bridges with other groups not only could shed light on structure–activity relationships but also generate variants that are resistant to aerobic oxidation and have better metabolic stability. Although the preliminary biological evaluation of this larger ring analog showed no antimicrobial activity at comparable concentrations to the natural lactin A2, the methodology described has found wide application for the preparation of rigidified peptidomimetics containing multiple carbocyclic rings. In fact, shortly after, Schiller *et al.* synthesized the dicarba analogs of the cyclic opioid peptide H-Tyr-*c*[D-Cys-Gly-Phe-D(or L)-Cys]NH_2 by RCM on Rink resin using [Ru]-**II** (20 mol%, reflux 48 hours) [55]. Among the dicarba analogs, the *trans* isomer of H-Tyr-*c*[D-Allylgly-Gly-Phe-L-Allylgly]NH_2 proved to be an excellent μ and δ agonist with subnanomolar potency for both receptors.

While the kinetic product of a metathesis generally is the *cis* alkene, as the reaction is reversible, equilibration typically leads to the thermodynamically most favorable alkene geometry. In macrocycles, it is not uncommon to obtain a mixture of alkene geometries. This has led many researchers to reduce the alkene by catalytic hydrogenation using palladium on charcoal after cleavage from the resin. Robinson *et al.* have shown that this reduction could perhaps be performed more conveniently on solid phase using Wilkinson's catalyst and medium hydrogen pressure (60 psi) [56]. As for many transition-metal-catalyzed reactions, microwave irradiation can be a particularly effective method to rapidly reach high temperatures resulting in short reaction times. This is also the case in the metathesis for peptide cyclization. For example, the dicarba analog of α-conotoxin was obtained in just 1 hour with microwave irradiation while 48 hours were required with conventional heating [57].

13.6 Cross-metathesis on Solid Phase

The practical and general implementation of CM [58] is challenging for alkenes of comparable reactivity, as a statistical mixture of homo and heterocoupling is typically observed. Blechert *et al.* quickly recognized that immobilization of one of the olefinic partners would provide a means to isolate the desired heterocoupling product albeit without preventing some homocoupling in solution. To this end, different olefins were reacted in the presence of [Ru]-**I** to tritylpolystyrene-bound *N*-Boc-*N*-allylglycinol and *N*-Boc-*C*-allylglycinol (Figure 13.10) [59]. Of course, the reaction was most productive with an excess of the alkene in solution, as the latter was partially consumed in the formation of a homodimer product, which in principle can react with the polymer-bound alkene albeit more slowly. It was also shown that CM could be performed on an allyl silane resin which offers mild acidic cleavage conditions (TFA results in rapid protodesilylation) in addition to cleavage with concomitant allylation using an acetal with a Lewis acid ($TiCl_4$) [60, 61]. CM has also been used in conjunction with ROM of strained ring systems. Cuny *et al.* capitalized on strained norbornene to synthesize highly functionalized

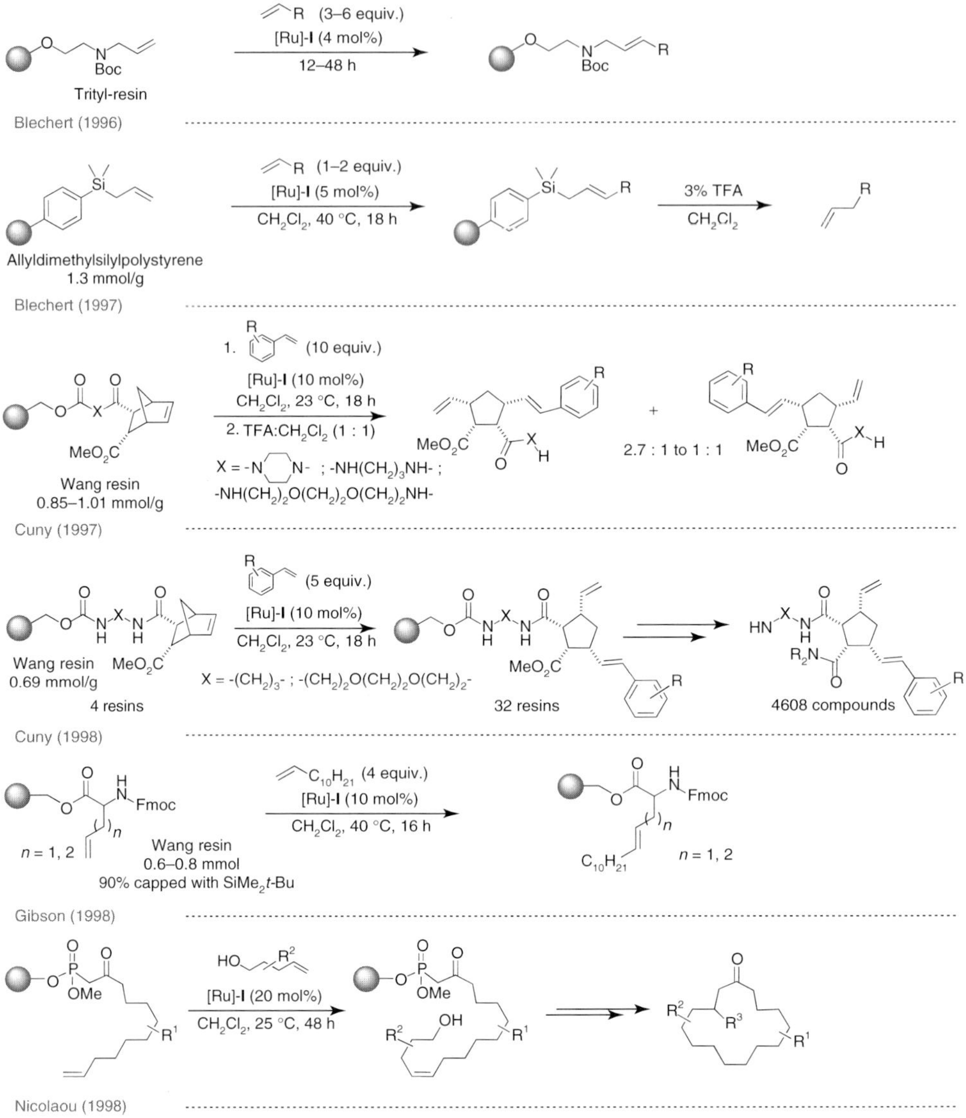

Figure 13.10 Pioneering examples of cross-metathesis on solid phase.

cyclopentanes after a ring opening/CM (Figure 13.10) [62]. In this case, the reaction was performed at room temperature with [Ru]-**I**. Under these conditions, metathesis with the terminal alkene produced in the ring-opening reaction is very slow, the ROM being the driving force for the reaction. The fact that the strained alkene is polymer bound reduces the efficacy of its polymerization and favors the CM pathway. However, this will of course depend to some extent on the loading of the resin. The unsymmetrical norbornenes generated two regioisomeric products [63]; nevertheless, the methodology was robust enough to be used in the generation of a library of 4608 highly functionalized cyclopentanes by using different diamine linkers and different terminal aryl olefins [64]. In an attempt to apply the CM for the derivatization of amino acids, Gibson *et al.* studied the reaction of polymer-bound unsaturated amino acids with different alkenes. The reaction of polymer-bound homoallylglycine (Hag) was achieved successfully for aryl- and alkyl-substituted alkenes when very low loading of Wang resin was used (about 0.06–0.1 mmol g^{-1}), although the efficiency of the reaction dropped dramatically for vinylglycine and dehydroalanine [65]. It is important to note that this study was performed with [Ru]-**I** and the lack of reaction in the latter two examples should be overcome using second-generation catalysts. Nicolaou *et al.* used a CM in a strategic bond-forming reaction to access muscone via a Horner–Wadsworth–Emmons cyclorelease using a ketophosphonate linker. In this case, the mildness and functional group compatibility of the CM were essential for this carbon–carbon bond formation. The CM between unprotected unsaturated alcohols and terminal alkenes using [Ru]-**I** proceeded in 60–70% yield [66].

Concurrently to the aforementioned efforts by Schmidt *et al.* using a cyclorelease reaction to cleave an oligosaccharide from the resin, Seeberger *et al.* relied on a CM under ethylene atmosphere to cleave the oligosaccharide (Figure 13.11). This latter linker is readily prepared from Merrifield resin and 4-octene-1,8-diol in one step, and affords the *n*-4-pentenyl glycosides which can be engaged in further glycosylations promoted by electrophilic reagents such as NIS. This linker has been applied to the synthesis of different polysaccharides [67–69], including the automated synthesis of antigens Gb-3 and Globo-H [70]. This linker has also been used for amino sugars such as the amino glycoside antibiotics and heparin. In this case, it was necessary to use [Ru]-**II** and pent-1-ene for the cleavage [71]. However, an alkenic linker is incompatible with two types of glycosyl donors (thioglycosides and of course *n*-pentenyl glycosides) because of the fact that both of them require an electrophilic activation which cannot be performed in presence of the olefinic double bond. To this end, it was shown that the alkene could be masked as a vicinal dibromide which was compatible with NIS-mediated glycosylation and that the latter vicinal dibromide functionality could be converted back to the alkene under the action of tetrabutylammonium iodide (TBAI). After regeneration of the olefin, the linker could be cleaved, as previously described, by CM with ethylene [72, 73]. Other examples of CM on solid phase catalyzed by [Ru]-**I** include the formation of stilbenoids [74]. The development of the more reactive second-generation catalysts proved particularly beneficial for CM. This reaction was applied to the solid-phase synthesis of a library of 3-(aryl)alkenyl β-lactams as cholesterol absorption inhibitors. The CM step was carried out in high yields using [Ru]-**II** (5 mol%), and the whole synthetic sequence gave good overall isolated yields and the *E*-alkenes exclusively [75]. In an alternative strategy to the aforementioned use of the RCM to access peptidic macrocycle surrogate of disulfide, Liskamp *et al.* used [Ru]-**II** to perform a CM between two allylglycine units which were subsequently macrolactamized en route to the DE fragment of nisin Z [76]. Having the bicyclic intermediate as a reference, the authors investigated

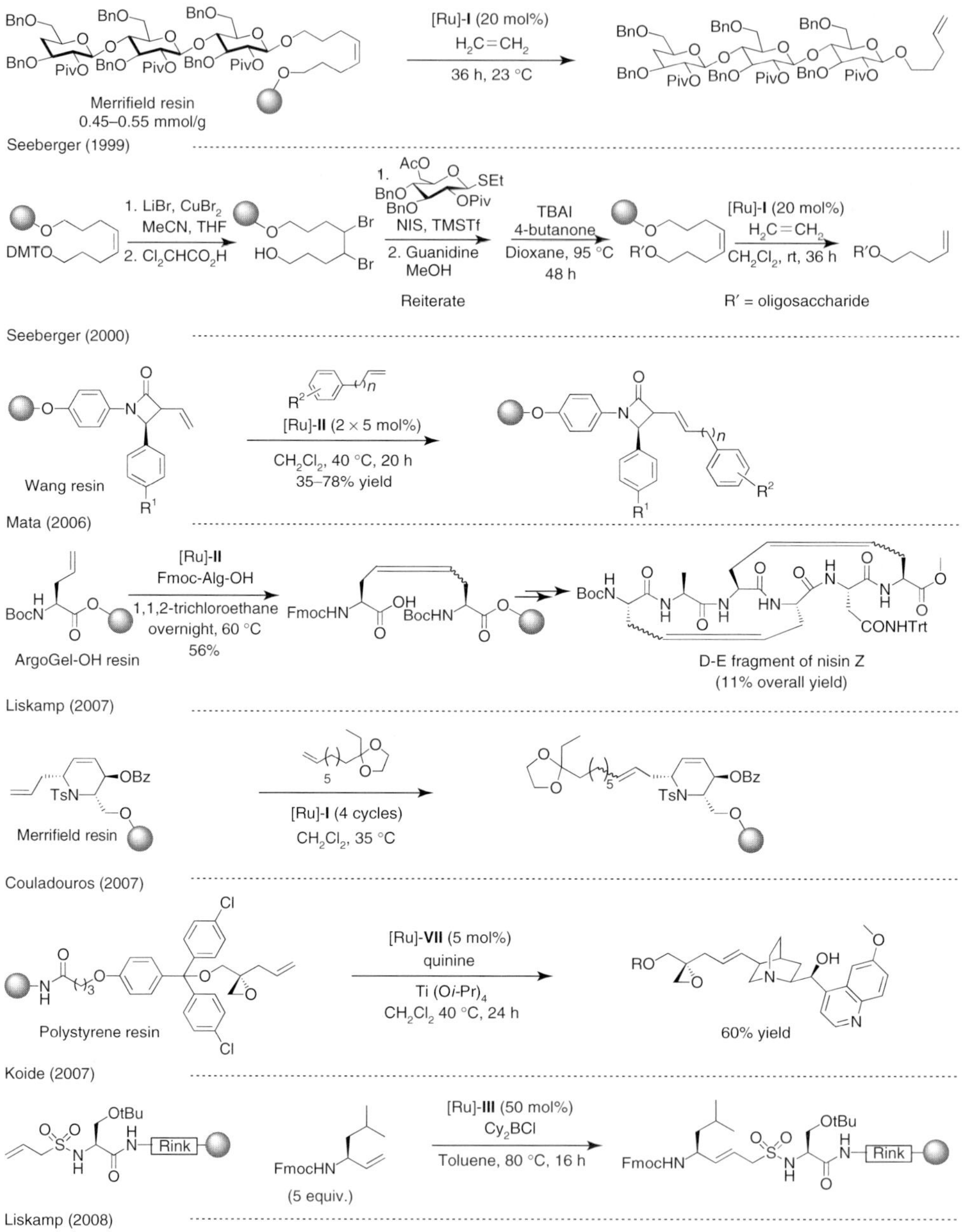

Figure 13.11 Selected examples of cross-metathesis on solid phase.

the more challenging prospect of obtaining the desired bicyclic system in one step by RCM of the linear peptide containing the four allyl glycine units, which proved feasible. The lower reactivity of first-generation catalyst can be an asset when performing a CM on a terminal alkene in the presence of an internal alkene. For the synthesis of prosophylline, a natural product which belongs to the family of the piperidine alkaloids known to have medicinal properties, Couladouros *et al.* relied on a CM of a terminal alkene in the presence of an endocyclic alkene. After extensive experimentation, it was found that four consecutive cycles at a fairly low temperature (35 °C) using [Ru]-**I** gave optimal yield [77]. Koide and Garner exploited the CM as a general method to derivatize natural products with the addition of an epoxide moiety [78]. Having a heteroatom in the allylic position can be problematic in a metathesis reaction, as it can coordinate to the ruthenium carbene intermediate thereby slowing the catalytic cycle. To this end, the addition of a Lewis acid such as $Ti(O\textit{i}Pr_4)$ to compete for chelation with the epoxide led to higher yields. In the latter case, introduction of a spacer between the trityl linker and the resin was critical for obtaining higher yields. It was also noted that [Ru]-**VII** [79] afforded better results. The optimized conditions (longer linker, [Ru]-**VII**, and Lewis acid) enable derivatization of complex natural products such as quinine. Another application of the CM is Liskamp's synthesis of alkene dipeptidosulfonamide as isosteres of the native peptide bond [80], in which a polymer-bound sulfonamide is reacted with readily available allylic amines. The reactions were found to require fairly high catalyst loading and the use of a Lewis acid (Cy_2BCl) to afford only modest yields. Nevertheless, the peptide incorporating this isostere was found to have improved biological activity thus warranting this method. Other examples of CM using second-generation catalysts also include the preparation of porphyrin–RGD peptide conjugate [81]. As it is the case for solution-phase chemistry, it has been noted that CM with α,β-conjugated systems performed better with [Ru]-**III** rather than [Ru]-**II** [82].

13.7
Ene–yne Metathesis on Solid Phase

The ene–yne metathesis, as its name implies, is a bond-forming and reorganization reaction between an alkyne and an alkene to yield a 1,3-diene. The reaction is mechanistically akin to the alkene metathesis and has been carried out both in the intermolecular and intramolecular version. The first examples of the reaction on solid phase were reported by Blechert *et al.* using the intermolecular version of the reaction. In the first example, a polymer-bound propargyl ester was reacted with a variety of highly functionalized alkenes in the presence of [Ru]-**I**, affording an allylic diene which could be released from the resin by taking advantage of the reactivity of allylic esters under palladium catalysis [83]. It is interesting to note that, as the cleavage is leveraged on the reactivity of allylic esters, only compounds having successfully participated in the ene–yne metathesis will be released. In the second example, the reaction was achieved between an allylic

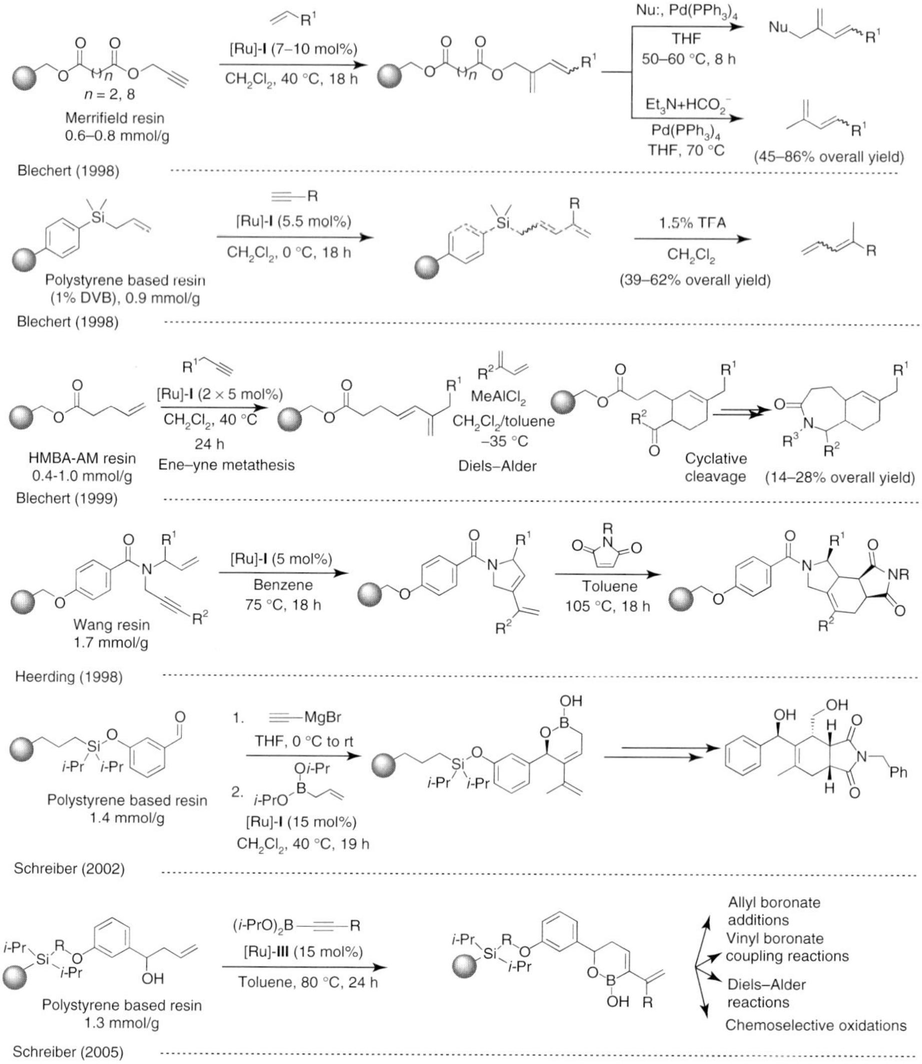

Figure 13.12 Ene–yne metathesis on solid phase.

silane polystyrene resin and several highly functionalized alkynes, thus affording polymer-bound dienes which could be released from the resin under mild acidic condition (protodesilylation) to afford an isomerized diene product as mixture of (*E*)- and (*Z*)-isomers. The broad scope of the reaction was demonstrated with a variety of substrates such as Fmoc-protected amino acids bearing an alkene ester and protected sugars (Figure 13.12) [84]. Blechert *et al.* recognized early on that the ene–yne reaction was particularly attractive for diversity-oriented synthesis. Indeed, it allowed bringing diversity in a complexity-building reaction. The resulting

1,3-diene product could indeed be engaged in further powerful complexity-building reactions such as a Diels–Alder or palladium cross-couplings. The application of an ene–yne metathesis followed by a Diels–Alder reaction was demonstrated a year later by Blechert *et al.* The product of the Diels–Alder was further harnessed as a handle for cyclization/cleavage with amines to afford octahydrobenzapinones with high purity [85]. The potential of intramolecular ene–yne metathesis coupled to a Diels–Alder reaction was elegantly used in a library synthesis by Heerding *et al.* for the preparation of 4200 isoindolines using split-and-mix synthesis. Analysis of a sample of the library showed that the final products were typically obtained in over 70% purity [86]. Schreiber *et al.* further elaborated the potential of this reaction sequence in the context of diversity-oriented synthesis by using boronic esters in the ene–yne annulations, thus providing an extra handle to divergent reaction pathways [87, 88]. In this case, it was observed that the efficacy of the reaction in solution could not be directly transferred to solid phase using highly loaded macrobeads. A significant difference in performance was indeed observed between [Ru]-**II** and [Ru]-**III** despite the fact that the key intermediate in the catalytic cycle was the same in both cases. It was also noted that marked improvement in the reaction could be obtained when the reaction was carried out under an ethylene atmosphere. This suggests that regeneration of the catalyst following the ene–yne annulation may be critical for good performance of the reaction. Indeed, if the intermediate from the ene–yne annulations has the potential to participate in a subsequent RCM thus releasing the catalyst in solution, the performance of both [Ru]-**II** and [Ru]-**III** was comparable.

13.8 Conclusion

The functional group tolerance of the ruthenium-based metathesis catalysts has had a tremendous impact on solid-phase organic synthesis. The efficacy of the reaction in solution generally translates directly to solid-phase transformation and its potential has been harnessed in a number of library syntheses, solid-phase syntheses of natural products, or diversity-oriented syntheses. It enables the use of chemically robust alkenes as linkers which can be cleaved by RCM or CM. It, of course, provides new manifolds of diversification in diversity-oriented synthesis as has been elegantly shown in landmark examples by Schreiber and Nelson. Another metathesis application of paramount importance is in peptide chemistry where solid-phase synthesis is omnipresent. The ability to stabilize secondary structures in short peptide motifs and replace pharmacologically unsuitable disulfide bonds or simply restrict the conformation of a peptidic library has already been successfully implemented in a number of important examples. The orthogonality of the metathesis reaction to peptide chemistry provides a really powerful tool in this regard.

Acknowledgments

We thank our collaborators who have contributed to our work in the area of diversity-oriented synthesis and solid-phase organic synthesis as well as the funding agencies (ANR, ARC, ERC, CNRS). While this review covers development of the olefin metathesis in solid-phase organic chemistry since the first reports in 1996, it is not intended to be comprehensive and we apologize for the inevitable arbitrary omissions.

References

1 Schrock, R.R. (2006) *Angew. Chem. Int. Ed.*, **45**, 3748–3759.
2 Grubbs, R.H. (2006) *Angew. Chem. Int. Ed.*, **45**, 3760–3765.
3 Chauvin, Y. (2006) *Angew. Chem. Int. Ed.*, **45**, 3740–3747.
4 Nicolaou, K.C., Bulger, P.G., and Sarlah, D. (2005) *Angew. Chem. Int. Ed. Engl.*, **44**, 4490–4527.
5 Burke, M.D. and Schreiber, S.L. (2004) *Angew. Chem. Int. Ed. Engl.*, **43**, 46–58.
6 Veerman, J.J.N., Van Maarseveen, J.H., Visser, G.M., Kruse, C.G., Schoemaker, H.E., Hiemstra, H., and Rutjes, F.P.J.T. (1998) *Eur. J. Org. Chem.*, **11**, 2583–2589.
7 van Maarseveen, J.H., den Hartog, J.A.J., Engelen, V., Finner, E., Visser, G., and Kruse, C.G. (1996) *Tetrahedron Lett.*, **37**, 8249–8252.
8 Ahn, Y.M., Yang, K., and Georg, G.I. (2001) *Org. Lett.*, **3**, 1411–1413.
9 Matsugi, M. and Curran, D.P. (2005) *J. Org. Chem.*, **70**, 1636–1642.
10 Pernerstorfer, J., Schuster, M., and Blechert, S. (1997) *Chem. Commun.*, 1949–1950.
11 Nicolaou, K.C., Winssinger, N., Pastor, J., Ninkovic, S., Sarabia, F., He, Y., Vourloumis, D., Yang, Z., Li, T., Giannakakou, P., and Hamel, E. (1997) *Nature*, **387**, 268–272.
12 Piscopio, A.D., Miller, J.F., and Koch, K. (1997) *Tetrahedron Lett.*, **38**, 7143–7146.
13 Piscopio, A.D., Miller, J.F., and Koch, K. (1999) *Tetrahedron*, **55**, 8189–8198.
14 Nicolaou, K.C., Vourloumis, D., Li, T., Pastor, J., Winssinger, N., He, Y., Ninkovic, S., Sarabia, F., Vallberg, H., Roschangar, F., King, N.P., Finlay, M.R.V., Giannakakou, P., Verdier-Pinard, P., and Hamel, E. (1997) *Angew. Chem. Int. Ed. Engl.*, **36**, 2097–2103.
15 Timmer, M.S., Verdoes, M., Sliedregt, L.A., van der Marel, G.A., van Boom, J.H., and Overkleeft, H.S. (2003) *J. Org. Chem.*, **68**, 9406–9411.
16 Moriggi, J.-D., Brown, L.J., Castro, J.L., and Brown, R.C.D. (2004) *Org. Biomol. Chem.*, **2**, 835–844.
17 Marsault, E., Hoveyda, H.R., Peterson, M.L., Saint-Louis, C., Landry, A., Vezina, M., Ouellet, L., Wang, Z., Ramaseshan, M., Beaubien, S., Benakli, K., Beauchemin, S., Deziel, R., Peeters, T., and Fraser, G.L. (2006) *J. Med. Chem.*, **49**, 7190–7197.
18 Morton, D., Leach, S., Cordier, C., Warriner, S., and Nelson, A. (2009) *Angew. Chem. Int. Ed.*, **48**, 104–109.
19 Luo, Z., Zhang, Q., Oderaotoshi, Y., and Curran, D.P. (2001) *Science*, **291**, 1766–1769.
20 Curran, D.P. (2004) *Handbook of Fluorous Chemistry*, John Wiley & Sons, 101–127.
21 Zhang, W. and Curran, D.P. (2006) *Tetrahedron*, **62**, 11837–11865.
22 Knerr, L. and Schmidt, R.R. (1999) *Synlett*, 1802–1804.
23 Knerr, L. and Schmidt, R.R. (2000) *Eur. J. Org. Chem.*, **15**, 2803–2808.
24 Brohm, D., Philippe, N., Metzger, S., Bhargava, A., Muller, O., Lieb, F., and Waldmann, H. (2002) *J. Am. Chem. Soc.*, **124**, 13171–13178.
25 Brohm, D., Metzger, S., Bhargava, A., Muller, O., Lieb, F., and Waldmann, H.

(2002) *Angew. Chem. Int. Ed. Engl.*, **41**, 307–311.
26 Timmer, M.S.M., Codee, J.D.C., Overkleeft, H.S., van Boom, J.H., and van der Marel, G.A. (2004) *Synlett*, 2155–2158.
27 Crowley, J.I., Harvey, T.B.III, and Rapoport, H. (1973) *J. Macromol. Sci., Chem.*, **7**, 1117–1126.
28 Pernerstorfer, J., Schuster, M., and Blechert, S. (1999) *Synthesis*, 138–144.
29 Lee, D., Sello, J.K., and Schreiber, S.L. (1999) *J. Am. Chem. Soc.*, **121**, 10648–10649.
30 Stanton, B.Z., Peng, L.F., Maloof, N., Nakai, K., Wang, X., Duffner, J.L., Taveras, K.M., Hyman, J.M., Lee, S.W., Koehler, A.N., Chen, J.K., Fox, J.L., Mandinova, A., and Schreiber, S.L. (2009) *Nat. Chem. Biol.*, **5**, 154–156.
31 Koide, K., Finkelstein, J.M., Ball, Z., and Verdine, G.L. (2001) *J. Am. Chem. Soc.*, **123**, 398–408.
32 Lee, D., Sello, J.K., and Schreiber, S.L. (2000) *Org. Lett.*, **2**, 709–712.
33 Arya, P., Durieux, P., Chen, Z.-X., Joseph, R., and Leek, D.M. (2004) *J. Comb. Chem.*, **6**, 54–64.
34 Khadem, S., Joseph, R., Rastegar, M., Leek, D.M., Oudatchin, K.A., and Arya, P. (2004) *J. Comb. Chem.*, **6**, 724–734.
35 Arya, P., Couve-Bonnaire, S., Durieux, P., Laforce, D., Kumar, R., and Leek, D.M. (2004) *J. Comb. Chem.*, **6**, 735–745.
36 Prakesch, M., Sharma, U., Sharma, M., Khadem, S., Leek, D.M., and Arya, P. (2006) *J. Comb. Chem.*, **8**, 715–734.
37 Reddy, P.T., Quevillon, S., Gan, Z., Forbes, N., Leek, D.M., and Arya, P. (2006) *J. Comb. Chem.*, **8**, 856–871.
38 Umarye, J.D., Lessmann, T., Garcia, A.B., Mamane, V., Sommer, S., and Waldmann, H. (2007) *Chem. Eur. J.*, **12,13**, 3305–3319.
39 Winssinger, N. and Barluenga, S. (2007) *Chem. Commun.*, 22–36.
40 Barluenga, S., Dakas, P.-Y., Ferandin, Y., Meijer, L., and Winssinger, N. (2006) *Angew. Chem. Int. Ed.*, **45**, 3951–3954.
41 Barluenga, S., Wang, C., Fontaine, J.G., Aouadi, K., Beebe, K., Tsutsumi, S., Neckers, L., and Winssinger, N. (2008) *Angew. Chem. Int. Ed. Engl.*, **47**, 4432–4435.
42 Blackwell, H.E., Clemons, P.A., and Schreiber, S.L. (2001) *Org. Lett.*, **3**, 1185–1188.
43 Liao, Y., Fathi, R., and Yang, Z. (2003) *J. Comb. Chem.*, **5**, 79–81.
44 Olenyuk, B., Jitianu, C., and Dervan, P.B. (2003) *J. Am. Chem. Soc.*, **125**, 4741–4751.
45 Blackwell, H.E. and Grubbs, R.H. (1998) *Angew. Chem. Int. Ed.*, **37**, 3281–3284.
46 Schafmeister, C.E., Po, J., and Verdine, G.L. (2000) *J. Am. Chem. Soc.*, **122**, 5891–5892.
47 Walensky, L.D., Kung, A.L., Escher, I., Malia, T.J., Barbuto, S., Wright, R.D., Wagner, G., Verdine, G.L., and Korsmeyer, S.J. (2004) *Science*, **305**, 1466–1470.
48 Bernal, F., Tyler, A.F., Korsmeyer, S.J., Walensky, L.D., and Verdine, G.L. (2007) *J. Am. Chem. Soc.*, **129**, 2456–2457.
49 Chapman, R.N., Dimartino, G., and Arora, P.S. (2004) *J. Am. Chem. Soc.*, **126**, 12252–12253.
50 Chapman, R.N. and Arora, P.S. (2006) *Org. Lett.*, **8**, 5825–5828.
51 Schmiedeberg, N. and Kessler, H. (2002) *Org. Lett.*, **4**, 59–62.
52 Derksen, D.J., Stymiest, J.L., and Vederas, J.C. (2006) *J. Am. Chem. Soc.*, **128**, 14252–14253.
53 Stymiest, J.L., Mitchell, B.F., Wong, S., and Vederas, J.C. (2003) *Org. Lett.*, **5**, 47–49.
54 Pattabiraman, V.R., Stymiest, J.L., Derksen, D.J., Martin, N.I., and Vederas, J.C. (2007) *Org. Lett.*, **9**, 699–702.
55 Berezowska, I., Chung, N.N., Lemieux, C., Wilkes, B.C., and Schiller, P.W. (2007) *J. Med. Chem.*, **50**, 1414–1417.
56 Whelan, A.N., Elaridi, J., Harte, M., Smith, S.V., Jackson, W.R., and Robinson, A.J. (2004) *Tetrahedron Lett.*, **45**, 9545–9547.
57 Robinson, A.J., Elaridi, J., Van Lierop, B.J., Mujcinovic, S., and Jackson, W.R. (2007) *J. Pept. Sci.*, **13**, 280–285.
58 Testero, S.A. and Mata, E.G. (2008) *J. Comb. Chem.*, **10**, 487–497.
59 Schuster, M., Pernerstorfer, J., and Blechert, S. (1996) *Angew. Chem. Int. Ed. Engl.*, **35**, 1979–1980.

60 Schuster, M., Lucas, N., and Blechert, S. (1997) *Chem. Commun.*, 823–824.

61 For a review covering the scope of the metathesis at the time: Grubbs, R.H. and Chang, S. (1998) *Tetrahedron*, **54**, 4413.

62 Cuny, G.D., Cao, J., and Hauske, J.R. (1997) *Tetrahedron Lett.*, **38**, 5237–5240.

63 Arjona, O., Csaky, A.G., and Plumet, J. (2000) *Synthesis*, **6**, 857–861.

64 Cao, J., Cuny, G.D., and Hauske, J.R. (1998) *Mol. Diversity*, **3**, 173–179.

65 Biagini, S.C.G., Gibson, S.E., and Keen, S.P. (1998) *J. Chem. Soc., Perkin Trans. 1*, 2485–2500.

66 Nicolaou, K.C., Pastor, J., Winssinger, N., and Murphy, F. (1998) *J. Am. Chem. Soc.*, **120**, 5132–5133.

67 Andrade, R.B., Plante, O.J., Melean, L.G., and Seeberger, P.H. (1999) *Org. Lett.*, **1**, 1811–1814.

68 Palmacci, E.R., Plante, O.J., Hewitt, M.C., and Seeberger, P.H. (2003) *Helv. Chim. Acta*, **86**, 3975–3990.

69 Castagner, B. and Seeberger, P.H. (2007) *Top. Curr. Chem.*, **278**, 289–309.

70 Werz, D.B., Castagner, B., and Seeberger, P.H. (2007) *J. Am. Chem. Soc.*, **129**, 2770–2771.

71 Kanemitsu, T. and Seeberger, P.H. (2003) *Org. Lett.*, **5**, 4541–4544.

72 Melean, L.G., Haase, W.-C., and Seeberger, P.H. (2000) *Tetrahedron Lett.*, **41**, 4329–4333.

73 Seeberger, P.H. and Haase, W.C. (2000) *Chem. Rev.*, **100**, 4349–4394.

74 Chang, S., Na, Y., Shin, H.J., Choi, E., and Jeong, L.S. (2002) *Tetrahedron Lett.*, **43**, 7445–7448.

75 Testero, S.A. and Mata, E.G. (2006) *Org. Lett.*, **8**, 4783–4786.

76 Ghalit, N., Kemmink, J., Hilbers, H.W., Versluis, C., Rijkers, D.T.S., and Liskamp, R.M.J. (2007) *Org. Biomol. Chem.*, **5**, 924–934.

77 Couladouros, E.A., Strongilos, A.T., and Neokosmidis, E. (2007) *Tetrahedron Lett.*, **48**, 8227–8229.

78 Garner, A.L. and Koide, K. (2007) *Org. Lett.*, **9**, 5235–5238.

79 Michrowska, A., Bujok, R., Harutyunyan, S., Sashuk, V., Dolgonos, G., and Grela, K. (2004) *J. Am. Chem. Soc.*, **126**, 9318–9325.

80 Brouwer, A.J., Elgersma, R.C., Jagodzinska, M., Rijkers, D.T.S., and Liskamp, R.M.J. (2008) *Bioorg. Med. Chem. Lett.*, **18**, 78–84.

81 Sol, V., Chaleix, V., Granet, R., and Krausz, P. (2008) *Tetrahedron*, **64**, 364–371.

82 Poeylaut-Palena, A.A., Testero, S.A., and Mata, E.G. (2008) *J. Org. Chem.*, **73**, 2024–2027.

83 Schuerer, S.C. and Blechert, S. (1998) *Synlett*, 166–168.

84 Schuster, M. and Blechert, S. (1998) *Tetrahedron Lett.*, **39**, 2295–2298.

85 Schurer, S.C. and Blechert, S. (1999) *Synlett*, 1879–1882.

86 Heerding, D.A., Takata, D.T., Kwon, C., Huffman, W.F., and Samanen, J. (1998) *Tetrahedron Lett.*, **39**, 6815–6818.

87 Micalizio, G.C. and Schreiber, S.L. (2002) *Angew. Chem. Int. Ed.*, **41**, 152–154.

88 Brittain, D.E.A., Gray, B.L., and Schreiber, S.L. (2005) *Chem. Eur. J.*, **11**, 5086–5093.

Index

Metathesis in Natural Product Synthesis: Strategies, Substrates and Catalysts. Edited by Janine Cossy, Stellios Arseniyadis, and Christophe Meyer

ISBN: 978-3-527-32440-8

t

u